T0211974

Lecture Notes in Computer Science 11989

More information about this series at http://www.springer.com/series/7407

Daniel Slamanig · Elias Tsigaridas ·
Zafeirakis Zafeirakopoulos (Eds.)

Mathematical Aspects of Computer and Information Sciences

8th International Conference, MACIS 2019
Gebze, Turkey, November 13–15, 2019
Revised Selected Papers

 Springer

Editors
Daniel Slamanig ⓘ
AIT Austrian Institute of Technology
Vienna, Austria

Zafeirakis Zafeirakopoulos ⓘ
Institute of Information Technologies
Gebze Technical University
Gebze, Turkey

Elias Tsigaridas
IMJ-PRG
Sorbonne University
Paris, France

ISSN 0302-9743   ISSN 1611-3349   (electronic)
Lecture Notes in Computer Science
ISBN 978-3-030-43119-8   ISBN 978-3-030-43120-4   (eBook)
https://doi.org/10.1007/978-3-030-43120-4

LNCS Sublibrary: SL1 – Theoretical Computer Science and General Issues

This Springer imprint is published by the registered company Springer Nature Switzerland AG
The registered company address is: Gewerbestrasse 11, 6330 Cham, Switzerland

Preface

Mathematical Aspects of Computer and Information Sciences (MACIS) is a series of biennial conferences focusing on research in mathematical and computational aspects of computing and information science. It is broadly concerned with algorithms, their complexity, and their embedding in larger logical systems. At the algorithmic level, there is a rich interplay between the numerical/algebraic/geometrical/topological axes. At the logical level, there are issues of data organization, interpretation, and associated tools. These issues often arise in scientific and engineering computation where we need experiments and case studies to validate or enrich the theory. At the application level, there are significant applications in the areas of mathematical cryptography, machine learning, and data analysis, and the various combinatorial structures and coding theory concepts that are used in a pivotal role in computing and information sciences. MACIS is interested in outstanding and emerging problems in all these areas. Previous MACIS conferences have been held in Beijing (2006, 2011), Paris (2007), Fukuoka (2009), Nanning (2013), Berlin (2015), and Vienna (2017). MACIS 2019 was held at the Gebze Technical University (GTU) located at the borders of Istanbul with Kocaeli, during November 13–15, 2019.

We are grateful to the track chairs and the Program Committee for their critical role in putting together a very successful technical program, especially under strict deadlines. We also wish to extend our gratitude to all MACIS 2019 conference participants – all of them contributed to making the conference a success. The conference would not have been possible without the hard work of the local organizing team: Gizem Süngü, Fatma Nur Esirci, Başak Karakaş, and Tülay Ayyıldız Akoğlu. We are grateful to GTU, in particular the Rector and the International Relations office of GTU, for offering the facilities where the conference took place, sponsoring the conference through the Scientific Research Programs Fund of the University, and providing material distributed to the participants. Moreover, we are grateful for the constant support of the Institute of Information Technologies of GTU. We also acknowledge support by the project 117F100 under the program 3501 of the Scientific and Technological Research Council of Turkey, by the project 118F321 under the program 2509 of the Scientific and Technological Research Council of Turkey, and the project NEMO under the program 2232 (International Leading Researchers Program) of the Scientific and Technological Research Council of Turkey. Finally we are grateful to Maplesoft for sponsoring the Best Early Stage Researcher Presentation Award. Last but not least, we are thankful to the three invited speakers, Matthias Beck (San Francisco State University, USA, and Free University of Berlin, Germany), Georg Fuchsbauer (Inria/ENS, France), and Agnes Szanto (North Carolina State University, USA), for honoring the conference with their participation and stimulating talks.

This volume contains 36 refereed papers (22 regular and 14 short papers) carefully selected out of 66 total submissions (48 regular, 18 short); thus, MACIS 2019 had an overall acceptance rate of 55%. The papers are organized in different categories corresponding to four tracks featured in the MACIS 2019 conference. The topics of the MACIS 2019 tracks cover a wide array of research areas, as follows:

Track 1: Algorithms and Foundations
Track Chairs: Chenqi Mou, Maximilian Jaroschek, Fadoua Ghourabi
Track 2: Security and Cryptography
Track Chairs: Alp Bassa, Olivier Blazy, Guénaël Renault
Track 3: Combinatorics, Codes, Designs and Graphs
Track Chairs: Michel Lavrauw, Liam Solus, Tınaz Ekim
Track 4: Data Modeling and Machine Learning
Track Chairs: Giorgos Kollias, Kaie Kubjas, Günce Orman
Tools and Software Track
Track Chairs: Matthew England, Vissarion Fisikopoulos, Ali Kemal Uncu

We wish to thank all the track chairs for their hard work in putting together these tracks. Last but not least, we thank the Springer management and production team for their support during the production of this volume.

January 2020

Daniel Slamanig
Elias Tsigaridas
Zafeirakis Zafeirakopoulos

Organization

General Chair

Zafeirakis Zafeirakopoulos   Gebze Technical University, Turkey

Program Committee Chairs

Daniel Slamanig   AIT Austrian Institute of Technology, Austria
Elias Tsigaridas   Inria, IMJ-PRG, France

Program Committee

Alp Bassa   Bosphorus University, Turkey
Olivier Blazy   University of Limoges, France
Türkü Özlüm Çelik   MPI MiS Leipzig, Germany
Tınaz Ekim   Bosphorus University, Turkey
Matthew England   Coventry University, UK
Vissarion Fisikopoulos   National University of Athens, Greece
Fadoua Ghourabi   Ochanomizu University, Japan
Maximilian Jaroschek   Technische Universität Wien, Austria
Giorgos Kapetanakis   University of Crete, Greece
Giorgos Kollias   IBM Research, USA
Kaie Kubjas   Aalto University, Finland
Kağan Kurşungöz   Sabanci University, Turkey
Michel Lavrauw   Sabanci University, Turkey
Chenqi Mou   Beihang University, China
Gunce Orman   Galatasaray University, Turkey
Veronika Pillwein   RISC, Austria
Mohan Ravichandran   Mimar Sinan Fine Arts University, Turkey
Guénaël Renault   Inria, École Polytechnique, France
Liam Solus   KTH, Sweden
Ali Kemal Uncu   RISC, Austria

Additional Reviewers

Per Alexandersson	Didem Gözüpek	Markus Schofnegger
Ibrahim Almakky	Antonio Jimenez-Pastor	Dimitris E. Simos
Carlos Améndola	Vassilios Kalantzis	Colin Stephen
Erchan Aptoula	Lukas Katthän	Ivan Tomasic
Kamal Bentahar	Florian Kohl	Pinar Uluer
Eliana Duarte	Stephan Krenn	Thibaut Verron
Dorian Florescu	Luke Oeding	Mario Werner
Ragnar Freij-Hollanti	Sebastian Ramacher	Juan Xu
Matteo Gallet	Georg Regensburger	Burcu Yilmaz
Oliver Gnilke	Elina Robeva	Nikolai Zamarashkin

MACIS Steering Committee

Dimitris E. Simos (Chair)	SBA Research, Austria
Ilias Kotsireas	Wilfrid Laurier University, Canada
Siegfried Rump	Hamburg University of Technology, Germany
Chee Yap	New York University, USA
Temur Kutsia	RISC, Johannes Kepler University, Austria
Johannes Blömer	Paderborn University, Germany

Contents

Algorithms and Foundations

Certified Hermite Matrices from Approximate Roots - Univariate Case   3
    Tulay Ayyildiz Akoglu and Agnes Szanto

On Parametric Border Bases.   10
    Yosuke Sato, Hiroshi Sekigawa, Ryoya Fukasaku,
    and Katsusuke Nabeshima

Reliable Computation of the Singularities of the Projection
in $\mathbb{R}^3$ of a Generic Surface of $\mathbb{R}^4$ 16
    Sény Diatta, Guillaume Moroz, and Marc Pouget

Evaluation of Chebyshev Polynomials on Intervals and Application
to Root Finding   35
    Viviane Ledoux and Guillaume Moroz

Proving Two Conjectural Series for $\zeta(7)$ and Discovering More
Series for $\zeta(7)$ 42
    Jakob Ablinger

Generalized Integral Dependence Relations.   48
    Katsusuke Nabeshima and Shinichi Tajima

Hilbert-Type Dimension Polynomials of Intermediate
Difference-Differential Field Extensions.   64
    Alexander Levin

Comprehensive *LU* Factors of Polynomial Matrices.   80
    Ana C. Camargos Couto, Marc Moreno Maza, David Linder,
    David J. Jeffrey, and Robert M. Corless

Sublinear Cost Low Rank Approximation via Subspace Sampling.   89
    Victor Y. Pan, Qi Luan, John Svadlenka, and Liang Zhao

CUR LRA at Sublinear Cost Based on Volume Maximization   105
    Qi Luan and Victor Y. Pan

New Practical Advances in Polynomial Root Clustering.   122
    Rémi Imbach and Victor Y. Pan

On the Chordality of Simple Decomposition in Top-Down Style.   138
    Chenqi Mou and Jiahua Lai

Automatic Synthesis of Merging and Inserting Algorithms on Binary Trees
Using Multisets in *Theorema* 153
   Isabela Drămnesc and Tudor Jebelean

Algebraic Analysis of Bifurcations and Chaos for Discrete
Dynamical Systems...................................... 169
   Bo Huang and Wei Niu

Security and Cryptography

Acceleration of Spatial Correlation Based Hardware Trojan Detection
Using Shared Grids Ratio................................ 187
   Fatma Nur Esirci and Alp Arslan Bayrakci

A Parallel GPU Implementation of SWIFFTX 202
   Metin Evrim Ulu and Murat Cenk

Computing an Invariant of a Linear Code 218
   Mijail Borges-Quintana, Miguel Ángel Borges-Trenard,
   Edgar Martínez-Moro, and Gustavo Torres-Guerrero

Generalized Secret Sharing Schemes Using N^μMDS Codes 234
   Sanyam Mehta and Vishal Saraswat

Exploiting Linearity of Modular Multiplication..................... 249
   Hamdi Murat Yıldırım

Combinatorics, Codes, Designs and Graphs

On a Weighted Spin of the Lebesgue Identity...................... 273
   Ali Kemal Uncu

Edge-Critical Equimatchable Bipartite Graphs..................... 280
   Yasemin Büyükçolak, Didem Gözüpek, and Sibel Özkan

Determining the Rank of Tensors in $\mathbb{F}_q^2 \otimes \mathbb{F}_q^3 \otimes \mathbb{F}_q^3$ 288
   Nour Alnajjarine and Michel Lavrauw

Second Order Balance Property on Christoffel Words 295
   Lama Tarsissi and Laurent Vuillon

IPO-Q: A Quantum-Inspired Approach to the IPO Strategy
Used in CA Generation 313
   Michael Wagner, Ludwig Kampel, and Dimitris E. Simos

A Fast Counting Method for 6-Motifs with Low Connectivity 324
   Taha Sevim, Muhammet Selçuk Güvel, and Lale Özkahya

LaserTank is NP-Complete.   333
  Per Alexandersson and Petter Restadh

Data Modeling and Machine Learning

Improved Cross-Validation for Classifiers that Make Algorithmic Choices
to Minimise Runtime Without Compromising Output Correctness.   341
  Dorian Florescu and Matthew England

A Numerical Efficiency Analysis of a Common Ancestor Condition   357
  Luca Carlini, Nihat Ay, and Christiane Görgen

Optimal Transport to a Variety.   364
  Türkü Özlüm Çelik, Asgar Jamneshan, Guido Montúfar,
  Bernd Sturmfels, and Lorenzo Venturello

SFV-CNN: Deep Text Sentiment Classification with Scenario
Feature Representation.   382
  Haoliang Zhang, Hongbo Xu, Jinqiao Shi, Tingwen Liu,
  and Jing Ya

Reinforcement Learning Based Interactive Agent for Personalized
Mathematical Skill Enhancement.   395
  Muhammad Zubair Islam, Kashif Mehmood, and Hyung Seok Kim

Common Vector Approach Based Image Gradients Computation
for Edge Detection   408
  Sahin Isik and Kemal Ozkan

Optimizing Query Perturbations to Enhance Shape Retrieval   422
  Bilal Mokhtari, Kamal Eddine Melkemi, Dominique Michelucci,
  and Sebti Foufou

Authorship Attribution by Functional Discriminant Analysis.   438
  Chahrazed Kettaf and Abderrahmane Yousfate

Tools and Software Track

An Overview of Geometry Plus Simulation Modules.   453
  Angelos Mantzaflaris

DD-Finite Functions Implemented in Sage   457
  Antonio Jiménez-Pastor

Author Index   463

Algorithms and Foundations

Certified Hermite Matrices from Approximate Roots - Univariate Case

Tulay Ayyildiz Akoglu[1]([⊠]) and Agnes Szanto[2]

[1] Karadeniz Technical University, Trabzon, Turkey
tulayaa@ktu.edu.tr
[2] North Carolina State University, Raleigh, NC, USA
aszanto@ncsu.edu

Abstract. Let $f_1, \ldots, f_m$ be univariate polynomials with rational coefficients and $\mathcal{I} := \langle f_1, \ldots, f_m \rangle \subset \mathbb{Q}[x]$ be the ideal they generate. Assume that we are given approximations $\{z_1, \ldots, z_k\} \subset \mathbb{Q}[i]$ for the common roots $\{\xi_1, \ldots, \xi_k\} = V(\mathcal{I}) \subseteq \mathbb{C}$. In this study, we describe a symbolic-numeric algorithm to construct a rational matrix, called *Hermite matrix*, from the approximate roots $\{z_1, \ldots, z_k\}$ and certify that this matrix is the true Hermite matrix corresponding to the roots $V(\mathcal{I})$. Applications of Hermite matrices include counting and locating real roots of the polynomials and certifying their existence.

Keywords: Symbolic–numeric computation · Approximate roots · Hermite matrices

1 Introduction

The development of numerical and symbolic techniques to solve systems of polynomial equations resulted in an explosion of applicability, both in term of the size of the systems efficiently solvable and the reliability of the output. Nonetheless, many of the results produced by numerical methods are not certified. In this paper, we show how to compute exact Hermite matrices from approximate roots of polynomials, and how to certify that these Hermite matrices are correct.

Hermite matrices and Hermite bilinear forms were introduced by Hermite in 1850 [7], and have many applications, including counting real roots [3,8,9] and locating them [2]. Assume that we are given the ideal $\mathcal{I} := \langle f_1, \ldots, f_m \rangle \subset \mathbb{Q}[x]$ generated by rational polynomials, and assume that $\dim_\mathbb{Q} \mathbb{Q}[x]/\mathcal{I} = k$. Hermite matrices have two kinds of definitions (see the precise formulation in Sect. 2.1):

1. The first definition of Hermite matrices uses the traces of k^2 multiplication matrices, each of them of size $k \times k$. The advantage of this definition is that it can be computed exactly, working with rational numbers only. The disadvantage is that it requires the computation of the traces of k^2 matrices.

T. A. Akoglu—partially supported by TUBITAK grant 119F211.
A. Szanto—partially supported by NSF grants CCF-1813340 and CCF-1217557.

D. Slamanig et al. (Eds.): MACIS 2019, LNCS 11989, pp. 3–9, 2020.
https://doi.org/10.1007/978-3-030-43120-4_1

2. The second definition uses symmetric functions of the k common roots of $\mathcal{I}$, counted with multiplicity. The advantage of this definition is that it gives a very efficient way to evaluate the entries of the Hermite matrix, assuming that we know the common roots of $\mathcal{I}$ exactly. The disadvantage is that we need to compute the common roots exactly, which may involve working in field extensions of $\mathbb{Q}$.

In this paper we propose to use the second definition to compute Hermite matrices, but instead of using exact roots, we use approximate roots that can be computed with numerical methods efficiently [6]. Once we obtain an approximate Hermite matrix, we use rational number reconstruction (RNR) to construct a matrix with rational entries of bounded denominators. Finally, we give a symbolic method which certifies that the rational Hermite matrix we computed is in fact the correct one, corresponding to the exact roots of $\mathcal{I}$.

Using RNR techniques on rational polynomial systems is not a new concept. A common approach is to use p-adic lifting or iterative refinement to build an approximate solution, then apply rational number reconstruction [13–15]. Peryl and Parrilo [11] used the approximate solutions as starting points for the computation of exact rational sum of squares decomposition of rational polynomials. RNR is also used to solve systems of linear equations and inequalities over the rational numbers [12]. Moreover, RNR can be used to construct the coefficients of the rational univariate representation of rational polynomial systems [1].

The novelty of this note and the difficulty of this problem is to certify the correctness of the Hermite matrix that we computed with the above heuristic approach. This part of the algorithm is purely symbolic. The main idea is to use the fact that companion matrices act like roots of the polynomials, so we can certify them, and then we use the famous Newton-Girard formulas [16] to connect the entries of the companion matrix with the entries of the Hermite matrix.

A natural question arises about the advantage of this hybrid symbolic–numeric approach over purely symbolic methods, for example by taking the gcd of the input polynomials and computing the symbolic Hermite matrix of the gcd using the definition with traces. In many cases, the input polynomials have much higher degree D than the number of common roots, so the bottleneck of the computation is computing the common roots or the gcd of the polynomials. Our approach computes numerically the roots of one polynomial with integer coefficients of size at most h, substitutes them into the other $m - 1$ polynomials to find the common roots, which can be done using $O((D^3 + hD^2) + hmD)$ binary operations up to logarithmic factors (c.f. [4,10]). On the other hand, computing the gcd of m degree D polynomials with integer coefficients of overall size $H \leq mh$ takes $O(mD^3)$ arithmetic operation with integers of size $O(D^4H)$. (c.f. [5]).

2   Preliminaries

2.1   Hermite Matrices

Let $f_1, \ldots, f_m \in \mathbb{Q}[x]$, $\mathcal{I} = \langle f_1, \ldots, f_m \rangle \subset \mathbb{Q}[x]$ and $k := \dim_{\mathbb{Q}} \mathbb{Q}[x]/\mathcal{I}$. Assume that (the residue classes of the polynomials in) $\mathcal{B} = \{1, x, \ldots, x^{k-1}\}$ form a basis for $\mathbb{Q}[x]/\mathcal{I}$. Note that all definitions in this section are valid for polynomials over $\mathbb{R}$ or $\mathbb{C}$, but in this note we only consider polynomials with rational coefficients.

In [3, Section 4.3.2] it is shown that the following two definitions of Hermite matrices are equivalent:

Definition 1. *Let $\xi_1, \xi_2, \ldots, \xi_k \in \mathbb{C}$ be the common roots of $\mathcal{I}$ (here each root is listed as many times as their multiplicity) and $g \in \mathbb{Q}[x]$. Then the **Hermite matrix** of $\mathcal{I}$ with respect to g is*

$$H_g := V_B^T G V_B \tag{1}$$

*where $V_B = [\xi_i^{j-1}]_{i,j=1,\ldots,k}$ is the Vandermonde matrix of the roots with respect to the basis $\mathcal{B}$ and G is an $k \times k$ diagonal matrix with $[G]_{ii} = g(\xi_i)$ for $i = 1, \ldots, k$. We will also need the **extended Hermite matrix** of $\mathcal{I}$ with respect to g*

$$H_g^+ := V_{B+}^T G V_{B+} \in \mathbb{Q}^{(k+1)\times(k+1)} \tag{2}$$

where $V_{B+} = [\xi_i^{j-1}]_{i=1,\ldots,k,\, j=1,\ldots,k+1} \in \mathbb{C}^{k\times(k+1)}$ is the Vandermonde matrix corresponding to $\mathcal{B}^+ := \{1, x, x^2, \ldots, x^k\}$.

Definition 1 gives the following formula for $g = 1$

$$H_1 = \left[\sum_{l=1}^{k} \xi_l^{i+j-2} \right]_{i,j=1,\ldots,k}. \tag{3}$$

The right hand side of (3) is the $(i+j-2)$-th power sum of the roots, which is an elementary symmetric function of the roots.

The second definition implies that the Hermite matrix has a Hankel structure and its entries are rational numbers.

Definition 2. *Let $\mathcal{I}$ as above and $g \in \mathbb{Q}[x]$. The Hermite matrix of $\mathcal{I}$ with respect of g is*

$$H_g := \left[\mathrm{Tr}(M_{gx^{i+j-2}}) \right]_{i,j=1}^{k},$$

where M_f denotes the matrix of the multiplication map $\mu_f : \mathbb{Q}[x]/\mathcal{I} \to \mathbb{Q}[x]/\mathcal{I}$, $\mu_f(p) := p \cdot f + \mathcal{I}$ in the basis $\mathcal{B}$.

2.2   Rational Number Reconstruction

Continued fractions are widely used for rational approximation purposes. Let z be a real number, one can compute the sequence of repeated quotients using continued fractions, yielding rational approximations for z. If the denominator is bounded, the following theorem guarantees the uniqueness of the rational approximation in case of existence.

Theorem 1. *[12] There exists a polynomial time algorithm which, for a given rational number z and a natural number B tests if there exists a pair of integers (p, q) with $1 \leq q \leq B$ and*

$$\left| z - \frac{p}{q} \right| < \frac{1}{2B^2}$$

if so, finds this unique pair of integers.

If we have a bound E for the absolute approximation error of z, then the denominator bound can be defined as $B := \left\lceil (2E)^{-1/2} \right\rceil$ to guarantee the uniqueness of a rational number within distance E from z with denominator at most B.

3   Construction and Certification of Hermite Matrices

In the following algorithm we assume that $\mathcal{I} = \langle f_1, \ldots, f_m \rangle$ is radical, i.e. if $k = \dim \mathbb{Q}[x]/\mathcal{I}$ then $V(\mathcal{I})$ has cardinality k. Our algorithm to construct and certify Hermite matrices from approximate roots is as follows.

Algorithm: Certified Univariate Hermite Matrix

- **Input:** $f_1, \ldots, f_m, g \in \mathbb{Q}[x]$; $k = \dim \mathbb{Q}[x]/\mathcal{I}$; $\{z_1, \ldots, z_k\} \subset \mathbb{Q}[i]$ approximate roots; a bound E on the absolute error of these approximate roots.
- **Output:** $H_g \in \mathbb{Q}^{k \times k}$ or Fail.
  1: Compute the approximate extended Hermite matrix

$$\tilde{H}_1^+ := \left[\sum_{l=1}^{k} z_l^{i+j-2} \right]_{i,j=1,\ldots,k+1} \in \mathbb{Q}[i]^{(k+1)\times(k+1)}.$$

  2: Use Rational Number Reconstruction for the real part of each entry $\tilde{H}_1^+$, using Theorem 1 with denominator bound for the (i, j)-th entry

$$B_{i,j} := \left\lceil (2k(i+j-2)EA^{i+j-3})^{-1/2} \right\rceil. \tag{4}$$

  Here A is an upper bound for the coordinates of the approximate roots. The resulting matrix is denoted by $H_1^+ \in \mathbb{Q}^{(k+1)\times(k+1)}$.
  3: $H_1 \leftarrow$ the first k rows and the first k columns of H_1^+ $H_1^k \leftarrow$ the first k rows and the last k columns of H_1^+.
  4: **If** H_1^+ has Hankel structure and $\operatorname{rank}(H_1) = \operatorname{rank}(H_1^+) = k$, **then**

$$M_x \leftarrow H_1^{-1} \cdot H_1^k$$

  else return Fail.

5: **If** M_x has a companion matrix shape and $f_i(M_x) = 0$ for $i = 1, \ldots, m$ **then** $p(x) \leftarrow$ charpol(M_x) **else** return Fail; **If** p is not square-free **then** return Fail. Otherwise M_x is the certified multiplication matrix by x in $\mathbb{Q}[x]/\mathcal{I}$.

6: Use the Newton–Girard formulas [16] with the coefficients of p to yield the d-th power sums of the roots of p for $d = 0, \ldots, 2k - 2$, as in (3). **If** each one matches to the corresponding entry of H_1, **then** it certifies H_1, **else** return Fail.

7: Once H_1 and M_x are certified, **return**

$$H_g \leftarrow H_1 \cdot g(M_x),$$

which is correct by $H_1 \cdot g(M_x) = (V^T V) \cdot (V^{-1} G V) = V^T G V = H_g$.

Note that if we do not give $k = \dim_{\mathbb{Q}} \mathbb{Q}[x]/\mathcal{I}$ as part of the input, the above algorithm only certifies that the output matrix H_g corresponds to a rational subvariety of $V(\mathcal{I})$, i.e. possibly a proper subset of $V(\mathcal{I})$ that is defined by rational polynomials.

We finish this note by describing a modification of the above algorithm for the case when $\mathcal{I}$ is not radical. In this case we return a certified Hermite matrix H_g corresponding to a rational component of the *radical* of $\mathcal{I}$, i.e. each common roots of $\mathcal{I}$ is counted with multiplicity one or zero. We still start with the same input, but $z_1, \ldots, z_k$ may have repetitions (or form clusters). In Step 4, instead of requiring H_1 to have rank k, we compute the companion matrix M_x using a maximal non-singular submatrix of H_1^+, which may have size smaller than k. In Step 6, we use the Newton–Girard formulas to define H_1, and return H_g defined as in Step 7, which may also have size smaller than k.

In future work, we plan to extend these results to multivariate and overdetermined polynomial systems.

4 Example

We demonstrate our algorithm on a simple example. Consider $f(x) = 16x^4 - 10x^2 + 1 \in \mathbb{Q}[x]$, with $g(x) = 1$. The exact roots of f are $1/\sqrt{2}, -1/\sqrt{2}, 1/2\sqrt{2}, -1/2\sqrt{2}$. We get the following approximate solutions using homotopy method in Maple: $z_1 = 0.7071067810, z_2 = -0.7071067810, z_3 = 0.3535533905, z_4 = -0.3535533905$. This solution has error bound $E := 10^{-8}$.

1: Compute the approximate extended Hankel matrix $\tilde{H}_1^+$ from z_1, z_2, z_3, z_4:

$$\tilde{H}_1^+ = \begin{bmatrix} 4.0 & -0.0000000007 & 1.2500000052 & -0.00000000026 & 0.5312500055 \\ -0.0000000007 & 1.2500000053 & -0.0000000002 & 0.5312500055 & -5.3363907043 \times 10^{-11} \\ 1.2500000052999999 & -0.0000000002 & 0.5312500055 & -5.4597088135 \times 10^{-11} & 0.2539062541 \\ -0.0000000002 & 0.5312500055 & -5.4597088135 \times 10^{-11} & 0.2539062542 & -9.3658008865 \times 10^{-12} \\ 0.5312500055 & -5.3363907043 \times 10^{-11} & 0.2539062541 & -9.3658008865 \times 10^{-12} & 0.1254882840 \end{bmatrix}.$$

2: Rationalize H_1^+, using $A = 0.8$ and $E = 10^{-8}$ and (4). This gives $B \cong 2700$ as upper bound for the denominators of each entry of the Hankel matrix H_1^+.

$$H_1^+ = \begin{bmatrix} 4 & 0 & \frac{5}{4} & 0 & \frac{17}{32} \\ 0 & \frac{5}{4} & 0 & \frac{17}{32} & 0 \\ \frac{5}{4} & 0 & \frac{17}{32} & 0 & \frac{65}{256} \\ 0 & \frac{17}{32} & 0 & \frac{65}{256} & 0 \\ \frac{17}{32} & 0 & \frac{65}{256} & 0 & \frac{257}{2048} \end{bmatrix}$$

3: Let H_1 be the first k rows and the first k columns of H_1^+, and H_1^k be the first k rows and the last k columns of H_1^+.
4: H_1^+ has Hankel structure and $\mathrm{rank}(H_1^+) = \mathrm{rank}(H_1) = 4$. Then

$$M_x = H_1^{-1} \cdot H_1^4 = \begin{bmatrix} 0 & 0 & 0 & -\frac{1}{16} \\ 1 & 0 & 0 & 0 \\ 0 & 1 & 0 & \frac{5}{8} \\ 0 & 0 & 1 & 0 \end{bmatrix}.$$

5: M_x has a companion matrix shape and $f(M_x) = 0$, then $p(x) := x^4 - \frac{5}{8}x^2 + \frac{1}{16}$ with $\gcd(p, p') = 1$ (square free). Thus we certified that M_x is the multiplication matrix by x in $\mathbb{Q}[x]/\langle f \rangle$.
6: We Newton–Girard formulas with the elementary symmetric functions: $e_0 = 1, e_1 = 0, e_2 = -\frac{5}{8}, e_3 = 0, e_4 = \frac{1}{16}$, which yields

$$\sum_{i=1}^{4} \xi_i^0 = 4, \sum_{i=1}^{4} \xi_i^2 = \frac{5}{4}, \sum_{i=1}^{4} \xi_i^4 = \frac{17}{32}, \sum_{i=1}^{4} \xi_i^6 = \frac{65}{256}, \sum_{i=1}^{4} \xi_i^8 = \frac{257}{2048},$$

and all odd power sums are zero. Each sum matches the corresponding entry, thus we certified H_1.
7: Since $g(x) = 1$, Return H_1.

References

1. Ayyildiz Akoglu, T., Hauenstein, J.D., Szanto, A.: Certifying solutions to overdetermined and singular polynomial systems over Q. J. Symb. Comput. **84**, 147–171 (2018)
2. Ayyildiz Akoglu, T.: Certifying solutions to polynomial systems over Q. Ph.D. thesis, North Carolina State University (2016)
3. Basu, S., Pollack, R., Roy, M.-F.: Algorithms in Real Algebraic Geometry. AACIM, vol. 10. Springer, Heidelberg (2006). https://doi.org/10.1007/3-540-33099-2
4. Becker, R., Sagraloff, M., Sharma, V., Yap, C.: A near-optimal subdivision algorithm for complex root isolation based on the pellet test and newton iteration. J. Symb. Comput. **86**, 51–96 (2018)

5. González-Vega, L.: On the complexity of computing the greatest common divisor of several univariate polynomials. In: Baeza-Yates, R., Goles, E., Poblete, P.V. (eds.) LATIN 1995. LNCS, vol. 911, pp. 332–345. Springer, Heidelberg (1995). https://doi.org/10.1007/3-540-59175-3_100

6. Bates, D.J., Hauenstein, J.D., Sommese, A.J., Wampler, C.W.: Numerically Solving Polynomial Systems with Bertini, vol. 25. SIAM, Philadelphia (2013)

7. Hermite, C.: Sur le nombre des racines d'une équation algébrique comprise entre des limites données. J. Reine Angew. Math. **52**, 39–51 (1850). Also in Oeuvres completes, vol. 1, pp. 397–414

8. Hermite, C.: Remarques sur le théorème de Sturm. CR Acad. Sci. Paris **36**(52–54), 171 (1853)

9. Hermite, C.: Extrait d'une lettre de Mr. Ch. Hermite de Paris à Mr. Borchardt de Berlin sur le nombre des racines d'une équation algébrique comprises entre des limites données. Journal für die reine und angewandte Mathematik **52**, 39–51 (1856)

10. Pan, V.Y.: Nearly optimal polynomial root-finders: the state of the art and new progress. arXiv:1805.12042v10 [cs.NA] (2019)

11. Peyrl, H., Parrilo, P.A.: Computing sum of squares decompositions with rational coefficients. Theoret. Comput. Sci. **409**(2), 269–281 (2008)

12. Schrijver, A.: Theory of Linear and Integer Programming. Wiley, New York (1998)

13. Steffy, D.E.: Exact solutions to linear systems of equations using output sensitive lifting. ACM Commun. Comput. Algebra **44**(3/4), 160–182 (2011)

14. Wan, Z.: An algorithm to solve integer linear systems exactly using numerical methods. J. Symb. Comput. **41**, 621–632 (2006)

15. Wang, X., Pan, V.Y.: Acceleration of Euclidean algorithm and rational number reconstruction. SIAM J. Comput. **32**(2), 548–556 (2003)

16. Weisstein, E.W.: Newton-Girard Formulas. From MathWorld-A Wolfram Web Resource. http://mathworld.wolfram.com/Newton-GirardFormulas.html

On Parametric Border Bases

Yosuke Sato[1(✉)], Hiroshi Sekigawa[1], Ryoya Fukasaku[2],
and Katsusuke Nabeshima[3]

[1] Tokyo University of Science, Tokyo, Japan
ysato@rs.kagu.tus.ac.jp, sekigawa@rs.tus.ac.jp
[2] Kyushu University, Fukuoka, Japan
fukasaku@math.kyushu-u.ac.jp
[3] Tokushima University, Tokushima, Japan
nabeshima@tokushima-u.ac.jp

Abstract. We study several properties of border bases of parametric polynomial ideals and introduce a notion of a minimal parametric border basis. It is especially important for improving the quantifier elimination algorithm based on the computation of comprehensive Gröbner systems.

Keywords: Parametric border basis · Comprehensive Gröbner system · Quantifier elimination

1 Introduction

We study properties of border bases of zero-dimensional parametric polynomial ideals. Main motivation of our work is to improve the CGS-QE algorithm introduced in [1]. It is a special type of a quantifier elimination (QE) algorithm which has a great effect on QE of a first order formula containing many equalities. The most essential part of the algorithm is to eliminate all existential quantifiers $\exists \bar{X}$ from the following basic first order formula:

$$\phi(\bar{A}) \wedge \exists \bar{X} \ (\bigwedge_{1 \leq i \leq s} f_i(\bar{A}, \bar{X}) = 0 \wedge \bigwedge_{1 \leq i \leq t} h_i(\bar{A}, \bar{X}) \geq 0) \tag{1}$$

with polynomials $f_1, \ldots, f_s, h_1, \ldots, h_t$ in $\mathbb{Q}[\bar{A}, \bar{X}]$ such that the parametric ideal $I = \langle f_1, \ldots, f_s \rangle$ is zero-dimensional in $\mathbb{C}[\bar{X}]$ for any specialization of the parameters $\bar{A} = A_1, \ldots, A_m$ satisfying $\phi(\bar{A})$, where $\phi(\bar{A})$ is a quantifier free formula consisting only of equality $=$ and disequality $\neq$. The algorithm computes a reduced comprehensive Gröbner system (CGS) $\mathcal{G} = \{(\mathcal{S}_1, G_1), \ldots, (\mathcal{S}_r, G_r)\}$ of the parametric ideal I on the algebraically constructible set $\mathcal{S} = \{\bar{a} \in \mathbb{C}^m | \phi(\bar{a})\}$, then applies the method of [9] with several improvements of [2–4,7]. One of the most important properties of the reduced CGS is that $\mathbb{C}[\bar{X}]/\langle f_1(\bar{X}, \bar{a}) \ldots, f_s(\bar{X}, \bar{a}) \rangle$ has an invariant basis $\{t \in T(\bar{X}) : t \nmid LT(g) \text{ for any } g \in G_i\}$ as a $\mathbb{C}$-vector space for every $\bar{a} \in \mathcal{S}_i$. It enables us to perform several uniform computations with parameters $\bar{A}$ for every $\bar{a} \in \mathcal{S}_i$. (More detailed descriptions can be found in [1].) In order to obtain a simple quantifier free formula, a compact representation

© Springer Nature Switzerland AG 2020
D. Slamanig et al. (Eds.): MACIS 2019, LNCS 11989, pp. 10–15, 2020.
https://doi.org/10.1007/978-3-030-43120-4_2

of a reduced CGS of I is desirable, minimizing the number r of the partition $\mathcal{S}_1, \ldots, \mathcal{S}_r$ of $\mathcal{S}$ is particularly important. Border bases are alternative tools for handling zero-dimensional ideals [5]. We have observed that the reduced CGS can be replaced with a parametric border basis in our algorithm. Since border bases have several nice properties which Gröbner bases do not possess, we can obtain a simpler quantifier free formula using a parametric border basis.

In this paper, we study border bases in parametric polynomial rings. We give a formal definition of a parametric border basis and show several properties which are important for improving the CGS-QE algorithm. Since our work is still on going and the paper is a short paper, we do not get deeply involved in the application of parametric border bases to QE.

The paper is organized as follows. In Sect. 2, we first give a quick review of a CGS for understanding the merit of our work, then give a formal definition of a parametric border basis. In Sect. 3, we introduce our main results together with a rather simple example for understanding our work. Numerical stability is one of the most important properties of border bases. In Sect. 4, we study this property in our setting. We follow the book [5] for the terminologies and notations concerning border bases.

2   Preliminary

In the rest of the paper, let $\mathbb{Q}$ and $\mathbb{C}$ denote the field of rational numbers and complex numbers, $\bar{X}$ and $\bar{A}$ denote some variables $X_1, \ldots, X_n$ and $A_1, \ldots, A_m$, $T(\bar{X})$ denote a set of terms in $\bar{X}$. For $t_1, t_2 \in T(\bar{X})$, $t_1 \mid t_2$ and $t_1 \nmid t_2$ denote that "t_2 is divisible by t_1" and "t_2 is not divisible by t_1" respectively. For a polynomial $f \in \mathbb{C}[\bar{A}, \bar{X}]$, regarding f as a member of a polynomial ring $\mathbb{C}[\bar{A}][\bar{X}]$ over the coefficient ring $\mathbb{C}[\bar{A}]$, its leading term and coefficient w.r.t. an admissible term order $\succ$ of $T(\bar{X})$ are denoted by $LT_{\succ}(f)$ and $LC_{\succ}(f)$ respectively. When $\succ$ is clear from context, they are simply denoted by $LT(f)$ and $LC(f)$.

2.1   Comprehensive Gröbner System

Definition 1. *For an algebraically constructible subset (ACS in short) $\mathcal{S}$ of $\mathbb{C}^m$, a finite set $\{\mathcal{S}_1, \ldots, \mathcal{S}_r\}$ of ACSs of $\mathbb{C}^m$ which satisfies $\cup_{i=1}^{r}\mathcal{S}_i = \mathcal{S}$ and $\mathcal{S}_i \cap \mathcal{S}_j = \emptyset (i \neq j)$ is called an algebraic partition of $\mathcal{S}$. Each $\mathcal{S}_i$ is called a segment.*

Definition 2. *Fix an admissible term order on $T(\bar{X})$. For a finite set $F \subset \mathbb{Q}[\bar{A}, \bar{X}]$ and an ACS $\mathcal{S}$ of $\mathbb{C}^m$, a finite set of pairs $\mathcal{G} = \{(G_1, \mathcal{S}_1), \ldots, (G_r, \mathcal{S}_r)\}$ with finite sets $G_1, \ldots, G_r$ of $\mathbb{Q}[\bar{A}, \bar{X}]$ satisfying the following properties is called a reduced comprehensive Gröbner system (CGS) of $\langle F \rangle$ on $\mathcal{S}$ with parameters $\bar{A}$. (When $\mathcal{S}$ is the whole space $\mathbb{C}^m$, "on $\mathbb{C}^m$" is usually omitted.)*

1. *$\{\mathcal{S}_1, \ldots, \mathcal{S}_r\}$ is an algebraic partition of $\mathcal{S}$.*
2. *For each i and $\bar{a} \in \mathcal{S}_i$, $G_i(\bar{a})$ is a reduced Gröbner basis of $\langle F(\bar{a}) \rangle \subset \mathbb{C}[\bar{X}]$, where $G_i(\bar{a}) = \{g(\bar{a}, \bar{X}) | g(\bar{A}, \bar{X}) \in G_i\}$ and $F(\bar{a}) = \{f(\bar{a}, \bar{X}) | f(\bar{A}, \bar{X}) \in F\}$.*
3. *For each i, $LC(g)(\bar{a}) \neq 0$ for every $g \in G_i$ and $\bar{a} \in \mathcal{S}_i$.*

Remark 3. *The set of leading terms of all polynomials of $G_i(\bar{a})$ is invariant for each $\bar{a} \in \mathcal{S}_i$. Hence, not only the dimension of the ideal $\langle G_i(\bar{a}) \rangle$ is invariant but also the $\mathbb{C}$-vector space $\mathbb{C}[\bar{X}]/\langle F(\bar{a}) \rangle$ has the same finite basis $\{t \in T(\bar{X}) :$ $t \nmid LT(g)$ for any $g \in G_i\}$ for every $\bar{a} \in \mathcal{S}_i$ when $\langle F(\bar{a}) \rangle$ is zero-dimensional.*

2.2 Border Bases in Parametric Polynomial Rings

Definition 4. *For a finite set $F \subset \mathbb{Q}[\bar{A}, \bar{X}]$ and an ACS $\mathcal{S}$ of $\mathbb{C}^m$ such that the ideal $\langle F(\bar{a}) \rangle$ is zero-dimensional for each $\bar{a} \in \mathcal{S}$, a finite set of triples $\mathcal{B} = \{(B_1, \mathcal{S}_1, \mathcal{O}_1), \dots, (B_r, \mathcal{S}_r, \mathcal{O}_r)\}$ with a finite set B_i of $\mathbb{Q}(\bar{A})[\bar{X}]$ and an order ideal $\mathcal{O}_i$ of $T(\bar{X})$ for each i satisfying the following properties is called a parametric border basis (PBB) of $\langle F \rangle$ on $\mathcal{S}$ with parameters $\bar{A}$. (When $\mathcal{S}$ is the whole space $\mathbb{C}^m$, "on $\mathbb{C}^m$" is usually omitted.)*

1. *$\{\mathcal{S}_1, \dots, \mathcal{S}_r\}$ is an algebraic partition of $\mathcal{S}$.*
2. *For each i, any denominator of a coefficient of an element of B_i does not vanish on $\mathcal{S}_i$.*
3. *For each i and $\bar{a} \in \mathcal{S}_i$, $B_i(\bar{a})$ is a $\mathcal{O}_i$-border basis of $\langle F(\bar{a}) \rangle \subset \mathbb{C}[\bar{X}]$.*

3 Properties of Parametric Border Bases

Consider the set $F = \{X^2 + \frac{1}{4}Y^2 - AXY + B - 1, \frac{1}{4}X^2 + Y^2 - BXY + A - 1\}$ of parametric polynomials in $\mathbb{Q}[A, B, X, Y]$ with parameters A and B, which is a similar but a little bit more complicated example than the one discussed in the book [5]. $\langle F(a, b) \rangle$ is zero-dimensional for every $(a, b) \in \mathbb{C}^2$. It has the following reduced CGS $\mathcal{G} = \{(G_1, \mathcal{S}_1), \dots, (G_7, \mathcal{S}_7)\}$ w.r.t. the lexicographic term order such that $X \succ Y$.

$G_1 = \{-5X^2 + 20BYX + 4, -5Y^2 - 20B + 4\}, \mathcal{S}_1 = \mathbb{V}(A - 4B),$

$G_2 = \{5X^2 - 4YX + 5B - 5, (5B - 1)YX - 5Y^2, (20B - 29)Y^3 + (-25B^3 + 35B^2 - 11B + 1)Y\},$
$\mathcal{S}_2 = \mathbb{V}(4A - B - 3) \setminus \{(\frac{4}{5}, \frac{1}{5}), (\frac{89}{80}, \frac{29}{20})\},$

$G_3 = \{16(A - 4B)(4A - B - 3)X + (-64A^2 + 272AB - 64B^2 - 225)Y^3 + (-64A^3 + (256B + 64)A^2 + (64B^2 - 320B - 240)A - 256B^3 + 256B^2 + 60B + 180)Y, (-64A^3 + 272A^2 - 64B^2 - 225)Y^4 + (-64A^3 + (256B + 64)A^2 + (64B^2 - 320B - 480)A - 256B^3 + 256B^2 + 120B + 360)Y^2 - 16(4A - B - 3)^2\},$
$\mathcal{S}_3 = \mathbb{C}^2 \setminus \mathcal{S}_1 \cup \mathcal{S}_2 \cup \mathcal{S}_4 \cup \dots \cup \mathcal{S}_7 = \mathbb{C}^2 \setminus \mathbb{V}((A - 4B)(4A - B - 3)(64A^2 - 272AB + 64B^2 + 225)),$

$G_4 = \{20X^2 + 9, Y\}, \mathcal{S}_4 = \{(\frac{89}{80}, \frac{29}{20})\},$

$G_5 = \{58Y^2 + 245, 35X - Y\}, \mathcal{S}_5 = \{(\frac{101}{20}, \frac{29}{20})\},$

$G_6 = \{60(20B - 29)X + ((400B^2 - 400B - 36)A - 1600B^3 + 1600B^2 + 519B - 375)Y, 15(((400B - 64)A - 64B - 425)Y^2 + 128((200B^2 - 80B - 42)A - 50B^3 + 20B^2 - 177B + 75)\},$
$\mathcal{S}_6 = \mathbb{V}(64A^2 - 272AB + 64B^2 + 225) \setminus \mathcal{S}_3 \cup \mathcal{S}_4 \cup \mathcal{S}_5,$

$G_7 = \{1\}, \mathcal{S}_7 = \mathbb{V}(-10881A - 10000B^3 + 8400B^2 + 9744B + 3925, 25A^2 - 17A + 25B^2 - 17B - 25, (400B - 64)A - 64B - 425) = \{(\alpha_1 + \beta_1 i, \alpha_1 - \beta_1 i), (\alpha_1 - \beta_1 i, \alpha_1 + \beta_1 i), (-\alpha_2 - \beta_2 i, -\alpha_2 + \beta_2 i), (\alpha_2 + \beta_2 i, -\alpha_2 - \beta_2 i)\} with \alpha_1 \fallingdotseq 1.16856, \beta_1 \fallingdotseq 0.266288, \alpha_2 \fallingdotseq 0.668559, \beta_2 \fallingdotseq 0.633712.$

Note that the $\mathbb{C}$-vector space $\mathbb{C}[X, Y]/\langle F(a, b) \rangle$ has dimension 4, 4, 4, 2, 2, 2 and 1 for $(a, b) \in \mathcal{S}_1, \mathcal{S}_2, \mathcal{S}_3, \mathcal{S}_4, \mathcal{S}_5, \mathcal{S}_6$ and $\mathcal{S}_7$ respectively. Even though $\mathcal{S}_1, \mathcal{S}_2$ and $\mathcal{S}_3$ are connected and the $\mathbb{C}$-vector space $\mathbb{C}[X, Y]/\langle F(a, b) \rangle$ has the same dimension 4 on $\mathcal{S}_1, \mathcal{S}_2$ and $\mathcal{S}_3$, we cannot glue them into a single segment as long as we use a reduced CGS. On the other hand, we can glue them into a single segment with the following PBB $\mathcal{B} = \{(B_1, \mathcal{S}'_1, \mathcal{O}_1), \dots, (B_5, \mathcal{S}'_5, \mathcal{O}_5)\}$.

$$B_1 = Y^2 + \tfrac{4(A-4B)}{15}XY + \tfrac{4}{15}(4A - B - 3), XY^2 + \tfrac{16(A-4B)(A-4B+3)}{64A^2 - 272AB + 64B^2 + 225}Y$$
$$+ \tfrac{60(4A-B-3)}{64A^2 - 272AB + 64B^2 + 225}X,$$
$$X^2 + \tfrac{4(B-4A)}{15}XY + \tfrac{4}{15}(4B - A - 3), X^2Y + \tfrac{16(B-4A)(B-4A+3)}{64A^2 - 272AB + 64B^2 + 225}X + \tfrac{60(4B-A-3)}{64A^2 - 272AB + 64B^2 + 225}Y,$$
$$S'_1 = S_1 \cup S_2 \cup S_3 = \mathbb{C}^2 \setminus \mathbb{V}(64A^2 - 272AB + 64B^2 + 225), \mathcal{O}_1 = \{1, X, Y, XY\},$$
$$B_2 = \{X^2 + \tfrac{9}{20}, Y, XY\}, S'_2 = S_4, \mathcal{O}_2 = \{1, X\},$$
$$B_3 = \{X - \tfrac{1}{35}Y, XY + \tfrac{7}{58}, Y^2 + \tfrac{245}{58}, \}, S'_3 = S_5 \mathcal{O}_3 = \{1, Y\},$$
$$B_4 = \{X + \tfrac{(400B^2 - 400B - 36)A - 1600B^3 + 1600B^2 + 519B - 375}{60(20B - 29)}Y,$$
$$XY - \tfrac{32((400B^2 - 400B - 36)A - 1600B^3 + 1600B^2 + 519B - 375)((200B^2 - 80B - 42)A - 50B^3 + 20B^2 - 177B + 75)}{225(20B - 29)((400B - 64)A - 64B - 425)},$$
$$Y^2 + \tfrac{128((200B^2 - 80B - 42)A - 50B^3 + 20B^2 - 177B + 75)}{15((400B - 64)A - 64B - 425)}\}, S'_4 = S_6, \mathcal{O}_4 = \{1, Y\},$$
$$B_5 = \{1\}, S'_5 = S_7, \mathcal{O}_5 = \emptyset.$$

Note also that $\mathbb{C}[X,Y]/\langle F(a,b)\rangle$ has the same dimension 2 on S'_2, S'_3 and S'_4. Even though S'_2, S'_4 and S'_3, S'_4 are connected, however, we cannot glue them into a single segment for both of them. The reason for S'_2, S'_4 is that $\langle F(a,b)\rangle$ has the only one order ideal $\mathcal{O}_2$ on $(a,b) \in S'_2$ (i.e., $(a,b) = (\tfrac{89}{80}, \tfrac{29}{20})$), while S'_4 contains a point $(\tfrac{29}{20}, \tfrac{89}{80})$ such that $\langle F(\tfrac{29}{20}, \tfrac{89}{80})\rangle$ has the only one order ideal $\mathcal{O}_4$ different from $\mathcal{O}_2$. The reason for S'_3, S'_4 is rather subtle. We cannot have a uniform parametric representation for both of B_3 and B_4. Those observations lead us to the following definition of a *minimal* PBB.

Definition 5. *A PBB* $\mathcal{B} = \{(B_1, S_1, \mathcal{O}_1), \ldots, (B_r, S_r, \mathcal{O}_r)\}$ *of* $\langle F\rangle$ *is said to be minimal if for any pair* (S_i, S_j) *of connected segments such that* $\mathbb{C}[\bar{X}]/\langle F(\bar{a}, \bar{X})\rangle$ *has the same dimension on them it satisfies either of the following:*

1. $\mathcal{O}_i \neq \mathcal{O}_j$, *but also* $\langle F(\bar{a})\rangle$ *does not possess a common order ideal on* $S_i \cup S_j$.
2. $\mathcal{O}_i = \mathcal{O}_j$ *and there exist no uniform parametric representation for both of* B_i *and* B_j *on* $S_i \cup S_j$.

Where "S_i and S_j are connected" means that $\overline{S_i} \cap \overline{S_j} \cap (S_i \cup S_j) \neq \emptyset$, $\overline{X}$ *denotes the Zariski closure of* X. *Intuitively,* S_i *and* S_j *are connected if and only if there exist two points* $\bar{a}_i \in S_i$ *and* $\bar{a}_j \in S_j$ *which are connected by a continuous path in* $S_i \cup S_j$.

Note that a Gröbner basis can be considered as a border basis with the naturally induced order ideal, we can convert a reduced CGS into a PBB using uniform parametric monomial reductions on each segment. Hence, we can compute a PBB of any given $\langle F\rangle$. Existence of a minimal PBB is also obvious, however, we have not obtained an effective algorithm yet. The reason is that we do not have an algorithm to decide whether the property 2 holds yet, while it is easy to check the property 1 using the (parametric) border division algorithm by B_i on S_i and by B_j on S_j. At this time, we have obtained the following results.

Lemma 6. *Let* $(B, S, \mathcal{O})$ *be a member of a PBB* $\mathcal{B}$ *of* $\langle F\rangle$ *such that* $S = \mathbb{C}^m \setminus \mathbb{V}(I)$ *for some ideal* $I \subset \mathbb{Q}[\bar{A}]$. *If there are other members* $(B_{n_1}, S_{n_1}, \mathcal{O}_{n_1}), \ldots, (B_{n_k}, S_{n_k}, \mathcal{O}_{n_k})$ *of* $\mathcal{B}$ *such that* $\mathbb{C}[\bar{X}]/\langle F(\bar{a}, \bar{X})\rangle$ *has the same dimension on* $S \cup S_{n_1} \cup \cdots \cup S_{n_k}$ *and* $\langle F(\bar{a}, \bar{X})\rangle$ *also has a unique order ideal* $\mathcal{O}'$ *on every* $\bar{a} \in S \cup S_{n_1} \cup \cdots \cup S_{n_k}$, *then we can compute a finite subset* B' *of* $\mathbb{Q}(\bar{A})[\bar{X}]$ *such that* $B'(\bar{a})$ *is a* $\mathcal{O}'$-*border basis of* $\langle F(\bar{a})\rangle$ *on* $S \cup S_{n_1} \cup \cdots \cup S_{n_k}$.

In the above example, by this lemma, we can glue $(G_1, S_1), (G_2, S_2), (G_3, S_3)$ into $(B_1, \mathcal{S}_1', \mathcal{O}_1)$ with $\mathcal{S}_1' = \mathcal{S}_1 \cup \mathcal{S}_2 \cup \mathcal{S}_3$ and the order ideal $\mathcal{O}_1$ induced from (G_1, S_1).

Lemma 7. *Let $(B_i, \mathcal{S}_i, \mathcal{O})$ and $(B_j, \mathcal{S}_j, \mathcal{O})$ be members of a PBB. If there exists $\bar{a} \in \mathcal{S}_i \cap \overline{\mathcal{S}_j}$ such that we cannot specialize some $t + h(\bar{A}, \bar{X}) \in B_j$ with $t \in \partial\mathcal{O}$ and $\bar{A} = \bar{a}$, then there exists no uniform parametric representation for B_i and B_j on $\mathcal{S}_i \cup \mathcal{S}_j$.*

In the above example, B_3 and B_4 do not have a uniform parametric representation since the denominator $60(20B - 29)$ of a coefficient of a polynomial in B_4 vanishes for $(A, B) = (\frac{101}{20}, \frac{29}{20}) \in \mathcal{S}_3' \cap \overline{\mathcal{S}_4'}$.

4   Stability of Parametric Border Basis

Numerical stability is one of the most important properties of border bases. We give a precise definition of the stability of a border basis of a parametric ideal as follows.

Definition 8. *Let F be a finite subset of $\mathbb{Q}[\bar{A}, \bar{X}]$ and $\mathcal{S}$ be a subset (not necessary to be algebraically constructible) of $\mathbb{C}^m$ such that the $\mathbb{C}$-vector space $\mathbb{C}[\bar{X}]/\langle F(\bar{a}, \bar{X}) \rangle$ has an invariant finite dimension for every $\bar{a} \in \mathcal{S}$. For $\bar{a} \in \mathcal{S}$ which is not an isolated point of $\mathcal{S}$, let $\langle F(\bar{a}, \bar{X}) \rangle$ have a $\mathcal{O}$-border basis $B = \{t_1 + g_1, \ldots, t_l + g_l\}$ with $\{t_1, \ldots, t_l\} = \partial\mathcal{O}$ and $g_1, \ldots, g_l \in \mathbb{C}[\bar{X}]$ for some order ideal $\mathcal{O} = \{s_1, \ldots, s_k\}$. If there exists an open neighborhood $\mathcal{S}' \subset \mathcal{S}$ of $\bar{a}$ such that $\langle F(\bar{c}, \bar{X}) \rangle$ has an invariant order ideal $\mathcal{O}$ together with a $\mathcal{O}$-border basis $\{t_1 + \phi_1^1(\bar{c})s_1 + \cdots + \phi_k^1(\bar{c})s_k, \ldots, t_l + \phi_1^l(\bar{c})s_1 + \cdots + \phi_k^l(\bar{c})s_k\}$ for each $\bar{c} \in \mathcal{S}'$ with mappings ϕ_j^i from $\mathcal{S}'$ to $\mathbb{C}$. (Note that it is uniquely determined.) In addition, if these mappings are continuous at $\bar{A} = \bar{a}$ that is $\lim_{\bar{c} \to \bar{a}} \phi_1^i(\bar{c})s_1 + \cdots + \phi_k^i(\bar{c})s_k = g_i$ for each $i = 1, \ldots, l$, then we say B is stable at $\bar{A} = \bar{a}$ in $\mathcal{S}$.*

Unfortunately, the stability property does not hold for some parametric ideal $\langle F(\bar{A}, \bar{X}) \rangle$.

Example 9. *Let $F = \{A(X - Y), AX^4 + X^2 + A - 1, AY^4 + Y^2 + A - 1\}$. $\mathbb{C}[X, Y]/\langle F(a) \rangle$ has dimension 4 for any $a \in \mathcal{S} = \mathbb{C}$. Possible order ideals of $\langle F(a) \rangle$ are $\{1, X, X^2, X^3\}$ and $\{1, Y, Y^2, Y^3\}$ for $a \neq 0$ but only $\{1, X, Y, XY\}$ for $a = 0$. Hence, the $\{1, X, Y, XY\}$-border basis B of $\langle F(0) \rangle$ is not stable at $A = 0$ in $\mathcal{S}$.*

In case a parametric ideal has an invariant order ideal in some connected region $\mathcal{S}$ its border basis seems to be stable at any point of $\mathcal{S}$, although we have not proved it yet.

Example 10. *For the example of the previous section, $\langle F(a, b, X, Y) \rangle$ has an order ideal $\{1, Y\}$ for every $(a, b) \in \mathcal{S}_3' \cup \mathcal{S}_4'$. As is mentioned at the end of previous section, we do not have a uniform parametric representation of the $\{1, Y\}$-border basis of $\langle F(a, b, X, Y) \rangle$ for every $(a, b) \in \mathcal{S}_3' \cup \mathcal{S}_4'$. It seems that the $\{1, Y\}$-border* *basis of* $\langle F(a,b,X,Y) \rangle$ *is not stable at* $(a,b) = (\frac{101}{20}, \frac{29}{20})$. *But it is actually stable at* $(A,B) = (\frac{101}{20}, \frac{29}{20})$ *in* $\mathcal{S}_3' \cup \mathcal{S}_4'$. *That is* $\frac{(400B^2 - 400B - 36)A - 1600B^3 + 1600B^2 + 519B - 375}{60(20B - 29)}$, $\frac{32((400B^2 - 400B - 36)A - 1600B^3 + 1600B^2 + 519B - 375)((200B^2 - 80B - 42)A - 50B^3 + 20B^2 - 177B + 75)}{225(20B - 29)((400B - 64)A - 64B - 425)}$ *and* $\frac{128((200B^2 - 80B - 42)A - 50B^3 + 20B^2 - 177B + 75)}{15((400B - 64)A - 64B - 425)}$ *converge to* $-\frac{1}{35}, -\frac{7}{58}$ *and* $\frac{245}{58}$ *as* $(A,B) \to (\frac{101}{20}, \frac{29}{20})$ *in* $\mathcal{S}_3' \cup \mathcal{S}_4'$.

5 Conclusion and Remarks

A *terrace* introduced in [8] is an ideal algebraic structure for a canonical representation of a comprehensive Gröbner system. It is the smallest commutative von Neumann regular ring extending $\mathbb{Q}[\bar{A}]$, meanwhile $\mathbb{Q}(\bar{A})$ is the smallest field extending $\mathbb{Q}[\bar{A}]$. If we are allowed to use this structure to represent coefficients of parametric polynomials, we can also similarly define a PBB and a minimal PBB. For the definition of a minimal PBB, we do not need the property 2, that is we always have $\mathcal{O}_i \neq \mathcal{O}_j$. Furthermore the better thing is that we can always compute it, though we have not tried to use it yet since the implementation of the structure of terrace is not very straightforward.

References

1. Fukasaku, R., Iwane, H., Sato, Y.: Real quantifier elimination by computation of comprehensive Gröbner systems. In: Proceedings of ISSAC 2015, pp. 173–180 (2015)
2. Fukasaku, R., Iwane, H., Sato, Y.: On the implementation of CGS real QE. In: Greuel, G.-M., Koch, T., Paule, P., Sommese, A. (eds.) ICMS 2016. LNCS, vol. 9725, pp. 165–172. Springer, Cham (2016). https://doi.org/10.1007/978-3-319-42432-3_21
3. Fukasaku, R., Sato, Y.: On real roots counting for non-radical parametric ideals. In: Blömer, J., Kotsireas, I.S., Kutsia, T., Simos, D.E. (eds.) MACIS 2017. LNCS, vol. 10693, pp. 258–263. Springer, Cham (2017). https://doi.org/10.1007/978-3-319-72453-9_18
4. Fukasaku, R., Iwane, H., Sato, Y.: On multivariate hermitian quadratic forms. Math. Comput. Sci. **13**(1–2), 79–93 (2019)
5. Kreuzer, M., Robbiano, L.: Computational Commutative Algebra 2. Springer, Heidelberg (2005). https://doi.org/10.1007/3-540-28296-3. Section 6.4 Border Bases
6. Montes, A.: The Gröbner Cover. ACM, vol. 27. Springer, Cham (2018). https://doi.org/10.1007/978-3-030-03904-2
7. Sato, Y., Fukasaku, R., Sekigawa, H.: On continuity of the roots of a parametric zero dimensional multivariate polynomial ideal. In: Proceedings of ISSAC 2018, pp. 359–365 (2018)
8. Suzuki, A., Sato, Y.: An alternative approach to comprehensive Gröbner bases. J. Symb. Comput. **36**(3–4), 649–667 (2003)
9. Weispfenning, V.: A new approach to quantifier elimination for real algebra. In: Caviness, B.F., Johnson, J.R. (eds.) Quantifier Elimination and Cylindrical Algebraic Decomposition. TEXTSMONOGR, pp. 376–392. Springer, Vienna (1998). https://doi.org/10.1007/978-3-7091-9459-1_20

Reliable Computation of the Singularities of the Projection in $\mathbb{R}^3$ of a Generic Surface of $\mathbb{R}^4$

Sény Diatta[1,2], Guillaume Moroz[2], and Marc Pouget[2(✉)]

[1] University Assane Seck of Ziguinchor, Ziguinchor, Senegal
`senydiatta@gmail.com`
[2] Université de Lorraine, CNRS, Inria, LORIA, 54000 Nancy, France
`{Guillaume.Moroz,Marc.Pouget}@inria.fr`

Abstract. Computing efficiently the singularities of surfaces embedded in $\mathbb{R}^3$ is a difficult problem, and most state-of-the-art approaches only handle the case of surfaces defined by polynomial equations. Let F and G be C^∞ functions from $\mathbb{R}^4$ to $\mathbb{R}$ and $\mathcal{M} = \{(x, y, z, t) \in \mathbb{R}^4 \mid F(x, y, z, t) = G(x, y, z, t) = 0\}$ be the surface they define. Generically, the surface $\mathcal{M}$ is smooth and its projection Ω in $\mathbb{R}^3$ is singular. After describing the types of singularities that appear generically in Ω, we design a numerically well-posed system that encodes them. This can be used to return a set of boxes that enclose the singularities of Ω as tightly as required. As opposed to state-of-the art approaches, our approach is not restricted to polynomial mapping, and can handle trigonometric or exponential functions for example.

1 Introduction

Consider two real analytic functions F, G defined in $\mathbb{R}^4$ and denote by $\mathcal{M}$ the smooth surface defined as the real common zeros of F and G. Let $\mathfrak{p}$ be the projection map from $\mathcal{M}$ to $\mathbb{R}^3$ along the direction $(0, 0, 0, 1)$ and Ω the image of $\mathcal{M}$ by $\mathfrak{p}$. The goal of this paper is to take advantage of the structure of the singularities of Ω and to present a regular system allowing to isolate them efficiently. Computing the singularities of such surfaces is fundamental for the reliable visualization of surfaces, and for problems that arise in fields such as mechanical design, control theory or biology.

The modern theory of singularities started with Whitney, Thom and Mather and the classification of singularities is an active research domain since then. Most of the literature focus on the local case of germs of functions, and only more recently the case of multigerms, that is taking into account the interplay of several points in the source space at once, attracted more attention, see e.g. [16] and references therein. Particularly relevant for our work is the case of functions from a surface to $\mathbb{R}^3$ which is studied in [7,8,13].

Unfortunately, these classifications do not lead directly to algorithms computing explicitly the singularities associated to varieties. Still, in [3], a numeric

© Springer Nature Switzerland AG 2020
D. Slamanig et al. (Eds.): MACIS 2019, LNCS 11989, pp. 16–34, 2020.
https://doi.org/10.1007/978-3-030-43120-4_3

approach is presented for computing the apparent contour of a function from the plane to itself. In [9], the authors proposed a reliable numeric algorithm to compute the singularities of the projection of smooth curves from $\mathbb{R}^3$ to $\mathbb{R}^2$, using a so-called *Ball system*. We generalize this approach to compute the singularities of the projection of smooth surfaces from $\mathbb{R}^4$ to $\mathbb{R}^2$.

After recalling some results from singularity and transversality theory in Sect. 2, we prove our first result on the types of singularities in Ω, the projection of a generic smooth compact surface, in Sect. 3. We prove in Sect. 4 that Equations (S-Ball) define a regular system that can be used to compute the set of singularities of Ω. Finally, in Sect. 5, we will illustrate our approach with the classical Whitney Umbrella, and with the computation of the singularities of a surface that cannot be handled by state-of-the-art method up to our knowledge.

Notation and Main Results
In the following, the surface $\mathcal{M}$ is a compact smooth 2-submanifold of $\mathbb{R}^4$ defined by the zero locus of the C^∞ functions F and G. We denote by $S_{compact}$ the subset of mappings in $C^\infty(\mathbb{R}^4, \mathbb{R}^2)$ that implicitly define a compact surface. With the coordinates (x, y, z, t) on $\mathbb{R}^4$, we denote $\mathfrak{p} : \mathcal{M} \to \mathbb{R}^3$ the projection along the t-axis, and Ω is the image of $\mathcal{M}$ by $\mathfrak{p}$. We call a plane in $\mathbb{R}^4$ vertical if it is parallel to the t-axis, that is it contains the vector $(0, 0, 0, 1)$. The tangent plane $\mathcal{P}$ of $\mathcal{M}$ at a point q is the set of vectors orthogonal to both $\nabla F(q) = (\partial_x F, \partial_y F, \partial_z F, \partial_t F)(q)$ and $\nabla G(q)$. Thus the tangent plane at q is vertical iff $\partial_t F(q) = \partial_t G(q) = 0$. We say that a property is *generic* if it is satisfied by a countable intersection of open dense sets of C^∞ mappings (see [4, §3.2.6]). The open sets we consider are given by the Whitney topology (as defined in [4, p.45] or [6, chap. II §3]) on the space of smooth maps $C^\infty(\mathbb{R}^4; \mathbb{R}^2)$, restricted to $S_{compact}$.

Our first result is a description of the generic singularities of Ω in terms of singularities of the projection map. We prove that Ω generically has only 3 kinds of singularities whose definition is given in [8], and recalled in Definition 3.

Theorem 1. *(Generic properties)*

1. *The surface defined by $F = G = 0$ is generically smooth.*
2. *The singularities of the projection in $\mathbb{R}^3$ of a generic compact surface of $\mathbb{R}^4$ is a curve $\mathcal{C}$ of double points having as singularities a discrete set of triple points and cross-caps.*

To compute the curve $\mathcal{C}$ of double points, a naive approach consists in duplicating the last variable, as in Equation (S-dble) of Sect. 4.1. However, this leads to a system that is not regular near the cross-caps. Thus, such an approach is not suitable for numerical solvers such as path continuation or subdivision algorithms.

Our second result shows that the computation of the singular curve $\mathcal{C}$ can be reduced to solving the regular system (S-Ball) of 4 equations in 5 variables. We call this system the *Ball system* as in [9] where the same approach was used for the projection of a space curve in the plane. We first define the operators S and D applied to a given smooth function A defined on $\mathbb{R}^4$.

$$
S.A(x, y, z, c, r) = \begin{cases} \frac{1}{2}(A(x, y, z, c + \sqrt{r}) + A(x, y, z, c - \sqrt{r})) & \text{if } r > 0 \\ A(x, y, z, c) & \text{if } r = 0, \end{cases}
$$

$$
D.A(x, y, z, c, r) = \begin{cases} \frac{1}{2\sqrt{r}}(A(x, y, z, c + \sqrt{r}) - A(x, y, z, c - \sqrt{r})) & \text{if } r > 0 \\ \partial_t A(x, y, z, c) & \text{if } r = 0. \end{cases}
$$

We then define the *Ball system* as

$$
\begin{cases} S.F(x, y, z, c, r) = 0 \\ S.G(x, y, z, c, r) = 0 \\ D.F(x, y, z, c, r) = 0 \\ D.G(x, y, z, c, r) = 0. \end{cases} \tag{S-Ball}
$$

Theorem 2. *(Computation of the singularities)*
Let $\mathcal{M} \subset \mathbb{R}^4$ be a compact surface solution of $F = G = 0$ that satisfies the generic properties of Theorem 1. Let $\mathcal{C}_{Ball}$ be the curve solution of the system (S-Ball).

1. *The points of $\mathcal{C}_{Ball}$ are regular points of System* (S-Ball).
2. *The projection of $\mathcal{C}_{Ball}$ to $\mathbb{R}^3$ is the singular locus $\mathcal{C}$ of Ω.*

A direct corollary of this theorem is that one can enclose the curve of singularities of Ω using state–of-the-art numerical algorithms such as the one presented in [11] for example.

2 Preliminaries

Before enumerating the different types of singularities that can appear on the projection in $\mathbb{R}^3$ of a generic surface of $\mathbb{R}^4$, we recall some basic definitions on regularity and transversality theory.

2.1 Regular, Critical and Singular Points

Definition 1. *(Regular and critical points of $\mathfrak{p}$)*

- **Regular point of $\mathfrak{p}$.** *A point $q \in \mathcal{M}$ is a* regular point of $\mathfrak{p}$ *when its derivative has full rank, that is* $\text{rank}(d\mathfrak{p})_q = 2$. *This is equivalent to say that the tangent plane to $\mathcal{M}$ at q is not vertical.*
- **Critical point of $\mathfrak{p}$.** *A point $q \in \mathcal{M}$ which is not a regular point of $\mathfrak{p}$ is called a* critical point of $\mathfrak{p}$. *Equivalently the tangent plane at q is vertical i.e. $\partial_t F(q) = \partial_t G(q) = 0$.*

Let P be a point of Ω, we say that a point $q \in \mathfrak{p}^{-1}(P) \subset \mathcal{M}$ is a regular (resp. singular) pre-image of P, if q is a regular (resp. critical) point of $\mathfrak{p}$.

Definition 2. *(Regular points of a variety or a system)*

- **Regular point of Ω.** *A point $P \in \Omega$ is a* regular point of Ω *if Ω is locally a 2-submanifold of $\mathbb{R}^3$, otherwise, it is a* singular point *of Ω.*

- **Regular solution of a system.** *A solution of a square system is regular if the Jacobian determinant does not vanish at this solution. When there are more variables than equations, one requires that the Jacobian matrix is full rank (i.e. the associated linear map is surjective) at the solution.*

For a point $P \in \Omega$ with pre-images $q_i \in \mathcal{M}$, we denote $\mathcal{P}_i$ the tangent plane of $\mathcal{M}$ at q_i and Π_i its image by $\mathfrak{p}$. We distinguish three types of singular points of Ω that are illustrated in Fig. 1.

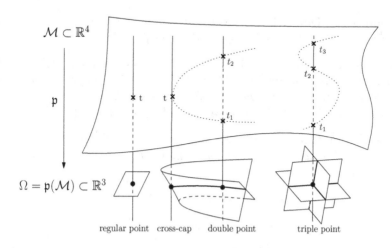

Fig. 1. Types of singularities of $\Omega = \mathfrak{p}(\mathcal{M})$ with their pre-images on $\mathcal{M}$

Definition 3. *(Singular points of Ω)*

- **Double point.** $P \in \Omega$ *is a double point if it has two regular pre-images q_1 and q_2 in $\mathcal{M}$ and $\Pi_1 \cap \Pi_2$ is a line. According to the classification in [8, Table 1], P is the image of a singularity of type A_0^2 of the mapping $\mathfrak{p}$.*
- **Triple point.** $P \in \Omega$ *is a triple point if it has three regular pre-images q_1, q_2 and q_3 and $\cap_{\{1 \le i \le 3\}} \Pi_i$ is a point. According to the classification in [8, Table 1], P is the image of a singularity of type A_0^3 of the mapping $\mathfrak{p}$.*
- **Cross-cap.** $P \in \Omega$ *is a cross-cap if it has one critical pre-image q in $\mathcal{M}$ and q is a singularity of type cross-cap of $\mathfrak{p}$ according to Definition 7. According to the classification in [8, Table 1], P is the image of a singularity of type S_0 of the mapping $\mathfrak{p}$.*

We use the following characterization of cross-caps in our particular setting. It is adapted from [12] and a private communication with David Mond, a proof is in the appendix.

Lemma 1 ([12]). *The projection $\mathfrak{p}$ has a singularity of type cross-cap iff the direction of projection is in the tangent plane and assuming wlog (indeed the surface can be parameterized by either (x,t), (y,t) or (z,t)) that $\mathcal{M}$ has a local parameterization of the form $(a(z,t), b(z,t), z, t)$, one has $\partial_{zt}a\partial_{tt}b - \partial_{tt}a\partial_{zt}b \ne 0$.*

2.2   Transversality and Genericity

For the results of Sect. 3, we introduce the relevant tools from singularity theory and in particular the notion of transversality.

Definition 4 ([4, **Definition 2.5.1**]). *Let E be a finite-dimensional vector space, the subspaces T and T' are transverse if $T + T' = E$.*

The notion of *transversality* extends to functions via the tangent map.

Definition 5 ([4, **Definition 3.7.1**]). *Let E, F be finite vector spaces, V and W be submanifolds of E and F respectively, and $f \in C^\infty(V; F)$.*

- *f is transverse to W at $q \in V$ if either $f(q)$ does not belong to W or $f(q)$ belongs to W and the image of the tangent space $T_q V$ by the tangent linear map $df(q)$ is transverse to the tangent space $T_{f(q)} W$.*
- *f is transverse to W if it is transverse to W at every point q of V.*

Definition 6 ([4, **§3.8.3**]). *Let r be a non-negative integer and E, F two finite-dimensional vector spaces. Let V be a submanifold of E and $f \in C^\infty(V; F)$. Then, the map*

$$j^r f : V \to J^r(V, F)$$
$$q \mapsto (q, f(q), f'(q), \ldots, f^{(r)}(q))$$

is called the r-jet of f and $J^r(V, F)$ is called the space of jets of order r of maps from V to F.

In our setting, $\mathfrak{p}$ is a mapping from $\mathcal{M}$ to $\mathbb{R}^3$. We denote by Σ^1 the submanifold of $J^1(\mathcal{M}, \mathbb{R}^3)$ of jets of corank 1, that is such that the linear map the jet defines from $T_q\mathcal{M}$ to $\mathbb{R}^3$ has corank 1 (with corank $= \min(\dim(\mathcal{M}), \mathbb{R}^3) -$rank $= 2-$rank). We then denote $\Sigma^1(\mathfrak{p}) = (j^1\mathfrak{p})^{-1}(\Sigma^1)$.

Definition 7 ([6, **Definition 4.5**]). *A point q of $\mathcal{M}$ is a cross-cap of $\mathfrak{p}$ if it is in $\Sigma^1(\mathfrak{p})$ and $j^1\mathfrak{p}$ is transverse to Σ^1 at q.*

We now state Thom's transversality theorem which is the main tool to determine the generic properties of projected surfaces (see Theorem 1).

Proposition 1 ([4, **Theorem 3.9.4**]). *Let E and F be two finite-dimensional vector spaces, with U an open set in E. Let r be an integer, and let W be a submanifold of $J^r(U; F)$. Then the set of maps $f \in C^\infty(U; F)$ such that $j^r f$ is transverse to W is a dense residual subset of $C^\infty(U; F)$. In other words, for generic f in $C^\infty(U; F)$ the map $j^r f$ is transverse to W. In addition, in this case, $(j^r f)^{-1}(W)$ is a submanifold of U of codimension equal to codim(W).*

Finally, we show that the subset $S_{compact}$ of mappings that define a compact set is open, such that a residual subset of C^∞ mappings is also a residual set in the set of mappings that define implicitly compact sets.

Lemma 2. $S_{compact}$ *is open in* $C^\infty(\mathbb{R}^4, \mathbb{R}^2)$ *equipped with the Whitney topology.*

Proof. If f_n is a sequence that converges toward f in $C^\infty(\mathbb{R}^4, \mathbb{R}^2)$, then according to [6, p.43], there exists a compact set $K \subset \mathbb{R}^4$ and an integer n such that $f_n(x) = f(x)$ for all $x \in \mathbb{R}^4 \setminus K$. This implies that $C^\infty(\mathbb{R}^4, \mathbb{R}^2) \setminus S_{compact}$ is a closed set, which concludes the proof. $\square$

3   Generic Properties of Projected Surfaces

In this section, we prove Theorem 1 describing the expected geometric structure of a projected surface, it is similar to [4, Prop. 4.7.8] for the apparent contour of a generic surface in $\mathbb{R}^3$.

Proof (of Theorem 1). First, we remark that if $\mathcal{M}$ is a smooth compact surface, and p is a point of Ω that has one regular pre-image q by $\mathfrak{p}$, then it is a regular point of Ω. Indeed by the regularity of q, there exists a neighbourhood U of q in $\mathcal{M}$ such that all points of U are regular for the projection $\mathfrak{p}$. Moreover, let us show that there exists a neighbourhood V of p such that $\mathfrak{p}^{-1}(V) \subset U$, then $\mathfrak{p}$ is an embedding between $\mathfrak{p}^{-1}(V)$ and V and thus p is a regular point of Ω. By contradiction, assume that for any neighbourhood V of p, $\mathfrak{p}^{-1}(V) \not\subset U$. Then one can construct a sequence $p_i \in \Omega$ converging to p such that $q_i = \mathfrak{p}^{-1}(p_i) \notin U$. By compacity of $\mathcal{M}$, one can assume that q_i converges to $q' \in \mathcal{M}$. By continuity of $\mathfrak{p}$, $\mathfrak{p}(q') = p$ and since p has a unique pre-image, one conclude that $q' = q$. This is in contradiction with the fact that the q_i are not in U which is a neighbourhood of q.

Using the Transversality Theorem 1 and its multijet version [4, Thm 3.9.7] we prove that generically: (a) $\mathcal{M}$ is smooth, (b) if a point of Ω has 2 pre-images by $\mathfrak{p}$ then it is a double point, (c) if a point of Ω has more than 2 pre-images then it is a triple point, (d) if a point p of Ω has a pre-image q and the tangent plane to $\mathcal{M}$ at q is vertical, then p is a cross-cap.

Let $\Delta_{(n)}(U)$ denote the subset of U^n consisting of n-tuples of pairwise distinct points and let $J^r_{(n)}(U, F)$ be the space of n-multijets of order r of maps from U to F (see [4, §3.9.6] for details). The idea is to express a geometric property as a submanifold of a jet space $J^r_{(n)}(U, F)$ such that the number of equations defining this submanifold coincides with its codimension. The transversality theorem then yields that generically the geometric property is satisfied on a submanifold of the original space with the same codimension. In particular when the codimension is larger than the dimension of the original space this means that the geometric property generically does not hold.

(a) Consider the jet of order 0:

$$j^0(F, G) : \ \mathbb{R}^4 \to J^0(\mathbb{R}^4, \mathbb{R}^2)$$
$$q \mapsto (q, F(q), G(q)).$$

The set $W = \{F(q) = G(q) = 0\}$ is a linear submanifold of $J^0(\mathbb{R}^4, \mathbb{R}^2)$ of codimension 2. The transversality theorem yields that, generically, the set

$\mathcal{M} = j^0(F, G)^{-1}(W)$ is a smooth surface, i.e. a 2-dimensional submanifold of $\mathbb{R}^4$.

(b) Let $q_i = (x_i, y_i, z_i, t_i)$ in $\mathbb{R}^4$. We consider the 2-multijet defined by:

$$j^1_{(2)}(F, G) : \ \Delta_{(2)}(\mathbb{R}^4) \to J^1_{(2)}(\mathbb{R}^4, \mathbb{R}^2)$$

$$(q_1, q_2) \mapsto (q_1, F(q_1), G(q_1), \nabla F(q_1), \nabla G(q_1),$$
$$q_2, F(q_2), G(q_2), \nabla F(q_2), \nabla G(q_2)).$$

The set $W = \{x_1 = x_2, y_1 = y_2, z_1 = z_2, F(q_1) = G(q_1) = F(q_2) = G(q_2) = 0\}$ is a linear submanifold of $J^1_{(2)}(\mathbb{R}^4, \mathbb{R}^2)$ of codimension 7. The transversality theorem yields that, generically, the set of pairs of distinct points of $\mathcal{M}$ that project to the same point of $\Omega = \mathfrak{p}(\mathcal{M})$ is a 1-dimensional submanifold of $\Delta_{(2)}(\mathbb{R}^4)$. In addition, generically, both points q_i are regular points of the projection $\mathfrak{p}$, since if it were not the case and q_1 were critical then this would add the two equations $\partial_t F(q_1) = \partial_t G(q_1) = 0$. This defines a 9-codimensional submanifold of $J^1_{(2)}(\mathbb{R}^4, \mathbb{R}^2)$ which pull back in $\Delta_{(2)}(\mathbb{R}^4)$ of dimension 8 must be void. Similarly, adding the condition that the tangent spaces Π_1 and Π_2 coincide would add two equations to W and thus generically does not hold.

(c) Consider the 3-multijet

$$j^0_{(3)}(F, G) : \ \Delta_{(3)}(\mathbb{R}^4) \to J^0_{(3)}(\mathbb{R}^4, \mathbb{R}^2)$$

$$(q_1, q_2, q_3) \mapsto (q_1, F(q_1), G(q_1), q_2, F(q_2), G(q_2), q_3, F(q_3), G(q_3))$$

The condition to have 3 points in $\mathcal{M}$ that project to the same point in Ω can be written in $J^0_{(3)}(\mathbb{R}^4, \mathbb{R}^2)$ as $\{x_1 = x_2 = x_3, y_1 = y_2 = y_3, z_1 = z_2 = z_3, F(q_i) = G(q_i) = 0, 1 \le i \le 3\}$ which is a submanifold of codimension 12, that is exactly the dimension of $\Delta_{(3)}(\mathbb{R}^4)$. By the transversality theorem, there is thus generically a discrete set of such points. In addition, extending this jet at order 1, the condition that the intersection of the tangent planes $\cap_{i=1}^3 \Pi_i$ is not a point or one of the points is critical for the projection would add other equations and thus this generically does not occur. Similarly, using a 4-multijet, one proves that there cannot be more than 3 distinct points projecting to the same point. The set of triple points of Ω is thus generically a discrete set.

(d) Consider the jet of order 1:

$$j^1(F, G) : \ \mathbb{R}^4 \to J^1(\mathbb{R}^4, \mathbb{R}^2)$$

$$q \mapsto (q, F(q), G(q), \nabla F(q), \nabla G(q))$$

The set critical points of $\mathfrak{p}$ can be written in $J^1(\mathbb{R}^4, \mathbb{R}^2)$ as $\{F(q) = G(q) = \partial_t F(q) = \partial_t G(q) = 0\}$ which is a submanifold of codimension 4, so that generically there is a discrete set of such points. To prove that these are generically cross-caps using Lemma 1, one has to use a jet of order 2 together with a local parameterization of $\mathcal{M}$ to see that with the additional condition $\partial_{zt}a\,\partial_{tt}b - \partial_{tt}a\,\partial_{zt}b = 0$ one defines a submanifold of codimension 5.

The conclusion is that, generically, the singular points of Ω have at most 3 pre-images. When there is only one pre-image, it is a critical point of $\mathfrak{p}$ and the point on Ω is a cross-cap. When there are 2 or 3 pre-images they are all regular points of $\mathfrak{p}$, and this gives a 1-dimensional curve of double points with a discrete set of triple points.   □

4 Computing the Singularities of the Projected Surface

Within this section, we assume the generic properties of Theorem 1 hold. The surface Ω is thus the disjoint union of regular points, double points, triple points and cross-caps.

4.1 Systems Encoding Singularities

We define the systems (S-dble), (S-tple) and (S-cros) to encode the singularities of the surface Ω in higher dimensional spaces. Figure 1 illustrates the geometry of these systems.

$$
\begin{cases}
F(x,y,z,t_1) = 0 \\
G(x,y,z,t_1) = 0 \\
F(x,y,z,t_2) = 0 \\
G(x,y,z,t_2) = 0 \\
\quad t_1 \neq t_2
\end{cases}
\text{(S-dble)}
\qquad
\begin{cases}
F(x,y,z,t_1) = 0 \\
G(x,y,z,t_1) = 0 \\
F(x,y,z,t_2) = 0 \\
G(x,y,z,t_2) = 0 \\
F(x,y,z,t_3) = 0 \\
G(x,y,z,t_3) = 0 \\
\quad t_i \neq t_j \ for \ i \neq j.
\end{cases}
\text{(S-tple)}
\qquad
\begin{cases}
F(x,y,z,t) = 0 \\
G(x,y,z,t) = 0 \\
\partial_t F(x,y,z,t) = 0 \\
\partial_t G(x,y,z,t) = 0
\end{cases}
\text{(S-cros)}
$$

One remarks that the solutions of system (S-dble) come in pairs by exchanging the t_1 and t_2 coordinates. Also a solution of system (S-tple) yields three pairs of solutions of system (S-dble). We will define in Sect. 4.3 the additional system (S-Ball) that gathers the double points, the triple points and the cross-caps. This subsection is devoted to the proof of the following theorem.

Theorem 3. *A point $p = (x,y,z) \in \Omega$ is a*

1. *Double point iff it has exactly two regular pre-images (x,y,z,t_1) and (x,y,z,t_2), and the Jacobian matrix associated to (S-dble) has maximum rank at (x,y,z,t_1,t_2).*
2. *Triple point iff it has three regular pre-images that give a regular solution of (S-tple).*
3. *Cross-cap iff it has one critical pre-image that is a regular solution of (S-cros).*

4.2   Regularity

We decompose the proof of Theorem 3 in several lemmas. We show that, generically, the double points are encoded by the system (S-dble) where its Jacobian has maximum rank (Lemma 3), the triple points are encoded by the regular solutions of system (S-tple) (Lemma 4) and the cross-caps are encoded by the regular solutions of system (S-cros) (Lemma 5).

Lemma 3. *[Theorem 3(1)] A point $P = (x, y, z)$ in Ω is a double point iff it has two regular pre-images $q_1 = (x, y, z, t_1)$ and $q_2 = (x, y, z, t_2)$, and the Jacobian matrix associated to (S-dble) has maximum rank at $\tilde{q} = (x, y, z, t_1, t_2)$.*

Proof. Let P be a double point of Ω with q_1, q_2 its regular preimages by $\mathfrak{p}$. Let $\mathcal{P}_1$ and $\mathcal{P}_2$ be the tangent planes to $\mathcal{M}$ at q_1 and q_2, and Π_1, Π_2 their projections. Since q_1, q_2 are regular points of $\mathfrak{p}$, $\mathcal{P}_1, \mathcal{P}_2$ are not vertical. The Jacobian matrix $\mathcal{J}_1$ associated to the system (S-dble) is

$$
\mathcal{J}_1 = \begin{pmatrix} \partial_x F_1 & \partial_y F_1 & \partial_z F_1 & \partial_{t_1} F_1 & 0 \\ \partial_x G_1 & \partial_y G_1 & \partial_z G_1 & \partial_{t_1} G_1 & 0 \\ \partial_x F_2 & \partial_y F_2 & \partial_z F_2 & 0 & \partial_{t_2} F_2 \\ \partial_x G_2 & \partial_y G_2 & \partial_z G_2 & 0 & \partial_{t_2} G_2 \end{pmatrix}
$$

with $F_i(x, y, z, t_1, t_2) = F(x, y, z, t_i)$ and $G_i(x, y, z, t_1, t_2) = G(x, y, z, t_i)$ for $i = 1, 2$.

We first show that if $\Pi_1 \cap \Pi_2$ is a line, then $\mathcal{J}_1$ has maximum rank at $\tilde{q}$. Consider two non-null vectors $u = (u_x, u_y, u_z, u_1, u_2)$ and $v = (v_x, v_y, v_z, v_1, v_2)$ in $Ker(\mathcal{J}_1(\tilde{q}))$, then we have

$$
\begin{cases} \nabla F(q_1) \cdot (u_x, u_y, u_z, u_1) = 0 \\ \nabla G(q_1) \cdot (u_x, u_y, u_z, u_1) = 0 \end{cases} \quad and \quad \begin{cases} \nabla F(q_2) \cdot (v_x, v_y, v_z, u_2) = 0 \\ \nabla G(q_2) \cdot (v_x, v_y, v_z, u_2) = 0. \end{cases}
$$

Since the tangent plane $\mathcal{P}_i$ to $\mathcal{M}$ at q_i is the set of vectors orthogonal to $\nabla F(q_i)$ and $\nabla G(q_i)$

$$
(u_x, u_y, u_z, u_1) \in \mathcal{P}_1 \quad and \quad (u_x, u_y, u_z, u_2) \in \mathcal{P}_2,
$$

which implies that $(u_x, u_y, u_z) \in \Pi_1 \cap \Pi_2$. Similarly $v \in Ker(\mathcal{J}_1(\tilde{q}))$ implies that $(v_x, v_y, v_z) \in \Pi_1 \cap \Pi_2$. Since $\Pi_1 \cap \Pi_2$ is a line, there exists $\lambda \in \mathbb{R}$ such that $\lambda(u_x, u_y, u_z) = (v_x, v_y, v_z)$.

If $\lambda = 0$ then $(v_x, v_y, v_z) = (0, 0, 0)$, which implies that the vector $(0, 0, 0, v_1)$ is in $\mathcal{P}_1$. This is not possible since q_1 is a regular point of $\mathfrak{p}$ and thus $\mathcal{P}_1$ is not vertical.

If $\lambda \neq 0$, since $\mathcal{P}_1$ is not vertical, at least one of the partial derivatives $\partial_t F$ or $\partial_t G$ is non-null at q_1. Without loss of generality, one can assume that $\partial_t F(q_1) = \partial_{t_1} F_1(\tilde{q}) \neq 0$. Thus $u, v \in Ker(\mathcal{J}_1(\tilde{q}))$ implies

$$
\begin{cases} u_x \partial_x F_1(\tilde{q}) + u_y \partial_y F_1(\tilde{q}) + u_z \partial_z F_1(\tilde{q}) + u_1 \partial_{t_1} F_1(\tilde{q}) = 0 \\ v_x \partial_x F_1(\tilde{q}) + v_y \partial_y F_1(\tilde{q}) + v_z \partial_z F_1(\tilde{q}) + v_1 \partial_{t_1} F_1(\tilde{q}) = 0. \end{cases}
$$

Multiplying the first line by λ and subtracting the second one where (v_x, v_y, v_z) is substituted by $\lambda(u_x, u_y, u_z)$ yields $(\lambda u_1 - v_1)\partial_{t_1} F_1(\tilde{q}) = 0$, thus $\lambda u_1 - v_1 = 0$ and finally $v_1 = \lambda u_1$.

Using the same approach at q_2 for the non-vertical tangent plane $\mathcal{P}_2$, one concludes that $v_2 = \lambda u_2$. So u and v are colinear vectors, thus $\dim(Ker(\mathcal{J}_1(\tilde{q}))) = 1$ and $\mathcal{J}_1(\tilde{q})$ has rank 4 which is maximal.

We now show the converse statement and thus assume that the Jacobian matrix is of maximum rank. By contradiction, if $\Pi_1 \cap \Pi_2$ is not a line, then it is a plane $\Pi := \Pi_1 = \Pi_2$. In this case, one can find two vectors (u_x, u_y, u_z) and (v_x, v_y, v_z) in Π that are linearly independent.

Let $u = (u_x, u_y, u_z, u_1, u_2)$ be such that (u_x, u_y, u_z, u_i) is the pre-image of (u_x, u_y, u_z) in $\mathcal{P}_i$. By definition of the tangent planes, one has

$$\begin{cases} \nabla F(q_1) \cdot (u_x, u_y, u_z, u_1) = 0 \\ \nabla G(q_1) \cdot (u_x, u_y, u_z, u_1) = 0 \end{cases} \quad and \quad \begin{cases} \nabla F(q_2) \cdot (v_x, v_y, v_z, u_2) = 0 \\ \nabla G(q_2) \cdot (v_x, v_y, v_z, u_2) = 0, \end{cases}$$

thus, u is in $Ker(\mathcal{J}_1(\tilde{q}))$. Similarly, let $v = (v_x, v_y, v_z, v_1, v_2)$ be such that (v_x, v_y, v_z, v_i) is the pre-image of (v_x, v_y, v_z) in $\mathcal{P}_i$, we also have that v is in $Ker(\mathcal{J}_1(\tilde{q}))$. Since the vectors (u_x, u_y, u_z) and (v_x, v_y, v_z) are linearly independent, the vectors u and v are also independent so $\dim(Ker(\mathcal{J}_1(\tilde{q}))) \geq 2$ and $\mathcal{J}_1(\tilde{q})$ is not of maximum rank. $\Pi_1 \cap \Pi_2$ is thus necessarily a line. $\square$

Lemma 4. *[Theorem 3(2)] A point $P = (x, y, z)$ in Ω is a triple point iff it has three regular pre-images $(x, y, z, t_i), i = 1, 2, 3$ and $\tilde{q} = (x, y, z, t_1, t_2, t_3)$ is a regular solution of the system (S-tple).*

Proof. Let P be a point in Ω with three regular pre-images q_1, q_2, q_3 by $\mathfrak{p}$. Let $\mathcal{P}_i$ be the tangent plane to $\mathcal{M}$ at q_i, note that $\mathcal{P}_i$ is not vertical since q_i is a regular point of $\mathfrak{p}$. The Jacobian matrix $\mathcal{J}_2$ associated to the system (S-tple) is

$$\mathcal{J}_2 = \begin{pmatrix} \partial_x F_1 & \partial_y F_1 & \partial_z F_1 & \partial_{t_1} F_1 & 0 & 0 \\ \partial_x G_1 & \partial_y G_1 & \partial_z G_1 & \partial_{t_1} G_1 & 0 & 0 \\ \partial_x F_2 & \partial_y F_2 & \partial_z F_2 & 0 & \partial_{t_2} F_2 & 0 \\ \partial_x G_2 & \partial_y G_2 & \partial_z G_2 & 0 & \partial_{t_2} G_2 & 0 \\ \partial_x F_3 & \partial_y F_3 & \partial_z F_3 & 0 & 0 & \partial_{t_3} F_3 \\ \partial_x G_3 & \partial_y G_3 & \partial_z G_3 & 0 & 0 & \partial_{t_3} G_3 \end{pmatrix}$$

with $F_i(x, y, z, t_1, t_2, t_3) = F(x, y, z, t_i)$ and $G_i(x, y, z, t_1, t_2, t_3) = G(x, y, z, t_i)$ for $i = 1, 2, 3$.

If $\mathcal{J}_2(\tilde{q})$ is not invertible, then there exists a non-zero vector $v = (v_x, v_y, v_z, v_1, v_2, v_3) \in Ker(\mathcal{J}_2(\tilde{q}))$. In other words, we have $\nabla F(q_i) \cdot (v_x, v_y, v_z, v_i) = \nabla G(q_i) \cdot (v_x, v_y, v_z, v_i) = 0$ and thus $(v_x, v_y, v_z, v_i) \in \mathcal{P}_i$. This implies that $(v_x, v_y, v_z) \in \cap_{i=1}^3 \Pi_i$ and on the other hand, since $\mathcal{P}_i$ is not vertical, this vector is non-null. We thus have that $\cap_{i=1}^3 \Pi_i$ is not a point.

Conversely, if $\cap_{i=1}^3 \Pi_i$ is not a point, then there exists a non-null vector $(v_x, v_y, v_z) \in \cap_{i=1}^3 \Pi_i$. Let $(v_x, v_y, v_z, v_i) \in \mathcal{P}_i$ be the pre-image of (v_x, v_y, v_z), we then have $\nabla F(q_i) \cdot (v_x, v_y, v_z, v_i) = \nabla G(q_i) \cdot (v_x, v_y, v_z, v_i) = 0$. In other words, $\mathcal{J}_2(\tilde{q}) \cdot (v_x, v_y, v_z, v_1, v_2, v_3) = 0$ and thus $\mathcal{J}_2(\tilde{q})$ is not invertible. $\square$

Lemma 5. *[Theorem 3(3)] A point $P = (x, y, z)$ in Ω is a cross-cap iff it has one critical pre-image that is a regular solution of (S-cros).*

Proof. First note that for a solution q of the system (S-cros), q is in $\mathcal{M}$ and $\partial_t F(q) = \partial_t G(q) = 0$, thus the tangent plane $\mathcal{P}$ to $\mathcal{M}$ at q is vertical which is the first condition for a cross-cap in Lemma 1.

Without loss of generality, one can assume the surface parameterized by the variables z and t. Indeed, $\nabla F(q)$ and $\nabla G(q)$ are independant so that there exists a 2×2 minor with non-null determinant. If we assume $\det \begin{pmatrix} \partial_x F(q) & \partial_y F(q) \\ \partial_x G(q) & \partial_y G(q) \end{pmatrix} \neq 0$ then, by the implicit function theorem, $\mathcal{M}$ is locally the image of a mapping $(z, t) \mapsto (a(z,t), b(z,t), z, t)$, with a and b two smooth functions. In other words, $\mathcal{M}$ is the zero locus of the functions

$$\begin{cases} \tilde{F}(x, y, z, t) = -x + a(z, t) \\ \tilde{G}(x, y, z, t) = -y + b(z, t). \end{cases}$$

The Jacobian matrix of the system (S-cros) using the functions $\tilde{F}$ and $\tilde{G}$ is then

$$\tilde{\mathcal{J}}_3 = \begin{pmatrix} -1 & 0 & \partial_z(a) & \partial_t(a) \\ 0 & -1 & \partial_z(b) & \partial_t(b) \\ 0 & 0 & \partial_{zt}(a) & \partial_{tt}(a) \\ 0 & 0 & \partial_{zt}(b) & \partial_{tt}(b) \end{pmatrix}$$

and its determinant reads as $\det(\tilde{\mathcal{J}}_3) = \partial_{zt}(a)\partial_{tt}(b) - \partial_{tt}(a)\partial_{zt}(b)$, which is precisely the quantity for the second condition of a cross-cap in Lemma 1. So we have just proved that P is a cross-cap iff $\det(\tilde{\mathcal{J}}_3) \neq 0$.

It remains to prove that $\det(\tilde{\mathcal{J}}_3) \neq 0$ iff $\det(\mathcal{J}_3) \neq 0$ where $\mathcal{J}_3$ is the Jacobian matrix associated to the system (S-cros):

$$\mathcal{J}_3 = \begin{pmatrix} \partial_x F & \partial_y F & \partial_z F & 0 \\ \partial_x G & \partial_y G & \partial_z G & 0 \\ \partial_{xt} F & \partial_{yt} F & \partial_{zt} F & \partial_{tt} F \\ \partial_{xt} G & \partial_{yt} G & \partial_{zt} G & \partial_{tt} G \end{pmatrix}.$$

We apply Hadamard's Lemma [4, Lemma 4.2.1] twice, first to $F(a+X, b+Y, z, t)$ with respect the variable X:

$$F(a + X, b + Y, z, t) - F(a, b + Y, z, t) = X g_1(X, b + Y, z, t) \qquad (1)$$

with

$$g_1(0, b + Y, z, t) = \partial_x F(a, b + Y, z, t) \qquad (2)$$

and then to $F(a, b + Y, z, t)$ with respect to the variable Y:

$$F(a, b + Y, z, t) - F(a, b, z, t) = Y g_2(Y, z, t)$$

with

$$g_2(0, z, t) = \partial_y F(a, b, z, t). \qquad (3)$$

By definition of the parametrization $(z,t) \mapsto (a(z,t), b(z,t), z, t)$, for any point on the surface $\mathcal{M}$ sufficiently close to q, $F(a(z,t), b(z,t), z, t) = 0$, thus equality (1) becomes

$$F(a + X, b + Y, z, t) = X g_1(X, b + Y, z, t) + Y g_2(Y, z, t) \qquad (4)$$

Now we set:

$$X = x - a(z,t)$$
$$Y = y - b(z,t).$$

Substituting X and Y in the relations (4), (2) and (3) yields

$$F(x, y, z, t) = -\tilde{F}(x, y, z, t) g_1(x + a, y, z, t) - \tilde{G}(x, y, z, t) g_2(b + y, z, t). \qquad (5)$$

In the same way, applying Hadamard's Lemma to $G(a + X, b + Y, z, t)$, there exist two smooth functions h_1 and h_2 such that

$$G(x, y, z, t) = -\tilde{F}(x, y, z, t) h_1(x + a, y, z, t) - \tilde{G}(x, y, z, t) h_2(b + y, z, t) \qquad (6)$$

with $h_1(0, y, z, t) = \partial_x G(a, y, z, t)$ and $h_2(0, z, t) = \partial_y G(a, b, z, t)$. We rewrite the relations (5) and (6) as $\begin{pmatrix} F \\ G \end{pmatrix} = -\underbrace{\begin{pmatrix} g_1 & g_2 \\ h_1 & h_2 \end{pmatrix}}_{\mathcal{A}} \begin{pmatrix} \tilde{F} \\ \tilde{G} \end{pmatrix} = \mathcal{A} \begin{pmatrix} \tilde{F} \\ \tilde{G} \end{pmatrix}$. Note that at

the point q, $\mathcal{A}(q) = -\begin{pmatrix} \partial_x F(q) & \partial_y F(q) \\ \partial_x G(q) & \partial_y G(q) \end{pmatrix}$ and by our assumption $\det \mathcal{A}(q) \neq 0$. Differentiating with respect to t yields

$$\begin{pmatrix} \partial_t F \\ \partial_t G \end{pmatrix} = \partial_t \mathcal{A} \begin{pmatrix} \tilde{F} \\ \tilde{G} \end{pmatrix} + \mathcal{A} \begin{pmatrix} \partial_t \tilde{F} \\ \partial_t \tilde{G} \end{pmatrix}.$$

We can thus rewrite the system for cross-caps as

$$\underbrace{\begin{pmatrix} F & G & \partial_t F & \partial_t G \end{pmatrix}^T}_{\mathcal{F}} = \underbrace{\begin{pmatrix} \mathcal{A} & 0 \\ \partial_t \mathcal{A} & \mathcal{A} \end{pmatrix}}_{\mathcal{N}} \underbrace{\begin{pmatrix} \tilde{F} & \tilde{G} & \partial_t \tilde{F} & \partial_t \tilde{G} \end{pmatrix}^T}_{\tilde{\mathcal{F}}}. \qquad (7)$$

The Jacobian determinants $\mathcal{J}_3 = \det(Jac(\mathcal{F}))$ and $\tilde{\mathcal{J}}_3 = \det(Jac(\tilde{\mathcal{F}}))$. The partial derivative of equation (7) with respect to any of the variables yields $\partial \mathcal{F} = \partial(\mathcal{N} \times \tilde{\mathcal{F}}) = \partial \mathcal{N} \times \tilde{\mathcal{F}} + \mathcal{N} \times \partial \tilde{\mathcal{F}}$, and since at the point q, $\tilde{\mathcal{F}}(q) = 0$, this simplifies to $\partial \mathcal{F}(q) = \mathcal{N}(q) \times \partial \tilde{\mathcal{F}}(q)$. At the point q, we thus have the equation $\mathcal{J}_3(q) = \mathcal{N}(q) \times \tilde{\mathcal{J}}_3(q)$, and since $\det \mathcal{N}(q) = \det \mathcal{A}(q)^2 \neq 0$ we conclude that

$$\det \mathcal{J}_3(q) \neq 0 \Leftrightarrow \det \tilde{\mathcal{J}}_3(q) \neq 0. \qquad (8)$$

□

4.3   Ball System

In this section, we show that the system (S-Ball) represents the solutions of (S-dble), (S-tple) and (S-cros) as regular solutions of a single system of equations via a change of variables. We call this system the Ball system as in [9] where the same approach was used for the projection of a space curve in the plane.

Lemma 6. *The projections in $\mathbb{R}^3$ of the solutions of the Ball system for $r \geq 0$ are the projections of the solutions of systems (S-dble), (S-tple) and (S-cros).*

Proof. Let (x, y, z, c, r) be a solution of the Ball system. If $r = 0$, the Ball system is exactly the system (S-cros). If $r > 0$, defining $t_1 = c - \sqrt{r}, t_2 = c + \sqrt{r}$, one can transform the Ball system into the system (S-dble) by multiplying the last two lines by $\sqrt{r}$ and adding or subtracting the two first lines by the last two ones. Finally by construction, the projection of the solutions of (S-tple) is included in the projection of the solutions of (S-dble), and thus in the projection of the solutions of the Ball system.   □

Lemma 7. *Let $P = (x, y, z)$ be a point in Ω.*

1. *P is a double point iff it has two regular pre-images (x, y, z, t_1) and (x, y, z, t_2) with $t_1 \neq t_2$ such that $(x, y, z, \frac{t_1+t_2}{2}, (\frac{t_1-t_2}{2})^2)$ is a regular solution of (S-Ball).*
2. *If P is a triple point, then it has three pre-images that give three regular solutions of (S-Ball).*
3. *If P is a cross-cap, then it has one critical pre-image (x, y, z, t) such that $(x, y, z, t, 0)$ is a regular solution of (S-Ball).*

To prove Lemma 7, we first note that $S.F, S.G, D.F$ and $D.G$ are smooth functions. The following lemma is a variation of [9, Lemma 6] to the case of functions of $\mathbb{R}^4$ that we state without proof.

Lemma 8. *If A is a real smooth function, then $S.A$ and $D.A$ are real smooth functions. Moreover, the derivatives of $S.A$ with respect to x, y, z, c, r are respectively $S.\partial_x A$, $S.\partial_y A$, $S.\partial_z A$, $S.\partial_t A$, $\frac{1}{2}D.\partial_t A$. The derivatives of $D.A$ with respect to x, y, z, c, r are respectively $D.\partial_x A$, $D.\partial_y A$, $D.\partial_z A$, $D.\partial_t A$ and $\frac{1}{2r}(S.\partial_t A - D.A)$ if $r > 0$ and $\frac{1}{6}\partial_{tt} A$ if $r = 0$.*

Proof (Proof of Lemma 7.). For the case $r > 0$, according to Lemma 8, the Jacobian of (S-Ball) is

$$\mathcal{J}_{(c,r>0)} = \begin{pmatrix} S.\partial_x F & S.\partial_y F & S.\partial_z F & S.\partial_t F & \frac{D.\partial_t F}{2} \\ S.\partial_x G & S.\partial_y G & S.\partial_z G & S.\partial_t G & \frac{D.\partial_t G}{2} \\ D.\partial_x F & D.\partial_y F & D.\partial_z F & D.\partial_t F & \frac{S.\partial_t F - D.F}{2r} \\ D.\partial_x G & D.\partial_y G & D.\partial_z G & D.\partial_t G & \frac{S.\partial_t G - D.G}{2r} \end{pmatrix}.$$

Let $q = (x, y, z, c, r)$ be a solution of the Ball system with $r > 0$, $\mathcal{J}_{(c,r>0)}$ can be simplified using the fact that $D.F(q) = D.G(q) = 0$. Denote $q_1 = (x, y, z, c + \sqrt{r})$ and $q_2 = (x, y, z, c - \sqrt{r})$ the two points of $\mathcal{M}$ solutions of (S-dble) according to Lemma 6. Applying to $\mathcal{J}_{(c,r>0)}$ successively the following transformations on its lines and columns: $\ell_3 \longleftarrow \sqrt{r} \times \ell_3, \ell_4 \longleftarrow \sqrt{r} \times \ell_4, c_5 \longleftarrow (2\sqrt{r})c_5, \ell_1 \longleftarrow \ell_1 + \ell_3, \ell_3 \longleftarrow \ell_1 - \ell_3, \ell_2 \longleftarrow \ell_2 + \ell_4, \ell_4 \longleftarrow \ell_2 - \ell_4$, one has:

$$\det \mathcal{J}_{(c,r>0)} = 0 \iff \det \begin{pmatrix} \partial_x F(q_1) & \partial_y F(q_1) & \partial_z F(q_1) & \partial_t F(q_1) & \partial_t F(q_1) \\ \partial_x G(q_1) & \partial_y G(q_1) & \partial_z G(q_1) & \partial_t G(q_1) & \partial_t G(q_1) \\ \partial_x F(q_2) & \partial_y F(q_2) & \partial_z F(q_2) & \partial_t F(q_2) & -\partial_t F(q_2) \\ \partial_x G(q_2) & \partial_y G(q_2) & \partial_z G(q_2) & \partial_t G(q_2) & -\partial_t G(q_2) \end{pmatrix} = 0$$

By changing again $c_4 \longleftarrow \frac{1}{2}(c_4 + c_5)$ and $c_5 \longleftarrow \frac{1}{2}(c_4 - c_5)$, we get

$$\det \mathcal{J}_{(c,r>0)} = 0 \iff \det \begin{pmatrix} \partial_x F(q_1) & \partial_y F(q_1) & \partial_z F(q_1) & \partial_t F(q_1) & 0 \\ \partial_x G(q_1) & \partial_y G(q_1) & \partial_z G(q_1) & \partial_t G(q_1) & 0 \\ \partial_x F(q_2) & \partial_y F(q_2) & \partial_z F(q_2) & 0 & \partial_t F(q_2) \\ \partial_x G(q_2) & \partial_y G(q_2) & \partial_z G(q_2) & 0 & \partial_t G(q_2) \end{pmatrix} = 0$$

The matrix on the right hand side is exactly that of the Jacobian of the system (S-dble), thus the lemma reduces to Lemma 3. In particular, this implies that both for double points and for triple points the solutions of the Ball system are regular.

For the case $r = 0$, the Ball system (S-Ball) coincides with the system (S-cros). In particular, according to Lemma 5, if P is cross-cap, the Jacobian of Equations (S-cros) is non-zero, which implies that the Jacobian matrix of the Ball system (S-Ball) is full rank. This implies that above cross-cap, the solution of the Ball system is regular. □

4.4   Algorithm

We developed a solver optimized for multivariate high degree polynomials called *voxelize* and available with GPL license ([14]). It is based on a classical bisection approach with an interval exclusion test that excludes the boxes that don't satisfy the input equations and inequalities ([15, Chapter 5] and references therein). For storing the set of boxes created during the subdivision, we use the Compressed Sparse Fiber data structure [2, 17], described in the literature as a generalization of the Compressed Sparse Row format. The main advantage of this data structure is that it allows us to efficiently evaluate a polynomial on a set of boxes appearing during the subdivision algorithm. More precisely, given a set S of K boxes in $\mathbb{R}^n$ arranged as a cube with $K = k^n$, evaluating a polynomial of degree d on S can be done in $O(d^n k + \cdots + dk^n)$ arithmetic operations. If $k > d$, this leads to $O(ndK)$ arithmetic operations.

5   Example

5.1   Whitney Umbrella

Our first example is the Whitney Umbrella. Its parametric equations are:
$x(u, v) = u$, $y(u, v) = v^2$ and $z(u, v) = uv$. Letting $F(x, y, z, t) = y - t^2$ and
$G(x, y, z, t) = z - xt$, the Whitney Umbrella is exactly the projection in $\mathbb{R}^3$ of
the surface defined by $F = G = 0$. The corresponding Ball system is:

$$\begin{cases} S.F = y - c^2 - r & = 0 \\ S.G = z - xc & = 0 \\ D.F = -2c & = 0 \\ D.G = -x & = 0 \end{cases}$$

Thus, substituting c by 0, we deduce that the set of singularities of the Whitney
Umbrella is defined by $x = 0$, $z = 0$ and $y = r \geq 0$.

Note that most state-of-the-art approaches start by computing the implicit
equation of the Whitney Umbrella: $P(x, y, z) = x^2 y - z^2 = 0$, and then compute
the singularities of this map as $P = \partial_x P = \partial_y P = \partial_z P = 0$. Unfortunately,
the solution to this system is $x = 0$ and $z = 0$, which adds a handle that is
not a singularity of the original surface. This is a known artifact that comes
from the Zariski closure of the original surface. Our method has the advantage
of returning the exact set of singularities of the Whitney Umbrella, without the
spurious handle.

5.2   Large Polynomials

Another advantage of our approach is that it is based on numerical methods, and
as such, it can compute the singularities of polynomial maps of high degrees. For
example, the polynomials in Equations (9) are generated randomly with degree
7. Computing SF, SG, DF, DG can be done quickly with a computer algebra
system. Then, using our subdivision solver *voxelize*, we enclosed the solutions
of the Ball system within the input box $x = [-0.35, 0.35]$, $y = [-0.35, 0.35]$, $z =
[0.4, 1.1]$, $c = [-5, 5]$, $r = [0, 5]$. Our result is displayed on Fig. 2, the red curve is
the projection in $\mathbb{R}^3$ of the boxes of $\mathbb{R}^5$ enclosing the Ball system, each box being
of size a factor 2^{-11} of the size of the input box. The surface $F = G = 0$ is also
enclosed by *voxelize* in boxes in $\mathbb{R}^4$, we then use a generalization to 4D of the
SurfaceNet approach [1,5] to compute a mesh that is eventually projected in $\mathbb{R}^3$
and displayed on the left of Fig. 2. On a quadcore Intel CPU i7-8650U, *voxelize*
running time was 11 seconds to enclose the Ball system and 8.5 s to enclose the
surface and compute its meshing.

$$F(x,y,z,t) = -x^7 + x^6y - 2x^5y^2 + x^4y^3 - 26x^3y^4 + 2x^2y^5 - 3y^7 - x^6z - x^5yz - x^4y^2z + 4x^3y^3z + 6xy^5z - y^6z$$
$$+ 3x^5z^2 + 17x^4yz^2 - 2x^2y^3z^2 - xy^4z^2 + y^5z^2 - 4x^2y^2z^3 + 2xy^3z^3 + y^4z^3 - 2x^3z^4 - 2x^2yz^4 - 3xy^2z^4$$
$$+ 2x^2z^5 + y^2z^5 - 5xz^6 - 6yz^6 + 3x^5yt + 5x^4y^2t - x^3y^3t + x^2y^4t - 2xy^5t - y^6t - 2x^5zt - x^4yzt$$
$$+ x^3y^2zt + x^2y^3zt + 2xy^4zt - 2y^5zt - 5x^4z^2t + x^2y^2z^2t + y^4z^2t + y^4z^2t - x^3z^3t + x^2yz^3t$$
$$- 2xy^2z^3t + 3y^3z^3t + 8x^2z^4t + 6xyz^4t + 2yz^5t - 2z^6t + 2x^5t^2 + xy^4t^2 + y^5t^2 - x^4zt^2 + 3x^3yzt^2$$
$$+ x^2y^2zt^2 + 2xy^3zt^2 - y^4zt^2 - x^3z^2t^2 - 2xy^2z^2t^2 + 3y^3z^2t^2 - 8x^2z^3t^2 - xyz^3t^2 + 2y^2z^3t^2 + xz^4t^2$$
$$+ 2yz^4t^2 + x^4t^3 - x^3yt^3 - 2x^2y^2t^3 - xy^3t^3 - 16y^4t^3 - x^3zt^3 - x^2yzt^3 + xy^2zt^3 + 6y^3zt^3 - 3x^2z^2t^3$$
$$+ xyz^2t^3 + 3y^2z^2t^3 + 2xz^3t^3 - yz^3t^3 - 4x^3t^4 + 2x^2yt^4 + 10xy^2t^4 + 14y^3t^4 + xyzt^4 - 2y^2zt^4 + xz^2t^4$$
$$- yz^2t^4 + 2z^3t^4 + 6xyt^5 + 2y^2t^5 - 4xzt^5 + 46yzt^5 + 29z^2t^5 - 6yt^6 - 5zt^6 - t^7 - x^6 + x^4y^2 + x^2y^4$$
$$+ 7xy^5 + 4y^6 - 8x^4yz - 373x^3y^2z + 15x^2y^3z - 2xy^4z + x^4z^2 - x^3yz^2 - x^2y^2z^2 + xy^3z^2 - 2y^4z^2$$
$$+ x^3z^3 + x^2yz^3 + xy^2z^3 + 3x^2z^4 + xyz^4 + xz^5 + 3yz^5 + 2z^6 - x^5t + 9x^4yt + x^3y^2t - 2x^2y^3t - xy^4t$$
$$+ 13x^4zt - x^3yzt + x^2y^2zt - 7xy^3zt + x^3z^2t - x^2yz^2t + xy^2z^2t + 3y^3z^2t - 4x^2z^3t + 2xyz^3t + y^2z^3t$$
$$- 3xz^4t - 6yz^4t + x^2y^2t^2 - xy^3t^2 - 2x^3zt^2 + 2x^2yzt^2 - 2y^3zt^2 - 6x^2z^2t^2 - 32xyz^2t^2 - xz^3t^2$$
$$- 5yz^3t^2 + z^4t^2 + x^3t^3 + 4x^2yt^3 - 3xy^2t^3 + y^3t^3 - x^2zt^3 - xyzt^3 - 8y^2zt^3 - xz^2t^3 - 84yz^2t^3$$
$$- z^3t^3 - 812x^2t^4 + xyt^4 + 2y^2t^4 + 2xzt^4 + yzt^4 - z^2t^4 - 2yt^5 + 10t^6 + x^4y + x^3y^2 + 29z^2y^3$$
$$+ xy^4 + 14y^5 - 4xy^3z + y^4z + 2x^3z^2 + x^2yz^2 + xy^2z^2 - y^3z^2 + 2x^2z^3 - xyz^3 - 6xz^4 + 4yz^4 + z^5$$
$$- x^5t + x^3yt - 2x^2y^2t + xy^3t - y^4t - x^3zt + xy^2zt - y^3zt + x^2z^2t + 28xyz^2t + 9y^2z^2t + xz^3t + x^3t^2$$
$$+ x^2yt^2 + 18xy^2t^2 - y^3t^2 - 2x^2zt^2 + 3xyzt^2 - 2y^2zt^2 + xz^2t^2 - 2z^3t^2 + 10x^2t^3 - xyt^3$$
$$+ y^2t^3 - xzt^3 - yzt^3 - xt^4 + yt^4 + 50zt^4 + t^5 - 67x^4 - x^3y + 5x^2y^2 + 17xy^3 - 2y^4 + 2x^3z + 5x^2yz$$
$$+ 4xy^2z + y^3z - xyz^2 + 10y^2z^2 - 2yz^3 + 4z^4 - 23x^3t + y^3t + x^2zt + 7y^2zt + 2yz^2t - x^2t^2 + 2xyt^2$$
$$- 9yt^2 - 4z^2t^2 + yt^3 - zt^3 - 4z^3 - 2x^2y + 2y^3 - 3x^2z + xyz + 2z^3 + x^2t + xyt - 5y^2t$$
$$- xzt + 2z^2t - 2xt^2 + yt^2 - zt^2 + t^3 - 4x^2 + 11y^2 - xz - 4yz - 2xt + zt - 2t^2 - x + 5y - 2t - 8$$

$$\text{(9a)}$$

$$G(x,y,z,t) = x^7 - 2x^6y - 2x^5y^2 + x^4y^3 - 7x^3y^4 + x^2y^5 + 6xy^6 + 2y^7 + x^6z - 11x^5yz - x^4y^2z - 2x^3y^3z - 2x^2y^4z$$
$$- xy^5z + 3y^6z - x^5z^2 + 13x^3y^2z^2 + x^2y^3z^2 - xy^4z^2 - y^5z^2 + x^3yz^3 - 8x^2y^2z^3 - 2xy^3z^3 - 2y^4z^3$$
$$+ 3x^2yz^4 + 10y^3z^4 - x^2z^5 - 3xyz^5 + y^2z^5 - xz^6 - 3yz^6 + z^7 + 2x^6t + x^5yt - 2x^4y^2t + x^3y^3t + 14x^2y^4t$$
$$- 2xy^5t + 2y^6t - x^4yzt + 3x^3y^2zt - x^2y^3zt - xy^4zt - y^5zt + x^4z^2t + x^3yz^2t + 7x^2y^2z^2t + 4xy^3z^2t$$
$$+ y^4z^2t + x^2yz^3t + xy^2z^3t - y^3z^3t + x^2z^4t - 10xyz^4t + yz^5t - 25z^6t + 5x^5t^2 + x^4yt^2 - 9x^3y^2t^2$$
$$+ 6x^2y^3t^2 + 2xy^4t^2 + y^5t^2 - x^4zt^2 + x^2y^2zt^2 + 7xy^3zt^2 + 2y^4zt^2 + 47x^2yz^2t^2 - 3xy^2z^2t^2 - y^3z^2t^2$$
$$- 2x^2z^3t^2 - 3xyz^3t^2 - 20y^2z^3t^2 - 2yz^4t^2 + z^5t^2 - x^4t^3 + 2x^3yt^3 + xy^2zt^3 + xyz^2t^3 + y^4t^3$$
$$- x^3zt^3 + x^2yzt^3 + 4xy^2zt^3 - y^3zt^3 + x^2z^2t^3 + xyz^2t^3 - xz^3t^3 + 10yz^3t^3 + 2z^4t^3 - 10x^3t^4 + y^3t^4$$
$$+ x^2zt^4 + 8xyzt^4 + y^2zt^4 - xz^2t^4 + 3yz^2t^4 + z^3t^4 + 2xyt^5 - y^2t^5 + xzt^5 + 2yzt^5 - z^2t^5 - xt^6$$
$$+ 3yt^6 - zt^6 + t^7 - 6x^6 + 2x^5y + 4x^4y^2 - 2x^3y^3 - 2y^4 + 3yz^5 - 3x^5z - 3x^2y^3z - xy^4z - y^5z - x^4z^2$$
$$- x^3yz^2 - 32x^2y^2z^2 + 18xy^3z^2 - 5x^3z^3 + 2x^2yz^3 - xy^2z^3 - x^2z^4 + 2xz^5 - 9yz^5 + 8x^4yt - 2x^3y^2t$$
$$+ xy^4t + y^5t - x^4zt - 7x^3yzt - xy^3zt + y^4zt + 3x^3z^2t - 2x^2yz^2t - xy^2z^2t + y^3z^2t + x^2z^3t - 8xyz^3t$$
$$+ 2y^3z^3t + xz^4t + yz^4t - 3x^4t^2 - x^3yt^2 + 5x^2y^2t + xy^3t^2 - y^4t^2 - x^3zt^2 - 4x^3yzt^2 + 4xy^2zt^2$$
$$+ 71y^3z^2t^2 + x^2z^2t^2 - y^2z^2t^2 - xz^3t^2 + yz^3t^2 + z^4t^2 - 5x^3yt^3 + 7x^2yt^3 + xy^2t^3 - 4y^3t^3 - 2x^2zt^3$$
$$+ xyzt^3 + y^2zt^3 + xz^2t^3 - yz^2t^3 + z^3t^3 + xyt^4 - y^2t^4 + 2yzt^4 - 4xt^5 - 7t^6 - 2x^5 - x^3y^2 - x^2y^3$$
$$- 9xy^4 - 2y^5 - 3x^4z + 12x^2y^2z - xy^3z + xy^4z - xyz^2 - 14y^3z^2 - x^2z^3 + 81xyz^3 - 2y^4z^3$$
$$+ 2x^5 + 2x^4t - 2x^3yt - x^2y^2t - xy^3t + 15y^4t - 7x^3zt + 5x^2yzt + 3y^3zt - 8x^2z^2t + 2xyz^2t - 461y^2z^2t$$
$$+ 2xz^3t - 44yz^3t + 6z^4t + 2x^2yt^2 + xy^2t^2 + y^3t^2 - x^2zt^2 - xyzt^2 + y^2zt^2 + 273xz^2t^2 + 56yz^2t^2$$
$$- x^2t^3 + 2xyt^3 - 2xzt^3 + 6yzt^3 + 5z^2t^3 - 4xt^4 - xt^4 - t^5 + 2x^4 - x^3y + x^2y^2 - y^4 + x^3yz - xyz$$
$$- xy^2z - x^2z^2 - y^2z^2 - xz^3 - yz^3 + 6z^4 - x^3t + xy^2t + y^3t + 4x^2zt - xyzt + yz^2t - 2z^3t + x^2t^2 + 3xyzt^2$$
$$- 22xzt^2 + yzt^2 + 17z^2t^2 - xt^3 - yt^3 - 2zt^3 + t^4 - 22x^3 - 5x^2y + 3xy^2 + 21y^3 - 8x^2z - 6y^2z + yz^2 - z^3$$
$$+ 7x^2t + 3y^2t - xzt - 12yzt - xt^2 + 3zt^2 + t^3 - x^2 + 3xy - 6xz - 2z^2 - xt + zt - 4t^2 - 2x - y - 9z + 2t - 1$$

$$\text{(9b)}$$

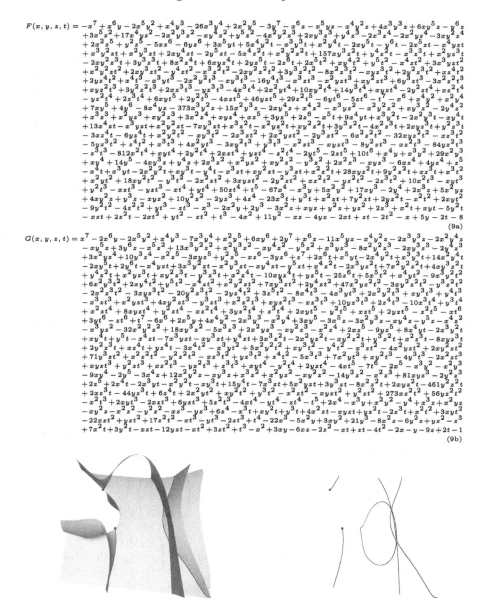

Fig. 2. Left: Singular surface Ω, projection in $\mathbb{R}^3$ of the smooth 2-manifold $\mathcal{M}$ of $\mathbb{R}^4$ defined by Equations (9). Right: Singular curve of Ω with cross-caps (blue) and triple points (green). (Color figure online)

6   Conclusion

As shown in the examples, our approach handles computation of singularities not handled by other state-of-the-art methods. Moreover, even though our approach cannot handle the computation of the singularities associated to any mapping, we showed that our approach works for almost all mappings.

With the systems we describe in Sect. 4.1, we could also compute the triple points and the cross-cap singularities. Note that in order to make this computation reliable, we need additional computation, not covered here, to ensure that we don't miss triple-points near cross-caps.

Finally, it is also possible to check the assumptions satisfied generically in Theorem 1 using a semi-algorithm that terminates if and only if the required conditions are satisfied, such an approach is exemplified in a close setting in [10].

Acknowledgment. We thanks David Mond for providing to us, via a private communication, this proof of the characterization of cusps.

7   Appendix: Proof of Lemma 1

Let $q \in \mathcal{M}$ be a cross-cap singularity of the projection $\mathfrak{p} : \mathcal{M} \mapsto \mathbb{R}^3$.

First, the condition $q \in \Sigma^1(\mathfrak{p})$ means that $d\mathfrak{p}(q)$ has corank 1. Since $\mathrm{rank}(\mathfrak{p})$ $= 2 - \mathrm{corank}(\mathfrak{p}) = 1$, the condition is also equivalent to $d\mathfrak{p}(q)$ has rank 1. In other words, the 2-dimensional tangent plane to $\mathcal{M}$ at q projects to a line, that is the direction of projection is in the tangent plane. Thus, the condition $q \in \Sigma^1(\mathfrak{p})$ of Definition 7 is equivalent to the first condition of Lemma 1: the direction of projection is in the tangent plane.

We now assume that the surface $\mathcal{M}$ is locally parameterized in a neighborhood of q by $(z,t) \mapsto (a(z,t), b(z,t), z, t)$, so that $\mathfrak{p}(z,t) = (a(z,t), b(z,t), z)$. The space $J^1(\mathcal{M}, \mathbb{R}^3)$ is thus locally equal to $U \times \mathbb{R}^3 \times L(\mathbb{R}^2, \mathbb{R}^3)$ where U is a subset of $\mathbb{R}^2$ and L stands for the space of linear mappings. The 1-jet of a mapping $(f_1(z,t), f_2(z,t), f_3(z,t)) : \mathcal{M} \mapsto \mathbb{R}^3$ is

$$\left((z,t), (f_1(z,t), f_2(z,t), f_3(z,t)), \begin{pmatrix} f_{1z} & f_{1t} \\ f_{2z} & f_{2t} \\ f_{3z} & f_{3t} \end{pmatrix} \right).$$

Σ^1 is the subset of $J^1(\mathcal{M}, \mathbb{R}^3)$ such that the matrix $\begin{pmatrix} f_{1z} & f_{1t} \\ f_{2z} & f_{2t} \\ f_{3z} & f_{3t} \end{pmatrix}$ has corank 1, that is has rank 1. Without loss of generality, if we assume $(f_{3z}, f_{3t}) \neq (0,0)$, Σ^1 is thus implicitly defined by the two equations: $\begin{vmatrix} f_{1z} & f_{1t} \\ f_{3z} & f_{3t} \end{vmatrix} = 0$ and $\begin{vmatrix} f_{2z} & f_{2t} \\ f_{3z} & f_{3t} \end{vmatrix} = 0$. One thus has $\Sigma^1 = \Phi^{-1}(0)$ with

$$\Phi : J^1(\mathcal{M}, \mathbb{R}^3) \to \mathbb{R}^2$$

$$\left((z,t), (f_1(z,t), f_2(z,t), f_3(z,t)), \begin{pmatrix} f_{1z} & f_{1t} \\ f_{2z} & f_{2t} \\ f_{3z} & f_{3t} \end{pmatrix} \right) \mapsto \begin{pmatrix} f_{1z}f_{3t} - f_{1t}f_{3z} \\ f_{2z}f_{3t} - f_{2t}f_{3z} \end{pmatrix}$$

According to [6, Lemma 4.3], $j^1\mathfrak{p}$ is transverse to Σ^1 at q iff $\Phi \cdot j^1\mathfrak{p}$ is a submersion at q. On the other hand, $\Phi \cdot j^1\mathfrak{p} = \Phi\left((z,t), (a(z,t), b(z,t), z), \begin{pmatrix} a_z & a_t \\ b_z & b_t \\ 1 & 0 \end{pmatrix} \right) = -(a_t, b_t)$. This mapping is a submersion iff its Jacobian $\begin{pmatrix} a_{zt} & a_{tt} \\ b_{zt} & b_{tt} \end{pmatrix}$ is full rank, that is $a_{zt}b_{tt} - a_{tt}b_{zt} \neq 0$ which is exactly the second condition of Lemma 1.

References

1. de Bruin, P.W., Vos, F.M., Post, F.H., Frisken-Gibson, S.F., Vossepoel, A.M.: Improving triangle mesh quality with SurfaceNets. In: Delp, S.L., DiGoia, A.M., Jaramaz, B. (eds.) MICCAI 2000. LNCS, vol. 1935, pp. 804–813. Springer, Heidelberg (2000). https://doi.org/10.1007/978-3-540-40899-4_83

2. Chou, S., Kjolstad, F., Amarasinghe, S.: Format abstraction for sparse tensor algebra compilers. Proc. ACM Program. Lang. 2(OOPSLA), 123:1–123:30 (2018). https://doi.org/10.1145/3276493

3. Delanoue, N., Lagrange, S.: A numerical approach to compute the topology of the apparent contour of a smooth mapping from R^2 to R^2. J. Comput. Appl. Math. 271, 267–284 (2014). https://doi.org/10.1016/j.cam.2014.03.032

4. Demazure, M.: Bifurcations and Catastrophes: Geometry of Solutions to Nonlinear Problems. UTX. Springer, Heidelberg (2000). https://doi.org/10.1007/978-3-642-57134-3

5. Gibson, S.F.F.: Constrained elastic surface nets: generating smooth surfaces from binary segmented data. In: Wells, W.M., Colchester, A., Delp, S. (eds.) MICCAI 1998. LNCS, vol. 1496, pp. 888–898. Springer, Heidelberg (1998). https://doi.org/10.1007/BFb0056277

6. Golubistky, M., Guillemin, V.: Stable Mappings and Their Singularities. GTM, vol. 14. Springer, New York (1973). https://doi.org/10.1007/978-1-4615-7904-5

7. Goryunov, V.V.: Local invariants of mappings of surfaces into three-space. In: Arnold, V.I., Gelfand, I.M., Retakh, V.S., Smirnov, M. (eds.) The Arnold-Gelfand Mathematical Seminars, pp. 223–255. Birkhäuser, Boston (1997). https://doi.org/10.1007/978-1-4612-4122-5_11

8. Hobbs, C.A., Kirk, N.P.: On the classification and bifurcation of multigerms of maps from surfaces to 3-space. Math. Scand. 89(1), 57–96 (2001). https://doi.org/10.7146/math.scand.a-14331

9. Imbach, R., Moroz, G., Pouget, M.: Numeric and certified isolation of the singularities of the projection of a smooth space curve. In: Kotsireas, I.S., Rump, S.M., Yap, C.K. (eds.) MACIS 2015. LNCS, vol. 9582, pp. 78–92. Springer, Cham (2016). https://doi.org/10.1007/978-3-319-32859-1_6

10. Imbach, R., Moroz, G., Pouget, M.: A certified numerical algorithm for the topology of resultant and discriminant curves. J. Symb. Comput. 80(Part 2), 285–306 (2017). https://doi.org/10.1016/j.jsc.2016.03.011

11. Martin, B., Goldsztejn, A., Granvilliers, L., Jermann, C.: Certified parallelotope continuation for one-manifolds. SIAM J. Numer. Anal. 51(6), 3373–3401 (2013). https://doi.org/10.1137/130906544

12. Mond, D.: Classification of certain singularities and applications to differential geometry. Ph.D. thesis, The University of Liverpool (1982)

13. Mond, D.: On the classification of germs of maps from $\mathbb{R}^2$ to $\mathbb{R}^3$. Proc. London Math. Soc. **s3–s50**(2), 333–369 (1985). https://doi.org/10.1112/plms/s3-50.2.333
14. Moroz, G.: Voxelize (2018–2019). https://gitlab.inria.fr/gmoro/voxelize. https://doi.org/10.5281/zenodo.3562432
15. Neumaier, A.: Interval Methods for Systems of Equations. Cambridge University Press, Cambridge (1990). https://doi.org/10.1017/CBO9780511526473
16. Sinha, R.O., Atique, R.W.: Classification of multigerms (from a modern viewpoint). Minicourse 3 of the School on Singularity Theory, 17–22 July 2016 (2016). www.worksing.icmc.usp.br/main_site/2016/minicourse3_notes.pdf
17. Smith, S., Karypis, G.: Tensor-matrix products with a compressed sparse tensor. In: Proceedings of the 5th Workshop on Irregular Applications: Architectures and Algorithms, IA3 2015, pp. 5:1–5:7. ACM (2015). https://doi.org/10.1145/2833179.2833183

Evaluation of Chebyshev Polynomials on Intervals and Application to Root Finding

Viviane Ledoux[1,2] and Guillaume Moroz[1(✉)]

[1] Université de Lorraine, CNRS, Inria, LORIA, F-54000 Nancy, France
viviane.ledoux@ens.fr, guillaume.moroz@inria.fr
[2] École Normale Supérieure, Paris, France

Abstract. In approximation theory, it is standard to approximate functions by polynomials expressed in the Chebyshev basis. Evaluating a polynomial f of degree n given in the Chebyshev basis can be done in $O(n)$ arithmetic operations using the Clenshaw algorithm. Unfortunately, the evaluation of f on an interval I using the Clenshaw algorithm with interval arithmetic returns an interval of width exponential in n. We describe a variant of the Clenshaw algorithm based on ball arithmetic that returns an interval of width quadratic in n for an interval of small enough width. As an application, our variant of the Clenshaw algorithm can be used to design an efficient root finding algorithm.

Keywords: Clenshaw algorithm · Chebyshev polynomials · Root finding · Ball arithmetic · Interval arithmetic

1 Introduction

Clenshaw showed in 1955 that any polynomial given in the form

$$p(x) = \sum_{i=0}^{n} a_i T_i(x) \qquad (1)$$

can be evaluated on a value x with a single loop using the following functions defined by recurrence:

$$u_k(x) = \begin{cases} 0 & \text{if } k = n+1 \\ a_n & \text{if } k = n \\ 2xu_{k+1}(x) - u_{k+2}(x) + a_k & \text{if } 1 \le k < n \\ xu_1(x) - u_2(x) + a_0 & \text{if } k = 0 \end{cases} \qquad (2)$$

such that $p(x) = u_0(x)$.

Unfortunately, if we use Eq. (2) with interval arithmetic directly, the result can be an interval of size exponentially larger than the input, as illustrated in Example 1.

© Springer Nature Switzerland AG 2020
D. Slamanig et al. (Eds.): MACIS 2019, LNCS 11989, pp. 35–41, 2020.
https://doi.org/10.1007/978-3-030-43120-4_4

Example 1. Let $\varepsilon > 0$ be a positive real number, and let x be the interval $[\frac{1}{2} - \varepsilon, \frac{1}{2} + \varepsilon]$ of width 2ε. Assuming that $a_n = 1$, we can see that u_{n-1} is an interval of width 4ε. Then by recurrence, we observe that u_{n-k} is an interval of width at least $4\varepsilon F_k$ where $(F_n)_{n \in \mathbb{N}}$ denotes the Fibonacci sequence, even if all $a_i = 0$ for $i < n$.

Note that the constant below the exponent is even higher when x is closer to 1. These numerical instabilities also appear with floating point arithmetic near 1 and -1 as analyzed in [4].

To work around the numerical instabilities near 1 and -1, Reinsch suggested a variant of the Clenshaw algorithm [4,7]. Let $d_n(x) = a_n$ and $u_n(x) = a_n$, and for k between 0 and $n - 1$, define $d_k(x)$ and $u_k(x)$ by recurrence as follows:

$$\begin{cases} d_k(x) = 2(x - 1)u_{k+1}(x) + d_{k+1}(x) + a_k \\ u_k(x) = d_k(x) + u_{k+1} \end{cases} \tag{3}$$

Computing $p(x)$ with this recurrence is numerically more stable near 1. However, this algorithm does not solve the problem of exponential growth illustrated in Example 1.

Our first main result is a generalization of Eq. 3 for any value in the interval $[-1, 1]$. This leads to Algorithm 1 that returns intervals with tighter radii, as analyzed in Lemma 2. Our second main result is the use of classical backward error analysis to derive Algorithm 2 which gives an even better radii. Then in Sect. 3 we use the new evaluation algorithm to design a root solver for Chebyshev series, detailed in Algorithm 3.

2   Evaluation of Chebyshev Polynomials on Intervals

2.1   Forward Error Analysis

In this section we assume that we want to evaluate a Chebyshev polynomial on the interval I. Let a be the center of I and r be its radius. Furthermore, let γ and $\overline{\gamma}$ be the 2 conjugate complex roots of the equation:

$$X^2 - 2aX + 1 = 0. \tag{4}$$

In particular, using Vieta's formulas that relate the coefficients to the roots of a polynomial, γ satisfies $\gamma + \overline{\gamma} = 2a$ and $\gamma\overline{\gamma} = 1$.

Let $z_n(x) = a_n$ and $u_n(x) = a_n$, and for k between 0 and $n - 1$, define $z_k(x)$ and $u_k(x)$ by recurrence as follows:

$$\begin{cases} z_k(x) = 2(x - a)u_{k+1}(x) + \gamma z_{k+1}(x) + a_k \\ u_k(x) = z_k(x) + \overline{\gamma}u_{k+1}(x) \end{cases} \tag{5}$$

Using Eq. (4), we can check that the u_k satisfies the recurrence relation $u_k(x) = 2xu_{k+1}(x) - u_{k+2}(x) + a_k$, such that $p(x) = xu_1(x) - u_2(x) + a_0$.

Let (e_k) and (f_k) be two sequences of positive real numbers. Let $\mathcal{B}_{\mathbb{R}}(a,r)$ and $\mathcal{B}_{\mathbb{R}}(u_k(a), e_k)$ represent the intervals $[a-r, a+r]$ and $[u_k(a) - e_k, u_k(a) + e_k]$. Let $\mathcal{B}_{\mathbb{C}}(z_k(a), f_k)$ be the complex ball of center $z_k(a)$ and radius f_k.

Our goal is to compute recurrence formulas on the e_k and the f_k such that:

$$\begin{cases} z_k(\mathcal{B}_{\mathbb{R}}(a,r)) \subset \mathcal{B}_{\mathbb{C}}(z_k(a), f_k) \\ u_k(\mathcal{B}_{\mathbb{R}}(a,r)) \subset \mathcal{B}_{\mathbb{R}}(u_k(a), e_k). \end{cases} \tag{6}$$

Lemma 1. *Let $e_n = 0$ and $f_n = 0$ and for $n > k \geq 1$:*

$$\begin{cases} f_k = 2r|u_{k+1}(a)| + 2re_{k+1} + f_{k+1} \\ e_k = \min(e_{k+1} + f_k, \frac{f_k}{\sqrt{1-a^2}}) \text{ if } |a| < 1 \text{ else } e_{k+1} + f_k \end{cases} \tag{7}$$

Then, (e_k) and (f_k) satisfy Eq. (6).

Proof (sketch). For the inclusion $z_k(\mathcal{B}_{\mathbb{R}}(a,r)) \subset \mathcal{B}_{\mathbb{C}}(z_k(a), f_k)$, note that γ has modulus 1, such that the radius of γz_{k+1} is the same as the radius of z_{k+1} when using ball arithmetics. The remaining terms bounding the radius of z_k follow from the standard rules of interval arithmetics.

For the inclusion $u_k(\mathcal{B}_{\mathbb{R}}(a,r)) \subset \mathcal{B}_{\mathbb{R}}(u_k(a), e_k)$, note that the error segment on u_k is included in the Minkowski sum of a disk of radius f_k and a segment of radius e_{k+1}, denoted by M. If θ is the angle of the segment with the horizontal, we have $\cos\theta = a$. We conclude that the intersection of M with a horizontal line is a segment of radius at most $\min(e_{k+1} + f_k, \frac{f_k}{\sqrt{1-a^2}})$.

Corollary 1. *Let $\mathcal{B}_{\mathbb{R}}(u,e) = \texttt{BallClenshawForward}((a_0, \ldots, a_n), a, r)$ be the result of Algorithm 1, then*

$$p(\mathcal{B}_{\mathbb{R}}(a,r)) \subset \mathcal{B}_{\mathbb{R}}(u,e)$$

Moreover, the following lemma bounds the radius of the ball returned by Algorithm 1.

Lemma 2. *Let $\mathcal{B}_{\mathbb{R}}(u,e) = \texttt{BallClenshawForward}((a_0, \ldots, a_n), a, r)$ be the result of Algorithm 1, and let M be an upper bound on $|u_k(a)|$ for $1 \leq k \leq n$. Assume that $\varepsilon_k < Mr$ for $1 \leq k \leq n$, then*

$$\begin{cases} e < 2Mn^2r & \text{if } \quad n < \frac{1}{2\sqrt{1-a^2}} \\ e < 9Mn\frac{r}{\sqrt{1-a^2}} & \text{if} \quad \frac{1}{2\sqrt{1-a^2}} \leq n < \frac{\sqrt{1-a^2}}{2r} \\ e < 2M\left[(1 + \frac{2r}{\sqrt{1-a^2}})^n - 1\right] & \text{if} \quad \frac{\sqrt{1-a^2}}{2r} < n \end{cases}$$

Proof (sketch). We distinguish 2 cases. First if $n < \frac{1}{2\sqrt{1-a^2}}$, we focus on the relation $e_k \leq e_{k+1} + f_k + Mr$, and we prove by descending recurrence that $e_k \leq 2M(n-k)^2r$ and $f_k \leq 2Mr(2(n-k-1)+1)$.

For the case $\frac{1}{2\sqrt{1-a^2}} \leq n$, we use the relation $e_k \leq \frac{f_k}{\sqrt{1-a^2}} + Mr$, that we substitute in the recurrence relation defining f_k to get $f_k \leq 2rM + \frac{2r}{\sqrt{1-a^2}}f_{k+1} + f_{k+1} + Mr\sqrt{1-a^2}$. We can check by recurrence that $f_k \leq \frac{3}{2}M\sqrt{1-a^2}\left[(1 + \frac{2r}{\sqrt{1-a^2}})^n - 1\right]$, which allows us to conclude for the case $\frac{\sqrt{1-a^2}}{2r} \leq n$. Finally, when $\frac{1}{2\sqrt{1-a^2}} \leq n < \frac{\sqrt{1-a^2}}{2r}$, we observe that $(1 + \frac{2r}{\sqrt{1-a^2}})^n - 1 \leq n\exp(1)\frac{2r}{\sqrt{1-a^2}}$ which leads to the bound for the last case.

Algorithm 1. Clenshaw evaluation algorithm, forward error

> **function** BALLCLENSHAWFORWARD$((a_0, \ldots, a_n), a, r)$
>   ▷ *Computation of the centers u_k*
> $u_{n+1} \leftarrow 0$
> $u_n \leftarrow a_n$
>   **for** k in $n-1, n-2, \ldots, 1$ **do**
> $u_k \leftarrow 2au_{k+1} - u_{k+2} + a_k$
> $\varepsilon_k \leftarrow$ bound on the rounding error for u_k
> $u_0 \leftarrow au_1 - u_2 + a_0$
> $\varepsilon_0 \leftarrow$ bound on the rounding error for u_0
>
>   ▷ *Computation of the radii e_k*
> $f_n \leftarrow 0$
> $e_n \leftarrow 0$
>   **for** k in $n-1, n-2, \ldots, 1$ **do**
> $f_k \leftarrow 2r|u_{k+1}| + 2re_{k+1} + f_{k+1}$
> $e_k \leftarrow min(e_{k+1} + f_k, \frac{f_k}{\sqrt{1-a^2}}) + \varepsilon_k$
> $f_0 \leftarrow r|u_1| + 2re_1 + f_1$
> $e_0 \leftarrow min(e_1 + f_0, \frac{f_0}{\sqrt{1-a^2}}) + \varepsilon_0$
>   **return** $\mathcal{B}_{\mathbb{R}}(u_0, e_0)$

2.2 Backward Error Analysis

In the literature, we can find an error analysis of the Clenshaw algorithm [3]. The main idea is to add the errors appearing at each step of the Clenshaw algorithm to the input coefficients. Thus the approximate result correspond to the exact result of an approximate input. Finally, the error bound is obtained as the evaluation of a Chebyshev polynomial. This error analysis can be used directly to derive an algorithm to evaluate a polynomial in the Chebyshev basis on an interval in Algorithm 2.

Lemma 3. *Let $e_n = 0$ and for $n > k \geq 1$:*

$$e_k = 2r|u_{k+1}(a)| + e_{k+1} \tag{8}$$

and $e_0 = r|u_1(a)| + e_1$. Then (e_k) satisfies $u_k(\mathcal{B}_{\mathbb{R}}(a, r)) \subset \mathcal{B}_{\mathbb{R}}(u_k(a), e_k)$.

Proof (sketch). In the case where the computations are performed without errors, D. Elliott [3, Equation (4.9)] showed that for $\gamma = \tilde{x} - x$ we have:

$$p(\tilde{x}) - p(x) = 2\gamma \sum_{i=0}^{n} u_i(\tilde{x})T_i(x) - \gamma u_1(\tilde{x})$$

In the case where $\tilde{x} = a$ and $x \in \mathcal{B}_{\mathbb{R}}(a, r)$ we have $\gamma \leq r$ and $|T(x)| \leq 1$ which implies $e_k \leq r|u_1(a)| + \sum_{i=2}^{n} 2r|u_i(a)|$.

Corollary 2. *Let* $\mathcal{B}_{\mathbb{R}}(u, e) = \texttt{BallClenshawBackward}((a_0, \ldots, a_n), a, r)$ *be the result of Algorithm 2, and let M be an upper bound on $|u_k(a)|$ for $1 \leq k \leq n$. Assume that $\varepsilon_k < Mr$ for $1 \leq k \leq n$, then $e < 3Mnr$.*

Algorithm 2. Clenshaw evaluation algorithm, backward error

---

    function BALLCLENSHAWBACKWARD$((a_0, \ldots, a_n), a, r)$
       ▷ *Computation of the centers u_k*
       $u_{n+1} \leftarrow 0$
       $u_n \leftarrow a_n$
       for k in $n-1, n-2, \ldots, 1$ **do**
          $u_k \leftarrow 2au_{k+1} - u_{k+2} + a_k$
          $\varepsilon_k \leftarrow$ bound on the rounding error for u_k
       $u_0 \leftarrow au_1 - u_2 + a_0$
       $\varepsilon_0 \leftarrow$ bound on the rounding error for u_0

       ▷ *Computation of the radii e_k*
       $e_n \leftarrow 0$
       for k in $n-1, n-2, \ldots, 1$ **do**
          $e_k \leftarrow e_{k+1} + 2r|u_{k+1}| + \varepsilon_k$
       $e_0 \leftarrow e_1 + r|u_1| + \varepsilon_0$
       return $\mathcal{B}_{\mathbb{R}}(u_0, e_0)$

---

3 Application to Root Finding

For classical polynomials, numerous solvers exist in the literature, such as those described in [5] for example. For polynomials in the Chebyshev basis, several approaches exist that reduce the problem to polynomial complex root finding [1], or complex eigenvalue computations [2] among other.

    In this section, we experiment a direct subdivision algorithm based on interval evaluation, detailed in Algorithm 3. This algorithm is implemented and publicly available in the software `clenshaw` [6].

    We applied this approach to Chebyshev polynomials whose coefficients are independently and identically distributed with the normal distribution with mean 0 and variance 1.

As illustrated in Fig. 1 our code performs significantly better than the classical companion matrix approach. In particular, we could solve polynomials of degree 90000 in the Chebyshev basis in less than 5 s and polynomials of degree 5000 in 0.043 s on a quad-core Intel(R) i7-8650U cpu at 1.9 GHz. For comparison, the standard numpy function chebroots took more than 65 s for polynomials of degree 5000. Moreover, using least square fitting on the ten last values, we observe that our approach has an experimental complexity closer to $\Theta(n^{1.67})$, whereas the companion matrix approach has a complexity closer to $\Theta(n^{2.39})$.

Algorithm 3. Subdivision algorithm for root finding

Require: $(a_0, \ldots, a_n)$ represents the Chebyshev polynomial approximating $f(x)$
$\quad\quad\quad (b_0, \ldots, b_n)$ represents the Chebyshev polynomial approximating $\frac{df}{dx}(x)$
Ensure: Res is a list of isolating intervals for the roots of f in $[-1,1]$
$\quad$ **function** SUBDIVIDECLENSHAW($(a_0, \ldots, a_n), (b_0, \ldots, b_n)$)
$\quad\quad \triangleright$ *Partition* $[-1,1]$ *in intervals where F either has constant sign or is monotonous*
$\quad\quad L \leftarrow [\mathcal{B}_\mathbb{R}(0,1)]$
$\quad\quad Partition \leftarrow []$
$\quad\quad$ **while** L is not empty **do**
$\quad\quad\quad \mathcal{B}_\mathbb{R}(a,r) \leftarrow$ **pop** the first element of L
$\quad\quad\quad \mathcal{B}_\mathbb{R}(f,s) \leftarrow$ BallClenshaw $((a_0, \ldots, a_n), a, r)$
$\quad\quad\quad \mathcal{B}_\mathbb{R}(df,t) \leftarrow$ BallClenshaw $((b_0, \ldots, b_n), a, r)$
$\quad\quad\quad$ **if** $f - s > 0$ **then**
$\quad\quad\quad\quad$ **append** the pair $(\mathcal{B}_\mathbb{R}(a,r),$ "*plus*") to $Partition$
$\quad\quad\quad$ **else if** $f + s < 0$ **then**
$\quad\quad\quad\quad$ **append** the pair $(\mathcal{B}_\mathbb{R}(a,r),$ "*minus*") to $Partition$
$\quad\quad\quad$ **else if** $g - s > 0$ or $g + s < 0$ **then**
$\quad\quad\quad\quad$ **append** the pair $(\mathcal{B}_\mathbb{R}(a,r),$ "*monotonous*") to $Partition$
$\quad\quad\quad$ **else**
$\quad\quad\quad\quad \mathcal{B}_1, \mathcal{B}_2 \leftarrow$ subdivide$\mathcal{B}_\mathbb{R}(a,r)$
$\quad\quad\quad\quad$ **append** $\mathcal{B}_1, \mathcal{B}_2$ to L

$\quad\quad \triangleright$ *Compute the sign of F at the boundaries*
$\quad\quad \mathcal{B}_\mathbb{R}(f,s) \leftarrow$ BallClenshaw $((a_0, \ldots, a_n), -1, 0)$
$\quad\quad$ **append** the pair $(\mathcal{B}_\mathbb{R}(-1,0), \texttt{sign}(\mathcal{B}_\mathbb{R}(f,s)))$ to $Partition$
$\quad\quad \mathcal{B}_\mathbb{R}(f,s) \leftarrow$ BallClenshaw $((a_0, \ldots, a_n), 1, 0)$
$\quad\quad$ **append** the pair $(\mathcal{B}_\mathbb{R}(1,0), \texttt{sign}(\mathcal{B}_\mathbb{R}(f,s)))$ to $Partition$

$\quad\quad \triangleright$ *Recover the root isolating intervals*
$\quad\quad Partition \leftarrow$ **sort** $Partition$
$\quad\quad Res \leftarrow$ the "monotonous" intervals of $Partition$
$\quad\quad\quad\quad$ such that the adjacent intervals have opposite signs
$\quad\quad$ **return** Res

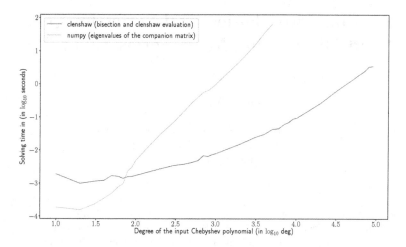

Fig. 1. Time for isolating the roots of a random Chebyshev polynomial, on a quad-core Intel(R) i7-8650U cpu at 1.9 GHz, with 16G of ram

References

1. Boyd, J.: Computing zeros on a real interval through chebyshev expansion and polynomial rootfinding. SIAM J. Numer. Anal. **40**(5), 1666–1682 (2002). https://doi.org/10.1137/S0036142901398325
2. Boyd, J.: Finding the zeros of a univariate equation: proxy rootfinders, chebyshev interpolation, and the companion matrix. SIAM Rev. **55**(2), 375–396 (2013). https://doi.org/10.1137/110838297
3. Elliott, D.: Error analysis of an algorithm for summing certain finite series. J. Aust. Math. Soc. **8**(2), 213–221 (1968). https://doi.org/10.1017/S1446788700005267
4. Gentleman, W.M.: An error analysis of Goertzel's (Watt's) method for computing Fourier coefficients. Comput. J. **12**(2), 160–164 (1969). https://doi.org/10.1093/comjnl/12.2.160
5. Kobel, A., Rouillier, F., Sagraloff, M.: Computing real roots of real polynomials ... and now for real! In: Proceedings of the ACM on International Symposium on Symbolic and Algebraic Computation, ISSAC 2016, pp. 303–310. ACM, New York (2016). https://doi.org/10.1145/2930889.2930937
6. Moroz, G.: Clenshaw 0.1, December 2019. https://doi.org/10.5281/zenodo.3571248, https://gitlab.inria.fr/gmoro/clenshaw
7. Oliver, J.: An error analysis of the modified Clenshaw method for evaluating Chebyshev and Fourier series. IMA J. Appl. Mathe. **20**(3), 379–391 (1977). https://doi.org/10.1093/imamat/20.3.379

Proving Two Conjectural Series for $\zeta(7)$ and Discovering More Series for $\zeta(7)$

Jakob Ablinger$^{(\boxtimes)}$

Research Institute for Symbolic Computation,
Johannes Kepler University, Linz, Austria
jakob.ablinger@risc.jku.at

Abstract. We give a proof of two identities involving binomial sums at infinity conjectured by Zhi-Wei Sun. In order to prove these identities, we use a recently presented method i.e., we view the series as specializations of generating series and derive integral representations. Using substitutions, we express these integral representations in terms of cyclotomic harmonic polylogarithms. Finally, by applying known relations among the cyclotomic harmonic polylogarithms, we derive the results. These methods are implemented in the computer algebra package `HarmonicSums`.

1 Introduction

In order to prove the two formulas (conjectured in [16])

$$\sum_{k=1}^{\infty} \frac{33H_k^{(5)} + 4/k^5}{k^2\binom{2k}{k}} = -\frac{45}{8}\zeta(7) + \frac{13}{3}\zeta(2)\zeta(5) + \frac{85}{6}\zeta(3)\zeta(4), \tag{1}$$

$$\sum_{k=1}^{\infty} \frac{33H_k^{(3)} + 8/k^3}{k^4\binom{2k}{k}} = -\frac{259}{24}\zeta(7) - \frac{98}{9}\zeta(2)\zeta(5) + \frac{697}{18}\zeta(3)\zeta(4), \tag{2}$$

where $H_k^{(a)} := \sum_{i=1}^{k}\frac{1}{i^a}$, we are going to use a method presented in [2], therefore we repeat some important definitions and properties (for more details we refer the interested reader to [4,7,12]). Let $\mathbb{K}$ be a field of characteristic 0. A function $f = f(x)$ is called *holonomic* (or *D-finite*) if there exist polynomials $p_d(x), p_{d-1}(x), \ldots, p_0(x) \in \mathbb{K}[x]$ (not all p_i being 0) such that the following holonomic differential equation holds:

$$p_d(x)f^{(d)}(x) + \cdots + p_1(x)f'(x) + p_0(x)f(x) = 0. \tag{3}$$

J. Ablinger—This work was supported by the Austrian Science Fund (FWF) grant SFB F50 (F5009-N15) and has received funding from the European Union's Horizon 2020 research and innovation programme under the Marie Skłodowska-Curie grant agreement No. 764850 "SAGEX".

D. Slamanig et al. (Eds.): MACIS 2019, LNCS 11989, pp. 42–47, 2020.
https://doi.org/10.1007/978-3-030-43120-4_5

A sequence $(f_n)_{n \geq 0}$ with $f_n \in \mathbb{K}$ is called holonomic (or *P-finite*) if there exist polynomials $p_d(n), p_{d-1}(n), \ldots, p_0(n) \in \mathbb{K}[n]$ (not all p_i being 0) such that the holonomic recurrence

$$p_d(n) f_{n+d} + \cdots + p_1(n) f_{n+1} + p_0(n) f_n = 0 \qquad (4)$$

holds for all $n \in \mathbb{N}$ (from a certain point on). In the following we utilize the fact that holonomic functions are precisely the generating functions of holonomic sequences: for a given holonomic sequence $(f_n)_{n \geq 0}$, the function defined by $f(x) = \sum_{n=0}^{\infty} f_n x^n$ (i.e., its generating function) is holonomic.

Note that given a holonomic recurrence for $(f_n)_{n \geq 0}$ it is straightforward to construct a holonomic differential equation satisfied by its generating function $f(x) = \sum_{n=0}^{\infty} f_n x^n$. For a recent overview of this holonomic machinery and further literature we refer to [12].

In the frame of the proofs we will deal with iterated integrals, hence we define

$$\mathrm{G}\left(f_1(\tau), f_2(\tau), \cdots, f_k(\tau); x\right) := \int_0^x f_1(\tau_1) \mathrm{G}\left(f_2(\tau), \cdots, f_k(\tau), \tau_1\right) d\tau_1,$$

where $f_1(x), f_2(x), \ldots, f_k(x)$ are hyperexponential functions. Note that $f(x)$ is called *hyperexponential* if $f'(x)/f(x) = q(x)$, where $q(x)$ is a rational function in x.

Another important class of iterated integrals that we will come across are the so called cyclotomic harmonic polylogarithms at cyclotomy 3 (compare [8]): let $m_i := (a_i, b_i) \in \{(0,0), (1,0), (3,0), (3,1)\}$ for $x \in (0,1)$ we define *cyclotomic polylogarithms at cyclotomy 3*:

$$\mathrm{H}(x) = 1,$$

$$\mathrm{H}_{m_1, \ldots, m_k}(x) = \begin{cases} \frac{1}{k!} (\log x)^k, & \text{if } m_i = (0,0) \\[2mm] \int_0^x \frac{y^{b_i}}{\Phi_{a_i}(y)} \mathrm{H}_{m_2, \ldots, m_k}(y) dy, & \text{otherwise,} \end{cases}$$

where $\Phi_a(x)$ denotes the ath cyclotomic polynomial, for instance $\Phi_1(x) = x - 1$ and $\Phi_3(x) = x^2 + x + 1$. We call k the weight of a cyclotomic polylogarithm and in case the limit exists we extend the definition to $x = 1$ and write

$$\mathrm{H}_{m_1, \ldots, m_k} := \mathrm{H}_{m_1, \ldots, m_k}(1) = \lim_{x \to 1} \mathrm{H}_{m_1, \ldots, m_k}(x).$$

Throughout this article we will write $0, 1, \lambda$ and μ for $(0,0), (1,0), (3,0)$, and $(3,1)$, respectively.

Note that cyclotomic polylogarithms evaluated at one posses a multitude of known relations, namely shuffle, stuffle, multiple argument, distribution and duality relations, for more details we refer to [6,8,10].

2   Proof of the Conjectures

In order to prove (1) and (2) we will apply the method described in [2] and hence we will make use of the command `ComputeGeneratingFunction` which is implemented in the package `HarmonicSums`[1][5]. Consider the sum left hand side of (1) and execute (note that in `HarmonicSums` $S[a, k] := \sum_{i=1}^{k} \frac{1}{i^a}$)

$$\textbf{ComputeGeneratingFunction} \left[\frac{33S[5, k] + 4/k^5}{k^2 \binom{2k}{k}}, x, \{n, 1, \infty\} \right]$$

which gives (after sending $x \to 1$)

$$\frac{4801781G(a, a; 1)}{73728} + \frac{451993G(0, a, a; 1)}{6144} + \frac{10193}{512}G(0, 0, a, a; 1)$$
$$+ \frac{363}{128}\sqrt{3}G(a, 0, a, a; 1) + \frac{1875}{128}G(0, 0, 0, a, a; 1) + \frac{363}{64}G(a, a, 0, a, a; 1)$$
$$+ \frac{37}{8}G(0, 0, 0, 0, a, a; 1) + \frac{33}{32}\sqrt{3}G(a, 0, 0, 0, a, a; 1) + \frac{37}{4}G(0, 0, 0, 0, 0, a, a; 1)$$
$$+ \frac{33}{16}G(a, a, 0, 0, 0, a, a; 1) + \frac{18937121G(a; 1)}{122880\sqrt{3}} - \frac{895605490019}{5573836800}, \qquad (5)$$

where 0 represents $1/\tau$ and $a := \sqrt{\tau}\sqrt{4 - \tau}$.
Internally `ComputeGeneratingFunction` splits the left hand side of (1) into

$$\sum_{k=1}^{\infty} x^k \frac{4}{k^7 \binom{2k}{k}} + \sum_{k=1}^{\infty} x^k \frac{33H_k^{(5)}}{k^2 \binom{2k}{k}} \qquad (6)$$

and computes the following two recurrences

$$0 = -(1 + k)^7 f(k) + 2(2 + k)^6(3 + 2k)f(1 + k),$$
$$0 = (1 + k)^2(2 + k)^6 f(k) - 2(2 + k)^2(3 + 2k)(5 + 2k)(55 + 75k + 40k^2$$
$$+ 10k^3 + k^4)f(1 + k) + 4(3 + k)^6(3 + 2k)(5 + 2k)f(2 + k),$$

satisfied by $\frac{4}{k^7 \binom{2k}{k}}$ and $\frac{33S[5,k]}{k^2 \binom{2k}{k}}$, respectively. Then it uses closure properties of holonomic functions to find the following differential equations

$$0 = f(x) + 3(-128 + 85x)f'(x) + x(-6906 + 3025x)f''(x)$$
$$+ 14x^2(-1541 + 555x)f^{(3)}(x) + 7x^3(-3112 + 993x)f^{(4)}(x)$$
$$+ 42x^4(-215 + 63x)f^{(5)}(x) + 2x^5(-841 + 231x)f^{(6)}(x)$$
$$+ 6x^6(-23 + 6x)f^{(7)}(x) + (-4 + x)x^7 f^{(8)}(x),$$
$$0 = 128f(x) + 8(-1650 + 2171x)f'(x) + 2\left(21870 - 164445x + 101876x^2\right)f''(x)$$
$$+ 2x\left(264850 - 761631x + 310438x^2\right)f^{(3)}(x)$$
$$+ 4x^2\left(354295 - 599492x + 183087x^2\right)f^{(4)}(x)$$
$$+ 2x^3\left(694988 - 826235x + 202454x^2\right)f^{(5)}(x)$$
$$+ 8x^4\left(76912 - 70638x + 14483x^2\right)f^{(6)}(x)$$

---

[1] The package `HarmonicSums` (Version 1.0 19/08/19) together with a `Mathematica` notebook containing the computations described here can be downloaded at https://risc.jku.at/sw/harmonicsums.

$$+ x^5 \left(135020 - 101534x + 17921x^2 \right) f^{(7)}(x)$$
$$+ x^6 \left(15020 - 9614x + 1491x^2 \right) f^{(8)}(x)$$
$$+ 2(-4 + x)x^7(-100 + 31x)f^{(9)}(x) + (-4 + x)^2 x^8 f^{(10)}(x),$$

satisfied by the first and the second sum in (6), respectively.

These differential equations are solved using the differential equation solver implemented in `HarmonicSums`[2]. This solver finds all solutions of holonomic differential equations that can be expressed in terms of iterated integrals over hyperexponential alphabets [4,7,11,14,15]; these solutions are called d'Alembertian solutions [9], in addition for differential equations of order two it finds all solutions that are Liouvillian [3,13,15].

Solving the differential equations, comparing initial values, summing the two results and sending $x \to 1$ leads to (5).

Since the iterated integrals in (5) only iterate over the integrands $1/\tau$ and $\sqrt{\tau}\sqrt{4 - \tau}$ we can use the substitution (compare [1, Section 3])

$$\tau \to (\tau - 1)^2/(1 + \tau + \tau^2)$$

to compute a representation in terms of cyclotomic harmonic polylogarithms at cyclotomy 3. This step is implemented in the command `SpecialGLToH` in `HarmonicSums` and executing this command leads to

$$- 3552 \mathrm{H}_{\lambda,\lambda,1,1,1,1,1} + 1776 \mathrm{H}_{\lambda,\lambda,1,1,1,1,\lambda} + 3552 \mathrm{H}_{\lambda,\lambda,1,1,1,1,\mu}$$
$$+ 1776 \mathrm{H}_{\lambda,\lambda,1,1,1,\lambda,1} - 3264 \mathrm{H}_{\lambda,\lambda,1,1,1,\lambda,\lambda} - 1776 \mathrm{H}_{\lambda,\lambda,1,1,1,\lambda,\mu}$$

$$\vdots$$

$$- 1776 \mathrm{H}_{\lambda,\lambda,\mu,\mu,\mu,\lambda,1} + 3264 \mathrm{H}_{\lambda,\lambda,\mu,\mu,\mu,\lambda,\lambda} + 1776 \mathrm{H}_{\lambda,\lambda,\mu,\mu,\mu,\lambda,\mu}$$
$$- 3552 \mathrm{H}_{\lambda,\lambda,\mu,\mu,\mu,\mu,1} + 1776 \mathrm{H}_{\lambda,\lambda,\mu,\mu,\mu,\mu,\lambda} + 3552 \mathrm{H}_{\lambda,\lambda,\mu,\mu,\mu,\mu,\mu},$$

where in total the expression consists of 243 cyclotomic polylogarithms.

Finally, we can use the command `ComputeCycloH1Basis[7,3]` to compute basis representation of the appearing cyclotomic harmonic polylogarithms. `ComputeCycloH1Basis` takes into account shuffle, stuffle, multiple argument, distribution and duality relations, for more details we refer to [6,8,10] and [1, Section 4]. Note that this is the computationally hardest part, since a linear system with more than 16000 variables and approximately 20000 equations has to be solved, however this has to be done only once. Applying these relations we find

$$- \frac{459}{4} \mathrm{H}_{0,0,1} \mathrm{H}_\lambda{}^4 - \frac{39}{2} \mathrm{H}_{0,0,0,0,1} \mathrm{H}_\lambda{}^2 + \frac{45}{8} \mathrm{H}_{0,0,0,0,0,0,1}, \tag{7}$$

---

[2] The `Mathematica` built-in differential equation solver was not sufficient to solve these differential equations. The implemented solver does not rely on the `Mathematica` built-in `DSolve`.

for which it is straightforward to verify that it is equal to the right hand side of (1) and hence this finishes the proof. Equivalently we find

$$\sum_{k=1}^{\infty} \frac{33H_k^{(3)} + 8/k^3}{k^4\binom{2k}{k}} = \frac{-6273}{20}H_\lambda{}^4H_{0,0,1} + 49H_\lambda{}^2H_{0,0,0,0,1} + \frac{259}{24}H_{0,0,0,0,0,0,1},$$

which is equal to the right hand side of (2).

3   More Identities

Using the same strategy it is possible to discover also other identities, here we list some of the additional identities that we found (with $c := \sum_{i=0}^{\infty} \frac{1}{(3i+1)^4}$):

$$\sum_{k=1}^{\infty} \frac{3H_k^{(2)} - 1/k^2}{k^5\binom{2k}{k}} = -\frac{205\zeta(7)}{18} + \frac{5\pi^2\zeta(5)}{18} + \frac{\pi^4\zeta(3)}{18} - \frac{\pi^7}{486\sqrt{3}} + \frac{\sqrt{3}c\pi^3}{8},$$

$$\sum_{k=1}^{\infty} \frac{11H_k^{(3)} + 8H_k^{(2)}/k}{k^4\binom{2k}{k}} = \frac{7337\zeta(7)}{216} + \frac{11\pi^2\zeta(5)}{81} + \frac{1417\pi^4\zeta(3)}{4860} - \frac{4\pi^7}{729\sqrt{3}} + \frac{c\pi^3}{\sqrt{3}},$$

$$\sum_{k=1}^{\infty} \frac{2H_k^{(5)} - H_k^{(3)}/k^2}{k^2\binom{2k}{k}} = -\frac{\zeta(7)}{72} + \frac{8\pi^2\zeta(5)}{81} - \frac{17\pi^4\zeta(3)}{4860}.$$

References

1. Ablinger, J.: Discovering and proving infinite binomial sums identities. J. Exp. Math. **26**, 62–71 (2017). arXiv: 1507.01703
2. Ablinger, J.: Discovering and proving infinite pochhammer sum identities. J. Exp. Math. 1–15 (2019). arXiv: 1902.11001
3. Ablinger, J.: Computing the inverse Mellin transform of holonomic sequences using Kovacic's algorithm. In: PoS RADCOR2017, vol. 69 (2017). arXiv: 1801.01039
4. Ablinger, J.: Inverse mellin transform of holonomic sequences. In: PoS LL 2016, vol. 067 (2016). arXiv: 1606.02845
5. Ablinger, J.: The package HarmonicSums: computer algebra and analytic aspects of nested sums. In: Loops and Legs in Quantum Field Theory - LL 2014 (2004). arXiv: 1407.6180
6. Ablinger, J., Blümlein, J., Schneider, C.: Generalized harmonic, cyclotomic, and binomial sums, their polylogarithms and special numbers. J. Phys. Conf. Ser. **523**, 012060 (2014). arxiv: 1310.5645
7. Ablinger, J., Blümlein, J., Raab, C.G., Schneider, C.: Iterated binomial sums and their associated iterated integrals. J. Math. Phys. Comput. **55**, 1–57 (2014). arXiv: 1407.1822
8. Ablinger, J., Blümlein, J., Schneider, C.: Harmonic sums and polylogarithms generated by cyclotomic polynomials. J. Math. Phys. **52**, 102301 (2011). arxiv: 1105.6063
9. Abramov, S.A., Petkovšek, M.: D'Alembertian solutions of linear differential and difference equations. In: Proceedings of ISSAC 1994. ACM Press (1994)

10. Blümlein, J., Broadhurst, D.J., Vermaseren, J.A.M.: The multiple zeta value data mine. Comput. Phys. Commun. **181**, 582–625 (2010). arXiv: 0907.2557
11. Bronstein, M.: Linear ordinary differential equations: breaking through the order 2 barrier. In: Proceedings of ISSAC 1992. ACM Press (1992)
12. Kauers, M., Paule, P.: The Concrete Tetrahedron. Text and Monographs in Symbolic Computation. Springer, Wien (2011). https://doi.org/10.1007/978-3-7091-0445-3
13. Kovacic, J.J.: An algorithm for solving second order linear homogeneous differential equations. J. Symb. Comput. **2**, 3–43 (1986)
14. Petkovšek, M.: Hypergeometric solutions of linear recurrences with polynomial coefficients. J. Symb. Comput. **14**, 243–264 (1992)
15. Hendriks, P.A., Singer, M.F.: Solving difference equations in finite terms. J. Symb. Comput. **27**, 239–259 (1999)
16. Sun, Z.-W.: List of conjectural series for powers of π and other constants. arXiv: 1102.5649

Generalized Integral Dependence Relations

Katsusuke Nabeshima[1]([✉]) and Shinichi Tajima[2]

[1] Graduate School of Technology, Industrial and Social Sciences,
Tokushima University, 2-1, Minamijosanjima-cho, Tokushima, Japan
nabeshima@tokushima-u.ac.jp
[2] Graduate School of Science and Technology, Niigata University,
8050, Ikarashi 2-no-cho, Nishi-ku, Niigata, Japan
tajima@emeritus.niigata-u.ac.jp

Abstract. A generalization of integral dependence relations in a ring of convergent power series is studied in the context of symbolic computation. Based on the theory of Grothendieck local duality on residues, an effective algorithm is introduced for computing generalized integral dependence relations. It is shown that, with the aid of local cohomology, generalized integral dependence relations in the ring of convergent power series can be computed in a polynomial ring. An extension of the proposed method to parametric cases is also discussed.

Keywords: Integral closure · Standard basis · Local cohomology

1 Introduction

Integral closure is an important concept in commutative algebra, number theory, algebraic geometry and singularity theory [17, 21]. In this paper, we consider the concept of integral dependence relations over an ideal from the point of view of complex analysis. We extend the classical concept of integral dependence relations and introduce a notion of generalized integral dependence relations in a ring of convergent power series. We study its basic properties and give an algorithm for computing these relations.

In [4], Kashiwara investigated Bernstein-Sato polynomials. In this paper, he used the concept of integral closure to prove the existence of good operators. In [22], Yano studied Bernstein-Sato polynomials and gave in particular several examples by utilizing integral dependence relations and some kinds of their generalizations. Note also that in [15], Scherk studied Gauss-Manin connections of isolated hypersurface singularities by using a sort of generalization of integral dependence relations for computing a saturation of Brieskorn lattice. These results therefore suggest that effective methods for computing integral dependence relations and their generalizations are desired in many applications.

This work has been partly supported by JSPS Grant-in-Aid for Scientific Research (C) (18K03214 and 18K03320).

In our previous paper [12], we gave an effective algorithm for computing integral numbers. We apply the same framework to generalized integral dependence relations, and study them in the context of symbolic computation. The key ideas of our approach are the use of local cohomology and computing ideal quotients in a polynomial ring.

This paper is organized as follows. Section 2 briefly reviews methods of solving extended ideal membership problems and algebraic local cohomology that are needed to construct an algorithm for computing generalized integral dependence relations. Section 3 gives the generalization of integral dependence relations and provides a new algorithm for computing generalized integral dependence relations. Moreover, Sect. 3 describes an extension of the proposed algorithm to parametric cases.

In this paper, we fix the following notations. The set of natural numbers $\mathbb{N}$ includes zero, and K is the field of rational numbers $\mathbb{Q}$ or the field of complex numbers $\mathbb{C}$. Let X be an open neighborhood of the origin O of the n-dimensional complex space $\mathbb{C}^n$ with coordinates $x = (x_1, \ldots, x_n)$ and let $\mathcal{O}_X$ be the sheaf on X of holomorphic functions, $\mathcal{O}_{X,O}$ the stalk at the origin of $\mathcal{O}_X$. We assume that s polynomials $f_1, \ldots, f_s$ in $K[x]$ satisfy

$$\{x \in X \,|\, f_1(x) = \cdots = f_s(x) = 0\} = \{O\}.$$

Let $\mathcal{I}_O$ be the ideal generated by $f_1, \ldots, f_s$ in $\mathcal{O}_{X,O}$. Let I be the ideal generated by $f_1, \ldots, f_s$ in the polynomial ring $K[x]$.

2 Preliminaries

Here we review methods for solving extended ideal membership problems and algebraic local cohomology.

2.1 Solving Extended Ideal Membership Problems

In [12] an algorithm for solving ideal membership problems of $\mathcal{I}_O$ is given, and in [9] an algorithm for solving extended ideal membership problems of $\mathcal{I}_O$ is given, too. Following [9] and [12], we briefly review two methods for solving the extended ideal membership problems. The underlining idea of these methods is the following lemma.

Lemma 1 ([12]). *Let h be a polynomial in $K[x]$. Then, $h \in \mathcal{I}_O$ if and only if there exists a polynomial u in the ideal quotient $I : \langle h \rangle$ such that $u \notin \mathfrak{m}$, where $I : \langle h \rangle = \{u \in K[x] \,|\, uh \in I\}$ is the ideal quotient in $K[x]$ and $\mathfrak{m} = \langle x_1, \ldots, x_n \rangle$ is the maximal ideal generated by $x_1, \ldots, x_n$.*

Suppose that $h \in K[x]$ and $h \in \mathcal{I}_O$. Then, there exists $u \in I : \langle h \rangle$ such that $u(O) \neq 0$. As $uh \in I$, there exist $p_1, p_2, \ldots, p_s \in K[x]$ such that $uh = p_1 f_1 + p_2 f_2 + \cdots + p_s f_s$. The condition $u(O) \neq 0$ implies that $\dfrac{p_i}{u}$ is an element of $\mathcal{O}_{X,O}$. That is, the extended ideal membership problem can be solved as follows:

$$h = \frac{p_1}{u} f_1 + \frac{p_2}{u} f_2 + \cdots + \frac{p_s}{u} f_s. \tag{1}$$

We review two methods to compute $p_1, p_2, \ldots, p_s \in K[x]$.

Method 1: The first method utilizes the extended Gröbner basis algorithm (for instance, see [1]) to compute the polynomials $p_1, p_2, \ldots, p_s$ of the Eq. (1). Let us fix a term order on the variables x and let $\{g_1, \ldots, g_r\}$ be a Gröbner basis of $I = \langle f_1, \ldots, f_s \rangle$ in $K[x]$. Then, the extended Gröbner basis algorithm outputs $a_{i1}, a_{i2}, \ldots, a_{is} \in K[x]$ that satisfy

$$g_i = a_{i1}f_1 + a_{i2}f_2 + \ldots + a_{is}f_s$$

where $1 \leq i \leq r$. As uh can be reduced to 0 by $\{g_1, \ldots, g_r\}$, polynomials $b_1, b_2, \ldots, b_r$ that satisfy

$$uh = b_1 g_1 + b_2 g_2 + \cdots + b_r g_r,$$

can be obtained by the division algorithm. Therefore,

$$uh = \left(\sum_{j=1}^{r} b_j a_{j1} \right) f_1 + \left(\sum_{j=1}^{r} b_j a_{j2} \right) + \cdots + \left(\sum_{j=1}^{r} b_j a_{js} \right) f_s,$$

namely, $p_i = \left(\sum_{j=1}^{r} b_j a_{ji} \right)$, for $1 \leq i \leq s$.

Method 2: The second method utilizes the syzygy computation. Let us consider the module M of syzygies of $uh, f_1, \ldots, f_s$. There exists an algorithm for computing the reduced Gröbner basis of M w.r.t. a POT (position over term) module order in $K[x]^{s+1}$. Let G be a Gröbner basis of M. As $uh \in \langle f_1, \ldots, f_s \rangle$ in $K[x]$, there exist a vector $(1, -a_1, -a_2, \ldots, -a_s)$ in G. This means

$$uh = a_1 f_1 + a_2 f_2 + \cdots + a_s f_s,$$

namely, $p_i = a_i$ for $1 \leq i \leq s$.

In Sect. 3, we will use the methods to compute generalized integral dependence relations.

2.2   Algebraic Local Cohomology and Standard Bases

Here we briefly review algebraic local cohomology and the relation between algebraic local cohomology and standard bases. The details are given in [11,19,20].

All local cohomology classes, in this paper, are algebraic local cohomology classes that belong to the set defined by

$$H_{[O]}^n(\mathcal{O}_X) = \lim_{k \to \infty} \mathrm{Ext}_{\mathcal{O}_{X,O}}^n (\mathcal{O}_{X,O}/\langle x_1, x_2, \ldots, x_n \rangle^k, \mathcal{O}_X)$$

where $\langle x_1, x_2, \ldots, x_n \rangle$ is the maximal ideal generated by $x_1, x_2, \ldots, x_n$. We represent an algebraic local cohomology class as a polynomial $\sum c_\lambda \xi^\lambda$ where $c_\lambda \in K$,

$\lambda \in \mathbb{N}^n$ and $\xi = (\xi_1, \xi_2, \ldots, \xi_n)$. (See [7,11,20].) For each $i \in \{1, 2, \ldots, n\}$, ξ_i corresponds to x_i. The multiplication by x^α is defined as

$$
x^\alpha * \xi^\lambda =
\begin{cases}
\xi^{\lambda - \alpha}, & \lambda_i \geq \alpha_i, i = 1, \ldots, n, \\
\\
0, & \text{otherwise,}
\end{cases}
$$

where $\alpha = (\alpha_1, \alpha_2, \ldots, \alpha_n) \in \mathbb{N}^n$ and $\lambda - \alpha = (\lambda_1 - \alpha_1, \ldots, \lambda_n - \alpha_n) \in \mathbb{N}^n$.

Let us fix a global term order $\prec$ on the variables ξ. For a given algebraic local cohomology class of the form

$$
\psi = c_\lambda \xi^\lambda + \sum_{\xi^{\lambda'} \prec \xi^\lambda} c_{\lambda'} \xi^{\lambda'} \quad (c_\lambda \neq 0),
$$

we call ξ^λ the *head term*, c_λ the *head coefficient* and $\xi^{\lambda'}$ the *lower terms*. We write the head term as $\mathrm{ht}(\psi)$, the set of terms of ψ as $\mathrm{Term}(\psi) = \{\xi^\kappa | \psi = \sum_{\kappa \in \mathbb{N}^n} c_\kappa \xi^\kappa, c_\kappa \neq 0, c_\kappa \in K\}$ and the set of lower terms of ψ as $\mathrm{LL}(\psi) = \{\xi^\kappa \in \mathrm{Term}(\psi) | \xi^\kappa \neq \mathrm{ht}(\psi)\}$. Let H be a finite subset of $H^n_{[O]}(\mathcal{O}_X)$. We write the set of head terms of H as $\mathrm{ht}(H)$ and the set of lower terms of H as $\mathrm{LL}(H) = \bigcup_{\psi \in H} \mathrm{LL}(\psi)$.

Definition 1 (Inverse orders). *Let $\prec$ be a local or global term order. Then, the inverse order $\prec^{-1}$ of $\prec$ is defined by*

$$
\xi^\alpha \prec \xi^\beta \iff x^\beta \prec^{-1} x^\alpha
$$

where $\alpha, \beta \in \mathbb{N}^n$.

Definition 2 (Minimal bases). *A basis $\{\xi^{\alpha_1}, \xi^{\alpha_2}, \ldots, \xi^{\alpha_\ell}\}$ for a monomial ideal is said to be minimal if no ξ^{α_i} in the basis divides other ξ^{α_j} for $i \neq j$, where $\alpha_1, \alpha_2, \ldots, \alpha_\ell \in \mathbb{N}^n$.*

Set

$$
H_F = \{\psi \in H^n_{[O]}(\mathcal{O}_X) | f_1 * \psi = f_2 * \psi = \cdots = f_s * \psi = 0\}
$$

and

$$
\mathrm{Ann}_{\mathcal{O}_{X,o}}(H_F) = \{g \in \mathcal{O}_{X,o} | g * \psi = 0, \forall \psi \in H_F\}.
$$

Note that since $\{x \in X | f_1(x) = \cdots = f_s(x) = 0\} = \{O\}$, H_F is a finite dimensional vector space. In [11,20], computation methods of a basis of the finite dimensional vector space H_F are introduced.

Let ξ^λ be a term. For a set T of terms in $K[\xi]$, we write the *neighbors* of T as $\mathrm{Neighbor}(T)$, i.e., $\mathrm{Neighbor}(T) = \{\xi_i \xi^\kappa | \xi^\kappa \in T, i\{1, 2, \ldots, n\}\}$.

Theorem 1 ([11]). *Using the same notation as above, let H be a basis of the vector space H_F such that for all $\psi \in H$, the head coefficient of ψ is 1, $\mathrm{ht}(\psi) \notin \mathrm{ht}(H \backslash \{\psi\})$ and $\mathrm{ht}(\psi) \notin \mathrm{LL}(H)$ w.r.t. a global term order $\prec$. Let Ψ be the minimal basis of the ideal generated by $\mathrm{Neighbor}(\mathrm{ht}(H)) \backslash \mathrm{ht}(H)$.*

Let $\psi = \xi^\lambda + \displaystyle\sum_{\xi^{\lambda'} \prec \xi^\lambda} c_{(\lambda,\lambda')} \xi^{\lambda'}$ in H where $c_{(\lambda,\lambda')} \in K$ and $\lambda, \lambda' \in \mathbb{N}^n$. The transfer SB_H *is defined by:*

$$
\begin{cases}
\mathrm{SB}_H(\xi^\alpha) = x^\alpha - \displaystyle\sum_{\xi^\kappa \in \mathrm{ht}(H)} c_{(\kappa,\alpha)} x^\kappa, & in \ \ K[x] \ if \ \xi^\alpha \in \mathrm{LL}(H), \\
\mathrm{SB}_H(\xi^\alpha) = x^\alpha, & in \ \ K[x] \ if \ \xi^\alpha \notin \mathrm{LL}(H),
\end{cases}
$$

where $\alpha, \kappa \in \mathbb{N}^n$.

  Then, $\mathrm{SB}_H(\Psi) = \{\mathrm{SB}_H(\xi^\lambda) | \xi^\lambda \in \Psi\}$ *is the reduced standard basis of the ideal* $\mathcal{I}_O$ *w.r.t. the local term order* $\prec^{-1}$ *in* $\mathcal{O}_{X,O}$. *Moreover,* $\mathcal{I}_O = \mathrm{Ann}_{\mathcal{O}_{X,O}}(H_F)$.

  In the next section, we will apply Theorem 1 for computing generalized integral dependence relations.

3   Generalized Integral Dependence Relations

In this section, we introduce the notion of a generalized integral dependence relation as an extension of the classical concept of integral dependence relation. We show its basic properties and give an algorithm for computing generalized integral dependence relations.

Definition 3 (Integral dependence relation). *Let $\mathcal{I}$ be an ideal in a ring R. An element $h \in R$ is said to be integral over $\mathcal{I}$ if there exists an integer ℓ and $a_i \in \mathcal{I}^i$, for $i = 1, 2, \ldots, \ell$, such that*

$$
h^\ell + a_1 h^{\ell-1} + a_2 h^{\ell-2} + \cdots + a_{\ell-1} h + a_\ell = 0.
$$

The smallest number ℓ that satisfies the equation above, is called integral number of h w.r.t. $\mathcal{I}$. The equation above is called an integral dependence relation of h over $\mathcal{I}$.

Remark. It is possible to check whether h is integral or not, by computing Hilbert-Samuel multiplicity of $\mathcal{I}$ and that of the ideal $(\mathcal{I}, h)$. An algorithm for computing Hilbert-Samuel multiplicity is given in [16].

  We generalize the concept of integral dependence relation as follows.

Definition 4 (Generalized integral dependence relation). *Let h be integral over $\mathcal{I}$, ℓ the integral number of h w.r.t. $\mathcal{I}$ and k a non-zero natural number with $k < \ell$. If there exists $b \in R$ and $a_i \in \mathcal{I}^i$, for $i = 1, 2, \ldots, k$, such that*

$$
bh^k + a_1 h^{k-1} + a_2 h^{k-2} + \cdots + a_{k-1} h + a_k = 0,
$$

then, we call the equation above a generalized integral dependence relation of h over $\mathcal{I}$.

We consider generalized integral dependence relations in the local ring $\mathcal{O}_{X,O}$.

Let h be integral over $\mathcal{I}_O$, ℓ the integral number of h w.r.t. $\mathcal{I}_O$ and k a non-zero natural number with $k < \ell$. Suppose that the following generalized integral dependence relation of h over $\mathcal{I}_O$ holds:

$$bh^k + a_1 h^{k-1} + a_2 h^{k-2} + \cdots + a_{k-1}h + a_k = 0 \tag{2}$$

where $b \in \mathcal{O}_{X,O}$, $a_i \in \mathcal{I}_O^i$ and $i = 1, \ldots, k$. Then,

$$bh^k = -a_1 h^{k-1} - a_2 h^{k-2} - \cdots - a_{k-1}h - a_k,$$

namely, b is a member of the ideal quotient

$$(\mathcal{I}_O h^{k-1} + \mathcal{I}_O^2 h^{k-2} + \cdots + \mathcal{I}_O^{k-1}h + \mathcal{I}_O^k) : \langle h^k \rangle.$$

Set

$$\mathscr{J}_k = \mathcal{I}_O h^{k-1} + \mathcal{I}_O^2 h^{k-2} + \cdots + \mathcal{I}_O^{k-1}h + \mathcal{I}_O^k,$$

then, a coefficient b of h^k in (2) is a member of $\mathscr{J}_k : \langle h^k \rangle$. Set

$$\mathscr{D}_k = \mathscr{J}_k : \langle h^k \rangle.$$

Lemma 2. *Let h be integral over $\mathcal{I}_O$ and ℓ the integral number of h w.r.t. $\mathcal{I}_O$. Then,*

$$\mathscr{D}_1 \subseteq \mathscr{D}_2 \subseteq \cdots \subseteq \mathscr{D}_\ell = \mathcal{O}_{X,O}.$$

Proof. For all $g \in \mathscr{D}_k$, we have $gh^k \in \mathscr{J}_k$. It is obviously $gh^{k+1} \in h\mathscr{J}_k$. Thus,

$$g \in h\mathscr{J}_k : \langle h^{k+1} \rangle.$$

Since $\mathscr{J}_k = \mathcal{I}_O h^{k-1} + \mathcal{I}_O^2 h^{k-2} + \cdots + \mathcal{I}_O^{k-1}h + \mathcal{I}_O^k$, we have $h\mathscr{J}_k \subset \mathscr{J}_{k+1}$. Hence,

$$g \in h\mathscr{J}_k : \langle h^{k+1} \rangle \subset \mathscr{J}_{k+1} : \langle h^{k+1} \rangle$$

so $\mathscr{D}_k \subseteq \mathscr{D}_{k+1}$. By the definition of the integral number, we have $1 \in \mathscr{J}_\ell : \langle h^\ell \rangle$. Therefore, $\mathscr{D}_\ell = \mathcal{O}_{X,O}$.   □

Next, we focus our attention on $K[x]$. For all $k \in \mathbb{N}$, let

$$J_k = Ih^{k-1} + I^2 h^{k-2} + \cdots + I^{k-1}h + I^k$$

and $Q_k = J_k : \langle h^k \rangle$ in $K[x]$.

We define H_{Q_k} to be the set of algebraic local cohomology classes in $H_{[O]}^n(\mathcal{O}_X)$ that is annihilated by the ideal Q_k, i.e.,

$$H_{Q_k} = \{\psi \in H_{[O]}^n(\mathcal{O}_X) | g * \psi = 0, \forall g \in Q_k\}.$$

We set

$$\mathrm{Ann}_{\mathcal{O}_{X,o}}(H_{Q_k}) = \{g \in \mathcal{O}_{X,o} | g * \psi = 0, \forall \psi \in H_{Q_k}\},$$

then we have the following lemma.

Lemma 3. *(i)* $\mathscr{Q}_k = \mathrm{Ann}_{\mathcal{O}_{X,O}}(H_{Q_k})$.
(ii) Let $b \in \mathrm{Ann}_{\mathcal{O}_{X,O}}(H_{Q_k})$. Then, there exists $u \in Q_k : \langle b \rangle \subset K[x]$ such that $u(O) \neq 0$.

Proof. (i) Let $J_k = J_{k,0} \cap J_k'$ be an ideal decomposition of J_k in $K[x]$ where $J_{k,0}$ is the primary ideal component at the origin O of the the ideal J_k. Then,

$$J_k : \langle h^k \rangle = (J_{k,0} : \langle h^k \rangle) \cap (J_k' : \langle h^k \rangle).$$

Since the common locus $\mathbb{V}(J_k)$ of the ideal J_k has an isolated point at O, $H_{Q_k} = H_{J_{k,0} : \langle h^k \rangle}$. Next, let us consider the ideal in $\mathcal{O}_{X,O}$. Then,

$$\mathcal{O}_{X,O}(J_{k,0} : \langle h^k \rangle) = \mathscr{J}_k : \langle h^k \rangle = \mathscr{Q}_k.$$

Thus,

$$H_{J_{k,0} : \langle h^k \rangle} = \{\psi \in H_{[O]}^n(\mathcal{O}_X) \mid q * \psi = 0, \forall q \in \mathscr{Q}_k\}.$$

By Grothendieck local duality theorem, $\mathrm{Ann}_{\mathcal{O}_{X,O}}(H_{Q_k}) = \mathscr{Q}_k$. (See [2, 10, 18].)
(ii) This follows from Lemma 1. □

As we described in Theorem 1, a standard basis of $\mathrm{Ann}_{\mathcal{O}_{X,O}}(H_{Q_k})$ can be obtained by computing a basis of the vector space H_{Q_k}. We can select b of Lemma 3 from the standard basis of $\mathrm{Ann}_{\mathcal{O}_{X,O}}(H_{Q_k})$. Let U be a Gröbner basis of $Q_k : \langle b \rangle$. Then, we can select an element u such that $u(O) \neq 0$ from the Gröbner basis U. Hence, $ub \in Q_k$, i.e., $ubh^k \in J_k \subset K[x]$. There exists $c_i \in I^i$ $(1 \leq i \leq k)$ such that

$$ubh^k = c_1 h^{k-1} + c_2 h^{k-2} + \cdots + c_{k-1}h + c_k. \tag{3}$$

As we mentioned in Sect. 2, the elements $c_1, \ldots, c_k$ can be obtained by solving the extended ideal membership problems. Therefore, as $u(O) \neq 0$, we can obtain the generalized integral dependence relation as follows

$$bh^k + \frac{-c_1}{u}h^{k-1} + \frac{-c_2}{u}h^{k-2} + \cdots + \frac{-c_{k-1}}{u}h + \frac{-c_k}{u} = 0, \quad \frac{-c_i}{u} \in \mathcal{I}_O^i,$$

where $1 \leq i \leq k$.
Set

$$F_i = \left\{ f_1^{\alpha_1} f_2^{\alpha_2} \cdots f_s^{\alpha_s} \;\middle|\; \sum_{j=1}^s \alpha_j = i, \alpha_1, \ldots, \alpha_s \in \mathbb{N} \right\}$$

and $F_i h^k = \{gh^k \mid g \in F_i\}$ where $i, k \in \mathbb{N}$.
Then, we have $I^i = \langle F_i \rangle$ and

$$J_k = \langle F_1 h^{k-1} \rangle + \langle F_2 h^{k-2} \rangle + \cdots + \langle F_{k-1}h \rangle + \langle F_k \rangle.$$

There exist d_g in $\mathcal{O}_{X,O}$ such that $-c_i = \sum_{g \in F_i} d_g g$.

Now, we are ready to introduce an algorithm for computing $u, b, -c_1, \ldots, -c_k$ of the Eq. (3) and each d_g.

Algorithm 1. (Generalized integral dependence relation)

Input: $h \in K[x]$, $\{f_1, \ldots, f_s\} \subset K[x]$: h is integral over $\mathcal{I}_O = \langle f_1, \ldots, f_s \rangle$ in $\mathcal{O}_{X,O}$.

Output:$L = \bigcup_{i=1}^{\ell} \{[i, L_i]\}$,

$$L_i = \{([b_{1i_0}, u_{1i_0}], v_{1i_1}, .., v_{1i_i}), ([b_{2i_0}, u_{2i_0}], v_{2i_1}, .., v_{2i_i}), .., ([b_{ri_0}, u_{ri_0}], v_{ri_1}, .., v_{ri_i})\}:$$

For each $[i, L_i]$, the set $\{b_{1i_0}, b_{2i_0}, \ldots, b_{ri_0}\}$ is a reduced standard basis of $\mathrm{Ann}_{\mathcal{O}_{X,O}}(H_{Q_i})$, $u_{ji_0}(O) \neq 0$ and

$$u_{ji_0} b_{ji_0} h^i + v_{ji_1} h^{i-1} + \cdots + v_{ji_{i-1}} h + v_{ji_i} = 0, \quad v_{ji_k} \in \mathcal{I}_O^{i-k}$$

where $Q_i = J_i : h^i \subset K[x]$, $H_{Q_i} = \{\psi \in H_{[O]}^n(\mathcal{O}_X)| g * \psi = 0, \forall g \in Q_i\}$, $1 \leq j \leq r$ and $1 \leq k \leq i$. The number ℓ is the integral number of h over $\mathcal{I}_O$.

BEGIN

$L \leftarrow \emptyset$; Allsb $\leftarrow \emptyset$; $J \leftarrow \emptyset$; $Sb \leftarrow \emptyset$; $k \leftarrow 1$;

while $Sb \neq \{1\}$ **do**

$F \leftarrow \{f_1^{\alpha_1} f_2^{\alpha_2} \cdots f_s^{\alpha_s} | \sum_j^s \alpha_j = k, \alpha_1, \ldots, \alpha_s \in \mathbb{N}\}$;

$J \leftarrow \{h \cdot g | g \in J\} \cup F$;

$Q \leftarrow$ Compute a basis of the ideal quotient $\langle J \rangle : h^k$ in $K[x]$;

$\Psi \leftarrow$ Compute a basis of the vector space $H_Q = \{\psi \in H_{[O]}^n(\mathcal{O}_X)| g * \psi = 0, \forall g \in Q\}$;

$Sb \leftarrow$ Compute the reduced standard basis of $\mathrm{Ann}_{\mathcal{O}_{X,O}}(\mathrm{Span}(\Psi))$;

$L_k \leftarrow \emptyset$;

while $Sb \neq \emptyset$ **do**

$b \leftarrow$ Select b from Sb; $Sb \leftarrow Sb \backslash \{b\}$;

if $b \notin$ Allsb **then**

$U \leftarrow$ Compute a basis of the ideal quotient $\langle Q \rangle : \langle b \rangle$ in $K[x]$;

$u \leftarrow$ Select u from U such that $u(O) \neq 0$;

$(v_1, \ldots, v_{k-1}) \leftarrow$ Compute v_j's that satisfy $ubh^k = \sum_{j=0}^{k-1} v_j h^j$

where $v_j = \sum_{gh^j \in J, h \nmid g} d_{jg} g$; $(\triangle 1)$

(Solve extended ideal membership problem of ubh^k w.r.t. J.)

$L_k \leftarrow L_k \cup \{([b, u], -v_1, \ldots, -v_{k-1})\}$;

Allsb $\leftarrow$ Allsb $\cup \{b\}$;

else

$[k', L_{k'}] \leftarrow$ Select $[k', L_{k'}]$ from L such that $([b, u], v_1, \ldots, v_{k'-1}) \in L_{k'}$;

$([b, u], v_1, \ldots, v_{k'-1}) \leftarrow$ Take $([b, u], v_1, \ldots, v_{k'-1})$ from $L_{k'}$;

$L_k \leftarrow L_k \cup \{([b, u], v_1, \ldots, v_{k'-1}, \underbrace{0, \ldots, 0}_{k-k' \text{elements}})\}$;

end-if

end-while

$L \leftarrow \{[k, L_k]\} \cup L$; $k \leftarrow k + 1$;

end-while

return L;

END

Theorem 2. *Algorithm 1 always terminates and outputs correctly.*

Proof. As h is integral over the zero-dimensional ideal $\mathcal{I}_O$, let us ℓ be its integral number. By Lemma 3, $\mathcal{Q}_k = \mathrm{Ann}_{\mathcal{O}_{X,o}}(H_Q)$, namely, $H_Q = \mathrm{Span}(\Psi)$ is a finite dimensional vector space. Thus, by Theorem 1, the reduced standard basis Sb of $\mathrm{Ann}_{\mathcal{O}_{X,o}}(H_Q)$ is a finite set of polynomials where $1 \leq k \leq \ell$. Therefore, the second **while-loop** stops after a finite number of iterations.

Since ℓ is the integral number, if $k = \ell$, then by Lemma 2 $\mathrm{Ann}_{\mathcal{O}_{X,o}}(\mathrm{Span}(\Psi)) = \langle 1 \rangle$, namely, the reduced standard basis is $\{1\}$. Hence, the first **while-loop** always stops after a finite number of iterations. Therefore, Algorithm 1 terminates.

If $b \notin A$llsb, the algorithm computes the relation

$$ubh^k - \sum_{j=0}^{k-1} \left(\sum_{gh^j \in J, h \nmid g} d_{jg}g \right) h^j = 0,$$

at ($\triangle 1$), by utilizing an algorithm for solving extended ideal membership problems. As we describe above, the relation is a generalized integral dependence relation or an integral dependence relation of h over $\mathcal{I}_O$. If $b \in A$llsb, then, for some $k' < k$, we already have a generalized integral dependence relation $ubh^{k'} + v_1 h^{k'-1} + \ldots + v_{k'-1} = 0$ of degree k'. Hence, a relation $h^{k-k'}(ubh^{k'} + v_1 h^{k'-1} + \cdots + v_{k'-1}) = 0$ over $\mathcal{I}_O$ of degree k follows directly from that of degree k'. □

Note that it is possible to return d_{jg}'s where $ubh^k = \sum_{j=0}^{k-1} \left(\sum_{gh^j \in J, h \nmid g} d_{jg}g \right) h^j$.

We illustrate Algorithm 1 with the following example.

Example 1. Let us consider $f = x^3z + y^6 + z^3$ and $h = y^4z$. Set $I = \langle \frac{\partial f}{\partial x}, \frac{\partial f}{\partial y}, \frac{\partial f}{\partial z} \rangle \subset K[x, y, z]$ and $\mathcal{I}_O = \langle \frac{\partial f}{\partial x}, \frac{\partial f}{\partial y}, \frac{\partial f}{\partial z} \rangle \subset \mathcal{O}_{X,O}$. The variables ξ, η, ζ correspond the variables x, y, z for algebraic local cohomology classes, respectively. The term order $\prec$ is the degree lexicographic order with $\zeta \prec \eta \prec \xi$.

We execute Algorithm 1 to get generalized integral dependence relations of h over $\mathcal{I}_O$.
▷ Case $k = 1$. Set $F = \{\frac{\partial f}{\partial x}, \frac{\partial f}{\partial y}, \frac{\partial f}{\partial z}\}$ and $J = F$ in $K[x, y, z]$.

1: The reduced Gröbner basis of $\langle J \rangle : \langle h \rangle \subset K[x, y, z]$ is $Q = \{x^2, z^2, y\}$.
2: A basis of the vector space $H_Q = \{\psi \in H^3_{[O]}(\mathcal{O}_X) | g * \psi = 0, \forall g \in Q\}$ is $\{1, \xi, \zeta, \xi\zeta\}$.
3: The reduced standard basis of $\mathrm{Ann}_{\mathcal{O}_{X,o}}(H_Q)$ w.r.t. $\prec^{-1}$ is Sb $= \{x^2, y, z^2\}$. Set $b_1 = x^2, b_2 = y, b_3 = z^2$ and Allsb $= S$b.
4-1: The reduced Gröbner basis of $\langle Q \rangle : \langle x_2 \rangle$ is $\{1\}$. Thus, set $u = 1$. As $ub_1 h \in \langle J \rangle = \langle \frac{\partial f}{\partial x}, \frac{\partial f}{\partial y}, \frac{\partial f}{\partial z} \rangle$, by utilizing a method of solving extended ideal membership problems, we obtain the following generalized integral dependence relation

$$-3x^2h + y^4(\tfrac{\partial f}{\partial x}) = 0. \tag{4}$$

4-2: Similarly, $\langle Q \rangle : \langle y \rangle = \langle Q \rangle : \langle z^2 \rangle = \langle 1 \rangle$. Set $u = 1$. As $ub_2h, ub_3h \in \langle J \rangle$, we obtain the following generalized integral dependence relations

$$-6yh + z(\tfrac{\partial f}{\partial y}) = 0, \quad 9z^2h + \left(-xy^4(\tfrac{\partial f}{\partial x}) + 3y^4z\tfrac{\partial f}{\partial z}\right) = 0. \tag{5}$$

▷ Case $k = 2$. Set $J = \{\tfrac{\partial f}{\partial x}h, \tfrac{\partial f}{\partial y}h, \tfrac{\partial f}{\partial z}h, (\tfrac{\partial f}{\partial x})^2, (\tfrac{\partial f}{\partial y})^2, (\tfrac{\partial f}{\partial z})^2, (\tfrac{\partial f}{\partial x})(\tfrac{\partial f}{\partial y}), (\tfrac{\partial f}{\partial x})(\tfrac{\partial f}{\partial z}),$ $(\tfrac{\partial f}{\partial y})(\tfrac{\partial f}{\partial z})\}$.

1: The reduced Gröbner basis of $\langle J \rangle : \langle h^2 \rangle \subset K[x, y, z]$ is $Q = \{x^2, y, z\}$. Set $b_1 = x^2, b_2 = y, b_3 = z$D

2: A basis of the vector space $H_Q = \{\psi \in H^3_{[O]}(\mathcal{O}_X) | g * \psi = 0, \forall g \in Q\}$ is $\{1, \xi\}$.

3: The reduced standard basis of $\mathrm{Ann}_{\mathcal{O}_{X,o}}(H_Q)$ w.r.t. $\prec^{-1}$ is $Sb = \{x^2, y, z\}$.

4-1: As $x^2, y \in Allsb$, we do not need to compute the relations. In fact, the generalized integral dependence relations

$$-3x^2h^2 + y^4(\frac{\partial f}{\partial x})h = 0, \quad -6yh^2 + z(\frac{\partial f}{\partial y})h = 0.$$

directly follow from (4), and (5).

4-2: The reduced Gröbner basis of $\langle Q \rangle : \langle z \rangle$ is $\{1\}$. Set $u = 1$. As $zh^2 \in \langle J \rangle$, by utilizing a method of solving extended ideal membership problems, we obtain the following generalized integral dependence relation

$$-54zh^2 + (-xy^3(\frac{\partial f}{\partial x})(\frac{\partial f}{\partial y}) + 3y^3z(\frac{\partial f}{\partial y})(\frac{\partial f}{\partial z})) = 0.$$

Renew $Allsb$ as $Allsb \cup \{z\}$.

▷ Cases $k = 3$. Set $J = \{\tfrac{\partial f}{\partial x}h^2, \tfrac{\partial f}{\partial y}h^2, \tfrac{\partial f}{\partial z}h^2, (\tfrac{\partial f}{\partial x})^2h, (\tfrac{\partial f}{\partial y})^2h, (\tfrac{\partial f}{\partial z})^2h, (\tfrac{\partial f}{\partial x})(\tfrac{\partial f}{\partial y})h,$ $(\tfrac{\partial f}{\partial x})(\tfrac{\partial f}{\partial z})h, (\tfrac{\partial f}{\partial y})(\tfrac{\partial f}{\partial z})h, (\tfrac{\partial f}{\partial x})^3, (\tfrac{\partial f}{\partial x})^2(\tfrac{\partial f}{\partial y}), (\tfrac{\partial f}{\partial x})^2(\tfrac{\partial f}{\partial z}), (\tfrac{\partial f}{\partial x})(\tfrac{\partial f}{\partial y})^2, (\tfrac{\partial f}{\partial x})(\tfrac{\partial f}{\partial z})^2,$ $(\tfrac{\partial f}{\partial x})(\tfrac{\partial f}{\partial y})(\tfrac{\partial f}{\partial z}), (\tfrac{\partial f}{\partial y})^3, (\tfrac{\partial f}{\partial y})^2(\tfrac{\partial f}{\partial z}), (\tfrac{\partial f}{\partial y})(\tfrac{\partial f}{\partial z})^2, (\tfrac{\partial f}{\partial z})^3\}$.

1: The reduced Gröbner basis of $\langle J \rangle : \langle h^3 \rangle$ is $Q = \{1\}$.

2: The reduced standard basis of $\mathrm{Ann}_{\mathcal{O}_{X,o}}(H_Q)$ is $Sb = \{1\}$. Thus, the integral number is 3. Set $u = 1$.

3-1: As $uh^3 \in \langle J \rangle$, we have the following integral dependence relation

$$h^3 + \frac{1}{324}(xy^2(\frac{\partial f}{\partial x})(\frac{\partial f}{\partial y})^2 - 3y^2z(\frac{\partial f}{\partial y})^2(\frac{\partial f}{\partial z})) = 0.$$

As $Sb = \{1\}$, we stop the computation.

We have implemented Algorithm 1 in the computer algebra system Risa/Asir [14].

4 Parametric Cases

We conclude this paper by considering the extension of Algorithm 1 to parametric cases.

Let $t = \{t_1, \ldots, t_m\}$ be variables such that $t \cap x = \emptyset$ and $\mathbb{C}[t][x]$ be a polynomial ring with coefficients in a polynomial ring $\mathbb{C}[t]$.

For $g_1, \ldots, g_r \in \mathbb{C}[t]$, $\mathbb{V}(g_1, \ldots, g_r) \subseteq \mathbb{C}^m$ denotes the affine variety of $g_1, \ldots, g_r$, i.e., $\mathbb{V}(g_1, \ldots, g_r) = \{\bar{t} \in \mathbb{C}^m | g_1(\bar{t}) = \cdots = g_r(\bar{t}) = 0\}$. We call an algebraic constructible set of a from $\mathbb{V}(g_1, \ldots, g_r) \backslash \mathbb{V}(g_1', \ldots, g_{r'}') \subseteq \mathbb{C}^m$ with $g_1, \ldots, g_r, g_1', \ldots, g_{r'}' \in \mathbb{C}[t]$, a stratum.

For every $\bar{t} \in \mathbb{C}^m$, the canonical specialization homomorphism $\sigma_{\bar{t}} : \mathbb{C}[t][x] \to \mathbb{C}[x]$ (or $\mathbb{C}[t] \to \mathbb{C}$) is defined as the map that substitutes t by $\bar{t}$ in $f(t, x) \in \mathbb{C}[t][x]$ (i.e., $\sigma_{\bar{t}}(f) = f(\bar{t}, x) \in \mathbb{C}[x]$). The image $\sigma_{\bar{t}}$ of a set F is denoted by $\sigma_{\bar{t}}(F) = \{\sigma_{\bar{t}}(f) | f \in F\} \subset \mathbb{C}[x]$.

Definition 5 (Comprehensive Gröbner system (CGS)). *Let $\prec$ be a term order on the variables x. Let F be a subset of $\mathbb{C}[t][x]$, $\mathbb{A}_1, \mathbb{A}_2, \ldots, \mathbb{A}_\nu$ strata in $\mathbb{C}^m$ and $G_1, G_2, \ldots, G_\nu$ subsets in $\mathbb{C}[t][x]$. A finite set $\mathcal{G} = \{(\mathbb{A}_1, G_1), (\mathbb{A}_2, G_2), \ldots, (\mathbb{A}_\nu, G_\nu)\}$ of pairs is called comprehensive Gröbner system on $\mathbb{A}_1 \cup \cdots \cup \mathbb{A}_\nu$ of $\langle F \rangle$ if for all $\bar{a} \in \mathbb{A}_i$, $\sigma_{\bar{a}}(G_i)$ is a Gröbner basis of $\langle \sigma_{\bar{a}}(F) \rangle$ in $\mathbb{C}[x]$ for each $i = 1, 2, \ldots, \nu$. We simply say $\mathcal{G}$ is a comprehensive Gröbner system of $\langle F \rangle$ if $\mathbb{A}_1 \cup \cdots \cup \mathbb{A}_\nu = \mathbb{C}^m$.*

We refer to [3,5,6] for algorithms and implementations of computing comprehensive Gröbner basis.

In order to extend Algorithm 1 to parametric cases, we need algorithms for computing comprehensive Gröbner systems of ideal quotients, parametric local cohomology classes and parametric standard bases. In [8,13], algorithms for computing comprehensive Gröbner systems of ideal quotients are given. In [7,11], algorithms for computing parametric local cohomology classes and parametric standard bases are given, too. Therefore, we are able to naturally extend Algorithm 1 to parametric cases.

Here we give an example for the parametric case.

Let us consider $f = x^2 y + y^5 + z^4 + y^4 z \in \mathbb{C}[x, y, z]$ (V_{18}^{*1} singularity) that defines an isolated singularity at the origin O in $\mathbb{C}^3$ and $h = xyz + t_1 y^3 z + t_2 y^3 z^2 \in (\mathbb{C}[t_1, t_2])[x, y, z]$ where t_1, t_2 are parameters. Set $I = \langle \frac{\partial f}{\partial x}, \frac{\partial f}{\partial y}, \frac{\partial f}{\partial z} \rangle$ in $\mathbb{C}[x, y, z]$, $\mathcal{I}_O = \langle \frac{\partial f}{\partial x}, \frac{\partial f}{\partial y}, \frac{\partial f}{\partial z} \rangle$ in $\mathcal{O}_{X,O}$ and $\prec$ is the degree lexicographic term order with $\zeta \prec \eta \prec \xi$.

▷ Case $k = 1$. Set $J = \{\frac{\partial f}{\partial x}, \frac{\partial f}{\partial y}, \frac{\partial f}{\partial z}\}$.

1-1: A comprehensive Gröbner system of $\langle J \rangle : \langle h \rangle$ w.r.t. the degree lexicographic term order with (x, y, z) is

$\{\mathbb{A}_1 = (\mathbb{V}(t_1, t_2), Q_{11} = \{1\}),$
$(\mathbb{A}_2 = \mathbb{V}(t_1) \backslash \mathbb{V}(t_1, t_2), Q_{12} = \{4y^2 + 25z, -16yz + 125z, 64z^2 + 625z, x\}),$
$(\mathbb{A}_3 = \mathbb{V}(64t_1 - 625t_2) \backslash \mathbb{V}(t_1, t_2), Q_{13} = \{5y^2 + 4yz, z^2, x\}),$
$(\mathbb{A}_4 = \mathbb{C}^2 \backslash \mathbb{V}(64t_1^2 - 625t_1 t_2), Q_{14} = \{16yz^2 - 125z^2, 64z^3 + 625z^2, 5y^2 + 4yz, x\})\}.$

We compute generalized integral dependence relations in each stratum.

1-2-1: If (t_1, t_2) belongs to the stratum $\mathbb{A}_1$, then the reduced standard basis of $\mathrm{Ann}_{\mathcal{O}_{X,O}}(H_{Q_{11}})$ w.r.t. $\prec^{-1}$ is $\{1\}$. Therefore, the integral number is 1. As the reduced Gröbner basis of $\langle Q_{11} \rangle : \langle 1 \rangle$ is $\{1\}$, an integral dependence relation of h over $\mathcal{I}_O$ is

- $h - \frac{1}{2}z\frac{\partial f}{\partial x} = 0$.

1-2-2: If (t_1, t_2) belongs to the stratum $\mathbb{A}_2$, then the reduced standard basis of $\mathrm{Ann}_{\mathcal{O}_{X,O}}(H_{Q_{12}})$ w.r.t. $\prec^{-1}$ is $Sb_{12} = \{x, y^2, z\}$. As $\langle Q_{12} \rangle : \langle x \rangle = \langle 1 \rangle$ and $\langle Q_{12} \rangle : \langle y^2 \rangle = \langle Q_{12} \rangle : \langle z \rangle = \langle x, 16y - 125, 64z + 625 \rangle$, set $u_{12} = 16y - 125$. (One can also select $64z + 625$ that has a constant term.) Then, generalized integral dependence relations are the following.

- $xh + ((xz + t_2z^2y^2)\frac{\partial f}{\partial x}) = 0$,

- $u_{12}y^2h + ((2x^2 - \frac{5}{2}t_2xy^3 + 2t_2xy^2z - \frac{25}{2}t_2xz^2 + \frac{125}{2}y^2z - 40z^3)\frac{\partial f}{\partial x} - (4xy + 4t_2y^3z - 25t_2z^2y + 20t_2z^3)\frac{\partial f}{\partial y} + (5t_2x^2 + 20xy + 20t_2y^3z)\frac{\partial f}{\partial z}) = 0$,

- $u_{12}zh + (\frac{25}{8}t_2xy^2 - \frac{5}{2}t_2xyz + 2t_2xz^2 - 8yz^2 + \frac{125}{2}z^2)\frac{\partial f}{\partial x} - (\frac{25}{4}t_2y^3 - 5t_2y^2z + 4t_2yz^2)\frac{\partial f}{\partial y} + (\frac{125}{4}t_2y^3)\frac{\partial f}{\partial z}) = 0$.

1-2-3: If (t_1, t_2) belongs to the stratum $\mathbb{A}_3$, then the reduced standard basis of $\mathrm{Ann}_{\mathcal{O}_{X,O}}(H_{Q_{13}})$ w.r.t. $\prec^{-1}$ is $Sb_{13} = \{x, y^3, z^2, yz + \frac{5}{4}y^2\}$. For all $b \in Sb_{13}$, the reduced Gröbner basis of $\langle Q_{13} \rangle : \langle b \rangle$ is $\{1\}$. Then, generalized integral dependence relations are the following.

- $xh - \frac{1}{2}(xz + t_2y^2z^2 + t_1y^2z)\frac{\partial f}{\partial x} = 0$,

- $y^3h + \frac{1}{128}(16x^2 + 20t_2xy^3 + 16t_2xy^2z + 125t_2xyz - 100t_1xz^2 - 320z^3)\frac{\partial f}{\partial x} + \frac{-1}{128}(32xy + 32t_2y^3z + 250t_2y^2z - 200t_2yz^2 + 160t_2z^3)\frac{\partial f}{\partial y} + \frac{1}{128}(40t_2x^2 + 160xy + 160t_2y^3z)\frac{\partial f}{\partial z} = 0$,

- $z^2h + \frac{1}{512}(-16t_2xy^3 - 125t_2xy^2 + 100t_2xyz - 80t_2xz^2 - 256z^3)\frac{\partial f}{\partial x} + \frac{1}{512}(250t_2y^3 - 200t_2 y^2z + 160t_2yz^2 - 128t_2z^3)\frac{\partial f}{\partial y} + \frac{1}{512}(32t_2x^2 - 1250t_2y^3)\frac{\partial f}{\partial z} = 0$,

- $(yz + \frac{5}{4}y^2)h + \frac{1}{8}(t_2xz^2 + t_1xz - 5y^2z - 4yz^2)\frac{\partial f}{\partial x} - \frac{1}{8}(2t_2yz^2 + 2t_1yz)\frac{\partial f}{\partial y} = 0$. Set $Allsb_3 = Sb_{13}$.

1-2-4: If (t_1, t_2) belongs to the stratum $\mathbb{A}_4$, then the reduced standard basis of $\mathrm{Ann}_{\mathcal{O}_{X,O}}(H_{Q_{14}})$ w.r.t. $\prec^{-1}$ is $Sb_{14} = \{x, y^3, z^2, yz + \frac{5}{4}y^2\}$. As $\langle Q_{14} \rangle : \langle x \rangle = \langle Q_{14} \rangle : \langle yz + \frac{5}{4}y^2 \rangle = \langle 1 \rangle$ and $\langle Q_{14} \rangle : \langle y^3 \rangle = \langle Q_{14} \rangle : \langle z^2 \rangle = \langle x, 16y - 125, 64z + 625 \rangle$, set $u_{14} = 16y - 125$. Then, generalized integral dependence relations are the following.

- $xh - \frac{1}{2}(xz + t_1y^2z + t_2y^2z^2)\frac{\partial f}{\partial x} = 0$,

- $u_{14}y^3h + \frac{-1}{8}(4t_2x^3 + 125x^2 - 16t_1y^3 + 100t_2yz^2 + 100t_1yz - 160t_2xz^3 - 80t_1xz^2 - 256z^4 - 2500z^3)\frac{\partial f}{\partial x} + \frac{-1}{8}(-8t_2x^2y - 250xy - 200t_2y^2z^2 - 200t_1y^2z + 160t_2yz^3 + 160t_1yz^2 - 128t_2z^4 - 128t_1z^3)\frac{\partial f}{\partial y} + \frac{-1}{8}(40t_2x^2y + 32t_2x^2z + 32t_1x^2 + 128xyz + 1250xy + 128t_2y^3z^3 + 128t_1yz^2)\frac{\partial f}{\partial z} = 0$,

- $u_{14}z^2h + \frac{1}{40}(125t_2xy^2z - 100t_2xyz^2 + 80t_2xz^3 + 125t_1xy^2 - 100t_1xyz + 80t_1xz^2 + 256z^4 - 16x^2y + 2500z^3)\frac{\partial f}{\partial x} + \frac{1}{40}(-250t_2y^3z + 200t_2y^2z^2 - 160t_2yz^3 - 250t_1y^3 + 200t_1y^2z - 160t_1yz^2 + 32xy^2)\frac{\partial f}{\partial y} + \frac{1}{40}(250t_2y^3z + 1250t_1y^3 - 160xy^2 - 128xyz)\frac{\partial f}{\partial z} = 0$,

- $(yz + \frac{5}{4}y^2)h - \frac{1}{8}(-t_1xz - t_2xz^2 + 5y^2z + 4yz^2)\frac{\partial f}{\partial x} + (2t_1yz + 2t_2yz^2)\frac{\partial f}{\partial y} = 0$.
  Set $Allsb_4 = Sb_{14}$.

▷ Case $k = 2$. Set $J = \{\frac{\partial f}{\partial x}h, \frac{\partial f}{\partial y}h, \frac{\partial f}{\partial z}h, (\frac{\partial f}{\partial x})^2, (\frac{\partial f}{\partial y})^2, (\frac{\partial f}{\partial z})^2, (\frac{\partial f}{\partial x})(\frac{\partial f}{\partial y}), (\frac{\partial f}{\partial x})(\frac{\partial f}{\partial z}), (\frac{\partial f}{\partial y})(\frac{\partial f}{\partial z})\}$.

2-1: A comprehensive Gröbner system of $\langle J\rangle : \langle h^2\rangle$ on $\mathbb{C}^2\backslash\mathbb{V}(t_1, t_2)$ is

$$\{(\mathbb{A}_2, Q_{22} = \{x, 16y - 125, 64z + 625\}),$$
$$(\mathbb{A}_3, Q_{23} = \{5y + 4z, z^2, x\}),$$
$$(\mathbb{A}_4, Q_{24} = \{64z^3 + 625z^2, 5y + 4z, x\})\}.$$

2-2-2: If (t_1, t_2) belongs to the stratum $\mathbb{A}_2$, then the reduced standard basis of $\mathrm{Ann}_{\mathcal{O}_{X,O}}(H_{Q_{21}})$ w.r.t. $\prec^{-1}$ is $\{1\}$. Therefore, the integral number is 2. As $\langle Q_{22}\rangle : \langle 1\rangle = \langle Q_{22}\rangle$, set $u_{22} = 16y - 125$. An integral dependence relation of h over $\mathcal{I}_O$ is

- $800u_{22}h^2 + v_{221}h + v_{222} = 0$ where
  $v_{221} = 75000t_2x^2z - 250000t_2y^4z - 300000t_2y^3z^2 + 500000t_2z^4$,
  $v_{222} = -125000t_2^2y^{11} - 100000t_2^2y^{10}z - 12800t_2^2y^7z^4 - 250000t_2^2y^7z^3 - 25000t_2^2x^2y^7 - 500000t_2^2y^3z^6 - 25000t_2^2x^2y^3z^3 - 25600t_2xy^5z^3 + 250000t_2xy^5z^2 + 500000t_2xy^4z^3 - 500000t_2xyz^5 + 5y^8 + 4y^7z + 75000t_2x^3yz^2 - 12800x^2y^3z^2 + 20y^4z^3 + 16y^3z^4 + x^2y^4 + 100000x^2y^2z^2 + 25000b^2y^3 + 4x^2z^3$.

2-2-3: If (t_1, t_2) belongs to the stratum $\mathbb{A}_3$, then the reduced standard basis of $\mathrm{Ann}_{\mathcal{O}_{X,O}}(H_{Q_{23}})$ w.r.t. $\prec^{-1}$ is $Sb_{23} = \{x, y^2, yz, z + \frac{5}{4}y\}$. Since $x \in Allsb_3$, its generalized integral dependence relation can be obtained from 1-2-3. For all $b \in Sb_{23}\backslash\{x\}$, the reduced Gröbner basis of $\langle Q_{23}\rangle : \langle b\rangle$ is $\{1\}$. Then, generalized integral dependence relations are the following.

- $-16384y^2h^2 + v_{231}h + v_{232} = 0$ where
  $v_{231} = (160000t_2^2y^8z^3 + 128000t_2^2y^7z^4 + 160000t_1t_2y^8z^2 + 128000t_1t_2y^7z^3 + 32000t_2^2x^2y^4z^3 + 32000t_1t_2x^2y^4z^2 + 160000t_2xy^6z^2 + 128000t_2xy^5z^3 + 32000t_2x^3y^2z^2)$,
  $v_{232} = (16384t_2^2y^8z^4 + 160000t_2^2y^8z^3 - 128000t_2^2y^7z^4 - 1250000t_2^2y^7z^3 - 32000t_2^2x^2y^4z^3 + 32768t_2y^6z^3 - 312500t_2^2x^2y^4z^2 + 160000t_2xy^6z^2 - 128000t_2xy^5z^3 - 32000t_2x^3y^2z^2 + 16384x^2y^4z^2)$.
- $327680yzh^2 + (-327680t_2^2y^7z^5 - 6400000t_2^2y^7z^4 - 31250000t_2^2y^7z^3 - 655360t_2xy^5z^4 - 6400000t_2xy^5z^3 - 327680x^2y^2z^3) = 0$.
- $327680(z + \frac{5}{4})h^2 + v_{233}h + v_{234} = 0$ where
  $v_{233} = -4000000t_2^2y^7z^3 - 3200000t_2^2y^6z^4 - 4000000t_1t_2y^7z^2 - 3200000t_1t_2y^6z^3 - 800000t_2^2x^2y^3z^3 - 819200t_2xy^5z^3 - 800000t_1t_2x^2y^3z^2 - 819200t_1xy^5z^2 - 4000000t_2xy^5z^2 - 3200000t_2xy^4z^3 - 800000t_2x^3yz^2 - 819200x^2y^3z^2$,
  $v_{234} = -409600t_2^2y^7z^4 - 327680t_2^2y^6z^5 - 4000000t_2^2y^7z^3 - 3200000t_2^2y^6z^4 + 800000t_2^2x^2y^3z^3 - 655360t_2xy^4z^4 + 7812500t_2^2x^2y^3z^2 + 4000000t_2xy^5z^2 - 3200000t_2xy^4z^3 + 800000t_2x^3yz^2 + 409600x^2y^3z^2 - 327680x^2y^2z^3$.

2-2-4: If (t_1, t_2) belongs to the stratum $\mathbb{A}_4$, then the reduced standard basis of $\mathrm{Ann}_{\mathcal{O}_{X,O}}(H_{Q_{24}})$ w.r.t. $\prec^{-1}$ is $S\mathrm{b}_{24} = \{x, y^2, yz, z + \frac{5}{4}y\}$. Since $x \in A\mathrm{llsb}_4$, its reduced integral dependence relation can be obtained from 1-2-4. As $\langle Q_{24} \rangle : \langle y^2 \rangle = \langle Q_{24} \rangle : \langle yz \rangle = \langle x, 16y - 125, 64z + 625 \rangle$ and $\langle Q_{24} \rangle : \langle z + \frac{5}{4}y \rangle = \langle 1 \rangle$, set $u_{24} = 16y - 125$. Then, generalized integral dependence relations are the following.

- $128u_{24}y^2h^2 + v_{241}h + v_{242} = 0$ where

  $v_{241} = 16000t_1t_2y^8z^3 - 20480t_1t_2y^6z^5 + 250000t_2^2y^7z^4 + 300000t_2^2y^6z^5 + 16000t_1^2y^8z^2 - 20480t_1^2y^6z^4 + 250000t_1t_2y^7z^3 + 300000t_1t_2y^6z^4 + 51200t_1t_2y^3z^7 - 500000t_2^2y^3z^7 + 51200t_1^2y^3z^6 + 3200t_1t_2x^2y^4z^3 - 5120t_1t_2x^2y^3z^4 - 500000t_1t_2y^3z^6 + 75000t_2^2x^2y^3z^4 + 3200t_2^2x^2y^4z^2 - 5120t_2^2x^2y^3z^3 + 750000t_1t_2x^2y^3z^3 + 16000t_1xy^6z^2 - 20480t_1xy^4z^4 + 250000t_2xy^5z^3 + 300000t_2xy^4z^4 + 51200t_1xyz^6 - 500000t_2xyz^6 + 3200t_1x^3y^2z^2 - 5120t_1x^3yz^3 + 75000t_2x^3yz^3$,

  $v_{242} = -2048b^2y^9z^4 - 4096t_1t_2y^9z^3 + 16000t_2^2y^8z^4 - 2048t_1^2y^9z^2 + 16000t_1t_2y^8z^3 + 20480t_1t_2y^6z^5 - 250000t_2^2y^7z^4 - 300000t_2^2y^6z^5 + 20480t_1^2y^6z^4 - 250000t_1t_2y^7z^3 - 300000t_1t_2y^6z^4 - 51200t_1t_2y^3z^7 + 500000t_2^2y^3z^7 - 4096t_2xy^7z^3 - 51200t_1^2y^3z^6 - 3200t_1t_2x^2y^4z^3 + 5120t_1t_2x^2y^3z^4 + 500000t_1t_2y^3z^6 - 4096t_1xy^7z^2 - 75000t_2^2x^2y^3z^4 + 32000t_2xy^6z^3 - 3200t_1^2x^2y^4z^2 + 5120t_1^2x^2y^3z^3 - 75000t_1t_2x^2y^3z^3 + 16000t_1xy^6z^2 + 20480t_1xy^4z^4 - 250000t_2xy^5z^3 - 300000t_2xy^4z^4 - 51200t_1xyz^6 + 500000t_2xyz^6 - 2048x^2y^5z^2 - 3200t_1x^3y^2z^2 + 5120t_1x^3yz^3 - 75000t_2x^3yz^3 + 16000x^2y^4z^2$.

- $12800u_{24}yzh^2 + v_{243}h + v_{244} = 0$ where

  $v_{243} = 3200000t_1t_2y^7z^4 + 3840000t_1t_2y^6z^5 - 31250000t_2^2y^7z^4 - 37500000t_2^2y^6z^5 + 320\ 0000t_1^2y^7z^3 + 3840000t_1^2y^6z^4 - 31250000t_1t_2y^7z^3 - 37500000t_1t_2y^6z^4 - 6400000t_1t_2y^3z^7 + 62500000t_2^2y^3z^7 - 6400000t_1^2y^3z^6 + 960000t_1t_2x^2y^3z^4 + 62500000t_1t_2y^3z^6 - 93750\ 00t_2^2x^2y^3z^4 + 960000t_1^2x^2y^3z^3 - 9375000t_1t_2x^2y^3z^3 + 3200000t_1xy^5z^3 + 3840000t_1x\ y^4z^4 - 31250000t_2xy^5z^3 - 37500000t_2xy^4z^4 - 6400000t_1xyz^6 + 62500000t_2xyz^6 + 960\ 000t_1x^3yz^3 - 9375000t_2x^3yz^3$,

  $v_{244} = (-204800t_2^2y^8z^5 - 409600t_1t_2y^8z^4 + 1600000t_2^2y^7z^5 - 204800t_1^2y^8z^3 - 3840000\ t_1t_2y^6z^5 + 31250000t_2^2y^7z^4 + 37500000t_2^2y^6z^5 - 1600000t_1^2y^7z^3 - 3840000t_1^2y^6z^4 + 312\ 50000t_1t_2y^7z^3 + 37500000t_1t_2y^6z^4 + 6400000t_1t_2y^3z^7 - 62500000t_2^2y^3z^7 - 409600t_2x\ y^6z^4 + 6400000t_1^2y^3z^6 - 960000t_1t_2x^2y^3z^4 - 62500000t_1t_2y^3z^6 - 409600t_1xy^6z^3 + 937\ 5000t_2^2x^2y^3z^4 + 3200000t_2xy^5z^4 - 960000t_1^2x^2y^3z^3 + 9375000t_1t_2x^2y^3z^3 - 3840000t_1\ xy^4z^4 + 31250000t_2xy^5z^3 + 37500000t_2xy^4z^4 + 6400000t_1xyz^6 - 62500000t_2xyz^6 - 204\ 800x^2y^4z^3 - 960000t_1x^3yz^3 + 9375000t_2x^3yz^3 + 1600000x^2y^3z^3$.

- $320(z + \frac{5}{4}y)h^2 + v_{245}h + v_{246} = 0$ where

  $v_{245} = -400t_2y^4z^2 - 320t_2y^3z^3 - 400t_1y^4z - 320t_1y^3z^2 - 80t_2x^2z^2 - 80t_1x^2z - 800x\ y^2z - 640xyz^2$,

  $v_{246} = 80t_2^2x^2y^3z^4 + 160t_1t_2x^2y^3z^3 + 400t_2xy^5z^3 + 320t_2xy^4z^4 + 80t_2^2x^2y^2z^2 + 400t_1xy^5z^2 + 320t_1xy^4z^3 + 80t_2x^3yz^3 + 80t_1x^3yz^2 + 400x^2y^3z^2 + 320x^2y^2z^3$.

▷ **Case $k = 3$.** $J = Ih^2 + I^2h + I^3$.

3-1: A comprehensive Gröbner system of $J : \langle h^3 \rangle$ on $\mathbb{C}^2 \backslash \mathbb{V}(t_1)$ is

$\{(\mathbb{V}(t_2) \backslash \mathbb{V}(t_1, t_2), Q_{31} = \{x, 16y - 125, 64z + 625\}),$
$(\mathbb{V}(64t_1 - 625t_2) \backslash \mathbb{V}(t_1, t_2), Q_{32} = \{1\}),$
$(\mathbb{C}^2 \backslash \mathbb{V}(64t_1^2t_2 - 625t_1t_2^2), Q_{33} = \{x, 16y - 125, 64z + 625\})\}.$

3-2: If (t_1, t_2) belongs to strata $\mathbb{V}(t_1)\backslash\mathbb{V}(t_1, t_2)$, $\mathbb{V}(64t_1 - 625t_2)\backslash\mathbb{V}(t_1, t_2)$ and $\mathbb{C}^2\backslash\mathbb{V}(64\ t_1 t_2^2 - 625t_1 t_2^2)$, then the reduced standard bases of $\mathrm{Ann}_{\mathcal{O}_{X,O}}(H_{Q_{31}})$, $\mathrm{Ann}_{\mathcal{O}_{X,O}}(H_{Q_{32}})$ and $\mathrm{Ann}_{\mathcal{O}_{X,O}}(H_{Q_{33}})$ are $\{1\}$. Therefore, the integral number is 3.

We omit the integral dependence relations of h over $\mathcal{I}_O$.

References

1. Becker, T., Weispfenning, V.: Gröbner Bases. Springer, New York (1993). https://doi.org/10.1007/978-1-4612-0913-3_6
2. Griffiths, P., Harris, J.: Principles of Algebraic Geometry. A Wiley-Interscience publication (1978)
3. Kapur, D., Sun, Y., Wang, D.: An efficient algorithm for computing a comprehensive Gröbner system of a parametric polynomial systems. J. Symb. Comput. **49**, 27–44 (2013)
4. Kashiwara, M.: B-functions and holonomic systems. Rationality of roots of B-functions. Invent. Math. **38**, 33–53 (1976–1977)
5. Nabeshima, K.: On the computation of parametric Gröbner bases for modules and syzygies. Jpn. J. Ind. Appl. Math. **27**, 217–238 (2010)
6. Nabeshima, K.: Stability conditions of monomial bases and comprehensive Gröbner systems. In: Gerdt, V.P., Koepf, W., Mayr, E.W., Vorozhtsov, E.V. (eds.) CASC 2012. LNCS, vol. 7442, pp. 248–259. Springer, Heidelberg (2012). https://doi.org/10.1007/978-3-642-32973-9_21
7. Nabeshima, K., Tajima, S.: On efficient algorithms for computing parametric local cohomology classes associated with semi-quasihomogeneous singularities and standard bases. In: Proceedings of the ISSAC 2014, pp. 351–358. ACM (2014)
8. Nabeshima, K., Tajima, S.: Computing logarithmic vector fields associated with parametric semi-quasihomogeneous hypersurface isolated singularities. In: Proceedings of the ISSAC 2015, pp. 291–298. ACM (2015)
9. Nabeshima, K., Tajima, S.: Solving extended ideal membership problems in rings of convergent power series via Gröbner bases. In: Kotsireas, I.S., Rump, S.M., Yap, C.K. (eds.) MACIS 2015. LNCS, vol. 9582, pp. 252–267. Springer, Cham (2016). https://doi.org/10.1007/978-3-319-32859-1_22
10. Nabeshima, K., Tajima, S.: Computing Tjurina stratifications of μ-constant deformations via parametric local cohomology systems. Appl. Algebra Eng. Commun. Comput. **27**, 451–467 (2016)
11. Nabeshima, K., Tajima, S.: Algebraic local cohomology with parameters and parametric standard bases for zero-dimensional ideals. J. Symb. Comput. **82**, 91–122 (2017)
12. Nabeshima, K., Tajima, S.: Solving parametric ideal membership problems and computing integral numbers in a ring of convergent power series via comprehensive Gröbner systems. Math. Comput. Sci. **13**, 185–194 (2019)
13. Nabeshima, K., Tajima, S.: Testing zero-dimensionality of varieties at a point. To appear in Mathematics in Computer Science. arXiv:1903.12365 [cs.SC] (2019)
14. Noro, M., Takeshima, T.: Risa/Asir - a computer algebra system. In: Proceedings of the ISSAC 1992, pp. 387–396. ACM (1992). http://www.math.kobe-u.ac.jp/Asir/asir.html

15. Scherk, J.: On the Gauss-Manin connection of an isolated hypersurface singularity. Math. Ann. **238**, 23–32 (1978)
16. Shibuta, F., Tajima, S.: An algorithm for computing the Hilbert-Samuel multiplicities and reductions of zero-dimensional ideal of Cohen-Macaulay local rings. J. Symb. Comput. **96**, 108–121 (2020)
17. Swanson, I., Huneke, C.: Integral Closure of Ideals. Rings, and Modules. Cambridge University Press, Cambridge (2006)
18. Tajima, S.: On polar varieties, logarithmic vector fields and holonomic D-modules. RIMS Kôkyûroku Bessatsu **40**, 41–51 (2013)
19. Tajima, S., Nakamura, Y.: Annihilating ideals for an algebraic local cohomology class. J. Symb. Comput. **44**, 435–448 (2009)
20. Tajima, S., Nakamura, Y., Nabeshima, K.: Standard bases and algebraic local cohomology for zero dimensional ideals. Adv. Stud. Pure Math. **56**, 341–361 (2009)
21. Vasconcelos, W.: Computational Methods in Commutative Algebra and Algebraic Geometry. Springer, Heidelberg (1998)
22. Yano, T.: On the theory of b-functions. Pub. Res. Inst. Math. Sci. **14**, 111–202 (1978)

Hilbert-Type Dimension Polynomials of Intermediate Difference-Differential Field Extensions

Alexander Levin[(⊠)]

The Catholic University of America, Washington, DC 20064, USA
levin@cua.edu
https://sites.google.com/a/cua.edu/levin

Abstract. Let K be an inversive difference-differential field and L a (not necessarily inversive) finitely generated difference-differential field extension of K. We consider the natural filtration of the extension L/K associated with a finite system η of its difference-differential generators and prove that for any intermediate difference-differential field F, the transcendence degrees of the components of the induced filtration of F are expressed by a certain numerical polynomial $\chi_{K,F,\eta}(t)$. This polynomial is closely connected with the dimension Hilbert-type polynomial of a submodule of the module of Kähler differentials $\Omega_{L^*|K}$ where L^* is the inversive closure of L. We prove some properties of polynomials $\chi_{K,F,\eta}(t)$ and use them for the study of the Krull-type dimension of the extension L/K. In the last part of the paper, we present a generalization of the obtained results to multidimensional filtrations of L/K associated with partitions of the sets of basic derivations and translations.

Keywords: Difference-differential field · Difference-differential module · Kähler differentials · Dimension polynomial

1 Introduction

Dimension polynomials associated with finitely generated differential field extensions were introduced by Kolchin in [4]; their properties and various applications can be found in his fundamental monograph [5, Chapter 2]. A similar technique for difference and inversive difference field extensions was developed in [7,8,12,13] and some other works of the author. Almost all known results on differential and difference dimension polynomials can be found in [6] and [10]. One can say that the role of dimension polynomials in differential and difference algebra is similar to the role of Hilbert polynomials in commutative algebra and algebraic geometry. The same can be said about dimension polynomials associated with difference-differential algebraic structures. They appear as generalizations of their differential and difference counterparts and play a key role in the

Supported by the NSF grant CCF-1714425.

D. Slamanig et al. (Eds.): MACIS 2019, LNCS 11989, pp. 64–79, 2020.
https://doi.org/10.1007/978-3-030-43120-4_7

study of dimension of difference-differential modules and extensions of difference-differential fields. Existence theorems, properties and methods of computation of univariate and multivariate difference-differential dimension polynomials can be found in [15], [6, Chapters 6 and 7], [14], [19] and [20].

In this paper we prove the existence and obtain some properties of a univariate dimension polynomial associated with an intermediate difference-differential field of a finitely generated difference-differential field extension (see Theorem 2 that can be considered as the main result of the paper). Then we use the obtained results for the study of the Krull-type dimension of such an extension. In particular, we establish relationships between invariants of dimension polynomials and characteristics of difference-differential field extensions that can be expressed in terms of chains of intermediate fields. In the last part of the paper we generalize our results on univariate dimension polynomials and obtain multivariate dimension polynomials associated with multidimensional filtrations induced on intermediate difference-differential fields. (Such filtrations naturally arise when one considers partitions of the sets of basic derivations and translations.) Note that we consider arbitrary (not necessarily inversive) difference-differential extensions of an inversive difference-differential field. In the particular case of purely differential extensions and in the case of inversive difference field extensions, the existence and properties of dimension polynomials were obtained in [11] and [13]. The main problem one runs into while working with a non-inversive difference (or difference-differential) field extension is that the translations are not invertible and there is no natural difference (respectively, difference-differential) structure on the associated module of Kähler differentials. We overcome this obstacle by considering such a structure on the module of Kähler differentials associated with the inversive closure of the extension. Finally, the results of this paper allow one to assign a dimension polynomial to a system of algebraic difference-differential equations of the form $f_i = 0$, $i \in I$ (f_i lie in the algebra of difference-differential polynomials $K\{y_1, \ldots, y_n\}$ over a ground field K) such that the difference-differential ideal P generated by the left-hand sides is prime and the solutions of the system should be invariant with respect to the action of a group G that commutes with basic derivations and translations. As in the case of systems of differential or difference equations, the dimension polynomial of such a system is defined as the dimension polynomial of the subfield of the difference-differential quotient field $K\{y_1, \ldots, y_n\}/P$ whose elements remain fixed under the action of G. Using the correspondence between dimension polynomials and Einstein's strength of a system of algebraic differential or difference equations established in [16] and [6, Chapter 6] (this characteristic of a system of PDEs governing a physical field was introduced in [1]), one can consider this dimension polynomial as an expression of the Einstein's strength of a system of difference-differential equations with group action.

2 Preliminaries

Throughout the paper $\mathbb{Z}$, $\mathbb{N}$ and $\mathbb{Q}$ denote the sets of all integers, all non-negative integers and all rational numbers, respectively. As usual, $\mathbb{Q}[t]$ will denote the ring of polynomials in one variable t with rational coefficients. By a ring we always mean an associative ring with a unity. Every ring homomorphism is unitary (maps unit onto unit), every subring of a ring contains the unity of the ring. Every module is unitary and every algebra over a commutative ring is unitary as well. Every field is supposed to have characteristic zero.

A *difference-differential ring* is a commutative ring R considered together with finite sets $\Delta = \{\delta_1, \ldots, \delta_m\}$ and $\sigma = \{\alpha_1, \ldots, \alpha_n\}$ of derivations and injective endomorphisms of R, respectively, such that any two mappings of the set $\Delta \bigcup \sigma$ commute. The elements of the set σ are called *translations* and the set $\Delta \bigcup \sigma$ will be referred to as a *basic set* of the difference-differential ring R, which is also called a Δ-σ-ring. We will often use prefix Δ-σ- instead of the adjective "difference-differential". If all elements of σ are automorphisms of R, we say that the Δ-σ-ring R is *inversive*. In this case we set $\sigma^* = \{\alpha_1, \ldots, \alpha_n, \alpha_1^{-1}, \ldots, \alpha_n^{-1}\}$ and call R a Δ-σ^*-ring.

If a Δ-σ-ring R is a field, it is called a *difference-differential field* or a Δ-σ-field. If R is inversive, we say that R is a Δ-σ^*-field.

In what follows, Λ will denote the free commutative semigroup of all power products $\lambda = \delta_1^{k_1} \ldots \delta_m^{k_m} \alpha_1^{l_1} \ldots \alpha_n^{l_n}$ where $k_i, l_j \in \mathbb{N}$ $(1 \leq i \leq m, 1 \leq j \leq n)$. Furthermore, Θ and T will denote the commutative semigroups of power products $\delta_1^{k_1} \ldots \delta_m^{k_m}$ and $\alpha_1^{l_1} \ldots \alpha_n^{l_n}$ $(k_i, l_j \in \mathbb{N})$, respectively. If $\lambda = \delta_1^{k_1} \ldots \delta_m^{k_m} \alpha_1^{l_1} \ldots \alpha_n^{l_n} \in \Lambda$, we define the order of λ as $\operatorname{ord} \lambda = \sum_{i=1}^{m} k_i + \sum_{j=1}^{n} l_j$ and set $\Lambda(r) = \{\lambda \in \Lambda \mid \operatorname{ord} \lambda \leq r\}$ for any $r \in \mathbb{N}$.

If the elements of σ are automorphisms, then Λ^* and Γ will denote the free commutative semigroup of all power products $\mu = \delta_1^{k_1} \ldots \delta_m^{k_m} \alpha_1^{l_1} \ldots \alpha_n^{l_n}$ with $k_i \in \mathbb{N}$, $l_j \in \mathbb{Z}$ and the free commutative group of power products $\gamma = \alpha_1^{l_1} \ldots \alpha_n^{l_n}$ with $l_1, \ldots, l_n \in \mathbb{Z}$, respectively. The order of such elements μ and γ are defined as $\operatorname{ord} \lambda = \sum_{i=1}^{m} k_i + \sum_{j=1}^{n} |l_j|$ and $\operatorname{ord} \gamma = \sum_{j=1}^{n} |l_j|$, respectively. We also set $\Lambda^*(r) = \{\mu \in \Lambda^* \mid \operatorname{ord} \mu \leq r\}$ $(r \in \mathbb{N})$.

A subring (ideal) S of a Δ-σ-ring R is said to be a difference-differential (or Δ-σ-) subring of R (respectively, difference-differential (or Δ-σ-) ideal of R) if S is closed with respect to the action of any operator of $\Delta \bigcup \sigma$. In this case the restriction of a mapping from $\Delta \bigcup \sigma$ on S is denoted by same symbol. If S is a Δ-σ-subring R, we also say that R is a Δ-σ-overring of S. If S is a Δ-σ-ideal of R and for any $\tau \in T$, the inclusion $\tau(a) \in S$ implies that $a \in S$, we say that the Δ-σ-ideal S is *reflexive* or that S is a Δ-σ^*-ideal of R.

If L is a Δ-σ-field and K a subfield of L which is also a Δ-σ-subring of L, then K is said to be a Δ-σ-subfield of L; L, in turn, is called a difference-differential (or Δ-σ-) field extension or a Δ-σ-overfield of K. In this case we also say that we have a Δ-σ-field extension L/K.

If R is a Δ-σ-ring and $S \subseteq R$, then the intersection of all Δ-σ-ideals of R containing the set S is, obviously, the smallest Δ-σ-ideal of R containing S. This ideal is denoted by $[S]$; as an ideal, it is generated by the set $\{\lambda(x) \mid x \in S, \lambda \in \Lambda\}$. If S is finite, $S = \{x_1, \ldots, x_k\}$, we say that the Δ-σ-ideal $I = [S]$ is finitely generated, write $I = [x_1, \ldots, x_k]$ and call $x_1, \ldots, x_k$ Δ-σ-generators of I.

If K is a Δ-σ-subfield of the Δ-σ-field L and $S \subseteq L$, then the intersection of all Δ-σ-subfields of L containing K and S is the unique Δ-σ-subfield of L containing K and S and contained in every Δ-σ-subfield of L with this property. It is denoted by $K\langle S\rangle$. If S is finite, $S = \{\eta_1, \ldots, \eta_s\}$ we write $K\langle\eta_1, \ldots, \eta_s\rangle$ for $K\langle S\rangle$ and say that this is a finitely generated Δ-σ-extension of K with the set of Δ-σ-generators $\{\eta_1, \ldots, \eta_s\}$. It is easy to see that $K\langle\eta_1, \ldots, \eta_s\rangle$ coincides with the field $K(\{\lambda\eta_i \,|\, \lambda \in \Lambda, 1 \leq i \leq s\})$. (If there might be no confusion, we often write $\lambda\eta$ for $\lambda(\eta)$ where $\lambda \in \Lambda$ and η is an element of a Δ-σ-ring.)

Let R_1 and R_2 be two difference-differential rings with the same basic set $\Delta \bigcup \sigma$. (More rigorously, we assume that there exist injective mappings of the sets Δ and σ into the sets of derivations and automorphisms of the rings R_1 and R_2, respectively, such that the images of any two elements of $\Delta \bigcup \sigma$ commute. We will denote the images of elements of $\Delta \bigcup \sigma$ under these mappings by the same symbols $\delta_1, \ldots, \delta_m, \alpha_1, \ldots, \alpha_n$). A ring homomorphism $\phi : R_1 \longrightarrow R_2$ is called a *difference-differential* (or Δ-σ-) *homomorphism* if $\phi(\tau a) = \tau\phi(a)$ for any $\tau \in \Delta \bigcup \sigma$, $a \in R$. It is easy to see that the kernel of such a mapping is a Δ-σ^*-ideal of R_1.

If R is a Δ-σ-subring of a Δ-σ-ring R^* such that the elements of σ act as automorphisms of R^* and for every $a \in R^*$ there exists $\tau \in T$ such that $\tau(a) \in R$, then the Δ-σ^*-ring R^* is called the *inversive closure* of R.

The proof of the following result can be obtained by mimicking the proof of the corresponding statement about inversive closures of difference rings, see [10, Proposition 2.1.7].

Proposition 1. *(i) Every Δ-σ-ring has an inversive closure.*
(ii) If R_1^* and R_2^* are two inversive closures of a Δ-σ-ring R, then there exists a Δ-σ-isomorphism of R_1^* onto R_2^* that leaves elements of R fixed.
(iii) If a Δ-σ-ring R is a Δ-σ-subring of a Δ-σ^*-ring U, then U contains an inversive closure of R.
(iv) If a Δ-σ-ring R is a field, then its inversive closure is also a field.

If K is an inversive difference-differential field and $L = K\langle\eta_1, \ldots, \eta_s\rangle$, then the inversive closure of L is denoted by $K\langle\eta_1, \ldots, \eta_s\rangle^*$. Clearly, this Δ-σ^*-field coincides with the field $K(\{\mu\eta_i | \mu \in \Lambda^*, 1 \leq i \leq s\})$.

Let R be a Δ-σ-ring and $U = \{u_i \,|\, i \in I\}$ a family of elements of some Δ-σ-overring of R. We say that the family U is Δ-σ-*algebraically dependent* over R, if the family $\{\lambda u_i \,|\, \lambda \in \Lambda, i \in I\}$ is algebraically dependent over R. Otherwise, the family U is said to be Δ-σ-*algebraically independent* over R.

If K is a Δ-σ-field and L a Δ-σ-field extension of K, then a set $B \subseteq L$ is said to be a Δ-σ-*transcendence basis* of L over K if B is Δ-σ-algebraically independent over K and every element $a \in L$ is Δ-σ-algebraic over $K\langle B\rangle$ (that is, the set $\{\lambda a \,|\, \lambda \in \Lambda\}$ is algebraically dependent over $K\langle B\rangle$). If L is a finitely generated Δ-σ-field extension of K, then all Δ-σ-transcendence bases of L over K are finite and have the same number of elements (the proof of this fact can be obtained by mimicking the proof of the corresponding properties of difference transcendence bases, see [10, Section 4.1]). In this case, the number of elements of any Δ-σ-transcendence basis is called the *difference-differential* (or Δ-σ-) *transcendence degree* of L over K (or the Δ-σ-transcendence degree of the extension L/K); it is denoted by Δ-σ-trdeg$_K L$.

The following theorem proved in [15] generalizes the Kolchin's theorem on differential dimension polynomial (see [5, Chapter II, Theorem 6]) and also the author's theorems on dimension polynomials of difference and inversive difference field extensions (see [10, Theorems 4.2.1 and 4.2.5]).

Theorem 1. *With the above notation, let* $L = K\langle\eta_1, \ldots, \eta_s\rangle$ *be a* Δ-σ-*field extension of a* Δ-σ-*field* K *generated by a finite set* $\eta = \{\eta_1, \ldots, \eta_s\}$. *Then there exists a polynomial* $\chi_{\eta|K}(t) \in \mathbb{Q}[t]$ *such that*

(i) $\chi_{\eta|K}(r) = \mathrm{trdeg}_K K(\{\lambda\eta_j \mid \lambda \in \Lambda(r), 1 \leq j \leq s\})$ *for all sufficiently large* $r \in \mathbb{Z}$ *(that is, there exists* $r_0 \in \mathbb{Z}$ *such that the equality holds for all* $r > r_0$*).*

(ii) $\deg \chi_{\eta|K} \leq m + n$ *and* $\chi_{\eta|K}(t)$ *can be written as* $\chi_{\eta|K}(t) = \sum_{i=0}^{m+n} a_i \binom{t+i}{i}$, *where* $a_i \in \mathbb{Z}$.

(iii) $d = \deg \chi_{\eta|K}, a_{m+n}$ *and* a_d *do not depend on the set of* Δ-σ-*generators* η *of* L/K *(*$a_{m+n} = 0$ *if* $d < m + n$*). Moreover,* $a_{m+n} = \Delta$-σ-trdeg$_K L$.

The polynomial $\chi_{\eta|K}(t)$ is called the Δ-σ-*dimension polynomial* of the Δ-σ-field extension L/K associated with the system of Δ-σ-generators η. We see that $\chi_{\eta|K}(t)$ is a polynomial with rational coefficients that takes integer values for all sufficiently large values of the argument. Such polynomials are called *numerical*; their properties are thoroughly described in [6, Chapter 2]. The invariants $d = \deg \chi_{\eta|K}$ and a_d (if $d < m + n$) are called the Δ-σ-*type* and *typical* Δ-σ-*transcendence degree* of L/K; they are denoted by Δ-σ-type$_K L$ and Δ-σ-t. trdeg$_K L$, respectively.

3 Dimension Polynomials of Intermediate Difference-Differential Fields. The Main Theorem

The following result is an essential generalization of Theorem 1. This generalization allows one to assign certain numerical polynomial to an intermediate Δ-σ-field of a Δ-σ-field extension L/K where K is an inversive Δ-σ-field. (We use the notation introduced in the previous section.)

Theorem 2. *Let* K *be an inversive* Δ-σ-*field with basic set* $\Delta \bigcup \sigma$ *where* $\Delta = \{\delta_1, \ldots, \delta_m\}$ *and* $\sigma = \{\alpha_1, \ldots, \alpha_n\}$ *are the sets of derivations and automorphisms of* K, *respectively. Let* $L = K\langle\eta_1, \ldots, \eta_s\rangle$ *be a* Δ-σ-*field extension of* K *generated by a finite set* $\eta = \{\eta_1, \ldots, \eta_s\}$. *Let* F *be an intermediate* Δ-σ-*field of the extension* L/K *and for any* $r \in \mathbb{N}$, *let* $F_r = F \bigcap K(\{\lambda\eta_j \mid \lambda \in \Lambda(r), 1 \leq j \leq s\})$. *Then there exists a numerical polynomial* $\chi_{K,F,\eta}(t) \in \mathbb{Q}[t]$ *such that*

(i) $\chi_{K,F,\eta}(r) = \mathrm{trdeg}_K F_r$ *for all sufficiently large* $r \in \mathbb{N}$;

(ii) $\deg \chi_{K,F,\eta} \leq m + n$ *and* $\chi_{K,F,\eta}(t)$ *can be written as* $\chi_{K,F,\eta}(t) = \sum_{i=0}^{m+n} c_i \binom{t+i}{i}$ *where* $c_i \in \mathbb{Z}$ *(*$1 \leq i \leq m + n$*).*

(iii) $d = \deg \chi_{K,F,\eta}(t), c_{m+n}$ and c_d do not depend on the set of Δ-σ-generators η of the extension L/K. Furthermore, $c_{m+n} = \Delta$-σ-trdeg$_K F$.

The polynomial $\chi_{K,F,\eta}(t)$ is called a Δ-σ-*dimension polynomial of the intermediate field F associated with the set of Δ-σ-generators η of L/K.*

The proof of Theorem 2 is based on properties of difference-differential modules and the difference-differential structure on the module of Kähler differentials considered below. Similar properties in differential and difference cases can be found in [2] and [10, Section 4.2], respectively.

Let K be a Δ-σ-field and Λ the semigroup of power products of basic operators introduced in Sect. 2. Let $\mathcal{D}$ denote the set of all finite sums of the form $\sum_{\lambda \in \Lambda} a_\lambda \lambda$ where $a_\lambda \in K$ (such a sum is called a Δ-σ-*operator* over K; two Δ-σ-operators are equal if and only if their corresponding coefficients are equal). The set $\mathcal{D}$ can be treated as a ring with respect to its natural structure of a left K-module and the relationships $\delta a = a\delta + \delta(a)$, $\alpha a = \alpha(a)\alpha$ for any $a \in K$, $\delta \in \Delta$, $\alpha \in \sigma$ extended by distributivity. The ring $\mathcal{D}$ is said to be the *ring of Δ-σ-operators* over K.

If $A = \sum_{\lambda \in \Lambda} a_\lambda \lambda \in \mathcal{D}$, then the number $\operatorname{ord} A = \max\{\operatorname{ord} \lambda \,|\, a_\lambda \neq 0\}$ is called the *order* of the Δ-σ-operator A. In what follows, we treat $\mathcal{D}$ as a filtered ring with the ascending filtration $(\mathcal{D}_r)_{r \in \mathbb{Z}}$ where $\mathcal{D}_r = 0$ if $r < 0$ and $\mathcal{D}_r = \{A \in \mathcal{D} \,|\, \operatorname{ord} A \leq r\}$ if $r \geq 0$.

Similarly, if a Δ-σ-field K is inversive and Λ^* is the semigroup defined in Sect. 2, then $\mathcal{E}$ will denote the set of all finite sums $\sum_{\mu \in \Lambda^*} a_\mu \mu$ where $a_\mu \in K$. Such a sum is called a Δ-σ^*-operator over K; two Δ-σ^*-operators are equal if and only if their corresponding coefficients are equal. Clearly, the ring $\mathcal{D}$ of Δ-σ-operators over K is a subset of $\mathcal{E}$. Moreover, $\mathcal{E}$ can be treated as an overring of $\mathcal{D}$ such that $\alpha^{-1}a = \alpha^{-1}(a)\alpha^{-1}$ for every $\alpha \in \sigma$, $a \in K$. This ring is called the *ring of Δ-σ^*-operators* over K.

The order of a Δ-σ^*-operator $B = \sum_{\mu \in \Lambda^*} a_\mu \mu$ is defined in the same way as the order of a Δ-σ-operator: $\operatorname{ord} B = \max\{\operatorname{ord} \mu \,|\, a_\mu \neq 0\}$. In what follows the ring $\mathcal{E}$ is treated as a filtered ring with the ascending filtration $(\mathcal{E}_r)_{r \in \mathbb{Z}}$ such that $\mathcal{E}_r = 0$ if $r < 0$ and $\mathcal{E}_r = \{B \in \mathcal{E} \,|\, \operatorname{ord} B \leq r\}$ if $r \geq 0$.

If K is a Δ-σ-field, then a *difference-differential module* over K (also called a Δ-σ-K-module) is a left $\mathcal{D}$-module M, that is, a vector K-space where elements of $\Delta \bigcup \sigma$ act as additive mutually commuting operators such that $\delta(ax) = a(\delta x) + \delta(a)x$ and $\alpha(ax) = \alpha(a)\alpha x$ for any $\delta \in \Delta$, $\alpha \in \sigma$, $x \in M$, $a \in K$. We say that M is a finitely generated Δ-σ-K-module if M is finitely generated as a left $\mathcal{D}$-module.

Similarly, if K is a Δ-σ^*-field, then an *inversive difference-differential module* over K (also called a Δ-σ^*-K-module) is a left $\mathcal{E}$-module (that is, a Δ-σ-K-module M with the action of elements of σ^* such that $\alpha^{-1}(ax) = \alpha^{-1}(a)\alpha^{-1}x$ for every $\alpha \in \sigma$). A Δ-σ^*-K-module M is said to be finitely generated if it is generated as a left $\mathcal{E}$-module by a finite set whose elements are called Δ-σ^*-generators of M.

If M is a Δ-σ-K-module (respectively, a Δ-σ^*-module, if K is a Δ-σ^*-field), then by a filtration of M we mean an exhaustive and separated filtration of

M as a $\mathcal{D}$- (respectively, $\mathcal{E}$-) module, that is, an ascending chain $(M_r)_{r\in\mathbb{Z}}$ of vector K-subspaces of M such that $\mathcal{D}_r M_s \subseteq M_{r+s}$ (respectively, $\mathcal{E}_r M_s \subseteq M_{r+s}$) for all $r, s \in \mathbb{Z}$, $M_r = 0$ for all sufficiently small $r \in \mathbb{Z}$, and $\bigcup_{r\in\mathbb{Z}} M_r = M$. A filtration $(M_r)_{r\in\mathbb{Z}}$ of a Δ-σ-K- (respectively, Δ-σ^*-K) module M is said to be *excellent* if every M_r is a finite dimensional vector K-space and there exists $r_0 \in \mathbb{Z}$ such that $M_r = \mathcal{D}_{r-r_0} M_{r_0}$ (respectively, $M_r = \mathcal{E}_{r-r_0} M_{r_0}$) for any $r \geq r_0$. Clearly, if M is generated as a $\mathcal{D}$- (respectively, $\mathcal{E}$-) module by elements $x_1, \ldots x_s$, then $\left(\sum_{i=1}^s \mathcal{D}_r x_i\right)_{r\in\mathbb{Z}}$ (respectively, $\left(\sum_{i=1}^s \mathcal{E}_r x_i\right)_{r\in\mathbb{Z}}$) is an excellent filtration of M; it is said to be the natural filtration associated with the set of generators $\{x_1, \ldots, x_s\}$.

If M' and M'' are Δ-σ-K- (respectively, Δ-σ^*-K-) modules, then a mapping $f : M' \to M''$ is said to be a Δ-σ-homomorphism if it is a homomorphism of $\mathcal{D}$- (respectively, $\mathcal{E}$-) modules. If M' and M'' are equipped with filtrations $(M'_r)_{r\in\mathbb{Z}}$ and $(M''_r)_{r\in\mathbb{Z}}$, respectively, and $f(M'_r) \subseteq M''_r$ for every $r \in \mathbb{Z}$, then f is said to be a Δ-σ-homomorphism of filtered Δ-σ-K- (respectively, Δ-σ^*-K-) modules.

The following two statements are direct consequences of [6, Theorem 6.7.3] and [6, Theorem 6.7.10], respectively.

Theorem 3. *With the above notation, let K be a Δ-σ-field, M a finitely generated Δ-σ-K-module, and $(M_r)_{r\in\mathbb{Z}}$ the natural filtration associated with some finite system of generators of M over the ring of Δ-σ-operators $\mathcal{D}$. Then there is a numerical polynomial $\phi(t) \in \mathbb{Q}[t]$ such that:*

(i) $\phi(r) = \dim_K M_r$ for all sufficiently large $r \in \mathbb{Z}$.

(ii) $\deg \phi \leq m + n$ and $\phi(t)$ can be written as $\phi(t) = \sum_{i=0}^{m+n} a_i \binom{t+i}{i}$ where

$a_0, \ldots, a_{m+n} \in \mathbb{Z}$.
(iii) $d = \deg \phi(t)$, a_n and a_d do not depend on the finite set of generators of the $\mathcal{D}$-module M the filtration $(M_r)_{r\in\mathbb{Z}}$ is associated with. Furthermore, a_{m+n} is equal to the Δ-σ-dimension of M over K (denoted by Δ-σ-$\dim_K M$), that is, to the maximal number of elements $x_1, \ldots, x_k \in M$ such that the family $\{\lambda x_i \mid \lambda \in \Lambda, 1 \leq i \leq k\}$ is linearly independent over K.

Theorem 4. *Let $f : M' \to M''$ be an injective homomorphism of filtered Δ-σ-K-modules M' and M'' with filtrations $(M'_r)_{r\in\mathbb{Z}}$ and $(M''_r)_{r\in\mathbb{Z}}$, respectively. If the filtration of M'' is excellent, then the filtration of M' is excellent as well.*

Proof of Theorem 2. Let $L = K\langle \eta_1, \ldots, \eta_s \rangle$ be a Δ-σ-field extension of a Δ-σ^*-field K. Let L^* be the inversive closure of L, that is, $L^* = K\langle \eta_1, \ldots, \eta_s \rangle^*$. Let $M = \Omega_{L^*|K}$, the module of Kähler differentials associated with the extension L^*/K. Then M can be treated as a Δ-σ^*-L^*-module where the action of the elements of $\Delta \bigcup \sigma^*$ is defined in such a way that $\delta(d\zeta) = d\delta(\zeta)$ and $\alpha(d\zeta) = d\alpha(\zeta)$ for any $\zeta \in L^*$, $\delta \in \Delta$, $\alpha \in \sigma^*$ (see [2] and [12, Lemma 4.2.8]).

For every $r \in \mathbb{N}$, let M_r denote the vector L^*-subspace of M generated by all elements $d\zeta$ where $\zeta \in K(\bigcup_{i=1}^s \Lambda^*(r)\eta_i)$. It is easy to check that $(M_r)_{r\in\mathbb{Z}}$ ($M_r = 0$

if $r < 0$) is the natural filtration of the Δ-σ^*-L^*-module M associated with the system of Δ-σ^*-generators $\{d\eta_1, \ldots, d\eta_s\}$.

Let F be any intermediate Δ-σ-field of L/K, $F_r = F \bigcap K(\{\lambda\eta_j \,|\, \lambda \in \Lambda(r), 1 \leq j \leq s\})$ $(r \in \mathbb{N})$ and $F_r = 0$ if $r < 0$. Let $\mathcal{E}$ and $\mathcal{D}$ denote the ring of Δ-σ^*-operators over L^* and the ring of Δ-σ-operators over L, respectively. Let N be the $\mathcal{D}$-submodule of M generated by all elements of the form $d\zeta$ with $\zeta \in F$ (by $d\zeta$ we always mean $d_{L^*|K}\zeta$). Furthermore, for any $r \in \mathbb{N}$, let N_r be the vector L-space generated by all elements $d\zeta$ with $\zeta \in F_r$ and $N_r = 0$ if $r < 0$.

It is easy to see that $(N_r)_{r\in\mathbb{Z}}$ is a filtration of the Δ-σ-L-module N, and if $M' = \sum_{i=1}^{s} \mathcal{D}d\eta_i$, then the embedding $N \to M'$ is a homomorphism of filtered $\mathcal{D}$-modules. (M' is considered as a filtered $\mathcal{D}$-module with the excellent filtration $(\sum_{i=1}^{s} \mathcal{D}_r d\eta_i)_{r\in\mathbb{Z}}$.) By Theorem 4, $(N_r)_{r\in\mathbb{Z}}$ is an excellent filtration of the $\mathcal{D}$-module N. Applying Theorem 3 we obtain that there exists a polynomial $\chi_{K,F,\eta}(t) \in \mathbb{Q}[t]$ such that $\chi_{K,F,\eta}(t)(r) = \dim_K N_r$ for all sufficiently large $r \in \mathbb{Z}$.

As it is shown in [17, Chapter V, Section 23], elements $\zeta_1, \ldots, \zeta_k \in L^*$ are algebraically independent over K if and only if the elements $d\zeta_1, \ldots, d\zeta_k$ are linearly independent over L^*. Thus, if $\zeta_1, \ldots, \zeta_k \in F_r$ $(r \in \mathbb{Z})$ are algebraically independent over K, then the elements $d\zeta_1, \ldots, d\zeta_k \in N_r$ are linearly independent over L^* and therefore over L. Conversely, if elements $dx_1, \ldots, dx_h$ $(x_i \in F_r$ for $i = 1, \ldots, h)$ are linearly independent over L, then $x_1, \ldots, x_h$ are algebraically independent over K. Otherwise, we would have a polynomial $f(X_1, \ldots, X_h) \in K[X_1, \ldots, X_h]$ of the smallest possible degree such that $f(x_1, \ldots, x_h) = 0$. Then $df(x_1, \ldots, x_h) = \sum_{i=1}^{h} \frac{\partial f}{\partial X_i}(x_1, \ldots, x_h)dx_i = 0$ where not all coefficients of dx_i are zeros (they are expressed by polynomials of degree less than $\deg f$). Since all the coefficients lie in L, we would have a contradiction with the linear independence of $dx_1, \ldots, dx_h$ over L.

It follows that $\dim_L N_r = \operatorname{trdeg}_K F_r$ for all $r \in \mathbb{N}$. Applying Theorem 3 we obtain the statement of Theorem 2. $\qquad\qquad\qquad\qquad\qquad\qquad\qquad\qquad\qquad\square$

Clearly, if $F = L$, then Theorem 2 implies Theorem 1. Note also that if an intermediate field F of a finitely generated Δ-σ-field extension L/K is not a Δ-σ-subfield of L, there might be no numerical polynomial whose values for sufficiently large integers r are equal to $\operatorname{trdeg}_K(F \bigcap K(\{\lambda\eta_j \,|\, \lambda \in \Lambda(r), 1 \leq j \leq s\}))$. Indeed, let $\Delta = \{\delta\}$ and $\sigma = \emptyset$. Let $L = K\langle y \rangle$, where the Δ-σ-generator y is Δ-σ-independent over K, and let $F = K(\delta^2 y, \ldots, \delta^{2k}y, \ldots)$. Then $\Lambda = \{\delta^i \,|\, i \in \mathbb{N}\}$, $\Lambda(r) = \{1, \delta, \ldots, \delta^r\}$, $F_r = F \bigcap K(\lambda y \,|\, \lambda \in \Lambda(r))$ and $\operatorname{trdeg}_K F_r = [\frac{r}{2}]$ (the integer part of $\frac{r}{2}$), which is not a polynomial of r. In this case, the function $\phi(r) = \operatorname{trdeg}_K F_r$ is a quasi-polynomial, but if one takes $F = K(\delta^2 y, \ldots, \delta^{2^k}y, \ldots)$, then $\operatorname{trdeg}_K F_r = [\log_2 r]$.

4   Type and Dimension of Difference-Differential Field Extensions

Let K be an inversive difference-differential (Δ-σ-) field with a basic set $\Delta \bigcup \sigma$ where $\Delta = \{\delta_1, \ldots, \delta_m\}$ and $\sigma = \{\alpha_1, \ldots, \alpha_n\}$ are the sets of derivations and automorphisms of K, respectively. Let $L = K\langle \eta_1, \ldots, \eta_s \rangle$ be a Δ-σ-field extension of K generated by a finite set $\eta = \{\eta_1, \ldots, \eta_s\}$. (We keep the notation introduced in Sect. 2.)

Let $\mathfrak{U}$ denote the set of all intermediate Δ-σ-fields of the extension L/K and

$$\mathfrak{B}_{\mathfrak{U}} = \{(F, E) \in \mathfrak{U} \times \mathfrak{U} \,|\, F \supseteq E\}.$$

Furthermore, let $\overline{\mathbb{Z}}$ denote the ordered set $\mathbb{Z} \bigcup \{\infty\}$ (where the natural order on $\mathbb{Z}$ is extended by the condition $a < \infty$ for any $a \in \mathbb{Z}$).

Proposition 2. *With the above notation, there exists a unique mapping $\mu_{\mathfrak{U}}$: $\mathfrak{B}_{\mathfrak{U}} \to \overline{\mathbb{Z}}$ such that*

(i) $\mu_{\mathfrak{U}}(F, E) \geq -1$ for any pair $(F, E) \in \mathfrak{B}_{\mathfrak{U}}$.
(ii) If $d \in \mathbb{N}$, then $\mu_{\mathfrak{U}}(F, E) \geq d$ if and only if $\mathrm{trdeg}_E F > 0$ and there exists an infinite descending chain of intermediate Δ-σ-fields

$$F = F_0 \supseteq F_1 \supseteq \cdots \supseteq F_r \supseteq \cdots \supseteq E \tag{1}$$

such that

$$\mu_{\mathfrak{U}}(F_i, F_{i+1}) \geq d - 1 \quad (i = 0, 1, \ldots). \tag{2}$$

Proof. In order to show the existence and uniqueness of the desired mapping $\mu_{\mathfrak{U}}$, one can just mimic the proof of the corresponding statement for chains of prime differential ideals given in [3, Section 1] (see also [11, Proposition 4.1] and [13, Section 4] where similar arguments were applied to differential and inversive difference field extensions, respectively). Namely, let us set $\mu_{\mathfrak{U}}(F, E) = -1$ if $F = E$ or the field extension F/E is algebraic. If $(F, E) \in \mathfrak{B}_{\mathfrak{U}}$, $\mathrm{trdeg}_E F > 0$ and for every $d \in \mathbb{N}$, there exists a chain of intermediate Δ-σ-fields (1) with condition (2), we set $\mu_{\mathfrak{U}}(F, E) = \infty$. Otherwise, we define $\mu_{\mathfrak{U}}(F, E)$ as the maximal integer d for which condition (ii) holds (that is, $\mu_{\mathfrak{U}}(F, E) \geq d$). It is clear that the mapping $\mu_{\mathfrak{U}}$ defined in this way is unique. $\square$

With the notation of the last proposition, we define the *type* of a Δ-σ-field extension L/K as the integer

$$\mathrm{type}(L/K) = \sup\{\mu_{\mathfrak{U}}(F, E) \,|\, (F, E) \in \mathfrak{B}_{\mathfrak{U}}\}. \tag{3}$$

and the *dimension* of the Δ-σ-extension L/K as the number
$\dim(L/K) = \sup\{q \in \mathbb{N} \,|\, \text{there exists a chain } F_0 \supseteq F_1 \supseteq \cdots \supseteq F_q \text{ such that } F_i \in \mathfrak{U} \text{ and}$

$$\mu_{\mathfrak{U}}(F_{i-1}, F_i) = \mathrm{type}(L/K) \quad (i = 1, \ldots, q)\}. \tag{4}$$

It is easy to see that for any pair of intermediate Δ-σ-fields of L/K such that $(F, E) \in \mathfrak{B}_{\mathfrak{U}}$, $\mu_{\mathfrak{U}}(F, E) = -1$ if and only if the field extension E/F is algebraic. It is also clear that if $\mathrm{type}(L/K) < \infty$, then $\dim(L/K) > 0$.

Proposition 3. *With the above notation, let F and E be intermediate Δ-σ-fields of a Δ-σ-field extension $L = K\langle \eta_1, \ldots, \eta_s \rangle$ generated by a finite set $\eta = \{\eta_1, \ldots, \eta_s\}$. Let $F \supseteq E$, so that $(F, E) \in \mathfrak{B}_\mathfrak{U}$. Then for any integer $d \geq -1$, the inequality $\mu_\mathfrak{U}(F, E) \geq d$ implies the inequality $\deg(\chi_{K,F,\eta}(t) - \chi_{K,E,\eta}(t)) \geq d$. ($\chi_{K,F,\eta}(t)$ and $\chi_{K,E,\eta}(t)$ are the Δ-σ-dimensions polynomials of the fields F and E associated with the set of Δ-σ-generators η of L/K.)*

Proof. We proceed by induction on d. Since $\deg(\chi_{K,F,\eta}(t) - \chi_{K,E,\eta}(t)) \geq -1$ for any pair $(F, E) \in \mathfrak{B}_\mathfrak{U}$ and $\deg(\chi_{K,F,\eta}(t) - \chi_{K,E,\eta}(t)) \geq 0$ if $\mathrm{trdeg}_E F > 0$, our statement is true for $d = -1$ and $d = 0$. (As usual we assume that the degree of the zero polynomial is -1.)

Let $d > 0$ and let the statement be true for all nonnegative integers less than d. Let $\mu_\mathfrak{U}(F, E) \geq d$ for some pair $(F, E) \in \mathfrak{B}_\mathfrak{U}$, so that there exists a chain of intermediate Δ-σ-fields (1) such that $\mu_\mathfrak{U}(F_i, F_{i+1}) \geq d - 1$ ($i = 0, 1, \ldots$). If $\deg(\chi_{K,F_i,\eta}(t) - \chi_{K,F_{i+1},\eta}(t)) \geq d$ for some $i \in \mathbb{N}$, then $\deg(\chi_{K,F,\eta}(t) - \chi_{K,E,\eta}(t)) \geq \deg(\chi_{K,F_i,\eta}(t) - \chi_{K,F_{i+1},\eta}(t)) \geq d$, so the statement of the proposition is true.

Suppose that $\deg(\chi_{K,F_i,\eta}(t) - \chi_{K,F_{i+1},\eta}(t)) = d - 1$ for every $i \in \mathbb{N}$, that is,

$$\chi_{K,F_i,\eta}(t) - \chi_{K,F_{i+1},\eta}(t) = \sum_{j=0}^{d-1} a_j^{(i)} \binom{t+j}{j} \text{ where } a_0^{(1)}, \ldots, a_{d-1}^{(i)} \in \mathbb{Z}, \ a_{d-1}^{(i)} > 0.$$

Then

$$\chi_{K,F,\eta}(t) - \chi_{K,F_{i+1},\eta}(t) = \sum_{k=0}^{i}(\chi_{K,F_k,\eta}(t) - \chi_{K,F_{k+1},\eta}(t)) = \sum_{j=0}^{d-1} b_j^{(i)} \binom{t+j}{j}$$

where $b_0^{(i)}, \ldots, b_{d-1}^{(i)} \in \mathbb{Z}$ and $b_{d-1}^{(i)} = \sum_{k=0}^{i} a_{d-1}^{(k)}$. Therefore, $b_{d-1}^{(0)} < b_{d-1}^{(1)} < \cdots$ and $\lim_{i \to \infty} b_{d-1}^{(i)} = \infty$. On the other hand, $\deg(\chi_{K,F,\eta}(t) - \chi_{K,F_{i+1},\eta}(t)) \leq \deg(\chi_{K,F,\eta}(t) - \chi_{K,E,\eta}(t))$. If $\deg(\chi_{K,F,\eta}(t) - \chi_{K,E,\eta}(t)) = d - 1$, that is,

$$\chi_{K,F,\eta}(t) - \chi_{K,E,\eta}(t) = \sum_{j=0}^{d-1} c_j \binom{t+j}{j} \text{ for some } c_0, \ldots, c_{d-1} \in \mathbb{Z}, \text{ then we would}$$

have $b_{d-1}^{(i)} < c_{d-1}$ for all $i \in \mathbb{N}$ contrary to the fact that $\lim_{i \to \infty} b_{d-1}^{(i)} = \infty$. Thus, $\deg(\chi_{K,F,\eta}(t) - \chi_{K,E,\eta}(t)) \geq d$, so the proposition is proved. $\qquad\square$

The following theorem provides a relationship between the introduced characteristics of a finitely generated Δ-σ-extension and the invariants of its Δ-σ-dimension polynomial introduced by Theorem 2.

Theorem 5. *Let K be an inversive difference-differential (Δ-σ-) field with basic set $\Delta \bigcup \sigma$ where $\Delta = \{\delta_1, \ldots, \delta_m\}$ and $\sigma = \{\alpha_1, \ldots, \alpha_n\}$ are the sets of derivations and automorphisms of K, respectively. Let L be a finitely generated Δ-σ-field extension of K. Then*

(i) $\mathrm{type}(L/K) \leq \Delta$-$\sigma$-$\mathrm{type}_K L \leq m + n$.
(ii) If Δ-σ-$\mathrm{trdeg}_K L > 0$, then $\mathrm{type}(L/K) = m+n$, $\dim(L/K) = \Delta$-σ-$\mathrm{trdeg}_K L$.
(iii) If Δ-σ-$\mathrm{trdeg}_K L = 0$, then $\mathrm{type}(L/K) < m + n$.

Proof. Let $\eta = \{\eta_1, \ldots, \eta_s\}$ be a system of Δ-σ-generators of L over K and for every $r \in \mathbb{N}$, let $L_r = K(\{\lambda\eta_i \,|\, \lambda \in \Lambda(r), 1 \leq i \leq s\})$. Furthermore, if F is any intermediate Δ-σ-field of the extension L/K, then F_r ($r \in \mathbb{N}$) will denote the field $F \bigcap L_r$. By Theorem 2, there is a polynomial $\chi_{K,F,\eta}(t) \in \mathbb{Q}[t]$ such that $\chi_{K,F,\eta}(r) = \text{trdeg}_K F_r$ for all sufficiently large $r \in \mathbb{N}$, $\deg \chi_{K,F,\eta} \leq m + n$, and this polynomial can be written as $\chi_{K,F,\eta}(t) = \sum_{i=1}^{m+n} a_i \binom{t+i}{i}$ where $a_0, \ldots, a_{m+n} \in \mathbb{Z}$ and $a_{m+n} = \Delta$-σ-$\text{trdeg}_K F$. Clearly, if E and F are two intermediate Δ-σ-fields of L/K and $F \supseteq E$, then $\chi_{K,F,\eta}(t) \geq \chi_{K,E,\eta}(t)$. (This inequality means that $\chi_F(r) \geq \chi_E(r)$ for all sufficiently large $r \in \mathbb{N}$. As it is first shown in [18], the set W of all differential dimension polynomials of finitely generated differential field extensions is well ordered with respect to this ordering. At the same time, as it is proved in [6, Chapter 2], W is also the set of all Δ-σ-dimension polynomials associated with finitely generated Δ-σ-field extensions).

Note that if $F \supseteq E$ and $\chi_{K,F,\eta}(t) = \chi_{K,E,\eta}(t)$, then the field extension F/E is algebraic. Indeed, if $x \in F$ is transcendental over E, then there exists $r_0 \in \mathbb{N}$ such that $x \in F_r$ for all $r \geq r_0$. Therefore, $\text{trdeg}_K F_r = \text{trdeg}_K E_r + \text{trdeg}_{E_r} F_r > \text{trdeg}_K E_r$ for all $r \geq r_0$ hence $\chi_{K,F,\eta}(t) > \chi_{K,E,\eta}(t)$ contrary to our assumption.

Since $\deg(\chi_{K,F,\eta}(t) - \chi_{K,E,\eta}(t)) \leq m + n$ for any pair $(F, E) \in \mathfrak{B}_{\mathfrak{u}}$, the last proposition implies that $\text{type}(L/K) \leq \Delta$-$\sigma$-$\text{type}_K L \leq m+n$. If Δ-σ-$\text{trdeg}_K L = 0$, then $\text{type}(L/K) \leq \Delta$-$\sigma$-$\text{type}_K L < m+n$. Thus, it remains to prove statement (ii) of the theorem.

Let Δ-σ-$\text{trdeg}_K L > 0$, let element $x \in L$ be Δ-σ-transcendental over K and let $F = K\langle x \rangle$. Clearly, in order to prove that $\text{type}(L/K) = m + n$ it is sufficient to show that $\mu_{\mathfrak{u}}(F, K) \geq m + n$. This inequality, in turn, immediately follows from the consideration of the following $m + n$ strictly descending chains of intermediate Δ-σ-fields of F/K.

$$F = K\langle x \rangle \supset K\langle \delta_1 x \rangle \supset K\langle \delta_1^2 x \rangle \supset \cdots \supset K\langle \delta_1^{i_1} x \rangle \supset K\langle \delta_1^{i_1+1} x \rangle \supset \cdots \supset K,$$
$$K\langle \delta_1^{i_1} x \rangle \supset K\langle \delta_1^{i_1+1} x, \delta_1^{i_1}\delta_2 x \rangle \supset K\langle \delta_1^{i_1+1} x, \delta_1^{i_1}\delta_2^2 x \rangle \supset \ldots K\langle \delta_1^{i_1+1} x, \delta_1^{i_1}\delta_2^{i_2} x \rangle \supset$$
$$K\langle \delta_1^{i_1+1} x, \delta_1^{i_1}\delta_2^{i_2+1} x \rangle \supset \cdots \supset K\langle \delta_1^{i_1+1} x \rangle,$$

$$\cdots$$

$$K\langle \delta_1^{i_1+1} x, \delta_1^{i_1+1}\delta_2^{i_2+1} x, \ldots, \delta_1^{i_1+1} \ldots \delta_{m-1}^{i_{m-1}+1} x, \delta_1^{i_1+1} \ldots \delta_{m-1}^{i_{m-1}+1}\delta_m^{i_m} x \rangle \supset K\langle \delta_1^{i_1+1} x,$$
$$\ldots, \delta_1^{i_1+1} \ldots \delta_{m-1}^{i_{m-1}+1} x, \delta_1^{i_1+1} \ldots \delta_{m-1}^{i_{m-1}+1}\delta_m^{i_m+1} x, \delta_1^{i_1+1} \ldots \delta_{m-1}^{i_{m-1}+1}\delta_m^{i_m}(\alpha_1 - 1)x \rangle \supset$$
$$\supset \cdots \supset K\langle \delta_1^{i_1+1} x, \ldots, \delta_1^{i_1+1} \ldots \delta_{m-1}^{i_{m-1}+1} x, \delta_1^{i_1+1} \ldots \delta_{m-1}^{i_{m-1}+1}\delta_m^{i_m}(\alpha_1 - 1)^2 x \rangle \supset$$
$$\supset \cdots \supset K\langle \delta_1^{i_1+1} x, \ldots, \delta_1^{i_1+1} \ldots \delta_{m-1}^{i_{m-1}+1} x, \delta_1^{i_1+1} \ldots \delta_{m-1}^{i_{m-1}+1}\delta_m^{i_m}(\alpha_1 - 1)^{i_{m+1}} x \rangle \supset$$
$$\cdots \supset K\langle \delta_1^{i_1+1} x, \ldots, \delta_1^{i_1+1} \ldots \delta_{m-1}^{i_{m-1}+1} x, \delta_1^{i_1+1} \ldots \delta_{m-1}^{i_{m-1}+1}\delta_m^{i_m}(\alpha_1 - 1)^{i_{m+1}+1} x \rangle \supset$$
$$\cdots \supset K\langle \delta_1^{i_1+1} x, \delta_1^{i_1+1}\delta_2^{i_2+1} x, \ldots, \delta_1^{i_1+1} \ldots \delta_{m-1}^{i_{m-1}+1} x, \delta_1^{i_1+1} \ldots \delta_{m-1}^{i_{m-1}+1}\delta_m^{i_m+1} x \rangle,$$

$$\cdots$$

$$K\langle \delta_1^{i_1+1}x,\dots,\delta_1^{i_1+1}\dots\delta_{m-1}^{i_{m-1}+1}\delta_m^{i_m+1}(\alpha_1-1)^{i_m+1+1}\dots(\alpha_{n-1}-1)^{i_m+n-1}x\rangle \supset$$

$$K\langle \delta_1^{i_1+1}x,\dots,\delta_1^{i_1+1}\dots\delta_{m-1}^{i_{m-1}+1}x,\delta_1^{i_1+1}\dots\delta_{m-1}^{i_{m-1}+1}\delta_m^{i_m}(\alpha_1-1)^{i_m+1+1}\dots$$

$$(\alpha_{n-1}-1)^{i_m+n-1+1}(\alpha_n-1)x\rangle \supset \dots \supset K\langle \delta_1^{i_1+1}x,\dots,\delta_1^{i_1+1}\dots\delta_{m-1}^{i_{m-1}+1}x,\delta_1^{i_1+1}\dots$$

$$\delta_{m-1}^{i_{m-1}+1}\delta_m^{i_m}(\alpha_1-1)^{i_m+1+1}\dots(\alpha_{n-1}-1)^{i_m+n-1+1}(\alpha_n-1)^{i_m+n}x\rangle \supset \dots \supset K\langle \delta_1^{i_1+1}x,$$

$$\dots,\delta_1^{i_1+1}\dots\delta_{m-1}^{i_{m-1}+1}x,\delta_1^{i_1+1}\dots\delta_{m-1}^{i_{m-1}+1}\delta_m^{i_m}(\alpha_1-1)^{i_m+1+1}\dots(\alpha_{n-1}-1)^{i_m+n-1+1}x\rangle.$$

These $m+n$ chains show that $\mu_{\mathfrak{U}}(F,K) \geq m+n$, hence $\mathrm{type}(L/K) = m+n$. Furthermore, if Δ-σ-$\mathrm{trdeg}_K L = k > 0$ and $x_1,\dots,x_k$ is a Δ-σ-transcendence basic of L over K, then every x_i $(2 \leq i \leq k)$ is a Δ-σ-independent over $K\langle x_1,\dots,x_{i-1}\rangle$. Therefore, the above chains show that $\mu_{\mathfrak{U}}(K\langle x_1\rangle, K) = \mu_{\mathfrak{U}}(K\langle x_1,x_2\rangle, K\langle x_1\rangle) = \dots = \mu_{\mathfrak{U}}(K\langle x_1,\dots,x_k\rangle, K\langle x_1,\dots,x_{k-1}\rangle) = m+n$, hence $\dim(L/K) \geq k = \Delta$-$\sigma$-$\mathrm{trdeg}_K L$.

In order to prove the opposite inequality, suppose that $F_0 \supseteq F_1 \supseteq \dots \supseteq F_p$ is an ascending chain of intermediate Δ-σ-fields of the extension L/K such that $\mu_{\mathfrak{U}}(F_i, F_{i+1}) = \mathrm{type}(L/K) = m+n$ for $i = 0,\dots,p-1$. Clearly, in order to prove our inequality, it is sufficient to show that $p \leq k$.

For every $i = 0,\dots,p$, the Δ-σ-dimension polynomial $\chi_{K,F_i,\eta}(t)$, whose existence is established by Theorem 2, can be written as $\chi_{K,F_i,\eta}(t) = \sum_{j=0}^{m+n} a_j^{(i)}\binom{t+j}{j}$ where $a_j^{(i)} \in \mathbb{Z}$ $(0 \leq i \leq p-1, 0 \leq j \leq m+n)$. Then

$$\chi_{K,F_0,\eta}(t) - \chi_{K,F_p,\eta}(t) = \sum_{i=1}^{p}(\chi_{K,F_{i-1},\eta}(t) - \chi_{K,F_i,\eta}(t)) = \sum_{i=1}^{p}\sum_{j=0}^{m+n}(a_j^{(i-1)} -$$

$$a_j^{(i)})\binom{t+j}{j} = (a_{m+n}^{(0)} - a_{m+n}^{(p)})\binom{t+m+n}{m+n} + o(t^{m+n})$$ where $o(t^{m+n})$ denotes a polynomial of degree at most $m+n-1$.

Since $\mu_{\mathfrak{U}}(F_i,F_{i+1}) = m+n$ $(0 \leq i \leq p-1)$, we have $\deg(\chi_{K,F_i,\eta}(t) - \chi_{K,F_{i+1},\eta}(t)) = m+n$ (see Proposition 3). Therefore, $a_{m+n}^{(0)} > a_{m+n}^{(1)} > \dots > a_{m+n}^{(p)}$, hence

$$a_{m+n}^{(0)} - a_{m+n}^{(q)} = \sum_{i=1}^{p}(a_{m+n}^{(i-1)} - a_{m+n}^{(i)}) \geq p.$$

On the other hand, $\chi_{K,F_0,\eta}(t) - \chi_{K,F_p,\eta}(t) \leq \chi_{K,L,\eta}(t) = \sum_{i=0}^{m+n} a_i\binom{t+i}{i}$ where $a_{m+n} = \Delta$-σ-$\mathrm{trdeg}_K L$. Therefore, $p \leq a_{m+n}^{(0)} - a_{m+n}^{(p)} \leq k = \sigma$-$\mathrm{trdeg}_K L$. This completes the proof of the theorem. $\square$

5 Multivariate Dimension Polynomials of Intermediate difference-Differential Field Extensions

In this section we present a result that generalizes both Theorem 2 and the theorem on multivariate dimension polynomial of a finitely generated differential field extension associated with a partition of the basic set of derivations, see [9, Theorem 4.6].

Let K be a difference-differential (Δ-σ-) field with basic sets $\Delta = \{\delta_1, \ldots, \delta_m\}$ and $\sigma = \{\alpha_1, \ldots, \alpha_n\}$ of derivations and automorphisms, respectively. Suppose that these sets are represented as the unions of p and q nonempty disjoint subsets, respectively ($p, q \geq 1$):

$$\Delta = \Delta_1 \bigcup \cdots \bigcup \Delta_p, \quad \sigma = \sigma_1 \bigcup \cdots \bigcup \sigma_q, \tag{5}$$

$\Delta_1 = \{\delta_1, \ldots, \delta_{m_1}\}, \Delta_2 = \{\delta_{m_1+1}, \ldots, \delta_{m_1+m_2}\}, \ldots, \Delta_p = \{\delta_{m_1+\cdots+m_{p-1}+1}, \ldots, \delta_m\}, \sigma_1 = \{\alpha_1, \ldots, \alpha_{n_1}\}, \sigma_2 = \{\alpha_{n_1+1}, \ldots, \alpha_{n_1+n_2}\}, \ldots, \sigma_q = \{\alpha_{n_1+\cdots+n_{q-1}+1}, \ldots, \alpha_n\}; (m_1 + \cdots + m_p = m; n_1 + \cdots + n_q = n).$

For any element $\lambda = \delta_1^{k_1} \ldots \delta_m^{k_m} \alpha_1^{l_1} \ldots \alpha_n^{l_n} \in \Lambda$ ($k_i, l_j \in \mathbb{N}$; we use the notation of Sect. 2), the *order of λ with respect to a set Δ_i* ($1 \leq i \leq p$) is defined as $\sum_{\mu=m_1+\cdots+m_{i-1}+1}^{m_1+\cdots+m_i} k_\mu$; it is denoted by $\mathrm{ord}_i \lambda$. (If $i = 1$, the last sum is replaced by $\sum_{\mu=1}^{m_1} k_\mu$.) Similarly, the order of λ with respect to a set σ_j ($1 \leq j \leq q$), denoted by $\mathrm{ord}'_j \lambda$, is defined as $\sum_{\nu=n_1+\cdots+n_{j-1}+1}^{n_1+\cdots+n_j} l_\nu$. (If $j = 1$, the last sum is $\sum_{\nu=1}^{n_1} l_\nu$.)

If $r_1, \ldots, r_{p+q} \in \mathbb{N}$, we set

$$\Lambda(r_1, \ldots, r_{p+q}) = \{\lambda \in \Lambda \mid \mathrm{ord}_i \lambda \leq r_i \, (1 \leq i \leq p) \text{ and } \mathrm{ord}'_j \lambda \leq r_{p+j}(1 \leq j \leq q)\}.$$

Furthermore, for any permutation $(j_1, \ldots, j_{p+q})$ of the set $\{1, \ldots, p + q\}$, let $<_{j_1,\ldots,j_{p+q}}$ be the lexicographic order on $\mathbb{N}^{p+q}$ such that $(r_1, \ldots, r_{p+q}) <_{j_1,\ldots,j_{p+q}} (s_1, \ldots, s_{p+q})$ if and only if either $r_{j_1} < s_{j_1}$ or there exists $k \in \mathbb{N}$, $1 \leq k \leq p + q$, such that $r_{j_\nu} = s_{j_\nu}$ for $\nu = 1, \ldots, k$ and $r_{j_{k+1}} < s_{j_{k+1}}$.

If $A \subseteq \mathbb{N}^{p+q}$, then A' will denote the set of all $(p + q)$-tuples $a \in A$ that are maximal elements of this set with respect to one of the $(p+q)!$ orders $<_{j_1,\ldots,j_{p+q}}$. Say, if $A = \{(1, 1, 1), (2, 3, 0), (0, 2, 3), (2, 0, 5), (3, 3, 1), (4, 1, 1), (2, 3, 3)\} \subseteq \mathbb{N}^3$, then $A' = \{(2, 0, 5), (3, 3, 1), (4, 1, 1), (2, 3, 3)\}$.

Theorem 6. *With the above notation, let F be an intermediate Δ-σ-field of a Δ-σ-field extension $L = K\langle\eta_1, \ldots, \eta_s\rangle$ generated by a finite family $\eta = \{\eta_1, \ldots, \eta_s\}$. Let partitions (5) be fixed and for any $r_1, \ldots, r_{p+q} \in \mathbb{N}^{p+q}$, let*

$$F_{r_1,\ldots,r_{p+q}} = F \bigcap K(\bigcup_{j=1}^{s} \Lambda(r_1, \ldots, r_{p+q})\eta_j).$$

Then there exists a polynomial in $p + q$ variables $\Phi_{K,F,\eta} \in \mathbb{Q}[t_1, \ldots, t_{p+q}]$ such that

(i) $\Phi_{K,F,\eta}(r_1, \ldots, r_{p+q}) = \mathrm{trdeg}_K K(\bigcup_{j=1}^{s} \Lambda(r_1, \ldots, r_{p+q})\eta_j)$

for all sufficiently large $(r_1, \ldots, r_{p+q}) \in \mathbb{N}^{p+q}$. *(That is, there exist* $r_1^{(0)}, \ldots, r_{p+q}^{(0)} \in \mathbb{N}$ *such that the equality holds for all* $(r_1, \ldots, r_{p+q}) \in \mathbb{N}^{p+q}$ *with* $r_i \geq r_i^{(0)}$, $1 \leq i \leq p+q$.);

(ii) $\deg_{t_i} \Phi_\eta \leq m_i$ $(1 \leq i \leq p)$, $\deg_{t_{p+j}} \Phi_\eta \leq n_j$ $(1 \leq j \leq q)$ *and* $\Phi_\eta(t_1, \ldots, t_{p+q})$ *can be represented as*

$$\Phi_\eta = \sum_{i_1=0}^{m_1} \cdots \sum_{i_p=0}^{m_p} \sum_{i_{p+1}=0}^{n_1} \cdots \sum_{i_{p+q}=0}^{n_q} a_{i_1 \ldots i_{p+q}} \binom{t_1 + i_1}{i_1} \cdots \binom{t_{p+q} + i_{p+q}}{i_{p+q}} \qquad (6)$$

where $a_{i_1 \ldots i_{p+q}} \in \mathbb{Z}$.

(iii) Let $E_\eta = \{(i_1, \ldots, i_{p+q}) \in \mathbb{N}^{p+q} \mid 0 \leq i_k \leq m_k$ *for* $k = 1, \ldots, p$, $0 \leq i_{p+j} \leq n_j$ *for* $j = 1, \ldots, q$, *and* $a_{i_1 \ldots i_{p+q}} \neq 0\}$. *Then* $d = \deg \Phi_\eta$, $a_{m_1 \ldots m_p n_1 \ldots n_q}$, *elements* $(k_1, \ldots, k_{p+q}) \in E'_\eta$, *the corresponding coefficients* $a_{k_1 \ldots k_{p+q}}$, *and the coefficients of the terms of total degree* d *do not depend on the choice of the set of* Δ-σ-*generators* η. *Furthermore,* $a_{m_1 \ldots m_p n_1 \ldots n_q} = \Delta$-$\sigma$-$\mathrm{trdeg}_K L$.

Proof. We will mimic the method of the proof of Theorem 2 using the results on multivariate dimension polynomials of Δ-σ-L-modules. Let $\mathcal{D}$ be the ring of Δ-σ-operators over L considered as a filtered ring with $(p+q)$-dimensional filtration $\{\mathcal{D}_{r_1, \ldots, r_{p+q}} \mid (r_1, \ldots, r_{p+q}) \in \mathbb{Z}^{p+q}\}$ where for any $r_1, \ldots, r_{p+q} \in \mathbb{N}^{p+q}$, $\mathcal{D}_{r_1, \ldots, r_{p+q}}$ is the vector L-subspace of $\mathcal{D}$ generated by $\Lambda(r_1, \ldots, r_{p+q})$, and $\mathcal{D}_{r_1, \ldots, r_{p+q}} = 0$ if at least one r_i is negative. If M is a Δ-σ-L-module, then a family $\{M_{r_1, \ldots, r_{p+q}} \mid (r_1, \ldots, r_{p+q}) \in \mathbb{Z}^{p+q}\}$ of vector K-subspaces of M is said to be a $(p+q)$-dimensional filtration of M if

(i) $M_{r_1, \ldots, r_{p+q}} \subseteq M_{s_1, \ldots, s_{p+q}}$ whenever $r_i \leq s_i$ for $i = 1, \ldots, p+q$.

(ii) $\bigcup_{(r_1, \ldots, r_{p+q}) \in \mathbb{Z}^{p+q}} M_{r_1, \ldots, r_{p+q}} = M$.

(iii) There exists $(r_1^{(0)}, \ldots, r_{p+q}^{(0)}) \in \mathbb{Z}^p$ such that $M_{r_1, \ldots, r_{p+q}} = 0$ if $r_i < r_i^{(0)}$ for at least one index i.

(iv) $\mathcal{D}_{r_1, \ldots, r_{p+q}} M_{s_1, \ldots, s_{p+q}} \subseteq M_{r_1+s_1, \ldots, r_{p+q}+s_{p+q}}$ for any $(p+q)$-tuples $(r_1, \ldots, r_{p+q}), (s_1, \ldots, s_{p+q}) \in \mathbb{Z}^{p+q}$,

If every vector L-space $M_{r_1, \ldots, r_{p+q}}$ is finite-dimensional and there exists an element $(h_1, \ldots, h_p) \in \mathbb{Z}^p$ such that $\mathcal{D}_{r_1, \ldots, r_{p+q}} M_{h_1, \ldots, h_{p+q}} = M_{r_1+h_1, \ldots, r_{p+q}+h_{p+q}}$ for any $(r_1, \ldots, r_{p+q}) \in \mathbb{N}^{p+q}$, the filtration $\{M_{r_1, \ldots, r_{p+q}} \mid (r_1, \ldots, r_{p+q}) \in \mathbb{Z}^{p+q}\}$ is called *excellent*. Clearly, if $z_1, \ldots, z_k$ is a finite system of generators of a Δ-σ-L-module M, then $\{\sum_{i=1}^{k} \mathcal{D}_{r_1, \ldots, r_{p+q}} z_i \mid (r_1, \ldots, r_{p+q}) \in \mathbb{Z}^{p+q}\}$ is an excellent $(p+q)$-dimensional filtration of M.

Let L^* be the inversive closure of L. As we have seen, the module of Kähler differentials $\Omega_{L^*|K}$ can be equipped with a structure of a Δ-σ^*-L-module such that $\beta(d\zeta) = d\beta(\zeta)$ for any $\zeta \in L^*$, $\beta \in \Delta \bigcup \sigma$ $(d = d_{L^*|K})$. Let M' denote a $\mathcal{D}$-submodule $\sum_{i=1}^{s} \mathcal{D} d\eta_i$ of M treated as a filtered $\mathcal{D}$-module with the natural $(p+q)$-dimensional filtration $\{M'_{r_1, \ldots, r_{p+q}} \mid (r_1, \ldots, r_{p+q}) \in \mathbb{Z}^{p+q}\}$ where $M'_{r_1, \ldots, r_{p+q}} = \sum_{i=1}^{s} \mathcal{D}_{r_1, \ldots, r_{p+q}} d\eta_i$. Let N be a $\mathcal{D}$-submodule of M' generated by all elements $d\zeta$ where $\zeta \in F$ and for any $r_1, \ldots, r_{p+q} \in \mathbb{N}$, let $N_{r_1, \ldots, r_{p+q}}$ be the vector L-space generated by all elements $d\zeta$ where $\zeta \in F_{r_1,\ldots,r_{p+q}}$. Setting $N_{r_1,\ldots,r_{p+q}} = 0$ if $(r_1,\ldots,r_{p+q}) \in \mathbb{Z}^{p+q} \setminus \mathbb{N}^{p+q}$, we get a $(p+q)$-dimensional filtration of the Δ-σ-L-module N, and the embedding $N \to M'$ becomes a homomorphism of $(p+q)$-filtered Δ-σ-L-modules. Now, one can mimic the proof of Theorem 3.2.8 of [12] to show that the filtration $\{N_{r_1,\ldots,r_{p+q}} | (r_1,\ldots,r_{p+q}) \in \mathbb{Z}^{p+q}\}$ is excellent. The result of Theorem 6 immediately follows from the fact that $\dim_L N_{r_1,\ldots,r_{p+q}} = \operatorname{trdeg}_K F_{r_1,\ldots,r_{p+q}}$ for all $(r_1,\ldots,r_{p+q}) \in \mathbb{N}^{p+q}$ (as it is mentioned in the proof of Theorem 2, a family $(\zeta_i)_{i \in I}$ of elements of L (in particular, of $F_{r_1,\ldots,r_{p+q}}$) is algebraically independent over K if and only if the family $(d\zeta_i)_{i \in I}$ is linearly independent over L) and the result of [12, Theorem 3.5.8] (it states that under the above conditions, there exists a polynomial $\Phi_{K,F,\eta}(t_1,\ldots,t_{p+q}) \in \mathbb{Q}[t_1,\ldots,t_{p+q}]$ such that $\Phi_\eta(r_1,\ldots,r_{p+q}) = \dim_L N_{r_1,\ldots,r_{p+q}}$ for all sufficiently large $(r_1,\ldots,r_{p+q}) \in \mathbb{Z}^{p+q}$ and $\Phi_{K,F,\eta}(t_1,\ldots,t_{p+q})$ satisfies conditions (ii) of Theorem 6. Statement (iii) of Theorem 6 can be obtained in the same way as statement (iii) of Theorem 2 of [13].)

References

1. Einstein, A.: The Meaning of Relativity. Appendix II (Generalization of Gravitation Theory), 4th edn, pp. 133–165. Princeton University Press, Princeton (1953)
2. Johnson, J.L.: Kähler differentials and differential algebra. Ann. Math. **89**(2), 92–98 (1969)
3. Johnson, J.L.: A notion on Krull dimension for differential rings. Comment. Math. Helv. **44**, 207–216 (1969)
4. Kolchin, E.R.: The notion of dimension in the theory of algebraic differential equations. Bull. Amer. Math. Soc. **70**, 570–573 (1964)
5. Kolchin, E.R.: Differential Algebra and Algebraic Groups. Academic Press, New York (1973)
6. Kondrateva, M.V., Levin, A.B., Mikhalev, A.V., Pankratev, E.V.: Differential and Difference Dimension Polynomials. Kluwer Academic Publishers, Dordrecht (1999)
7. Levin, A.B.: Characteristic polynomials of filtered difference modules and difference field extensions. Russ. Math. Surv. **33**(3), 165–166 (1978)
8. Levin, A.B.: Characteristic polynomials of inversive difference modules and some properties of inversive difference dimension. Russ. Math. Surv. **35**(1), 217–218 (1980)
9. Levin, A.B.: Gröbner bases with respect to several orderings and multivariable dimension polynomials. J. Symbolic Comput. **42**(5), 561–578 (2007)
10. Levin, A.B.: Difference Algebra. Springer, New York (2008)
11. Levin, A.B.: Dimension polynomials of intermediate fields and Krull-type dimension of finitely generated differential field extensions. Math. Comput. Sci. **4**(2–3), 143–150 (2010)
12. Levin, A.: Multivariate dimension polynomials of inversive difference field extensions. In: Barkatou, M., Cluzeau, T., Regensburger, G., Rosenkranz, M. (eds.) AADIOS 2012. LNCS, vol. 8372, pp. 146–163. Springer, Heidelberg (2014). https://doi.org/10.1007/978-3-642-54479-8_7
13. Levin, A.: Dimension polynomials of intermediate fields of inversive difference field extensions. In: Kotsireas, I.S., Rump, S.M., Yap, C.K. (eds.) MACIS 2015. LNCS, vol. 9582, pp. 362–376. Springer, Cham (2016). https://doi.org/10.1007/978-3-319-32859-1_31

14. Levin, A.B.: Multivariate difference-differential polynomials and new invariants of difference-differential field extensions. In: Proceedings of ISSAC 2013, Boston, MA, pp. 267–274 (2013)
15. Levin, A.B., Mikhalev A.V.: Difference-Differential Dimension Polynomials. Moscow State University, VINITI, No. 6848-B 88, pp. 1–64 (1988)
16. Mikhalev, A.V., Pankratev, E.V.: Differential dimension polynomial of a system of differential equations. Algebra (Collection of Papers). Moscow State University, Moscow, pp. 57–67 (1980)
17. Morandi, P.: Fields and Galois Theory. Springer, New York (1996)
18. Sit, W.: Well-ordering of certain numerical polynomials. Trans. Amer. Math. Soc. **212**, 37–45 (1975)
19. Zhou, M., Winkler, F.: Computing difference-differential dimension polynomials by relative Gröbner bases in difference-differential modules. J. Symbolic Comput. **43**(10), 726–745 (2008)
20. Zhou, M., Winkler, F.: Gröbner bases in difference-differential modules and difference-differential dimension polynomials. Sci. China, Ser. A Math. **51**(9), 1732–1752 (2008)

Comprehensive *LU* Factors of Polynomial Matrices

Ana C. Camargos Couto[1] , Marc Moreno Maza[1], David Linder[2],
David J. Jeffrey[1(✉)] , and Robert M. Corless[1]

[1] ORCCA, University of Western Ontario, London, ON, Canada
djeffrey@uwo.ca
[2] Maplesoft, Waterloo, ON, Canada

Abstract. The comprehensive LU decomposition of a parametric matrix consists of a case analysis of the LU factors for each specialization of the parameters. Special cases can be discontinuous with respect to the parameters, the discontinuities being triggered by zero pivots encountered during factorization. For polynomial matrices, we describe an implementation of comprehensive LU decomposition in MAPLE, using the `RegularChains` package.

Keywords: Parametric linear algebra · LU decomposition · Regular chains

1 Introduction

Decomposing a matrix A into lower and upper triangular factors L and U is one of the fundamental operations in linear algebra. It is implemented in MAPLE's `LinearAlgebra` package as `LUDecomposition`. For polynomial matrices, the function takes the usual Computer Algebra option of returning only a generic factorization. Thus, for example,

$$A_1 = \begin{bmatrix} 1-x & 2 & 3 \\ 2-x & 5 & 6 \\ x & 3 & 2 \end{bmatrix} = \begin{bmatrix} 1 & 0 & 0 \\ \frac{x-2}{x-1} & 1 & 0 \\ \frac{-x}{x-1} & \frac{5x-3}{3x-1} & 1 \end{bmatrix} \begin{bmatrix} 1-x & 2 & 3 \\ 0 & \frac{3x-1}{x-1} & \frac{3x}{x-1} \\ 0 & 0 & -\frac{2}{3x-1} \end{bmatrix}. \tag{1}$$

The special cases $x = 1, 3/2$ make the elements singular. The importance of re-computing singular cases is established in [5], and, in the context of differential elimination, in [7]. We remark that special cases for LU factoring do not always occur when pivots are zero, because sometimes alternative pivots can lead to the same factoring. Indeed special cases can be *exactly detected* by Maple's existing `LUDecomposition` function through a special syntax implementing the algorithm of [5], which is not the default because its output is not just a simple answer, as we discuss below. See [8] for an example of the syntax. Because re-computation

D. Slamanig et al. (Eds.): MACIS 2019, LNCS 11989, pp. 80–88, 2020.
https://doi.org/10.1007/978-3-030-43120-4_8

is necessary in those special cases by that method, which allows comprehensive computation but is not itself comprehensive, we do not directly compare our present implementation to that syntax.

Symbolic computing in the presence of parameters has been the subject of discussion over many years [1]. Early systems, such as Macsyma, often asked a user interactively for information regarding a parameter, while other approaches used provisos, case analyses, error messages, etc. An important distinction is that between *comprehensive* approaches and *generic* approaches. In a comprehensive approach, a system will attempt to identify and compute all possible special cases, in contrast to a generic approach which selects one expression, implying conditions (which may not be stated) on the parameters.

Comprehensive solutions have been defined and used in several areas of mathematics. In algebraic geometry, a comprehensive Gröbner Basis was defined in [6], and a comprehensive triangular system based on regular chains was defined in [9]. In the MAPLE package DEtools, the rifsimp program offers a casesplit option, which is equivalent to a comprehensive analysis. A comprehensive solution of linear systems was presented in [14]. Computer Algebra systems have tended to avoid comprehensive results for several reasons. First, there is the difficulty of continuing a computation using a comprehensive result; secondly, there has been a fear that the number of cases will multiply exponentially and overwhelm the system. Although this could happen, there are many problems for which a comprehensive solution is possible and desirable.

2 Preliminaries

The implementation is based on MAPLE's RegularChains library, which we briefly describe in this section. The notion of a *regular chain*, introduced independently in [2] and [4], is closely related to that of a triangular decomposition of a polynomial system. Broadly speaking, a *triangular decomposition* of a polynomial system S is a set of simpler (in a precise sense) polynomial systems $S_1, \ldots, S_e$ such that a point p is a solution of S if, and only if, p is a solution of (at least) one of the systems $S_1, \ldots, S_e$.

If one wishes to describe all the solutions of S, those simpler systems are required to be regular chains. We refer to [3,10] for a formal presentation on the concepts of a regular chain.

Multivariate Polynomials. Let $\mathbb{K}$ be a field. If $\mathbb{K}$ is an ordered field, then we assume that it is a real closed field such as the field $\mathbb{R}$ of real numbers. Otherwise, we assume that $\mathbb{K}$ is algebraically closed, like the field $\mathbb{C}$ of complex numbers. Let $X_1 < \cdots < X_s$ be $s \geqslant 1$ ordered variables. We denote by $\mathbb{K}[X_1, \ldots, X_s]$ the ring of polynomials in the variables $X_1, \ldots, X_s$ with coefficients in $\mathbb{K}$. For a non-constant polynomial $p \in \mathbb{K}[X_1, \ldots, X_s]$, the greatest variable in p is called the *main variable* of p, denoted by mvar(p), and the leading coefficient of p w.r.t. mvar(p) is called the *initial* of p, denoted by init(p).

Regular Chains. A set R of non-constant polynomials in $\mathbb{K}[X_1, \ldots, X_s]$ is called a *triangular set*, if for all $p, q \in R$ with $p \neq q$ we have mvar(p) $\neq$ mvar(q).

A variable X_i is said to be *free* w.r.t. R if there exists no $p \in R$ such that $\mathrm{mvar}(p)=X_i$. For a nonempty triangular set R, we define the *saturated ideal* $\mathrm{sat}(R)$ of R to be the ideal$(R){:}h_R^\infty$, where h_R is the product of the initials of the polynomials in R. The saturated ideal of the empty triangular set is defined as the trivial ideal $\langle 0 \rangle$. From now on, R denotes a triangular set of $\mathbb{K}[X_1, \ldots, X_s]$. The ideal $\mathrm{sat}(R)$ has several properties, and in particular it is unmixed [11]. We denote its height, that is, the number of polynomials in R, by e, thus $\mathrm{sat}(R)$ has dimension $s - e$. Let $X_{i_1} < \cdots < X_{i_e}$ be the main variables of the polynomials in R. We denote by r_j the polynomial of R whose main variable is X_{i_j} and by h_j the initial of r_j. Thus h_R is the product $h_1 \cdots h_e$. We say that R is a *regular chain* whenever R is empty or, $\{r_1, \ldots, r_{e-1}\}$ is a regular chain and h_e is regular modulo the saturated ideal $\mathrm{sat}(\{r_1, \ldots, r_{e-1}\})$.

Constructible Sets. Let $F \subset \mathbb{K}[X_1, \ldots, X_s]$ be a set of polynomials and $g \in \mathbb{K}[X_1, \ldots, X_s]$ be a polynomial. We denote by $V(F) \subseteq \mathbb{K}^s$ the *zero set* or *affine variety* of F, that is, the set of points in the affine space $\mathbb{K}^s$ at which every polynomial $f \in F$ vanishes. If F consists of a single polynomial f, we write $V(f)$ instead of $V(F)$. We call a *constructible set* any subset of $\mathbb{K}^s$ of the form $V(F)\backslash V(g)$. Let $R \subset \mathbb{K}[X_1, \ldots, X_s]$ be a regular chain and let $h \in \mathbb{K}[X_1, \ldots, X_s]$ be a polynomial. We say that the pair $[R, h]$ is a *regular system* whenever h is regular modulo $\mathrm{sat}(R)$ and $V(h_R) \subseteq V(h)$ holds. We write $Z(R, h)$ for $V(R) \setminus V(h)$. One should observe that for a regular system $[R, h]$ the zero set $Z(R, h)$ is necessarily not empty. Regular systems provide an encoding for constructible sets. More precisely, there exists a finite family $\mathcal{T}$ of regular systems $[R_1, h_1], \ldots, [R_e, h_e]$ of $\mathbb{K}[X_1, \ldots, X_s]$ such that

$$V(F) \setminus V(g) \;=\; Z(R_1, h_1) \;\cup\; \cdots \;\cup\; Z(R_e, h_e).$$

We call $\mathcal{T}$ a *triangular decomposition* of the constructible set $V(F)\backslash V(g)$. Encoding constructible sets with regular systems has another benefit. It leads to efficient algorithms for performing set-theoretic operations on constructible sets; see [9]. These operations, as well as the above mentioned triangular decomposition algorithms, are part of the `RegularChains` library [12,13] distributed with the MAPLE CAS.

3   Comprehensive LU Method

We consider the LU factoring of matrices with multivariate polynomial entries, using partial pivoting. The pivots are analysed with the `RegularChains` library in MAPLE. Care is taken to identify cases where zero pivots do not, after all, lead to distinct LU factors. Considering that constructible sets represent the solution set of a polynomial, if there exists cases where the pivot is zero in a step of LU decomposition, constructible sets are used to represent their equations. Subsequently, these equations are used to express the constraints of validity of each solution branch (e.g.: $x = 3/2$ for the example shown in Sect. 1)

For the decomposition to be comprehensive (i.e., span all possible scenarios), we need to conduct the row reductions on all possible *unique* cases that

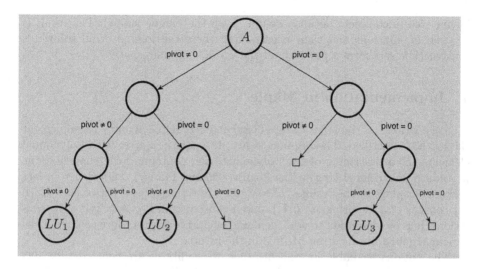

Fig. 1. Steps towards the comprehensive solution. Each root-to-leaf path represents a distinct LU decomposition of A. On each step, the calculation is split between two potential branches. Square nodes represents non-unique, and therefore dropped, cases.

may arise. Therefore, at each step, a pivot's constructible sets are analysed. Let CS_1 express the solution set of *pivot* $\neq 0$. This constructible set can be built with the GeneralConstruct command from the RegularChains library in MAPLE:

```
>> GeneralConstruct([],[A(k,k)],R);
```

Where the first and second arguments express equations and inequations to build the constructible set from, and the third argument is a polynomial ring. In order to build CS_1, we would give GeneralConstruct one inequation that represents the condition *pivot* $\neq 0$, and no equations. If CS_1 is nonempty, there are cases in which the natural matrix pivot can be used for the row reduction operation; so this operation is recorded in a branch (*pivot* $\neq 0$ in Fig. 1). Let CS_2 express the solution set of an inequation *pivot* $= 0$. Similarly to the other case, CS_2 can be built with the GeneralConstruct command:

```
>> GeneralConstruct([A(k,k)],[],R);
```

However, in this case, an equation *pivot* $= 0$ is passed as argument to the function, and no inequations are used. If CS_2 is non-empty, we look for an alternative pivot in the same column that has an empty CS_2 (this way guaranteeing that there will be no cases where division by zero is possible). The alternative pivot is used to build the permutation matrix and the original CS_2 value is saved, so that we can keep track of the exception case conditions. The alternative operation is recorded in a second branch (*pivot* $= 0$ in Fig. 1).

This process is repeated iteratively on each step of the LU factoring, every time splitting the result in two possible cases, and this way forming an incomplete binary tree (incomplete because we only keep the unique leaves). The result is a group of solutions and their constraints, where the joining of all solution's constructible sets form a partition of the variable domain space.

4   Implementation in Maple

We have written a Maple procedure ComprehensiveLU(A,R,opt) to implement the method described. The arguments are A, a square matrix with polynomial elements, R, a descriptor of the polynomial ring containing the elements (the procedure PolynomialRing in the RegularChains library), and opt, to select different displays of the results. The results are returned as a list of lists. Each list consists of a factoring (P, L, U), and a constructible set, specifying the conditions. It is our intention to add the ComprehensiveLU procedure as part of the LinearAlgebra library from MAPLE in the future.

The options available for the printing of conditions are constructible sets (the default), prettyprinting and programmable. To present some examples below, we have unpacked the output using the prettyprinting option, for easier reading. We have also confined our examples to small matrices with only a few polynomial entries. In each example, the first case corresponds to the generic result, and equals the result returned by the Maple command LinearAlgebra[LUDecomposition].

We return to the introductory example (1).

$$
\begin{bmatrix} 1-x & 2 & 3 \\ 2-x & 5 & 6 \\ x & 3 & 2 \end{bmatrix} = \begin{bmatrix} 1 & 0 & 0 \\ \frac{x-2}{x-1} & 1 & 0 \\ \frac{-x}{x-1} & \frac{5x-3}{3x-1} & 1 \end{bmatrix} \begin{bmatrix} 1-x & 2 & 3 \\ 0 & \frac{3x-1}{x-1} & \frac{3x}{x-1} \\ 0 & 0 & -\frac{2}{3x-1} \end{bmatrix}, \quad \begin{cases} x-1 \neq 0 , \\ 3x-1 \neq 0 . \end{cases} \tag{2}
$$

The permutation matrix is I and is omitted. The two conditions are not returned by Maple. The special cases are

$$
\begin{bmatrix} 1 & 0 & 0 \\ 0 & 0 & 1 \\ 0 & 1 & 0 \end{bmatrix} \begin{bmatrix} 1 & 0 & 0 \\ 1/2 & 1 & 0 \\ 5/2 & 0 & 1 \end{bmatrix} \begin{bmatrix} 2/3 & 2 & 3 \\ 0 & 2 & -1/2 \\ 0 & 0 & -3/2 \end{bmatrix}, \text{when} \begin{cases} 3x-1=0 . \end{cases}
$$

and

$$
\begin{bmatrix} 0 & 0 & 1 \\ 0 & 1 & 0 \\ 1 & 0 & 0 \end{bmatrix} \begin{bmatrix} 1 & 0 & 0 \\ 1 & 1 & 0 \\ 0 & 1 & 1 \end{bmatrix} \begin{bmatrix} 1 & 3 & 2 \\ 0 & 2 & 4 \\ 0 & 0 & -1 \end{bmatrix}, \text{ when } \begin{cases} x-1=0 . \end{cases}
$$

A multivariate example shows how the number of conditions increases as the number of parameters increases.

$$
A = \begin{bmatrix} a & 2b & 3 \\ d & -2 & 6 \\ 7 & 3 & 2 \end{bmatrix} . \tag{3}
$$

The generic case (also returned by Maple without conditions) is

$$A = \begin{bmatrix} 1 & 0 & 0 \\ \frac{d}{a} & 1 & 0 \\ \frac{7}{a} & \frac{-3a+14b}{2bd+2a} & 1 \end{bmatrix} \begin{bmatrix} a & 2b & 3 \\ 0 & \frac{-2bd-2a}{a} & \frac{-3d+6a}{a} \\ 0 & 0 & \frac{(4d-84)b+22a-9d-42}{2bd+2a} \end{bmatrix}, \begin{cases} a \neq 0 , \\ a + bd \neq 0 . \end{cases} \quad (4)$$

There are 3 special cases, and it is interesting to note that they uncover additional constraints.

$$A = \begin{bmatrix} 1 & 0 & 0 \\ 0 & 0 & 1 \\ 0 & 1 & 0 \end{bmatrix} \begin{bmatrix} 1 & 0 & 0 \\ \frac{-7}{bd} & 1 & 0 \\ -1/b & 0 & 1 \end{bmatrix} \begin{bmatrix} -bd & 2b & 3 \\ 0 & \frac{3d+14}{d} & \frac{2bd+21}{bd} \\ 0 & 0 & \frac{6b+3}{b} \end{bmatrix}, \begin{cases} a + bd = 0 \\ b \neq 0 \\ d \neq 0 \end{cases} \quad (5)$$

The second case is

$$A = \begin{bmatrix} 0 & 0 & 1 \\ 0 & 1 & 0 \\ 1 & 0 & 0 \end{bmatrix} \begin{bmatrix} 1 & 0 & 0 \\ d/7 & 1 & 0 \\ 0 & \frac{-14b}{3d+14} & 1 \end{bmatrix} \begin{bmatrix} 7 & 3 & 2 \\ 0 & -\frac{3d}{7} - 2 & -\frac{2d}{7} + 6 \\ 0 & 0 & \frac{-4bd+84b+9d+42}{3d+14} \end{bmatrix}, \begin{cases} a = 0 \\ 3d + 14 \neq 0 \end{cases}$$

$$(6)$$

Lastly,

$$A = \begin{bmatrix} 0 & 0 & 1 \\ 1 & 0 & 0 \\ 0 & 1 & 0 \end{bmatrix} \begin{bmatrix} 1 & 0 & 0 \\ 0 & 1 & 0 \\ -2/3 & 0 & 1 \end{bmatrix} \begin{bmatrix} 7 & 3 & 2 \\ 0 & 2b & 3 \\ 0 & 0 & 22/3 \end{bmatrix}, \begin{cases} a = 0 , \\ 3d + 14 = 0 \end{cases} \quad (7)$$

4.1 Efficiency

The implementation uses the `RegularChains` library, which is more efficient at performing polynomial arithmetic than the older `LUDecomposition` procedure. In order to perform comparison tests, we created a set of input matrices with polynomial elements up to degree 5, and measured the computation time for the LU factoring of each configuration. To ensure a fair efficiency analysis, the comparison test restricts the `ComprehensiveLU` program to computing only the generic case, in order to keep it comparable with the MAPLE library. The results are shown in Fig. 2. See below the script for the efficiency comparison test:

```
cf := proc(d) randpoly([x_1, x_2, x_3], dense, degree = d); end;
t1 := []: t2 := []: xd := []:
for d from 1 to 2 do      # coeff. degree
  m := 3; n:= 3;   ## order of the matrix
  xd := [op(xd),d];
  A := Matrix([[cf(d), cf(d), cf(d)], [1, cf(d), cf(d)],
```

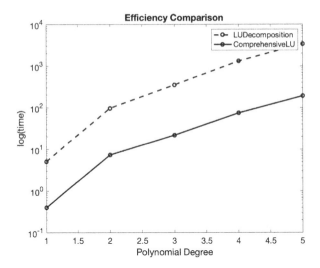

Fig. 2. Computation time vs polynomial degree for LUDecomposition of Maple and the present ComprehensiveLU. Recursion levels were introduced, with ComprehensiveLU being restricted to computing only the generic case.

```
[cf(d), 2, 4]]):
R := PolynomialRing([x_3, x_2, x_1]):
t_1 :=  x_1^n + x_1 + 1;    t_2 :=  x_2^2  - x_1 - 1;
t_3 :=  x_3^2 - x_2 - 1;
cs := GeneralConstruct([t_1, t_2, t_3], [], R):
t := time(): ComprehensiveLU(A, R, cs);
t1 := [op(t1),time() - t]:
printf("with \%g degree polynomials, ComprehensiveLU took
      \%g seconds to compute the result \n",d, time() - t);
a_1:=RootOf(t_1, x_1): a_2:=RootOf(y^2 - a_1 + 1, y):
a_3:=RootOf(z^2 - a_2 + 1, z):
B := eval(A, [x_1 = a_1, x_2 = a_2, x_3 = a_3]);
t := time():   LUDecomposition(B);
t2 := [op(t2),time() - t]:
printf("with \%g degree polynomials, LUDecomposition took
      \%g seconds to compute the result \n",d, time() - t);
end do:
```

The experiment consists of an LU factoring of 3×3 matrices with random polynomials. In order to make the computations algebraically challenging, recurrence levels were established to define the polynomial variables, which obey additional polynomial relationships with highest degree equal to 3. The experiment script loops over values of the random polynomial degree d ranging from 1 to 5 and records the time it took both algorithms to compute the final result in each iteration. The plot in Fig. 2 illustrates the comparison findings.

5   Conclusion

In parametric linear algebra, LU decomposition can be a discontinuous operation if the pivots encountered throughout the factorization are polynomials with roots defined in the problem's domain space. The discontinuity equations define special cases that we carefully consider in this project.

Our aim is to provide a comprehensive tool for computing LU factors of parametric matrices in MAPLE. We have shown that the main existing procedure in Maple's `LinearAlgebra` library, `LUDecomposition`, can only decompose generic cases of parametric matrices in an explicit way. Therefore, our algorithmic procedure `ComprehensiveLU` can be seen as a complement to the existing library function.

References

1. Corless, R.M., Jeffrey, D.J.: Well... it isn't quite that simple. SIGSAM Bull. **26**(3), 2–6 (1992)
2. Kalkbrener, M.: Three Contributions to Elimination Theory. Johannes Kepler University, Linz (1991)
3. Aubry, P., Lazard, D., Moreno Maza, M.: On the theories of triangular sets. J. Symb. Comp. **28**(1–2), 105–124 (1999)
4. Yang, L., Zhang, J.: Searching dependency between algebraic equations: an algorithm applied to automated reasoning. International Atomic Energy Agency, IC/89/263, Miramare, Trieste, Italy (1991)
5. Corless, R.M., Jeffrey, D.J.: The Turing factorization of a rectangular matrix. SIGSAM Bull. **31**(3), 20–30 (1997)
6. Weispfenning, V.: Comprehensive grobner bases. J. Symbolic Comput. **14**, 1–29 (1992)
7. Reid, G.: Algorithms for reducing a system of PDEs to standard form, determining the dimension of its solution space and calculating its Taylor series solution. Eur. J. Appl. Math. **2**, 293–318 (1991)
8. Jeffrey, D.J., Corless R.M.: Linear algebra in Maple. In: Hogben, L. (ed) Chapter 89 in the CRC Handbook of Linear Algebra, 2nd ed. Chapman & Hall/CRC (2013)
9. Chen, C., Golubitsky, O., Lemaire, F., Moreno Maza, M., Pan, W.: Comprehensive triangular decomposition. In: Ganzha, V.G., Mayr, E.W., Vorozhtsov, E.V. (eds.) CASC 2007. LNCS, vol. 4770, pp. 73–101. Springer, Heidelberg (2007). https://doi.org/10.1007/978-3-540-75187-8_7
10. Chen, C., Moreno Maza, M.: Algorithms for computing triangular decomposition of polynomial systems. J. Symb. Comput. **47**(6), 610–642 (2012)
11. Boulier, F., Lemaire, F., Moreno Maza, M.: Well Known Theorems on Triangular Systems and the D5 Principle. In: Dumas, J.-G. et al. (eds.) Proceedings of Transgressive Computing 2006, Granada, Spain (2006)
12. Chen, C., et al.: Solving semi-algebraic systems with the RegularChains library in Maple. In: Raschau, S. (ed.) Proceedings of the Fourth International Conference on Mathematical Aspects of Computer Science and Information Sciences (MACIS 2011), pp. 38–51 (2011)

13. Lemaire, F., Moreno Maza, M., Xie, Y.: The RegularChains library in Maple 10. In: Kotsireas, I.S. (ed.) Proceedings of Maple Summer Conference 2005, Waterloo, Canada (2005)
14. Sit, W.Y.: An algorithm for solving parametric linear systems. J. Symb. Comp. **13**, 353–394 (1992)

Sublinear Cost Low Rank Approximation via Subspace Sampling

Victor Y. Pan[1(✉)], Qi Luan[2], John Svadlenka[3], and Liang Zhao[1,3]

[1] Department of Computer Science,
Lehman College of the City University of New York, Bronx, NY 10468, USA
victor.pan@lehman.cuny.edu
[2] Program in Mathematics, The Graduate Center of the City University of New
York, New York, NY 10036, USA
qi_luan@yahoo.com
[3] Program in Computer Science, The Graduate Center of the City University of New
York, New York, NY 10036, USA
jsvadlenka@gradcenter.cuny.edu, lzhao1@gc.cuny.edu
http://comet.lehman.cuny.edu/vpan/

Abstract. Low Rank Approximation (LRA) of a matrix is a hot
research subject, fundamental for Matrix and Tensor Computations and
Big Data Mining and Analysis. Computations with LRA can be performed at *sublinear cost*, that is, by using much fewer memory cells and
arithmetic operations than an input matrix has entries. Although every
sublinear cost algorithm for LRA fails to approximate the worst case
inputs, we prove that our sublinear cost variations of a popular subspace
sampling algorithm output accurate LRA of a large class of inputs.

Namely, they do so with a high probability (*whp*) for a random
input matrix that admits its LRA. In other papers we propose and analyze other sublinear cost algorithms for LRA and Linear Least Sqaures
Regression. Our numerical tests are in good accordance with our formal
results.

Keywords: Low-rank approximation · Sublinear cost · Subspace
sampling

2000 Math. Subject Classification: 65Y20 · 65F30 · 68Q25 ·
68W20 · 15A52

1 Introduction

LRA Background. Low rank approximation (LRA) of a matrix is a hot
research area of Numerical Linear Algebra (NLA) and Computer Science (CS)
with applications to fundamental matrix and tensor computations and Data
Mining and Analysis (see surveys [HMT11, M11, KS17], and [CLO16]). Matrices from Big Data (e.g., unfolding matrices of multidimensional tensors) are
frequently so immense that realistically one can access only a tiny fraction of

© Springer Nature Switzerland AG 2020
D. Slamanig et al. (Eds.): MACIS 2019, LNCS 11989, pp. 89–104, 2020.
https://doi.org/10.1007/978-3-030-43120-4_9

their entries, although quite typically these matrices admit their LRA (cf. (1) in Sect. 2). One can operate with such matrices *at sublinear computational cost*, that is, by using much fewer memory cells and arithmetic operations than an input matrix has entries, but can we compute LRA at sublinear cost? Yes and no. No, because every sublinear cost LRA algorithm fails even on the small input families of Appendix B. Yes, because our sublinear cost variations of a popular *subspace sampling* algorithm output accurate LRA for a large class of input.

Let us provide some details.

Subspace sampling algorithms compute LRA of a matrix M by using auxiliary matrices FM, MH or FMH for random multipliers F and H, commonly called *test matrices* and having smaller sizes. The output LRA are nearly optimal whp provided that F and H are Gaussian, Rademacher's, SRHT or SRFT matrices;[1] furthermore the algorithms consistently output accurate LRA in their worldwide application with these and some other random multipliers F and H, all of which, however, are multiplied by M at superlinear cost (see [TYUC17, Section 3.9], [HMT11, Section 7.4], and the bibliography therein).

Our modifications are deterministic. They use fixed sparse orthogonal (e.g., *subpermutation*) multipliers[2] F and H, run at sublinear cost, and whp output reasonably close *dual LRA*, i.e., LRA of a random input admitting LRA; we deduce our error estimates under three distinct models of random matrix computations in Sections 4.1 – 4.3. Unlike the customary randomized algorithms of [HMT11], [M11], [KS17], which perform at superlinear cost and which whp output close LRA of *any matrix* that admits LRA, our deterministic algorithms run at sublinear cost and whp output close LRA of *many such matrices* and in a sense most of them. Namely we prove that whp they output close LRA of a random input matrix that admits LRA.

How meaningful are our results? Our definitions of three classes of random matrices of low numerical rank are quite natural for various real world applications of LRA, but are odd for some other ones, as is the case with any definition of that kind. In spite of such odds, however, our formal study is in good accordance with our numerical tests for both synthetic and real world inputs, some from [HMT11]. Surely it is not realistic to assume that an input matrix is random, but we can randomize it by means of pre-processing of an input with random multipliers and then apply our results. Moreover, empirically such a randomized pre-processing and sublinear cost pre-processing with proper sparse multipliers consistently give similar results.

Our upper bounds on the output error of LRA of an $m \times n$ matrix of numerical rank r exceed the optimal error bound by a factor of $\sqrt{\min\{m,n\}r}$, but if the optimal bound is small enough we can apply two algorithms for iterative

---

[1] Here and hereafter *"Gaussian matrices"* stands for "Gaussian random matrices" (see Definition 1). "SRHT and SRFT" are the acronyms for "Subsample Random Hadamard and Fourier transforms". Rademacher's are the matrices filled with iid variables, each equal to 1 or -1 with probability $1/2$.

[2] Subpermutation matrices are full-rank submatrices of permutation matrices.

refinement of LRA, proposed in [PLa], running at sublinear cost, and reasonably efficient according to the results of numerical tests in [PLa].

As we discussed earlier, any sublinear cost LRA algorithm (and ours are no exception) fails on some families of hard inputs, but our analysis and tests show that the class of such inputs is narrow. We conjecture that it shrinks fast if we recursively apply the same algorithm with new multipliers; in Sect. 5 we comment of some heuristic recipes for these recursive processes; our numerical tests consistently confirm their efficiency.

Impact of Our Study, Its Extensions and By-Products

(i) Our duality approach is efficient for some fundamental matrix computations besides LRA: [PQY15, PZ17a], and [PZ17b] formally support empirical efficiency of dual Gaussian elimination with no pivoting, while [LPb] proposes a dual sublinear cost deterministic modification of Sarlós' randomized algorithm of 2006 and then proves that whp it outputs nearly optimal solution of the important problem of *Linear Least Squares Regression (LLSR)* for random input, and consequently for a large class of inputs – in a sense for most of them. This formal study turned out to be in very good accordance with the results of our extensive tests with synthetic and real world inputs.

(ii) In the paper [PLa] we proposed, analyzed, and tested new sublinear cost algorithms for refinement of a crude but reasonably close LRA.

(iii) In [LPa] and [PLSZa] we proved that popular Cross-Approximation LRA algorithms running at sublinear cost as well as our simplified sublinear cost variations of these algorithms output accurate solution of dual LRA whp, and we also devised a sublinear cost algorithm for transformation of any LRA into its special form of CUR LRA, which is particularly memory efficient.

(iv) Our acceleration of LRA can be immediately extended to the acceleration of Tensor Train Decomposition because it is reduced to recursive computation of LRA of unfolding matrices. Likewise our results can be readily extended to Tucker Decomposition of tensors because Tucker Decomposition is essentially LRA of unfolding matrices of a tensor. Extension to CP Decomposition of Tensors, however, remains a challenge.

(v) In [LPa] we also extended our progress by devising deterministic and practically promising algorithm that at sublinear cost computes accurate LRA for a symmetric positive semidefinite matrix admitting LRA.

Related Works. LRA has huge bibliography; see, e.g., [M11, HMT11, KS17]. The papers [PLSZ16] and [PLSZ17] have provided the first formal support for dual accurate randomized LRA at sublinear cost (they call sublinear cost algorithms *superfast*). The earlier papers [PQY15, PLSZ16, PZ17a], and [PZ17b] studied duality for other fundamental matrix computations besides LRA, and we have already cited extension of our progress in [PLa], [LPa] and [LPb].

Organization of the Paper. In Sect. 2 we recall random sampling for LRA. In Sects. 3 and 4 we estimate output errors of our dual LRA algorithms running at sublinear cost. In Sect. 5 we generate multipliers for both pre-processing and sampling. Appendix A is devoted to background on matrix computations. In Appendix B we specify some small families of inputs on which any sublinear cost LRA algorithm fails. Because of size limitation for this paper we leave to [PLSZb] various details, our historical comments, the test results, and some proofs, in particular the proofs of Theorems 5 and 6.

Some Definitions. The concepts "large", "small", "near", "close", "approximate", "ill-" and "well-conditioned", are usually quantified in the context. "$\ll$" and "$\gg$" mean "much less than" and "much greater than", respectively. *"Flop"* stands for "floating point arithmetic operation"; *"iid"* for "independent identically distributed". In context a *"perturbation of a matrix"* can mean a perturbation having a small relative norm. $\mathbb{R}^{p \times q}$ denotes the class of $p \times q$ real matrices. We assume dealing with real matrices throughout, and so the Hermitian transpose M^* of M turns into transpose M^T, but our study can be readily extended to complex matrices; see some relevant results about complex Gaussian matrices in [E88, CD05, ES05], and [TYUC17].

2   Four Known Subspace Sampling Algorithms

Hereafter $|| \cdot ||$ and $|| \cdot ||_F$ denote the spectral and the Frobenius matrix norms, respectively; $| \cdot |$ can denote either of them. M^+ denotes the Moore – Penrose pseudo inverse of M.

Next we devise a sublinear cost algorithm for LRA XY of matrix M such that

$$M = XY + E, \ ||E||/||M|| \leq \epsilon, \tag{1}$$

for pairs of matrices X of size $m \times r$ and Y of size $r \times n$, a matrix norm $|| \cdot ||$, and a small tolerance ϵ.

Algorithm 1. *Range Finder* (see Remark 1).

INPUT: *An $m \times n$ matrix M and a target rank r.*
OUTPUT: *Two matrices $X \in \mathbb{R}^{m \times l}$ and $Y \in \mathbb{R}^{l \times m}$ defining an LRA $\tilde{M} = XY$..*
INITIALIZATION: *Fix an integer l, $r \leq l \leq n$, and an $n \times l$ test matrix (multiplier) H of rank l.*
COMPUTATIONS:
  1. *Compute the $m \times l$ matrix MH.*
  2. *Fix a nonsingular matrix $T^{-1} \in \mathbb{R}^{l \times l}$ and output the matrix $X :=$ $MHT^{-1} \in \mathbb{R}^{m \times l}$.*
  3. *Output an $l \times n$ matrix $Y := \operatorname{argmin}_V |XV - M| = X^+ MT$.*

Remark 1. Let $\operatorname{rank}(FM) = k$. Then $XY = MH(MH)^+ M$ independently of the choice of T^{-1}, but a proper choice of a nonsingular matrix T numerically stabilizes the algorithm. For $l > r \geq \operatorname{nrank}(MH)$ the matrix MH is ill-conditioned,[3] but let Q and R be the factors of the thin QR factorization

---

[3] $\operatorname{nrank}(W)$ denotes *numerical rank of W* (see Appendix A.1).

of MH, choose $T := R$, and observe that $X = MHT^{-1} = Q$ is an orthogonal matrix. $X = MHT^{-1}$ is also an orthogonal matrix if $T = R\Pi$ and if R and Π are factors of a rank-revealing QRΠ factorization of MH.

Column Subspace Sampling turns into *Column Subset Selection* in the case of a subpermutation matrix H.

Algorithm 2. *Transposed Range Finder* (see Remark 2).

INPUT: *As in Algorithm 1.*
OUTPUT: *Two matrices $X \in \mathbb{R}^{k \times n}$ and $Y \in \mathbb{R}^{m \times k}$ defining an LRA $\tilde{M} = YX$.*
INITIALIZATION: *Fix an integer k, $r \le k \le m$, and a $k \times m$ test matrix (multiplier) F of full numerical rank k.*
COMPUTATIONS:

   1. *Compute the $k \times m$ matrix FM.*
   2. *Fix a nonsingular $k \times k$ matrix S^{-1}; then output $k \times n$ matrix $X := S^{-1}FM$.*
   3. *Output an $m \times k$ matrix $Y := \operatorname{argmin}_V |VX - M|$.*

Row Subspace Sampling turns into random *Row Subset Selection* in the case of a subpermutation matrix F.

Remark 2. $Y = M(S^{-1}FM)^+$ and $YX = M(FM)^+FM$ independently of the choice of S^{-1} if $\operatorname{rank}(FM) = l$, but a proper choice of S numerically stabilizes the algorithm. For $k > r \ge \operatorname{nrank}(FMH)$ the matrix FMH is ill-conditioned, but $S^{-1}FM$ is orthogonal if $S = L$, $X := Q = L^{-1}FM$, $Y := Q^*M$, and L and Q are the factors of the thin LQ factorization of FM.

The following algorithm combines row and column subspace sampling. In the case of the identity matrix S it turns into the algorithm of [TYUC17, Section 1.4], whose origin can be traced back to [WLRT08].

Algorithm 3. *Row and Column Subspace Sampling* (see Remark 3).

INPUT: *As in Algorithm 1.*
OUTPUT: *Two matrices $X \in \mathbb{R}^{m \times k}$ and $Y \in \mathbb{R}^{k \times m}$ defining an LRA $\tilde{M} = XY$.*
INITIALIZATION: *Fix two integers k and l, $r \le k \le m$ and $r \le l \le n$; fix two test matrices (multipliers) $F \in \mathbb{R}^{k \times m}$ and $H \in \mathbb{R}^{n \times l}$ of full numerical ranks and two nonsingular matrices $S \in \mathbb{R}^{k \times k}$ and $T \in \mathbb{R}^{l \times l}$.*
COMPUTATIONS:

   1. *Output the matrix $X = MHT^{-1} \in \mathbb{R}^{m \times l}$.*
   2. *Compute the matrices $U := S^{-1}FM \in \mathbb{R}^{k \times n}$ and $W := S^{-1}FX \in \mathbb{R}^{m \times l}$.*
   3. *Output the $l \times n$ matrix $Y := \operatorname{argmin}_V |W^+V - U|$.*

Remark 3. $YX = MH(FMH)^+FM$ independently of the choice of the matrices S^{-1} and T^{-1} if the matrix FMH has full rank $\min\{k, l\}$, but a proper choice of S and T numerically stabilizes the computations of the algorithm. For $\min\{k, l\} > r \ge \operatorname{nrank}(FMH)$ the matrix FMH is ill-conditioned, but we can make it orthogonal by properly choosing the matrices S^{-1} and T^{-1}.

Remark 4. By applying Algorithm 3 to the transpose matrix M^* we obtain **Algorithm 4**, which begins with column subspace sampling followed by row subspace sampling. Our study of Algorithms 1 and 3 for input M actually covers Algorithms 2 and 4 as well.

Next we estimate the output errors of Algorithm 1 for any input; then extend these estimates to the output of Algorithm 3, at first for any input and then for random inputs.

3   Deterministic Error Bounds for Sampling Algorithms

Suppose that we are given matrices MHT^{-1} and $S^{-1}FM$. We can perform Algorithm 3 at arithmetic cost in $O(kln)$, which is sublinear if $kl \ll m$. Furthermore let $k^2 \ll m$ and $l^2 \ll n$. Then for proper deterministic choice of sparse (e.g., subpermutation) matrices S and T we can also compute the matrices MHT^{-1} and $S^{-1}FM$ at sublinear cost and thus complete computations of entire Algorithm 3 at sublinear cost. In this case we cannot ensure any reasonable accuracy of the output LRA for a worst case input and even for small input families of Appendix B, but we are going to prove that the output of that deterministic algorithm is quite accurate whp for random input and therefore for a large class of inputs, which is in good accordance with the results of our tests with synthetic and real world inputs.

We deduce some auxiliary deterministic output error bounds for any fixed input matrix in this section and refine them for random input under our probabilistic models in the next section. It turned out that the output error bounds are dominated at the stage of performing Range Finder because in Sect. 3.2 we rather readily bound additional impact of pre-processing with multipliers F and $S^{-1}F$.

3.1   Deterministic Error Bounds for Range Finder

Theorem 1 [HMT11, Theorem 9.1]. *Suppose that Algorithm 1 has been applied to a matrix M with a multiplier H and let*

$$C_1 = V_1^* H, \ C_2 = V_2^* H, \tag{2}$$

$$M = \begin{pmatrix} U_1 \, \Sigma_1 & V_1^* \\ U_2 & \Sigma_2 \, V_2^* \end{pmatrix}, \ M_r = U_1 \Sigma_1 V_1^*, \ \text{and} \ M - M_r = U_2 \Sigma_2 V_2^* \tag{3}$$

be SVDs of the matrices M, its rank-r truncation M_r, and $M - M_r$, respectively. $[\Sigma_2 = O$ and $XY = M$ if $\mathrm{rank}(M) = r$. The columns of V_1^* span the top right singular space of M.] Then

$$|M - XY|^2 \le |\Sigma_2|^2 + |\Sigma_2 C_2 C_1^+|^2. \tag{4}$$

Notice that $|\Sigma_2| = \bar{\sigma}_{r+1}(M)$, $|C_2| \le 1$, and $|\Sigma_2 C_2 C_1^+| \le |\Sigma_2| \, |C_2| \, |C_1^+|$ and obtain

$$|M - XY| \le (1 + |C_1^+|^2)^{1/2} \bar{\sigma}_{r+1}(M) \ \text{for} \ C_1 = V_1^* H. \tag{5}$$

It follows that the output LRA is optimal up to a factor of $(1 + |C_1^+|^2)^{1/2}$.

Next we deduce an upper bound on the norm $|C_1^+|$ in terms of $||((MH)_r)^+||$, $||M||$, and $\eta := 2\sigma_{r+1}(M) \ ||((MH)_r)^+||$.

Corollary 1. *Under the assumptions of Theorem 1 let the matrix $M_r H$ have full rank r. Then*

$$|(M_rH)^+|/|M_r^+| \leq |C_1^+| \leq |(M_rH)^+| \ |M_r| \leq |(M_rH)^+| \ |M|.$$

Proof. Deduce from (2) and (3) that $M_r H = U_1 \Sigma_1 C_1$. Hence $C_1 = \Sigma_1^{-1} U_1^* M_r H$.

Recall that the matrix $M_r H$ has full rank r, apply Lemma 2, recall that U_1 is an orthogonal matrix, and obtain $|(M_rH)^+|/|\Sigma_1^{-1}| \leq |C_1^+| \leq |(M_rH)^+| \ |\Sigma_1|$.

Substitute $|\Sigma_1| = |M_r|$ and $|\Sigma_1^{-1}| = |M_r^+|$ and obtain the corollary.

Corollary 2. *See [PLSZb]. Under the assumptions of Corollary 1 let*

$$\eta := 2\sigma_{r+1}(M) \ ||((MH)_r)^+|| < 1, \ \eta' := \frac{2\sigma_{r+1}(M)}{1-\eta} \ ||((MH)_r)^+|| < 1.$$

Then

$$\frac{1-\eta'}{||M_r^+||} \ ||((MH)_r)^+|| \leq ||C_1^+ \ || \leq \frac{||M||}{1-\eta} \ ||((MH)_r)^+||.$$

For a given matrix MH we compute the norm $||((MH)_r)^+||$ at sublinear cost if $l^2 \ll n$. If also some reasonable upper bounds on $||M||$ and $\sigma_{r+1}(M)$ are known, then Corollary 2 implies *a posteriori estimates* for the output errors of Algorithm 1.

3.2 Deterministic Impact of Pre-multiplication on the Errors of LRA

It turned out that the impact of pre-processing with multipliers $S^{-1}F$ into the output error bounds is dominated at the stage of Range Finder.

Lemma 1. [The impact of pre-multiplication on LRA errors.] *Suppose that Algorithm 3 outputs a matrix XY for $Y = (FX)^+ FM$ and that $m \geq k \geq l = \mathrm{rank}(X)$. Then*

$$M - XY = W(M - XX^+M) \text{ for } W = I_m - X(FX)^+F, \tag{6}$$

$$|M - XY| \leq |W| \ |M - XX^+M|, \ |W| \leq |I_m| + |X| \ |F| \ |(XF)^+|. \tag{7}$$

Proof. Recall that $Y = (FX)^+ FM$ and notice that $(FX)^+ FX = I_l$ if $k \geq l = \mathrm{rank}(FX)$. Therefore $Y = X^+M + (FX)^+F(M - XX^+M)$. Consequently (6) and (7) hold.

We bounded the norm $|M - XX^+M|$ in the previous subsection; next we bound the norms $|(FX)^+|$ and $|W|$ of the matrices FX and W, computed at sublinear cost for $kl \ll n$, a fixed orthogonal X, and proper choice of sparse F.

Theorem 2. *[P00, Algorithm 1] for a real $h > 1$ applied to an $m \times l$ orthogonal matrix X performs $O(ml^2)$ flops and outputs an $l \times m$ subpermutation matrix F such that $||(FX)^+|| \leq \sqrt{(m-l)lh^2 + 1}$, and $||W|| \leq 1 + \sqrt{(m-l)lh^2 + 1}$, for $W = I_m + X(FX)^+ F$ of (6) and any fixed $h > 1$; $||W|| \approx \sqrt{ml}$ for $m \gg l$ and $h \approx 1$.*

[P00, Algorithm 1] outputs $l \times m$ matrix F. One can strengthen deterministic bounds on the norm $|W|$ by computing proper $k \times m$ subpermutation matrices F for k of at least order l^2.

Theorem 3. *For k of at least order l^2 and a fixed orthogonal multiplier X compute a $k \times m$ subpermutation multiplier F by means of deterministic algorithms by Osinsky, running at sublinear cost and supporting [O18, equation (1)]. Then $||W|| \leq 1 + ||(FX)^+|| = O(l)$ for W of (6).*

4 Accuracy of Sublinear Cost Dual LRA Algorithms

Next we estimate the output errors of Algorithm 1 for a fixed orthogonal matrix H and two classes of random inputs of low numerical rank, in particular for perturbed factor-Gaussian inputs of Definition 2. These estimates formally support the observed accuracy of Range Finder with various dense multipliers (see [HMT11, Section 7.4], and the bibliography therein), but also with sparse multipliers, with which Algorithms 3 and 4 run at sublinear cost.[4] We extend these upper estimates for output accuracy to variations of Algorithm 3 that run at sublinear cost; then we extend them to Algorithm 4 by means of transposition of an input matrix. This study involves the norms of a Gaussian matrix and its pseudo inverse, whose estimates we recall in Appendix A.4.

Hereafter $\overset{d}{=}$ denotes equality in probability distribution.

Definition 1. *A matrix is Gaussian if its entries are iid Gaussian (normal) variables. We let $\mathcal{G}^{p \times q}$ denote a $p \times q$ Gaussian matrices, and define random variables $\nu_{p,q} \overset{d}{=} |G|$, $\nu_{sp,p,q} \overset{d}{=} ||G||$, $\nu_{F,p,q} \overset{d}{=} ||G||_F$, $\nu_{p,q}^+ \overset{d}{=} |G^+|$, $\nu_{sp,p,q}^+ \overset{d}{=} ||G^+||$, and $\nu_{F,p,q}^+ \overset{d}{=} ||G^+||_F$, for a $p \times q$ random Gaussian matrix G. $[\nu_{p,q} \overset{d}{=} \nu_{q,p}$ and $\nu_{p,q}^+ \overset{d}{=} \nu_{q,p}^+$, for all pairs of p and q.]*

Theorem 4 [Non-degeneration of a Gaussian Matrix]. *Let $F \overset{d}{=} \mathcal{G}^{r \times p}$, $H \overset{d}{=} \mathcal{G}^{q \times r}$, $M \in \mathbb{R}^{p \times q}$ and $r \leq \text{rank}(M)$. Then the matrices F, H, FM, and MH have full rank r with probability 1.*

Assumption 1. *We simplify the statements of our results by assuming that a Gaussian matrix has full rank and ignoring the probability 0 of its degeneration.*

In Theorems 5 and 6 of the next subsections we state our error estimates, which we prove in [PLSZb].

---

[4] We defined Algorithm 4 in Remark 4

4.1 Errors of Range Finder for a Perturbed Factor-Gaussian Input

Assumption 2. *Suppose that $\tilde{M} = AB$ is a right $m \times n$ factor Gaussian matrix of rank r, $H = U_H \Sigma_H V_H^*$ is a $n \times l$ test matrix, and let $\theta = \frac{e\sqrt{l}(\sqrt{n}+\sqrt{r})}{l-r}$ be a constant. Here and hereafter $e := 2.71828182\ldots$. Define random variables $\nu = ||B||$ and $\mu = ||(BU_H)^+||$, and recall that $\nu \overset{d}{=} \nu_{sp,r,n}$ and $\mu \overset{d}{=} \nu_{sp,r,l}^+$.*

Theorem 5. [Errors of Range Finder for a perturbed factor-Gaussian matrix.] *Under Assumption 2, let $\phi = \left(\nu\mu||H^+||\right)^{-1} - 4\alpha||H||$, and let $M = \tilde{M} + E$ be a right factor Gaussian with perturbation such that*

$$\alpha := \frac{||E||_F}{(\sigma_r(M) - \sigma_{r+1}(M))} \leq \min\left(0.2, \frac{\xi}{8\kappa(H)\theta}\right) \tag{8}$$

where $0 < \xi < 2^{-0.5}$. Apply Algorithm 1 to M with a test matrix (multiplier) H. Then

$$||M - XY||^2 \leq \left(1 + \phi^{-2}\right)\sigma_{r+1}^2(M) \text{ and}$$

$$||M - XY|| \leq \left(1 + 2||H^+||\theta/\xi\right)\sigma_{r+1}(M) \tag{9}$$

with a probability no less than $1 - 2\sqrt{\xi}$. If $r \ll l$, then $\theta \approx e\sqrt{n/l}$, implying that the coefficient of $\sigma_{r+1}(M)$ on the right hand side of (9) is close to

$$1 + \frac{2e||H^+||}{\xi}\sqrt{n/l} = O(\sqrt{n/l}).$$

4.2 Output Errors of Range Finder Near a Matrix with a Random Singular Space

Next we state similar estimates under an alternative randomization model for dual LRA.

Theorem 6 [Errors of Range Finder for an input with a random singular space]. *Let the matrix V_1 in Theorem 1 be the $n \times r$ Q factor in a QR factorization of a normalized $n \times r$ Gaussian matrix G and let the multiplier $H = U_H \Sigma_H V_H^*$ be any $n \times l$ matrix of full rank $l \geq r$.*

(i) Then for random variables $\nu = |G|$ and $\mu = |G^T U_H|$, it holds that

$$|M - XY|/\bar{\sigma}_{r+1}(M) \leq \phi_{r,l,n} := (1 + (\nu\mu|H^+|)^2)^{1/2}.$$

(ii) For $n \geq l \geq r + 4 \geq 6$, with a probability at most $1 - 2\sqrt{\xi}$ it holds that

$$\phi_{sp,r,l,n}^2 \leq 1 + \xi^{-2}\, e^2\, ||H^+||^2 \left(\frac{\sqrt{l}(\sqrt{n}+\sqrt{r})}{l-r}\right)^2$$

and

$$\phi_{F,r,l,n}^2 \leq 1 + \xi^{-2}\, r^2\, ||H^+||_F^2\, \frac{n}{l-r-1}.$$

Here $||H^+|| = 1$ and $||H^+||_F = \sqrt{l}$ if the matrix H is orthogonal.

Bound the output errors of Algorithms 3 and 4 by combining the estimates of this section and Sect. 3.2 and by transposing an input matrix M.

4.3 Impact of Pre-multiplication in the Case of Gaussian Noise

Next deduce randomized estimates for the impact of pre-multiplication in the case where an input matrix M includes considerable additive white Gaussian noise,[5] which is a classical representation of natural noise in information theory, is widely adopted in signal and image processing, and in many cases properly represents the errors of measurement and rounding (cf. [SST06]).

Theorem 7. *Suppose that two matrices $F \in \mathbb{R}^{k \times m}$ and $H \in \mathbb{R}^{n \times l}$ are orthogonal where $k \geq 2l + 2$, $l \geq 2$ and $k, l < \min(m, n)$, $A \in \mathbb{R}^{m \times n}$, λ_E is a positive scalar,*

$$M = A + E, \quad \frac{1}{\lambda_E} E \stackrel{d}{=} \mathcal{G}^{m \times n}, \tag{10}$$

and $W = I_m - MH(FMH)^+ F$ (cf. (6) for $X = MH$). Then

$$\mathbb{E}\left(\frac{\|W\|_F - \sqrt{m}}{\lambda_E \|M\|_F}\right) \leq \sqrt{\frac{l}{k - 2l - 1}} \quad and \quad \mathbb{E}\left(\frac{\|W\| - 1}{\lambda_E \|M\|}\right) \leq \frac{e\sqrt{k - l}}{k - 2l}. \tag{11}$$

Proof. Assumption (10) and Lemma 5 together imply that FEH is a scaled Gaussian matrix: $\frac{1}{\lambda_E} FEH \stackrel{d}{=} \mathcal{G}^{k \times l}$. Hence $FMH = FAH + \lambda_E G_{k,l}$. Apply Theorem 10 and obtain

$$\mathbb{E} \|(FMH)^+\| \leq \lambda_E \frac{e\sqrt{k - l}}{k - 2l} \quad and \quad \mathbb{E} \|(FMH)^+\|_F \leq \lambda_E \sqrt{\frac{l}{k - 2l - 1}}$$

Recall from (6) that $|W| \leq |I_m| + |(FMH)^+| \, |M|$ since the multipliers F and H are orthogonal, and thus

$$\mathbb{E}|W| \leq |I_m| + |M| \cdot \mathbb{E} |(FMH)^+|.$$

Substitute equations $\|I_m\|_F = \sqrt{m}$ and $\|I_m\| = 1$ and claim (iii) of Theorem 12 and obtain (11).

Remark 5. For $k = l = \rho$, $S = T = I_k$, subpermutation matrices F and H, and a nonsingular matrix FMH, Algorithms 3 and 4 output LRA in the form CUR where $C \in \mathbb{R}^{m \times \rho}$ and $R \in \mathbb{R}^{\rho \times n}$ are two submatrices made up of ρ columns and ρ rows of M and $U = (FMH)^{-1}$. [PLSZa] extends our current study to devising and analyzing algorithms for the computation of such CUR LRA in the case where k and l are arbitrary integers not exceeded by ρ.

---

[5] Additive white Gaussian noise is statistical noise having a probability density function (PDF) equal to that of the Gaussian (normal) distribution.

5 Multiplicative Pre-processing for LRA

We proved that sublinear cost variations of Algorithms 3 and 4 whp output accurate LRA of a random input. In the real world computations input matrices are not random, but we can randomize them by multiplying them by random matrices.

Algorithms 1–4 output accurate LRA whp if such multipliers are Gaussian, SRHT, SRFT or Rademacher's (cf. [HMT11, Sections 10 and 11], [T11]. Multiplication by these matrices runs at a superlinear cost, and our heuristic recipe is to apply these algorithms with a small variety of sparse multipliers F_i and/or H_i, $i = 1, 2, \ldots$, with which computational cost becomes sublinear, and then to monitor the accuracy of the output LRA by applying the criteria of the previous section, [PLa], and/or [PLSZa].

Various families of sparse multipliers have been proposed, extensively tested in [PLSZ16] and [PLSZ17], and turned out to be nearly as efficient as Gaussian multpliers according to these tests. One can readily complement these families with subpermutation matrices and, say, sparse quasi Rademacher's multipliers (see [PLSZa]) and then combine these basic multipliers together into their orthogonalized sums, products or other lower degree polynomials (cf. [HMT11, Remark 4.6]).

Acknowledgements. We were supported by NSF Grants CCF–1116736, CCF–1563942, CCF–1733834 and PSC CUNY Award 69813 00 48.

Appendix

A Background on Matrix Computations

A.1 Some Definitions

- An $m \times n$ matrix M is *orthogonal* if $M^*M = I_n$ or $MM^* = I_m$.
- For $M = (m_{i,j})_{i,j=1}^{m,n}$ and two sets $\mathcal{I} \subseteq \{1, \ldots, m\}$ and $\mathcal{J} \subseteq \{1, \ldots, n\}$, define the submatrices $M_{\mathcal{I},:} := (m_{i,j})_{i\in\mathcal{I};j=1,\ldots,n}$, $M_{:,\mathcal{J}} := (m_{i,j})_{i=1,\ldots,m;j\in\mathcal{J}}$, and $M_{\mathcal{I},\mathcal{J}} := (m_{i,j})_{i\in\mathcal{I};j\in\mathcal{J}}$.
- rank(M) denotes the *rank* of a matrix M.
- $\operatorname{argmin}_{|E| \leq \epsilon|M|} \operatorname{rank}(M + E)$ is the ϵ-rank(M) it is *numerical rank*, nrank(M), if ϵ is small in context.
- Write $\sigma_j(M) = 0$ for $j > r$ and obtain M_r, the *rank-r truncation* of M.
- $\kappa(M) = ||M|| \, ||M^+||$ is the spectral *condition number* of M.

A.2 Auxiliary Results

Next we recall some relevant auxiliary results (we omit the proofs of two well-known lemmas).

Lemma 2 [The norm of the pseudo inverse of a matrix product]. *Suppose that* $A \in \mathbb{R}^{k \times r}$, $B \in \mathbb{R}^{r \times l}$ *and the matrices* A *and* B *have full rank* $r \leq \min\{k, l\}$. *Then* $|(AB)^+| \leq |A^+| \, |B^+|$.

Lemma 3 (The norm of the pseudo inverse of a perturbed matrix, [B15, Theorem 2.2.4]). *If* $\mathrm{rank}(M + E) = \mathrm{rank}(M) = r$ *and* $\eta = ||M^+|| \, ||E|| < 1$, *then*

$$\frac{1}{\sqrt{r}}||(M + E)^+|| \leq ||(M + E)^+|| \leq \frac{1}{1 - \eta}||M^+||.$$

Lemma 4 (The impact of a perturbation of a matrix on its singular values, [GL13, Corollary 8.6.2]). *For* $m \geq n$ *and a pair of* $m \times n$ *matrices* M *and* $M + E$ *it holds that*

$$|\sigma_j(M + E) - \sigma_j(M)| \leq ||E|| \text{ for } j = 1, \dots, n.$$

Theorem 8 (The impact of a perturbation of a matrix on its top singular spaces, [GL13, Theorem 8.6.5]). *Let* $g =: \sigma_r(M) - \sigma_{r+1}(M) > 0$ *and* $||E||_F \leq 0.2g$. *Then for the left and right singular spaces associated with the* r *largest singular values of the matrices* M *and* $M + E$, *there exist orthogonal matrix bases* $B_{r,\mathrm{left}}(M)$, $B_{r,\mathrm{right}}(M)$, $B_{r,\mathrm{left}}(M + E)$, *and* $B_{r,\mathrm{right}}(M + E)$ *such that*

$$\max\{||B_{r,\mathrm{left}}(M+E) - B_{r,\mathrm{left}}(M)||_F, ||B_{r,\mathrm{right}}(M+E) - B_{r,\mathrm{right}}(M)||_F\} \leq \frac{4||E||_F}{g}.$$

For example, if $\sigma_r(M) \geq 2\sigma_{r+1}(M)$, which implies that $g \geq 0.5 \, \sigma_r(M)$, and if $||E||_F \leq 0.1 \, \sigma_r(M)$, then the upper bound on the right-hand side is approximately $8||E||_F/\sigma_r(M)$.

A.3 Gaussian and Factor-Gaussian Matrices of Low Rank and Low Numerical Rank

Lemma 5 [Orthogonal invariance of a Gaussian matrix]. *Suppose that* k, m, *and* n *are three positive integers,* $k \leq \min\{m, n\}$, $G_{m,n} \overset{d}{=} \mathcal{G}^{m \times n}$, $S \in \mathbb{R}^{k \times m}$, $T \in \mathbb{R}^{n \times k}$, *and* S *and* T *are orthogonal matrices. Then* SG *and* GT *are Gaussian matrices.*

Definition 2 [Factor-Gaussian matrices]. *Let* $r \leq \min\{m, n\}$ *and let* $\mathcal{G}_{r,B}^{m \times n}$, $\mathcal{G}_{A,r}^{m \times n}$, *and* $\mathcal{G}_{r,C}^{m \times n}$ *denote the classes of matrices* $G_{m,r}B$, $AG_{r,n}$, *and* $G_{m,r}CG_{r,n}$, *respectively, which we call left, right, and two-sided factor-Gaussian matrices of rank* r, *respectively, provided that* $G_{p,q}$ *denotes a* $p \times q$ *Gaussian matrix,* $A \in \mathbb{R}^{m \times r}$, $B \in \mathbb{R}^{r \times n}$, *and* $C \in \mathbb{R}^{r \times r}$, *and* A, B *and* C *are well-conditioned matrices of full rank* r.

Theorem 9. *The class* $\mathcal{G}_{r,C}^{m \times n}$ *of two-sided* $m \times n$ *factor-Gaussian matrices* $G_{m,r}\Sigma G_{r,n}$ *does not change if in its definition we replace the factor* C *by a well-conditioned diagonal matrix* $\Sigma = (\sigma_j)_{j=1}^r$ *such that* $\sigma_1 \geq \sigma_2 \geq \cdots \geq \sigma_r > 0$.

Proof. Let $C = U_C \Sigma_C V_C^*$ be SVD. Then $A = G_{m,r} U_C \overset{d}{=} \mathcal{G}^{m \times r}$ and $B = V_C^* G_{r,n} \overset{d}{=} \mathcal{G}^{r \times n}$ by virtue of Lemma 5, and so $G_{m,r} C G_{r,n} = A \Sigma_C B$ for $A \overset{d}{=} \mathcal{G}^{m \times r}$, $B \overset{d}{=} \mathcal{G}^{r \times n}$, and A independent from B.

Definition 3. The relative norm of a perturbation of a Gaussian matrix *is the ratio of the perturbation norm and the expected value of the norm of the matrix (estimated in Theorem 11).*

We refer to all three matrix classes above as *factor-Gaussian matrices of rank r*, to their perturbations within a relative norm bound ϵ as *factor-Gaussian matrices of ϵ-rank r*, and to their perturbations within a small relative norm as *factor-Gaussian matrices of numerical rank r* to which we also refer as *perturbations of factor-Gaussian matrices.*

Clearly $||(A\Sigma)^+|| \le ||\Sigma^{-1}|| \, ||A^+||$ and $||(\Sigma B)^+|| \le ||\Sigma^{-1}|| \, ||B^+||$ for a two-sided factor-Gaussian matrix $M = A\Sigma B$ of rank r of Definition 2, and so whp such a matrix is both left and right factor-Gaussian of rank r.

Theorem 10. *Suppose that λ is a positive scalar, $M_{k,l} \in \mathbb{R}^{k \times l}$ and G a $k \times l$ Gaussian matrix for $k - l \ge l + 2 \ge 4$. Then, we have*

$$\mathbb{E} \, ||(M_{k,l} + \lambda G)^+|| \le \frac{\lambda e \sqrt{k-l}}{k - 2l} \quad and \quad \mathbb{E} \, ||(M_{k,l} + \lambda G)^+||_F \le \lambda \sqrt{\frac{l}{k - 2l - 1}}$$

Proof. Let $M_{k,l} = U\Sigma V^*$ be *full SVD* such that $U \in \mathbb{R}^{k \times k}$, $V \in \mathbb{R}^{l \times l}$, U and V are orthogonal matrices, $\Sigma = (D \mid O_{l,k-l})^*$, and D is an $l \times l$ diagonal matrix. Write $W_{k,l} := U^*(M_{k,l} + \lambda G)V$ and observe that $U^* M_{k,l} V = \Sigma$ and $U^* G V = \begin{bmatrix} G_1 \\ G_2 \end{bmatrix}$ is a $k \times l$ Gaussian matrix by virtue of Lemma 5. Hence

$$\sigma_l(W_{k,l}) = \sigma_l\left(\begin{bmatrix} D + \lambda G_1 \\ \lambda G_2 \end{bmatrix} \right) \ge \max\{\sigma_l(D + \lambda G_1), \lambda \sigma_l(G_2)\},$$

and so $|W_{k,l}^+| \le \min\{|(D + \lambda G_1)^+|, |\lambda G_2^+|\}$. Recall that $G_1 \overset{d}{=} \mathcal{G}^{l \times l}$ and $G_2 \overset{d}{=} \mathcal{G}^{k-l \times l}$ are independent, and now Theorem 10 follows because $|(M_{k,l} + \lambda G_{k,l})^+| = |W_{k,l}^+|$ and by virtue of claim (iii) and (iv) of Theorem 12.

A.4 Norms of a Gaussian Matrix and Its Pseudo Inverse

$\Gamma(x) = \int_0^\infty \exp(-t) t^{x-1} dt$ denotes the Gamma function.

Theorem 11 [Norms of a Gaussian matrix. See [DS01, Theorem II.7] and our Definition 1].

(i) Probability$\{\nu_{\text{sp},m,n} > t + \sqrt{m} + \sqrt{n}\} \le \exp(-t^2/2)$ for $t \ge 0$, $\mathbb{E}(\nu_{\text{sp},m,n}) \le \sqrt{m} + \sqrt{n}$.

(ii) $\nu_{F,m,n}$ is the χ-function, with $\mathbb{E}(\nu_{F,m,n}) = mn$ and probability density $\frac{2x^{n-i} \exp(-x^2/2)}{2^{n/2} \Gamma(n/2)}$.

Theorem 12 [Norms of the pseudo inverse of a Gaussian matrix (see Definition 1)].

(i) Probability $\{\nu_{\text{sp},m,n}^+ \geq m/x^2\} < \frac{x^{m-n+1}}{\Gamma(m-n+2)}$ for $m \geq n \geq 2$ and all positive x,

(ii) Probability $\{\nu_{F,m,n}^+ \geq t\sqrt{\frac{3n}{m-n+1}}\} \leq t^{n-m}$ and Probability $\{\nu_{\text{sp},m,n}^+ \geq t\frac{e\sqrt{m}}{m-n+1}\} \leq t^{n-m}$ for all $t \geq 1$ provided that $m \geq 4$,

(iii) $\mathbb{E}((\nu_{F,m,n}^+)^2) = \frac{n}{m-n-1}$ and $\mathbb{E}(\nu_{\text{sp},m,n}^+) \leq \frac{e\sqrt{m}}{m-n}$ provided that $m \geq n+2 \geq 4$,

(iv) Probability $\{\nu_{\text{sp},n,n}^+ \geq x\} \leq \frac{2.35\sqrt{n}}{x}$ for $n \geq 2$ and all positive x, and furthermore $||M_{n,n} + G_{n,n}||^+ \leq \nu_{n,n}$ for any $n \times n$ matrix $M_{n,n}$ and an $n \times n$ Gaussian matrix $G_{n,n}$.

Proof. See [CD05, Proof of Lemma 4.1] for claim (i), [HMT11, Proposition 10.4 and equations (10.3) and (10.4)] for claims (ii) and (iii), and [SST06, Theorem 3.3] for claim (iv).

Theorem 12 implies reasonable probabilistic upper bounds on the norm $\nu_{m,n}^+$ even where the integer $|m - n|$ is close to 0; whp the upper bounds of Theorem 12 on the norm $\nu_{m,n}^+$ decrease very fast as the difference $|m - n|$ grows from 1.

B   Small Families of Hard Inputs for Sublinear Cost LRA

Any sublinear cost LRA algorithm fails on the following small families of LRA inputs.

Example 1. Let $\Delta_{i,j}$ denote an $m \times n$ matrix of rank 1 filled with 0s except for its (i,j)th entry filled with 1. The mn such matrices $\{\Delta_{i,j}\}_{i,j=1}^{m,n}$ form a family of δ-*matrices*. We also include the $m \times n$ null matrix $O_{m,n}$ filled with 0s into this family. Now any fixed sublinear cost algorithm does not access the (i,j)th entry of its input matrices for some pair of i and j. Therefore it outputs the same approximation of the matrices $\Delta_{i,j}$ and $O_{m,n}$, with an undetected error at least $1/2$. Arrive at the same conclusion by applying the same argument to the set of $mn + 1$ small-norm perturbations of the matrices of the above family and to the $mn + 1$ sums of the latter matrices with any fixed $m \times n$ matrix of low rank. Finally, the same argument shows that a posteriori estimation of the output errors of an LRA algorithm applied to the same input families cannot run at sublinear cost.

The example actually covers randomized LRA algorithms as well. Indeed suppose that with a positive constant probability an LRA algorithm does not access K entries of an input matrix with a positive constant probability. Apply this algorithm to two matrices of low rank whose difference at all these K entries is equal to a large constant C. Then, clearly, with a positive constant probability the algorithm has errors at least $C/2$ at at least $K/2$ of these entries. The paper [LPa] shows, however, that accurate LRA of a matrix that admits sufficiently close

LRA can be computed at sublinear cost in two successive Cross-Approximation (C-A) iterations (cf. [GOSTZ10]) provided that we avoid choosing degenerating initial submatrix, which is precisely the problem with the matrix families of Example 1. Thus we readily compute close LRA if we recursively perform C-A iterations and avoid degeneracy at some C-A step.

References

[B15]   Björck, Å.: Numerical Methods in Matrix Computations. TAM, vol. 59. Springer, Cham (2015). https://doi.org/10.1007/978-3-319-05089-8

[CD05]   Chen, Z., Dongarra, J.J.: Condition numbers of Gaussian random matrices, SIAM. J. Matrix Anal. Appl. **27**, 603–620 (2005)

[CLO16] Cichocki, C., Lee, N., Oseledets, I., Phan, A.-H., Zhao, Q., Mandic, D.P.: Tensor networks for dimensionality reduction and large-scale optimization: part 1 low-rank tensor decompositions. Found. Trends® Mach. Learn. **9**(4–5), 249–429 (2016)

[DS01]   Davidson, K.R., Szarek, S.J.: Local operator theory, random matrices, and banach spaces. In: Johnson, W.B., Lindenstrauss, J., (eds.) Handbook on Geometry of Banach Spaces, pp. 317–368, North Holland (2001)

[E88]   Edelman, A.: Eigenvalues and condition numbers of random matrices. SIAM J. Matrix Anal. Appl. **9**(4), 543–560 (1988)

[ES05]   Edelman, A., Sutton, B.D.: Tails of condition number distributions. SIAM J. Matrix Anal. Appl. **27**(2), 547–560 (2005)

[GL13]   Golub, G.H., Van Loan, C.F.: Matrix Computations, fourth edition. The Johns Hopkins University Press, Baltimore (2013)

[GOSTZ10] Goreinov, S., Oseledets, I., Savostyanov, D., Tyrtyshnikov, E., Zamarashkin, N.: How to find a good submatrix. In: *Matrix Methods: Theory, Algorithms, Applications,*(dedicated to the Memory of Gene Golub, edited by V. Olshevsky and E. Tyrtyshnikov), pp. 247–256. World Scientific Publishing, New Jersey (2010)

[HMT11] Halko, N., Martinsson, P.G., Tropp, J.A.: Finding structure with randomness: probabilistic algorithms for constructing approximate matrix decompositions. SIAM Rev. **53**(2), 217–288 (2011)

[KS17]   Kishore Kumar, N., Schneider, J.: Literature survey on low rank approximation of matrices. Linear Multilinear Algebra **65**(11), 2212–2244 (2017). arXiv:1606.06511v1 [math.NA] 21 June 2016

[LPa]   Luan, Q., Pan, V.Y.: CUR LRA at sublinear cost based on volume maximization, In: Salmanig, D. et al. (eds.) MACIS 2019, LNCS 11989, pp. xx–yy. Springer, Switzerland (2020). https://doi.org/10.1007/978-3-030-43120-49. arXiv:1907.10481 (2019)

[LPb]   Luan, Q., Pan, V.Y., Randomized approximation of linear least squares regression at sublinear cost. arXiv:1906.03784, 10 June 2019

[M11]   Mahoney, M.W.: Randomized algorithms for matrices and data. Found. Trends Mach. Learn. **3**, 2 (2011)

[O18]   Osinsky, A.: Rectangular maximum volume and projective volume search algorithms. arXiv:1809.02334, September 2018

[P00]   Pan, C.-T.: On the existence and computation of rank-revealing LU factorizations. Linear Algebra Appl. **316**, 199–222 (2000)

[PLa] Pan, V.Y., Luan, Q.: Refinement of low rank approximation of a matrix at sublinear cost. arXiv:1906.04223, 10 June 2019

[PLSZ16] Pan, V.Y., Luan, Q., Svadlenka, J., Zhao, L.: Primitive and Cynical Low Rank Approximation, Preprocessing and Extensions. arXiv 1611.01391, 3 November 2016

[PLSZ17] Pan, V.Y., Luan, Q., Svadlenka, J., Zhao, L.: Superfast Accurate Low Rank Approximation. Preprint, arXiv:1710.07946, 22 October 2017

[PLSZa] Pan, V.Y., Luan, Svadlenka, Q., Zhao, L.: CUR Low Rank Approximation at Sublinear Cost. arXiv:1906.04112, 10 June 2019

[PLSZb] Pan, V.Y., Luan, Q., Svadlenka, J., Zhao, L.: Low rank approximation at sublinear cost by means of subspace sampling. arXiv:1906.04327, 10 June 2019

[PQY15] Pan, V.Y., Qian, G., Yan, X.: Random multipliers numerically stabilize Gaussian and block Gaussian elimination: proofs and an extension to low-rank approximation. Linear Algebra Appl. **481**, 202–234 (2015)

[PZ17a] Pan, V.Y., Zhao, L.: New studies of randomized augmentation and additive preprocessing. Linear Algebra Appl. **527**, 256–305 (2017)

[PZ17b] Pan, V.Y., Zhao, L.: Numerically safe Gaussian elimination with no pivoting. Linear Algebra Appl. **527**, 349–383 (2017)

[SST06] Sankar, A., Spielman, D., Teng, S.-H.: Smoothed analysis of the condition numbers and growth factors of matrices. SIMAX **28**(2), 446–476 (2006)

[T11] Tropp, J.A.: Improved analysis of subsampled randomized Hadamard transform. Adv. Adapt. Data Anal. **3**(1–2), 115–126 (2011). (Special issue "Sparse Representation of Data and Images")

[TYUC17] Tropp, J.A., Yurtsever, A., Udell, M., Cevher, V.: Practical sketching algorithms for low-rank matrix approximation. SIAM J. Matrix Anal. Appl. **38**, 1454–1485 (2017)

[WLRT08] Woolfe, F., Liberty, E., Rokhlin, V., Tygert, M.: A fast randomized algorithm for the approximation of matrices. Appl. Comput. Harmonic. Anal. **25**, 335–366 (2008)

CUR LRA at Sublinear Cost Based on Volume Maximization

Qi Luan[1] and Victor Y. Pan[2,3(✉)]

[1] Mathematics, The Graduate Center of the City University of New York,
New York, NY 10036, USA
qi_luan@yahoo.com
[2] Computer Science and Mathematics,The Graduate Center of the City
University of New York, New York, NY 10036, USA
[3] Computer Science, Lehman College of the City University of New York,
Bronx, NY 10468, USA
victor.pan@lehman.cuny.edu
http://comet.lehman.cuny.edu/vpan/

Abstract. A matrix algorithm runs at *sublinear cost* if it uses much fewer memory cells and arithmetic operations than the input matrix has entries. Such algorithms are indispensable for Big Data Mining and Analysis, where input matrices are so immense that one can only access a small fraction of all their entries. Typically, however, such matrices admit their Low Rank Approximation *(LRA)*, which one can access and process at sublinear cost. Can, however, we compute LRA at sublinear cost? Adversary argument shows that no algorithm running at sublinear cost can output accurate LRA of worst case input matrices or even of the matrices of small families of our Appendix A, but we prove that some sublinear cost algorithms output a reasonably close LRA of a matrix W if (i) this matrix is sufficiently close to a low rank matrix or (ii) it is a Symmetric Positive Semidefinite *(SPSD)* matrix that admits LRA. In both cases supporting algorithms are deterministic and output LRA in its special form of CUR LRA, particularly memory efficient. The design of our algorithms and the proof of their correctness rely on the results of extensive previous study of CUR LRA in Numerical Linear Algebra using volume maximization. In case (i) we apply Cross-Approximation *(C-A)* iterations, running at sublinear cost and computing accurate LRA worldwide for more than a decade. We provide the first formal support for this long-known empirical efficiency assuming non-degeneracy of the initial submatrix of at least one C-A iteration. We cannot ensure non-degeneracy at sublinear cost for a worst case input but prove that it holds with a high probability *(whp)* for any initialization in the case of a random or randomized input. Empirically we can replace randomization with sparse multiplicative preprocessing of an input matrix, performed at sublinear cost. In case (ii) we make no additional assumptions about the input class of SPSD matrices admitting LRA or about initialization of our sublinear cost algorithms for CUR LRA, which promise to be practically valuable. We hope that proper combination of our deterministic techniques with randomized LRA methods, popular among Computer Science researchers, will lead them to further progress in LRA.

© Springer Nature Switzerland AG 2020
D. Slamanig et al. (Eds.): MACIS 2019, LNCS 11989, pp. 105–121, 2020.
https://doi.org/10.1007/978-3-030-43120-4_10

Keywords: Low Rank Approximation (LRA) · CUR LRA · Sublinear cost · Symmetric Positive Semidefinite (SPSD) matrices · Cross-Approximation (C-A) · Maximal volume

2000 Math. Subject Classification: 65Y20 · 65F30 · 68Q25 · 15A52

1   Introduction

1.1. LRA Problem. An $m \times n$ matrix W admits its close approximation of rank at most r if and only if the matrix W has *numerical rank* at most r (and then we write $\mathrm{nrank}(W) \le r$), that is,

$$W = AB + E, \ ||E||/||W|| \le \epsilon, \tag{1.1}$$

for $A \in \mathbb{C}^{m \times r}$, $B \in \mathbb{C}^{r \times n}$, a matrix norm $|| \cdot ||$, and a small tolerance ϵ. Such an LRA approximates the mn entries of W by using $(m + n)r$ entries of A and B. This is a crucial benefit in applications of LRA to Big Data Mining and Analysis, where the size mn of an input matrix is usually immense, and one can only access a tiny fraction of its mn entries. Quite typically, however, such matrices admit LRA of (1.1) where $(m+n)r \ll mn$. (Hereafter $a \ll b$ and $b \gg a$ mean that the ratio $|a/b|$ is small in context.)

Can we, however, compute close LRA *at sublinear cost*, that is, by using much fewer memory cells and flops than an input matrix has entries? Based on adversary argument one can prove that no algorithm running at sublinear cost can output close LRA of the worst case inputs and even of the matrices of small families of our Appendix A, but for more than a decade Cross-Approximation (C-A) iterations, running at sublinear cost, have been routinely computing close LRA worldwide. Moreover they output LRA in its special form of CUR LRA (see Sect. 2), which is particularly memory efficient and is defined by a proper choice of a submatrix G of W, said to be a generator of CUR LRA or a *CUR generator*.

1.2. Our First Main Result. The main result of Part I of our paper, made up of Sects. 2–5, provides partial formal support for this empirical phenomenon.

Let us elaborate. Let $\sigma_j(M)$ denote the jth largest singular value of a matrix M, which is the minimal distance from M to a matrix of rank $j + 1$ in spectral norm. Suppose that C-A iterations are applied to an $m \times n$ matrix W that admits a sufficiently close LRA (1.1). Let W_i and V_i denote the input and output submatrices of W at the ith C-A iteration for $i = 1, 2, \ldots$ and let $||\cdot||$ denote the spectral or Frobenius matrix norm. Then we prove (see Corollary 3 and Remark 3) that the approximation error norm $||W - V_{i+1}||$ is within a factor f from optimal, which is reasonably bounded unless the ratio $\sigma_r(W_i)/\sigma_r(W)$ is small.

Our proof relies on Theorems 1 and 2, recalled from [OZ18], which extend long study traced back to [CI94, GE96, GTZ97, GT01] and which bound the output errors of CUR LRA in term of maximization of the volume $v_2(G)$ or r-projective volume $v_{2,r}(G)$ of a CUR generator G (see Definition 1 for these concepts).

The ratio $\sigma_r(W_i)/\sigma_r(W)$ is small where one applies C-A iterations to a worst case input matrix, but one can prove that it is not small whp where an input matrix of small numerical rank is random or randomized by means of its pre-

and post-multiplying by random multipliers. Empirically the ratio tends to be not small even where an input matrix of small numerical rank is pre-processed with any fixed rather than random orthogonal multipliers, and in particular at sublinear cost for proper sparse multipliers. The above error factor f can be considered a price for obtaining CUR LRA at sublinear cost, but if the ratio $\sigma_{r+1}(W)/\sigma_r(W)$ is small enough, we can iteratively refine LRA at sublinear cost by means of our algorithms of [PLa].

1.3. Our Results About CUR LRA of SPSD Matrices. Our novel sublinear cost algorithm computes reasonably close CUR LRA of any SPSD matrix admitting LRA. Then again we devise and analyze our algorithm based on the cited link of the error bounds of an output CUR LRA and maximization of the volume or r-projective volume of a CUR generator, and we can reapply our comments on deviation from optimum and iterative refinement of the output.

1.4. Earlier Works. Our results of Part I appeared in [PLSZ16, Section 5] and [PLSZ17, Part II] together with various results on LRA of random input matrices.[1] Our progress in Part II has been inspired by the results of [OZ18] and [CKM19]. Section 1.4 of [LPa] covers relevant earlier works in more details.

1.5. Organization of Our Paper. We define CUR LRA and C-A iterations in the next section. We devote Sect. 3 to background material on matrix volumes, their maximization and its impact on LRA. In Sect. 4 we recall C-A iterations and in Sect. 5 prove that they output reasonably close LRA of a matrix having sufficiently low numerical rank. These sections make up Part I of our paper, while Sects. 6–8 make up its Part II. In Sect. 6 we state our main results for SPSD inputs. We prove the correctness of our algorithms in Sect. 7 and [LPa] and estimate their complexity in Sect. 8. In the Appendix we recall the relevant definitions and auxiliary results and specify small matrix families that are hard for LRA at sublinear cost.

Part I. CUR LRA by Means of C-A Iterations

2 Background: CUR LRA

We use basic definitions for matrix computations recalled in Appendix B. We simplify our presentation by confining it to the case of real matrices, but the extension to the case of complex matrices is straightforward.

   CUR LRA of a matrix W of numerical rank at most r is defined by three matrices C, U, and R, with C and R made up of l columns and k rows of W,

---

[1] The papers [PLSZ16], unsuccessfully submitted to ACM STOC 2017 and widely circulated at that time, and [PLSZ17] provided the first formal support for LRA at sublinear cost, which they called "superfast" LRA. Their approach has extended to LRA the earlier study in [PQY15, PZ17a], and [PZ17b] of randomized Gaussian elimination with no pivoting and other fundamental matrix computations. It was followed by sublinear cost randomized LRA algorithms of [MW17].

respectively, $U \in \mathbb{C}^{l \times k}$ said to be the *nucleus* of CUR LRA,

$$0 < r \leq k \leq m, \ r \leq l \leq n, \ kl \ll mn, \tag{2.1}$$

$$W = CUR + E, \text{ and } ||E||/||W|| \leq \epsilon, \text{ for a small tolerance } \epsilon > 0. \tag{2.2}$$

CUR LRA is a special case of LRA of (1.1) where $k = l = r$ and, say, $A = LU$, $B = R$. Conversely, given LRA of (1.1) one can compute CUR LRA of (2.2) at sublinear cost (see [PLa] and [PLSZa]).

Define a *canonical* CUR LRA as follows.

(i) Fix two sets of columns and rows of W and define its two submatrices C and R made up of these columns and rows, respectively.

(ii) Define the $k \times l$ submatrix $W_{k,l}$ made up of all common entries of C and R, and call it a *CUR generator*.

(iii) Compute its rank-r truncation $W_{k,l,r}$ by setting to 0 all its singular values, except for the r largest ones.

(iv) Compute the Moore–Penrose pseudo inverse $U =: W_{k,l,r}^{+}$ and call it the *nucleus* of CUR LRA of the matrix W (cf. [DMM08, OZ18]); see an alternative choice of a nucleus in [MD09]).

$W_{r,r} = W_{r,r,r}$, and if a CUR generator $W_{r,r}$ is nonsingular, then $U = W_{r,r}^{-1}$.

3 Background: Matrix Volumes

3.1 Definitions and Hadamard's Bound

Definition 1. *For three integers k, l, and r such that $1 \leq r \leq \min\{k, l\}$, define the volume $v_2(M) := \prod_{j=1}^{\min\{k,l\}} \sigma_j(M)$ and r-projective volume $v_{2,r}(M) := \prod_{j=1}^{r} \sigma_j(M)$ of a $k \times l$ matrix M such that $v_{2,r}(M) = v_2(M)$ if $r = \min\{k, l\}$, $v_2^2(M) = \det(MM^*)$ if $k \geq l$; $v_2^2(M) = \det(M^*M)$ if $k \leq l$, $v_2^2(M) = |\det(M)|^2$ if $k = l$.*

Definition 2. *The volume of a $k \times l$ submatrix $W_{\mathcal{I},\mathcal{J}}$ of a matrix W is h-maximal over all $k \times l$ submatrices if it is maximal up to a factor of h. The volume $v_2(W_{\mathcal{I},\mathcal{J}})$ is column-wise (resp. row-wise) h-maximal if it is h-maximal in the submatrix $W_{\mathcal{I},:}$ (resp. $W_{:,\mathcal{J}}$). The volume of a submatrix $W_{\mathcal{I},\mathcal{J}}$ is column-wise (resp. row-wise) locally h-maximal if it is h-maximal over all submatrices of W that differ from the submatrix $W_{\mathcal{I},\mathcal{J}}$ by a single column (resp. single row). Call volume (h_c, h_r)-maximal if it is both column-wise h_c-maximal and row-wise h_r-maximal. Likewise define locally (h_c, h_r)-maximal volume. Write maximal instead of 1-maximal and $(1,1)$-maximal in these definitions. Extend all of them to r-projective volumes.*

For a $k \times l$ matrix $M = (m_{ij})_{i,j=1,1}^{k,l}$ write $\mathbf{m}_j := (m_{ij})_{i=1}^{k}$ and $\bar{\mathbf{m}}_i := ((m_{ij})_{j=1}^{l})^*$ for all i and j. For $k = l = r$ recall *Hadamard's bound*

$$v_2(M) = |\det(M)| \leq \min \left\{ \prod_{j=1}^{r} ||\mathbf{m}_j||, \ \prod_{i=1}^{r} ||\bar{\mathbf{m}}_j^*||, \ r^{r/2} \max_{i,j=1}^{r} |m_{ij}|^r \right\}.$$

3.2   The Impact of Volume Maximization on CUR LRA

The estimates of the two following theorems in the Chebyshev matrix norm $||\cdot||_C$ increased by a factor of $\sqrt{mn}$ turn into estimates in the Frobenius norm $||\cdot||_F$ (see (B.3)).

Theorem 1 [OZ18].[2] *Suppose $r := \min\{k,l\}$, $W_{\mathcal{I},\mathcal{J}}$ is the $k \times l$ CUR generator, $U = W_{\mathcal{I},\mathcal{J}}^+$ is the nucleus of a canonical CUR LRA of an $m \times n$ matrix W, $E = W - CUR$, $h \geq 1$, and the volume of $W_{\mathcal{I},\mathcal{J}}$ is locally h-maximal, that is,*

$$h \, v_2(W_{\mathcal{I},\mathcal{J}}) = \max_B v_2(B)$$

where the maximum is over all $k \times l$ submatrices B of the matrix W that differ from $W_{\mathcal{I},\mathcal{J}}$ in at most one row and/or column. Then

$$||E||_C \leq h \, f(k,l) \, \sigma_{r+1}(W) \quad \text{for} \quad f(k,l) := \sqrt{\frac{(k+1)(l+1)}{|l-k|+1}}.$$

Theorem 2 [OZ18]. *Suppose that $W_{k,l} = W_{\mathcal{I},\mathcal{J}}$ is a $k \times l$ submatrix of an $m \times n$ matrix W, $U = W_{k,l,r}^+$ is the nucleus of a canonical CUR LRA of W, $E = W - CUR$, $h \geq 1$, and the r-projective volume of $W_{\mathcal{I},\mathcal{J}}$ is locally h-maximal, that is,*

$$h \, v_{2,r}(W_{\mathcal{I},\mathcal{J}}) = \max_B v_{2,r}(B)$$

where the maximum is over all $k \times l$ submatrices B of the matrix W that differ from $W_{\mathcal{I},\mathcal{J}}$ in at most one row and/or column. Then

$$||E||_C \leq h \, f(k,l,r) \, \sigma_{r+1}(W) \quad \text{for} \quad f(k,l,r) := \sqrt{\frac{(k+1)(l+1)}{(k-r+1)(l-r+1)}}.$$

Corollary 1. *Suppose that $BW = (BU|BV)$ for a nonsingular matrix B and that the submatrix U is h-maximal in the matrix $W = (U|V)$. Then the submatrix BU is h-maximal in the matrix BW.*

Remark 1. Theorems 1 and 2 have been stated in [OZ18] under assumptions that the matrix $W_{\mathcal{I},\mathcal{J}}$ has (globally) h-maximal volume or r-projective volume, respectively, but their proofs in [OZ18] support the above extensions to the case of locally maximal volume and r-projective volume.

4   C-A Iterations

C-A iterations recursively apply two auxiliary Subalgorithms $\mathcal{A}$ and $\mathcal{B}$ (see Algorithm 1).

---

[2] The theorem first appeared in [GT01, Corollary 2.3] in the special case where $k = l = r$ and $m = n$.

Given a 4-tuple of integers k, l, p, and q such that $r \le k \le p$ and $r \le l \le q$ subalgorithm $\mathcal{A}$ is applied to a $p \times q$ matrix and computes its $k \times l$ submatrix whose volume or projective volume is maximal up to a fixed factor $h \ge 1$ among all its $k \times l$ submatrices. For simplicity first consider the case where $k = l = p = q = r$ (see Fig. 1, borrowed from [PLSZa]).

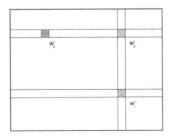

Fig. 1. The three successive C-A steps output three striped matrices.

Subalgorithm $\mathcal{B}$ verifies whether the error norm of the CUR LRA built on a fixed CUR generator is within a fixed tolerance τ (see [PLa] for some verification recipes).

5   CUR LRA by Means of C-A Iterations

We can apply C-A steps by choosing deterministic algorithms of [GE96] for Subalgorithm $\mathcal{A}$. In this case mq and pn memory cells and $O(mq^2)$ and $O(p^2n)$ flops are involved in "vertical" and "horizontal" C-A iterations, respectively. They run at sublinear cost if $p^2 = o(m)$ and $q^2 = o(n)$ and output submatrices having h-maximal volumes for h being a low degree polynomial in $m + n$. Every iteration outputs a matrix that has locally h-maximal volume in a "vertical" or "horizontal" submatrix, and the hope is to obtain globally $\bar{h}$-maximal submatrix (for reasonably bounded $\bar{h}$) when maximization is performed recursively in alternate directions.

Of course, the contribution of C-A step is nil where it is applied to a $p \times q$ input whose volume is 0 or nearly vanishes compared to the target maximum, but the consistent success of C-A iterations in practice suggests that in a small number of loops such a degeneration is regularly avoided.

In the next subsection we show that already two successive C-A iterations output a CUR generator having h-maximal volume (for any $h > 1$) if these iterations begin at a $p \times q$ submatrix of W that shares its rank $r > 0$ with W. By continuity of the volume the result is extended to small perturbations of such matrices within a norm bound estimated in Theorem 13. In Sect. 5.2 we extend these results to the case where r-projective volume rather than the volume of a CUR generator is maximized. (Theorem 2 shows benefits of such a maximization.) In Sect. 5.3 we summarize our study in this section and comment on the estimated and empirical performance of C-A iterations.

5.1   Volume of the Output of a C-A Loop

By comparing SVDs of the matrices W and W^+ obtain the following lemma.

Algorithm 1. C-A Iterations

Input: $W \in \mathbb{C}^{m \times n}$, four positive integers r, k, l, and ITER; a number $\tau > 0$.

Output: A CUR LRA of W with an error norm at most τ or FAILURE.

Initialization: Fix a submatrix W_0 made up of l columns of W and obtain an initial set $\mathcal{I}_0$.

Computations:
for $i = 1, 2, \ldots$, ITER do
  if i is even **then**
    "Horizontal" C-A step:
    1. Let $R_i := W_{\mathcal{I}_{i-1},:}$ be a $p \times n$ submatrix of W.
    2. Apply Subalgorithm $\mathcal{A}$ for $q = n$ to R_i and obtain a $k \times l$
    submatrix $W_i = W_{\mathcal{I}_{i-1}, \mathcal{J}_i}$.
  else
    "Vertical" C-A step:
    1. Let $C_i := W_{:,\mathcal{J}_{i-1}}$ be an $m \times q$ submatrix of W.
    2. Apply Subalgorithm $\mathcal{A}$ for $p = m$ to C_i and obtain a $k \times l$
    submatrix $W_i = W_{\mathcal{I}_i, \mathcal{J}_{i-1}}$.
  end if
  Apply subalgorithm $\mathcal{B}$ and obtain E, the error bound of CUR LRA built
  on the generator W_i.
  if $E \leq \tau$ **then**
    return CUR LRA built on the generator W_i.
  end if
end for
return Failure

Lemma 1. $\sigma_j(W)\sigma_{\mathrm{rank}(W)+1-j}(W^+) = 1$ for all matrices W and all subscripts j, $j \leq \mathrm{rank}(W)$.

Corollary 2. $v_2(W)v_2(W^+) = 1$ and $v_{2,r}(W)v_{2,r}(W_r^+) = 1$ for all matrices W of full rank and all integers r such that $1 \leq r \leq \mathrm{rank}(W)$.

We are ready to prove that a $k \times l$ submatrix of rank r that has (h, h')-locally maximal nonzero volume in a rank-r matrix W has hh'-maximal volume globally in W, that is, over all $k \times l$ submatrices of W.

Theorem 3. Suppose that the volume of a $k \times l$ submatrix $W_{\mathcal{I},\mathcal{J}}$ is nonzero and (h, h')-maximal in a matrix W for $h \geq 1$ and $h' \geq 1$ where $\mathrm{rank}(W) = r = \min\{k, l\}$. Then this volume is hh'-maximal over all its $k \times l$ submatrices of the matrix W.

Proof. The matrix $W_{\mathcal{I},\mathcal{J}}$ has full rank because its volume is nonzero.

Fix any $k \times l$ submatrix $W_{\mathcal{I}',\mathcal{J}'}$ of the matrix W, recall that $W = CUR$, and obtain that

$$W_{\mathcal{I}',\mathcal{J}'} = W_{\mathcal{I}',\mathcal{J}} W_{\mathcal{I},\mathcal{J}}^{+} W_{\mathcal{I},\mathcal{J}'}.$$

If $k \leq l$, then first apply claim (iii) of Theorem 14 for $G := W_{\mathcal{I}',\mathcal{J}}$ and $H := W_{\mathcal{I},\mathcal{J}}^{+}$; then apply claim (i) of that theorem for $G := W_{\mathcal{I}',\mathcal{J}} W_{\mathcal{I},\mathcal{J}}^{+}$ and $H := W_{\mathcal{I},\mathcal{J}'}$ and obtain that

$$v_2(W_{\mathcal{I}',\mathcal{J}}) = v_2(W_{\mathcal{I}',\mathcal{J}} W_{\mathcal{I},\mathcal{J}}^{+} W_{\mathcal{I},\mathcal{J}'}) \leq v_2(W_{\mathcal{I}',\mathcal{J}}) v_2(W_{\mathcal{I},\mathcal{J}}^{+}) v_2(W_{\mathcal{I},\mathcal{J}'}).$$

If $k > l$ deduce the same bound by applying the same argument to the matrix equation

$$W_{\mathcal{I}',\mathcal{J}'}^{T} = W_{\mathcal{I},\mathcal{J}'}^{T} W_{\mathcal{I},\mathcal{J}}^{+T} W_{\mathcal{I}',\mathcal{J}}^{T}.$$

Combine this bound with Corollary 2 for W replaced by $W_{\mathcal{I},\mathcal{J}}$ and deduce that

$$v_2(W_{\mathcal{I}',\mathcal{J}'}) = v_2(W_{\mathcal{I}',\mathcal{J}} W_{\mathcal{I},\mathcal{J}}^{+} W_{\mathcal{I},\mathcal{J}'}) \leq v_2(W_{\mathcal{I}',\mathcal{J}}) v_2(W_{\mathcal{I},\mathcal{J}'})/v_2(W_{\mathcal{I},\mathcal{J}}). \quad (5.1)$$

Recall that the matrix $W_{\mathcal{I},\mathcal{J}}$ is (h, h')-maximal and conclude that

$$h v_2(W_{\mathcal{I},\mathcal{J}}) \geq v_2(W_{\mathcal{I},\mathcal{J}'}) \text{ and } h' v_2(W_{\mathcal{I},\mathcal{J}}) \geq v_2(W_{\mathcal{I}',\mathcal{J}}).$$

Substitute these inequalities into the above bound on the volume $v_2(W_{\mathcal{I}',\mathcal{J}'})$ and obtain that $v_2(W_{\mathcal{I}',\mathcal{J}'}) \leq h h' v_2(W_{\mathcal{I},\mathcal{J}})$.

5.2   From Maximal Volume to Maximal r-Projective Volume

Recall that the CUR LRA error bound of Theorem 1 is strengthened when we shift to Theorem 2, that is, maximize r-projective volume for $r < k = l$ rather than the volume. Next we reduce maximization of r-projective volume of a CUR generators to volume maximization.

Corollary 1 implies the following lemma.

Lemma 2. *Let M and N be a pair of $k \times l$ submatrices of a $k \times n$ matrix and let Q be a $k \times k$ unitary matrix. Then $v_2(M)/v_2(N) = v_2(QM)/v_2(QN)$, and if $r \leq \min\{k, l\}$ then also $v_{2,r}(M)/v_{2,r}(N) = v_{2,r}(QM)/v_{2,r}(QN)$.*

The submatrices R' and $\begin{pmatrix} R' \\ O \end{pmatrix}$ of R of Algorithm 2 have maximal volume and maximal r-projective volume in the matrix R, respectively, by virtue of Theorem 14 and because $v_2(R) = v_{2,r}(R) = v_{2,r}(R')$. Therefore the submatrix $W_{:,\mathcal{J}}$ has maximal r-projective volume in the matrix W by virtue of Lemma 2.

Remark 2. By transposing a horizontal input matrix W and interchanging the integers m and n and the integers k and l we extend the algorithm to computing a $k \times l$ submatrix of maximal or nearly maximal r-projective volume in an $m \times l$ matrix of rank r.

Algorithm 2. From maximal volume to maximal r-projective volume

Input: Four integers k, l, n, and r such that $0 < r \leq k \leq n$ and $r \leq l \leq n$; a $k \times n$ matrix W of rank r; a black box algorithm that finds an $r \times l$ submatrix having locally maximal volume in an $r \times n$ matrix of full rank r.

Output: A column set $\mathcal{J}$ such that $W_{:,\mathcal{J}}$ has maximal r-projective volume in W.

Computations:

1. Compute a rank-revealing QRP factorization $W = QRP$, where Q is a unitary matrix, P is a permutation matrix, $R = \begin{pmatrix} R' \\ O \end{pmatrix}$, and R' is an $r \times n$ matrix. (See [GL13, Sections 5.4.3 and 5.4.4] and [GE96].)

2. Compute an $r \times l$ submatrix $R'_{:,\mathcal{J}}$ of R' having maximal volume.

return $\mathcal{J}'$ such that $P : \mathcal{J}' \longrightarrow \mathcal{J}$.

5.3 Complexity and Accuracy of a Two-Step C-A Loop

The following theorem summarizes our study in this section.

Theorem 4. *Given five integers k, l, m, n, and r such that $0 < r \leq k \leq m$ and $r \leq l \leq n$, suppose that two successive C-A steps (say, based on the algorithms of [GE96]) combined with Algorithm 2 have been applied to an $m \times n$ matrix W of rank r and have output $k \times l$ submatrices W_1' and $W_2' = W_{\mathcal{I}_2,\mathcal{J}_2}$ with nonzero r-projective column-wise locally h-maximal and nonzero r-projective row-wise locally h'-maximal volumes, respectively. Then the submatrix W_2' has $h'h$-maximal r-projective volume in the matrix W.*

By combining Theorems 1, 2, and 4 we obtain the following corollary.

Corollary 3. *Under the assumptions of Theorem 4 apply a two-step C-A loop to an $m \times n$ matrix W of rank r and suppose that both its C-A steps output $k \times l$ submatrices having nonzero r-projective column-wise and row-wise locally h-maximal volumes (see Remark 3 below). Build a canonical CUR LRA on a CUR generator $W_2' = W_{k,l}$ of rank r output by the second C-A step. Then*

(i) the computation of this CUR LRA by using the auxiliary algorithms of [GE96] involves $(m + n)r$ memory cells and $O((m + n)r^2)$ flops[3] and

(ii) the error matrix E of the output CUR LRA satisfies the bound $\|E\|_C \leq g(k,l,r) \, \bar{h} \, \sigma_{r+1}(W)$ for $\bar{h}$ of Theorem 4 and $g(k,l,r)$ denoting the functions $f(k,l)$ of Theorem 1 or $f(k,l,r)$ of Theorem 2. In particular $\|E\|_C \leq 2hh'\sigma_2(W)$ for $k = l = r = 1$.

Remark 3. Theorem 13 enables us to extend Theorem 4 and Corollary 3 to the case of an input matrix W of numerical rank r if the input matrix of any C-A

---

[3] For $r = 1$ an input matrix turns into a vector of dimension m or n, and then we compute its absolutely maximal coordinate just by applying $m - 1$ or $n - 1$ comparisons, respectively (cf. [O17]).

step shares its numerical rank with W. This is fulfilled whp for a random matrix W that admits LRA (see our full paper, arXiv:1907.10481).

Part II. CUR LRA for SPSD Matrices

6   CUR LRA of SPSD Matrices: Two Main Results

For SPSD matrices we can a little improve our estimates of Theorem 13 by applying Wielandt–Hoffman theorem (see [GL13, Theorem 8.6.4]), but we are going to compute reasonably close CUR LRA of an SPSD matrix at sublinear cost with no restriction on its distance from a low rank matrix.

Theorem 5 (Main Result 1). *Suppose that $A \in \mathbb{R}^{n \times n}$ is an SPSD matrix, r and n are two positive integers, $r < n$, ξ is a positive number, and $\mathcal{I}$ is the output of Algorithm 6. Write $C := A_{:,\mathcal{I}}$, $U := A_{\mathcal{I},\mathcal{I}}^{-1}$, and $R := A_{\mathcal{I},:}$. Then*

$$||A - CUR||_C \le (1 + \xi)(r + 1)\sigma_{r+1}(A). \tag{6.1}$$

Furthermore Algorithm 6 runs at an arithmetic cost in $O(nr^4 \log r)$.

Theorem 6 (Main Result 2, *proven in [LPa], due to size limitation for this paper*). *Suppose that $A \in \mathbb{R}^{n \times n}$ is an SPSD matrix, r, K and n are three positive integers such that $r < K < n$, ξ is a positive number, and $\mathcal{I}$ is the output of Algorithm 6. Write $C := A_{:,\mathcal{I}}$, $U := (A_{\mathcal{I},\mathcal{I}})_r^+$, and $R := A_{\mathcal{I},:}$. Then*

$$||A - CUR||_C \le (1 + \xi)\frac{K + 1}{K - r + 1}\sigma_{r+1}(A). \tag{6.2}$$

In particular, let $K = cr - 1$ for $c > 1$. Then

$$||A - CUR||_C \le (1 + \frac{1}{c - 1})(1 + \xi)\sigma_{r+1}(A). \tag{6.3}$$

Furthermore Algorithm 6 runs at an arithmetic cost in $O(r^2 K^4 n + r K^4 n \log n) = O((r + \log n)K^4 n)$, which turns into $O((r + \log n)n^5)$ in case of a constant $c > 1$.

7   Proof of Main Result 1

Theorem 7 *(Adapted from [OZ18, Thm. 6] and [GT01, Thm. 2.1]). Suppose that $W \in \mathbb{R}^{(r+1) \times (r+1)}$,*

$$W = \begin{bmatrix} A & b \\ c^T & d \end{bmatrix},$$

and $A \in \mathbb{R}^{r \times r}$ has maximal volume among all $r \times r$ submatrices of W. Then

$$\frac{v_2(W)}{v_2(A)} \le (1 + r)\sigma_{r+1}(W). \tag{7.1}$$

Algorithm 3. Greedy Column Subset Selection [CM09].

Input: $A \in \mathbb{R}^{m \times n}$ an a positive integer $K < n$.
Output: $\mathcal{I}$.
  Initialize $\mathcal{I} = \{\}$.
  $M^1 \leftarrow A$.
  for t = 1, 2, ..., K **do**
    $i \leftarrow \arg\max_{a \in [n]} ||M^t_{:,a}||$
    $\mathcal{I} \leftarrow \mathcal{I} \cup \{i\}$.
    $M^{t+1} \leftarrow M^t - ||M^t_{:,i}||^{-2}(M^t_{:,i})(M^t_{:,i})^T M^t$
  end for
  return $\mathcal{I}$.

Hereafter $[n]$ denotes the set of n integers $\{1, 2, \ldots, n\}$, and $|\mathcal{T}|$ denotes the cardinality (the number of elements) of a set $\mathcal{T}$.

The theorem is readily deduced from the following result.

Theorem 8 *(Cf. [CKM19]). Suppose that W is an $n \times n$ SPSD matrix and $\mathcal{I}$ and $\mathcal{J}$ are two sets of integers in $[n]$ having the same cardinality. Then $v_2\left(W_{\mathcal{I},\mathcal{J}}\right)^2 \leq v_2\left(W_{\mathcal{I},\mathcal{I}}\right) v_2\left(W_{\mathcal{J},\mathcal{J}}\right)$.*

Theorem 8 shows that the maximal volume submatrix M of an SPSD matrix A can be chosen to be principal. This can be exploited to greatly reduce the cost of searching for the maximal volume submatrix. As pointed out in [CKM19] and implied in [CM09] searching for a maximal volume submatrix in a general matrix or even in an SPSD matrix is NP hard and therefore is impractical for inputs of even moderately large size. [CKM19] proposed to search for a submatrix with a large volume by means of algorithm that is equivalent to **Gaussian Elimination with Complete Pivoting** (Algorithm 4). Such a submatrix, however, only guarantees an upper bound of $4^r \sigma_{r+1}(A)$ on the Chebyshev error norm for the output CUR LRA (see the definition of Chebyshev's norm in Appendix B).

Next we seek a principal submatrix $A_{\mathcal{I},\mathcal{I}}$ having maximal volume in every matrix $A_{\mathcal{S},\mathcal{S}}$ such that $\mathcal{S} \supset \mathcal{I}$ and $|\mathcal{S}| = |\mathcal{I}| + 1$. Such a submatrix generates a CUR LRA with Chebyshev error norm bound $(r + 1)\sigma_{r+1}(A)$, thus considerably improving the aforementioned exponential bound. According to the following theorem, we arrive at such a submatrix $A_{\mathcal{I},\mathcal{I}}$ by recursively replacing a single index in an initial set $\mathcal{I}$.

Theorem 9. *Suppose that $A \in \mathbb{R}^{n \times n}$ is an SPSD matrix, $\mathcal{I}$ is an index set, and $0 < |\mathcal{I}| = r < n$. Let $v_2\left(A_{\mathcal{I},\mathcal{I}}\right) \geq v_2\left(A_{\mathcal{J},\mathcal{J}}\right)$ for any index set $\mathcal{J}$ where $|\mathcal{J}| = r$, and $\mathcal{J}$ only differs from $\mathcal{I}$ at a single element. Then $A_{\mathcal{I},\mathcal{I}}$ is a maximal volume submatrix of $A_{\mathcal{S},\mathcal{S}}$ for any superset $\mathcal{S}$ of $\mathcal{I}$ lying in $[n]$ and such that $|\mathcal{S}| = r + 1$.*

Proof. Apply [CKM19] Thm. 1 to such an SPSD matrix $A_{\mathcal{S},\mathcal{S}}$ and obtain that there exists a subset $\mathcal{I}'$ of $\mathcal{S}$ such that $|\mathcal{I}'| = r$ and $A_{\mathcal{I}',\mathcal{I}'}$ is a maximal volume submatrix of $A_{\mathcal{S},\mathcal{S}}$. $v_2\left(A_{\mathcal{I},\mathcal{I}}\right) \geq v_2\left(A_{\mathcal{I}',\mathcal{I}'}\right)$ since $\mathcal{I}'$ and $\mathcal{I}$ differs at most at a single element, and this proves the theorem.

Algorithm 4. An SPSD Matrix: Gaussian Elimination with Complete
Pivoting (cf. [B00] and [CKM19]).

Input: An SPSD matrix $A \in \mathbb{R}^{n \times n}$ and a positive integer $K < n$.
Output: $\mathcal{I}$.
   Initialize $R \leftarrow A$, and $\mathcal{I} = \{\}$.
   for t = 1, 2, ..., K **do**
      $i_t \leftarrow \arg \max_{j \in [n]} |R_{j,j}|$.
      $\mathcal{I} \leftarrow \mathcal{I} \cup \{i_t\}$.
      $R \leftarrow R - R_{:,i_t} \cdot r_{i_t,i_t}^{-1} \cdot R_{i_t,:}$.
   end for
   return $\mathcal{I}$.

The papers [GT01] and [OZ18] have considerably relaxed the condition that
the generator $A_{\mathcal{I},\mathcal{I}}$ is a maximal volume submatrix: if $v_2 (A_{\mathcal{I},\mathcal{I}})$ is increased by a
factor of $h > 1$ from maximal, then the error bound only increases by at most the
same factor of h. In the case of SPSD inputs, we extend this relaxation further to
$A_{\mathcal{I},\mathcal{I}}$ having **close-to-maximal** volume among "nearby" principal submatrices.

Theorem 10. *For an SPSD matrix $A \in \mathbb{R}^{n \times n}$, a positive integer $r < n$, and
a positive number ξ, let $\mathcal{I} \subset [n]$ be an index set and let $|\mathcal{I}| = r$. Suppose that
$(1 + \xi) v_2 (A_{\mathcal{I},\mathcal{I}}) \geq v_2 (A_{\mathcal{J},\mathcal{J}})$ for any subset $\mathcal{J}$ of $[n]$ such that $|\mathcal{J}| = r$ and $\mathcal{J}$
differs from $\mathcal{I}$ at one element. Then*

$$\|A - A_{:,\mathcal{I}} A_{\mathcal{I},\mathcal{I}}^{-1} A_{\mathcal{I},:}\|_C \leq (1 + \xi)(r + 1)\sigma_{r+1}(A). \qquad (7.2)$$

If $v_2 (A_{\mathcal{I},\mathcal{I}})$ is increased by at most a factor of $1 + \xi$ each time when we replace
an index in $\mathcal{I}$, then Algorithm 6 would not run into infinite loop due to rounding
to machine precision. Furthermore, Theorem 10 guarantees that the accuracy is
mostly preserved, that is, upon termination, the returned index set $\mathcal{I}$ satisfies
inequality (7.2).

Let t denote the number of times a single index in $\mathcal{I}$ is replaced. In the
following, we show that t is bounded by $O(r \log r)$, if the initial set $\mathcal{I}_0$ is greedily
chosen in Algorithm 3.

Theorem 11 *(Adapted from [CM09] Thm. 10). For a matrix $C \in \mathbb{R}^{m \times n}$ and a
positive integer $r < n$, let Algorithm 3 with input C and r output a set $\mathcal{I}$. Then*

$$v_2 (C_{:,\mathcal{I}}) \geq \frac{1}{r!} \max_{\mathcal{S} \subset [n]:|\mathcal{S}|=r} v_2 (C_{:,\mathcal{S}}). \qquad (7.3)$$

Theorem 12. *For an SPSD matrix $A \in \mathbb{R}^{n \times n}$ and a positive integer $r < n$, let
Algorithm 4 with inputs A and r output a set $\mathcal{I}$. Then*

$$v_2 (A_{\mathcal{I},\mathcal{I}}) \geq \frac{1}{(r!)^2} \max_{\mathcal{S} \subset [n]:|\mathcal{S}|=r} v_2 (A_{\mathcal{S},\mathcal{S}}). \qquad (7.4)$$

Corollary 4. *For an SPSD matrix $A \in \mathbb{R}^{n \times n}$, a positive integer $r < n$, and a
positive number ξ, Algorithm 6 calls Algorithm 5 at most $O(r \log r)$ times.*

Algorithm 5. Index Swap

Input: An SPSD matrix $A \in \mathbb{R}^{n \times n}$, a set $\mathcal{I} \in [n]$, a positive integer $r \leq |\mathcal{I}|$, and a positive number ξ.

Output: $\mathcal{J}$

    Compute $v_{2,r}(A_{\mathcal{I},\mathcal{I}})$

    for all $i \in \mathcal{I}$ **do**

      $\mathcal{I}' \leftarrow \mathcal{I} - \{i\}$

      for all $j \in [n] - \mathcal{I}$ **do**

        $\mathcal{J} \leftarrow \mathcal{I}' \cup \{j\}$

        Compute $v_{2,r}(A_{\mathcal{J},\mathcal{J}})$

        if $v_{2,r}(A_{\mathcal{J},\mathcal{J}})/v_{2,r}(A_{\mathcal{I},\mathcal{I}}) > 1 + \xi$ **then**

          return $\mathcal{J}$

        end if

      end for

    end for

    return $\mathcal{I}$

---

8   Complexity Analysis

In this section, we estimate the time complexity of performing the Main Algorithm (Algorithm 6) in the case of both $r = K$ and $r < K$. The cost of finding the initial set $\mathcal{I}_0$ by means of Algorithm 4 is $O(nK^2)$. Let t denote the number of iterations and let $c(r, K)$ denote the arithmetic cost of performing Algorithm 5 with parameters r and K. Then the complexity is in $O(nK^2 + t \cdot c(r, K))$.

In the case of $r = K$, Corollary 4 implies that $t = O(r \log r)$. Algorithm 5 may need up to nr comparisons of $v_2(A_{\mathcal{I},\mathcal{I}})$ and $v_2(A_{\mathcal{J},\mathcal{J}})$. Since $\mathcal{I}$ and $\mathcal{J}$ differs at most at one index, we compute $v_2(A_{\mathcal{J},\mathcal{J}})$ faster by using small rank update of

---

Algorithm 6. Main Algorithm

Input: An SPSD matrix $A \in \mathbb{R}^{n \times n}$, two positive integers K and r such that $r \leq K < n$, and a positive number ξ.

Output: $\mathcal{I}$

    $\mathcal{I} \leftarrow$ Algorithm $4(A, K)$

    while TRUE do

      $\mathcal{J} \leftarrow$ Algorithm $5(A, \mathcal{I}, r, \xi)$

      if $\mathcal{J} = \mathcal{I}$ **then**

        BREAK

      else

        $\mathcal{I} \leftarrow \mathcal{J}$

      end if

    end while

    return $\mathcal{I}$.

$A_{\mathcal{I},\mathcal{I}}$ instead of computing from the scratch; this saves a factor of k. Therefore $c(r,r) = O(r^3 n)$, and the time complexity of the Main Algorithm is $O(nr^4 \log r)$.

In the case of $r < K$, according to [GE96, Theorem 7.2] and [CM09, Theorem 10], t increases slightly to $O(r^2 + r \log n)$, and if $v_{2,r}(A_{\mathcal{J},\mathcal{J}})$ is computed by using SVD, then $c(r, K) = O(K^4 n)$, and the time complexity of the Main Algorithm is $O(r^2 K^4 n + r K^4 n \log n)$.

Acknowledgements. Our research has been supported by NSF Grants CCF-1116736, CCF-1563942, and CCF-133834 and PSC CUNY Award 69813 00 48. We also thank A. Cortinovis, A. Osinsky, N. L. Zamarashkin for pointers to their papers [CKM19] and [OZ18], S. A. Goreinov for reprints, of his papers, and E. E. Tyrtyshnikov for pointers to the bibliography and the challenge of formally supporting empirical power of C-A algorithms.

Appendix

A   Small Families of Hard Inputs for Sublinear Cost LRA

Any sublinear cost LRA algorithm fails on the following small input families.

Example 1. Define a family of $m \times n$ matrices of rank 1 (we call them δ-*matrices*):

$$\{\Delta_{i,j}, \ i = 1, \ldots, m; \ j = 1, \ldots, n\}.$$

Also include the $m \times n$ null matrix $O_{m,n}$ into this family. Now fix any sublinear cost algorithm; it does not access the (i,j)th entry of its input matrices for some pair of i and j. Therefore it outputs the same approximation of the matrices $\Delta_{i,j}$ and $O_{m,n}$, with an undetected error at least $1/2$. Apply the same argument to the set of $mn + 1$ small-norm perturbations of the matrices of the above family and to the $mn + 1$ sums of the latter matrices with any fixed $m \times n$ matrix of low rank. Finally, the same argument shows that a posteriori estimation of the output errors of an LRA algorithm applied to the same input families cannot run at sublinear cost.

This example actually covers randomized LRA algorithms as well. Indeed suppose that with a positive constant probability an LRA algorithm does not access K entries of an input matrix. Apply this algorithm to two matrices of low rank whose difference at all these K entries is equal to a large constant C. Then, clearly, with a positive constant probability the algorithm has errors at least $C/2$ at at least $K/2$ of these entries.

B   Definitions for Matrix Computations and a Lemma

Next we recall some basic definitions for matrix computations (cf. [GL13]).

$\mathbb{C}^{m \times n}$ is the class of $m \times n$ matrices with complex entries.

I_s denotes the $s \times s$ identity matrix. $O_{q,s}$ denotes the $q \times s$ matrix filled with zeros.

$\mathrm{diag}(B_1, \ldots, B_k) = \mathrm{diag}(B_j)_{j=1}^{k}$ denotes a $k \times k$ block diagonal matrix with diagonal blocks $B_1, \ldots, B_k$.

$(B_1 \mid \ldots \mid B_k)$ and $(B_1, \ldots, B_k)$ denote a $1 \times k$ block matrix with blocks $B_1, \ldots, B_k$.

W^T and W^* denote the transpose and the Hermitian transpose of an $m \times n$ matrix $W = (w_{ij})_{i,j=1}^{m,n}$, respectively. $W^* = W^T$ if the matrix W is real.

For two sets $\mathcal{I} \subseteq \{1, \ldots, m\}$ and $\mathcal{J} \subseteq \{1, \ldots, n\}$ define the submatrices

$$W_{\mathcal{I},:} := (w_{i,j})_{i \in \mathcal{I}; j=1,\ldots,n}, W_{:,\mathcal{J}} := (w_{i,j})_{i=1,\ldots,m; j \in \mathcal{J}}, \ W_{\mathcal{I},\mathcal{J}} := (w_{i,j})_{i \in \mathcal{I}; j \in \mathcal{J}}. \tag{B.1}$$

An $m \times n$ matrix W is *unitary* (also *orthogonal* when real) if $W^*W = I_n$ or $WW^* = I_m$.

Compact SVD of a matrix W, hereafter just *SVD*, is defined by the equations

$$W = S_W \Sigma_W T_W^*,$$
$$\text{where } S_W^* S_W = T_W^* T_W = I_\rho, \ \Sigma_W := \mathrm{diag}(\sigma_j(W))_{j=1}^{\rho}, \ \rho = \mathrm{rank}(W), \tag{B.2}$$

$\sigma_j(W)$ denotes the jth largest singular value of W for $j = 1, \ldots, \rho$; $\sigma_j(W) = 0$ for $j > \rho$.

$||W|| = ||W||_2$, $||W||_F$, and $||W||_C$ denote spectral, Frobenius, and Chebyshev norms of a matrix W, respectively, such that (see [GL13, Section 2.3.2 and Corollary 2.3.2])

$$||W|| = \sigma_1(W), \ ||W||_F^2 := \sum_{i,j=1}^{m,n} |w_{ij}|^2 = \sum_{j=1}^{\mathrm{rank}(W)} \sigma_j^2(W), \ ||W||_C := \max_{i,j=1}^{m,n} |w_{ij}|,$$

$$||W||_C \le ||W|| \le ||W||_F \le \sqrt{mn} \, ||W||_C, \ ||W||_F^2 \le \min\{m,n\} \, ||W||^2. \tag{B.3}$$

$W^+ := T_W \Sigma_W^{-1} S_W^*$ is the Moore–Penrose pseudo inverse of an $m \times n$ matrix W.

$$||W^+|| \sigma_r(W) = 1 \tag{B.4}$$

for a full rank matrix W.

A matrix W has ϵ-*rank* at most $r > 0$ for a fixed tolerance $\epsilon > 0$ if there is a matrix W' of rank r such that $||W' - W|| / ||W|| \le \epsilon$. We write $\mathrm{nrank}(W) = r$ and say that a matrix W has *numerical rank* r if it has ϵ-rank r for a small ϵ.

Lemma 3. *Let* $G \in \mathbb{C}^{k \times r}$, $\Sigma \in \mathbb{C}^{r \times r}$ *and* $H \in \mathbb{C}^{r \times l}$ *and let the matrices* G, H *and* Σ *have full rank* $r \le \min\{k, l\}$. *Then* $||(G \Sigma H)^+|| \le ||G^+|| \, ||\Sigma^+|| \, ||H^+||$.

A proof of this well-known result is included in [LPa].

C The Volume and r-Projective Volume of a Perturbed Matrix

Theorem 13. *Suppose that* W' *and* E *are* $k \times l$ *matrices,* $\mathrm{rank}(W') = r \le \min\{k, l\}$, $W = W' + E$, *and* $||E|| \le \epsilon$. *Then*

$$\left(1 - \frac{\epsilon}{\sigma_r(W)}\right)^r \le \prod_{j=1}^{r}\left(1 - \frac{\epsilon}{\sigma_j(W)}\right) \le \frac{v_{2,r}(W)}{v_{2,r}(W')} \le \prod_{j=1}^{r}\left(1 + \frac{\epsilon}{\sigma_j(W)}\right) \le \left(1 + \frac{\epsilon}{\sigma_r(W)}\right)^r. \tag{C.1}$$

If $\min\{k,l\} = r$, then $v_2(W) = v_{2,r}(W)$, $v_2(W') = v_{2,r}(W')$, and

$$\left(1 - \frac{\epsilon}{\sigma_r(W)}\right)^r \le \frac{v_2(W)}{v_2(W')} = \frac{v_{2,r}(W)}{v_{2,r}(W')} \le \left(1 + \frac{\epsilon}{\sigma_r(W)}\right)^r. \qquad (C.2)$$

Proof. Bounds (C.1) follow because a perturbation of a matrix within a norm bound ϵ changes its singular values by at most ϵ (see [GL13, Corollary 8.6.2]). Bounds (C.2) follow because $v_2(M) = v_{2,r}(M) = \prod_{j=1}^r \sigma_j(M)$ for any $k \times l$ matrix M with $\min\{k,l\} = r$, in particular for $M = W'$ and $M = W = W' + E$.

If the ratio $\frac{\epsilon}{\sigma_r(W)}$ is small, then $\left(1 - \frac{\epsilon}{\sigma_r(W)}\right)^r = 1 - O\left(\frac{r\epsilon}{\sigma_r(W)}\right)$ and $\left(1 + \frac{\epsilon}{\sigma_r(W)}\right)^r = 1 + O\left(\frac{r\epsilon}{\sigma_r(W)}\right)$, which shows that the relative perturbation of the volume is amplified by at most a factor of r in comparison to the relative perturbation of the r largest singular values.

D   The Volume and r-Projective Volume of a Matrix Product

Theorem 14 *(Cf. [OZ18]).* [Examples 2 and 3 below show some limitations on the extension of the theorem.]

Suppose that $W = GH$ for an $m \times q$ matrix G and a $q \times n$ matrix H. Then

(i) $v_2(W) = v_2(G)v_2(H)$ if $q = \min\{m,n\}$; $v_2(W) = 0 \le v_2(G)v_2(H)$ if $q < \min\{m,n\}$.

(ii) $v_{2,r}(W) \le v_{2,r}(G)v_{2,r}(H)$ for $1 \le r \le q$,

(iii) $v_2(W) \le v_2(G)v_2(H)$ if $m = n \le q$.

Example 2. If G and H are unitary matrices and if $GH = O$, then $v_2(G) = v_2(H) = v_{2,r}(G) = v_{2,r}(H) = 1$ and $v_2(GH) = v_{2,r}(GH) = 0$ for all $r \le q$.

Example 3. If $G = (1 \mid 0)$ and $H = \mathrm{diag}(1,0)$, then $v_2(G) = v_2(GH) = 1$ and $v_2(H) = 0$.

References

[B00]   Bebendorf, M.: Approximation of boundary element matrices. Numer. Math. **86**(4), 565–589 (2000)

[CI94]   Chandrasekaran, S., Ipsen, I.: On rank revealing QR factorizations. SIAM J. Matrix Anal. Appl. **15**, 592–622 (1994)

[CKM19]   Cortinovis, A., Kressner, D., Massei, S.: MATHICSE technical report: on maximum volume submatrices and cross approximation for symmetric semidefinite and diagonally dominant matrices. MATHICSE, 12 February 2019

[CM09]   Çivril, A., Magdon-Ismail, M.: On selecting a maximum volume submatrix of a matrix and related problems. Theor. Comput. Sci. **410**(47–49), 4801–4811 (2009)

[DMM08] Drineas, P., Mahoney, M.W., Muthukrishnan, S.: Relative-error CUR matrix decompositions. SIAM J. Matrix Anal. Appl. **30**(2), 844–881 (2008)

[GE96] Gu, M., Eisenstat, S.C.: An efficient algorithm for computing a strong rank revealing QR factorization. SIAM J. Sci. Comput. **17**, 848–869 (1996)

[GL13] Golub, G.H., Van Loan, C.F.: Matrix Computations, 4th edn. The Johns Hopkins University Press, Baltimore (2013)

[GT01] Goreinov, S.A., Tyrtyshnikov, E.E.: The maximal-volume concept in approximation by low rank matrices. Contemp. Math. **208**, 47–51 (2001)

[GTZ97] Goreinov, S.A., Tyrtyshnikov, E.E., Zamarashkin, N.L.: A theory of pseudo-skeleton approximations. Linear Algebra Appl. **261**, 1–21 (1997)

[LPa] Luan, Q., Pan, V.Y.: Low rank approximation of a matrix at sublinear cost, 21 July 2019. arXiv:1907.10481

[MD09] Mahoney, M.W., Drineas, P.: CUR matrix decompositions for improved data analysis. Proc. Natl. Acad. Sci. USA **106**, 697–702 (2009)

[MW17] Musco, C., Woodruff, D.P.: Sublinear time low-rank approximation of positive semidefinite matrices. In: IEEE 58th FOCS, pp. 672–683 (2017)

[O17] Osinsky, A.I.: Probabilistic estimation of the rank 1 cross approximation accuracy, submitted on 30 June 2017. arXiv:1706.10285

[OZ18] Osinsky, A.I., Zamarashkin, N.L.: Pseudo-skeleton approximations with better accuracy estimates. Linear Algebra Appl. **537**, 221–249 (2018)

[PLa] Pan, V.Y., Luan, Q.: Refinement of low rank approximation of a matrix at sub-linear cost, submitted on 10 June 2019. arXiv:1906.04223

[PLSZ16] Pan, V.Y., Luan, Q., Svadlenka, J., Zhao, L.: Primitive and cynical low rank approximation, preprocessing and extensions, submitted on 3 November 2016. arXiv:1611.01391v1

[PLSZ17] Pan, V.Y., Luan, Q., Svadlenka, J., Zhao, L.: Superfast accurate approximation of low rank matrices, submitted on 22 October 2017. arXiv:1710.07946v1

[PLSZa] Pan, V.Y., Luan, Q., Svadlenka, J., Zhao, L.: CUR low rank approximation at sub-linear cost, submitted on 10 June 2019. arXiv:1906.04112

[PQY15] Pan, V.Y., Qian, G., Yan, X.: Random multipliers numerically stabilize Gaussian and block Gaussian elimination: proofs and an extension to low-rank approximation. Linear Algebra Appl. **481**, 202–234 (2015)

[PZ17a] Pan, V.Y., Zhao, L.: New studies of randomized augmentation and additive preprocessing. Linear Algebra Appl. **527**, 256–305 (2017)

[PZ17b] Pan, V.Y., Zhao, L.: Numerically safe Gaussian elimination with no pivoting. Linear Algebra Appl. **527**, 349–383 (2017)

New Practical Advances in Polynomial Root Clustering

Rémi Imbach[1](✉) and Victor Y. Pan[2]

[1] Courant Institute of Mathematical Sciences, New York University,
New York, USA
remi.imbach@nyu.edu
[2] Lehman College and the Graduate Center, City University of New York,
New York, USA
victor.pan@lehman.cuny.edu
https://cims.nyu.edu/~imbach/, http://comet.lehman.cuny.edu/vpan/

Abstract. We report an ongoing work on clustering algorithms for complex roots of a univariate polynomial p of degree d with real or complex coefficients. As in their previous best subdivision algorithms our root-finders are robust even for multiple roots of a polynomial given by a black box for the approximation of its coefficients, and their complexity decreases at least proportionally to the number of roots in a region of interest (ROI) on the complex plane, such as a disc or a square, but we greatly strengthen the main ingredient of the previous algorithms. We build the foundation for a new counting test that essentially amounts to the evaluation of a polynomial p and its derivative p', which is a major benefit, e.g., for sparse polynomials p. Moreover with evaluation at about $\log(d)$ points (versus the previous record of order d) we output correct number of roots in a disc whose contour has no roots of p nearby. Our second and less significant contribution concerns subdivision algorithms for polynomials with real coefficients. Our tests demonstrate the power of the proposed algorithms.

1 Introduction

We seek complex roots of a degree d univariate polynomial p with real or complex coefficients. For a while the user choice for this problem has been (the package MPsolve) based on *e.g.* Erhlich-Aberth (simultaneous Newton-like) iterations. Their empirical global convergence (right from the start) is very fast, but its formal support is a long-known challenge, and the iterations approximate the roots in a fixed region of interest (ROI) about as slow as all complex roots.

In contrast, for the known algorithms subdividing a ROI, e.g., box, the cost of root-finding in a ROI decreases at least proportionally to the number of roots in it. Some recent subdivision algorithms have a proved nearly optimal complexity,

Rémi's work is supported by NSF Grants # CCF-1563942 and # CCF-1708884.
Victor's work is supported by NSF Grants # CCF-1116736 and # CCF-1563942 and by PSC CUNY Award 698130048.

© Springer Nature Switzerland AG 2020
D. Slamanig et al. (Eds.): MACIS 2019, LNCS 11989, pp. 122–137, 2020.
https://doi.org/10.1007/978-3-030-43120-4_11

are robust in the case of root clusters and multiple roots, and their implementation in [IPY18] a little outperforms MPsolve for ROI containing only a small number of roots, which is an important benefit in many computational areas.

The Local Clustering Problem. For a complex set $\mathcal{S}$, $\text{Zero}(\mathcal{S}, p)$, or sometimes $\text{Zero}(\mathcal{S})$, stands for the roots of p in $\mathcal{S}$. $\#(\mathcal{S}, p)$ (or $\#(\mathcal{S})$) stands for the number of roots of p in $\mathcal{S}$. Here and hereafter the roots are counted with their multiplicity.

We consider boxes (that is, squares with horizontal and vertical edges, parallel to coordinate axis) and discs $D(c, r) = \{z \text{ s.t. } |z - c| \le r\}$ on the complex plane. For such a box (resp. disc) $\mathcal{S}$ and a positive δ we denote by $\delta\mathcal{S}$ its concentric δ-dilation. We call a disc Δ an *isolator* if $\#(\Delta) > 0$ and call it *natural* isolator if in addition $\#(\Delta) = \#(3\Delta)$. A set $\mathcal{R}$ of roots of p is called a *natural cluster* if there exists a natural isolator Δ with $\text{Zero}(\mathcal{R}) = \text{Zero}(\Delta)$. The Local Clustering Problem (LCP) is to compute natural isolators for natural clusters together with the sum of multiplicities of roots in the clusters:

Local Clustering Problem (LCP):
Given: a polynomial $p \in \mathbb{C}[z]$, a ROI $B_0 \subset \mathbb{C}$, $\epsilon > 0$
Output: a set of pairs $\{(\Delta^1, m^1), \ldots, (\Delta^\ell, m^\ell)\}$ where:
  - the Δ^j's are pairwise disjoint discs of radius $\le \epsilon$,
  - $m^j = \#(\Delta^j, p) = \#(3\Delta^j, p)$ and $m^j > 0$ for $j = 1, \ldots, \ell$,
  - $\text{Zero}(B_0, p) \subseteq \bigcup_{j=1}^\ell \text{Zero}(\Delta^j, p) \subseteq \text{Zero}(2B_0, p)$.

The basic tool of the nearly optimal subdivision algorithm of [BSS+16] for the LCP is the T^*-*test* for counting the roots of p in a complex disc (with multiplicity). It relies on Pellet's theorem, involves approximations of the coefficients of p, and applies shifting and scaling the variable z and Dandelin-Gräffe's root-squaring iterations. [IPY18] describes high-level improvement of this test, and Ccluster[1], a C implementation of [BSS+16].

Our Contributions. Our new counting test, the P^*-*test*, for a pair of complex c and positive r computes the number s_0 of roots of p in a complex disc Δ centered at c with radius r. If the boundary $\partial\Delta$ contains no roots of p, then

$$s_0 = \frac{1}{2\pi\mathbf{i}} \int_{\partial\Delta} \frac{p'(z)}{p(z)} dz, \text{ for } \mathbf{i} = \sqrt{-1}, \tag{1}$$

by virtue of Cauchy's theorem. By following [Sch82] and [Pan18], we approximate s_0 by s_0^* obtained by evaluating p'/p on q points on the boundary $\partial\Delta$ within the error bound $|s_0 - s_0^*|$ in terms of q and the relative width of a root-free annulus around $\partial\Delta$. Namely if $\#(\frac{1}{2}\Delta) = \#(2\Delta)$ then for $q = \lceil \log_2(d+4) + 2 \rceil$ we recover exact value of s_0 from s_0^*.

---

[1] https://github.com/rimbach/Ccluster.

Table 1. Running times in seconds of `Ccluster`, new and old versions, for computing clusters of roots in a small ROI (local) and a ROI containing all the roots, and `MPsolve`.

	Ccluster local				Ccluster global				MPsolve
	#Clus	t_{old}	t_{new}	t_{old}/t_{new}	#Clus	t_{old}	t_{new}	t_{old}/t_{new}	t
Mignotte$_{128}$	1	0.05	0.02	2.49	127	5.00	1.81	2.75	0.02
Mignotte$_{256}$	1	0.16	0.05	2.82	255	31.8	10.7	2.95	0.07
Mignotte$_{383}$	1	0.32	0.11	2.74	382	79.7	26.8	2.97	0.17
Mandelbrot$_7$	1	0.18	0.06	2.92	127	7.17	2.88	2.48	0.06
Mandelbrot$_8$	0	0.39	0.11	3.38	255	40.6	15.1	2.69	0.39
Mandelbrot$_9$	5	3.08	0.91	3.37	511	266	97.1	2.74	3.20

A usual practice to ensure condition $\#(\frac{1}{2}\Delta) = \#(2\Delta)$ when knowing a $\rho > 1$ so that $\#(\frac{1}{\rho}\Delta) = \#(\rho\Delta)$ is to apply Dandelin-Gräffe's iterations. In the test we propose here, such root-squaring operations can be applied implicitly by doubling the number q of evaluation points.

We give an *effective*[2] (*i.e.* implementable) description of this P^*-test, which involves no coefficients of p and can be applied to a polynomial p represented by a black box for its evaluation. For sparse polynomials and polynomials defined by recursive process such as Mandelbrot's polynomials (see [BF00], or Eq. (3) below), the test is particularly efficient and the resulting acceleration of the clustering algorithm of [BSS+16] is particularly strong.

Our second (and less significant) contribution applies to polynomials with real coefficients: the roots of such polynomials are either real or appear in complex conjugated pairs. As a consequence, one can recover all the roots in a ROI B_0 containing $\mathbb{R}$ from the ones with positive imaginary parts. We show how to improve a subdivision scheme by leveraging of the latter property.

Every polynomial p and its product $p\overline{p}$ with its complex conjugate $\overline{p}$ belongs to this class and has additional property that the multiplicity of its real roots is even, but we do not assume the latter restriction.

We implemented and tested our improvements in `Ccluster`. For polynomials with real coefficients that are sparse or can be evaluated by a fast procedure, we achieved a 2.5 to 3 fold speed-up as shown in Table 1 by columns t_{old}/t_{new}. When the ROI contains only a few solutions, `Ccluster` is, thanks to those improvements, a little more efficient than `MPsolve` (compare columns `Ccluster` local, t_{new} and `MPsolve` in Table 1). We give details on our experiments below.

Implementation and Experiments. All the timings shown in this article are sequential times in seconds on a Intel(R) Core(TM) i7-7600U CPU @ 2.80 GHz machine with Linux. `MPsolve` is called with the command `mpsolve`

---

[2] By effective, we refer to the pathway proposed in [XY19] to describe algorithms in three levels: abstract, interval, effective.

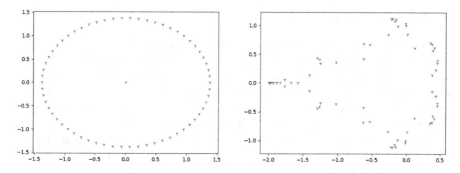

Fig. 1. Left: 63 clusters of roots for a Mignotte polynomial of degree 64. Right: Clusters of roots for the Mandelbrot polynomial of degree 63.

-as -Gi -o16 -j1[3]. Table 1 shows comparative running times of Ccluster and MPsolve on two families of polynomials, Mignotte and Mandelbrot's polynomials, with real coefficients, defined below. Columns t_{new} (resp. t_{old}) show timings of Ccluster with (resp. without) the improvements described in this paper. Columns #Clus show the number of clusters found by two versions. We used both versions of Ccluster with $\epsilon = 2^{-53}$. Ccluster global refers to the ROI $[-500, 500] + i[-500, 500]$, that contains all the roots of the tested polynomials; Ccluster local refers to an ROI containing only a few solutions. We used $[-0.5, 0.5] + i[-0.5, 0.5]$ for Mignotte's polynomials and $[-0.25, 0.25] + i[-0.25, 0.25]$ for Mandelbrot's polynomials.

The Mignotte's polynomial of degree d and parameter $a = 14$ is:

$$\texttt{Mignotte}_d(z) = z^d - 2(2^a z - 1)^2 \tag{2}$$

It has a cluster of two roots near the origin whose separation is near the theoretical minimum separation bound. It is sparse and can be evaluated very fast. We define the Mandelbrot's polynomial as $\texttt{Mandelbrot}_1(z) = 1$ and

$$\texttt{Mandelbrot}_k(z) = z\texttt{Mandelbrot}_{k-1}(z)^2 + 1 \tag{3}$$

$\texttt{Mandelbrot}_k(z)$ has degree $2^k - 1$. It can be evaluated with a straight line program. The 63 clusters of roots of $\texttt{Mandelbrot}_6(z)$ and $\texttt{Mignotte}_{64}(z)$ are depicted in Fig. 1.

Structure of the Paper. Our paper is organized as follows: in Sect. 2 we describe our P^*-test. In Sect. 3 we apply it to speeding up a clustering algorithm. In Sect. 4 we cover our root-finder for polynomials with real coefficients. Section 5 presents the results of our improvements. In the rest of the present section, we recall the related work and the clustering algorithm of [BSS+16].

---

[3] MPsolve tries to isolate the roots unless the escape bound 10^{-16} is reached.

1.1   Previous Works

Univariate polynomial root-finding is a long-standing and still actual problem; it is intrinsically linked to numerical factorization of a polynomial into the product of its linear factors. The algorithms of [Pan02] support record and nearly optimal bounds on the Boolean complexity of the solution of both problems of factorization and root-finding. The cost bound of the factorization is smaller by a factor of d, and both bounds differ from respective information lower bound by at most a polylogarithmic factor in the input size and in the bound on the required output precision. Root-finder supporting such bit complexity bounds are said to be nearly optimal. The algorithms of [Pan02] are involved and have never been implemented. User's choice has been for a while the package of subroutines MPsolve (see [BF00] and [BR14]), based on simultaneous Newton-like (*i.e.* Ehrlich-Aberth) iterations. These iterations converge to all roots simultaneously with cubic convergence rate, but only locally, that is, near the roots; empirically they converge very fast also globally, right from the start, although formal support for this empirical behavior is a long-known research challenge. Furthermore these iterations compute a small number of roots in a ROI not much faster than all roots.

In contrast, recent approaches based on subdivision (as well as the algorithms of [Pan02]) compute the roots in a fixed ROI at a cost that decreases at least proportionally to the number of roots. Near-optimal complexity has been achieved both for the real case (see [PT13,PT16,SM16] that combines the Descartes rule of signs with Newton's iterations and its implementation in [KRS16]) and the complex case. In the complex case [BSSY18] similarly combines counting test based on Pellet's theorem with complex version of the QIR algorithm, which in turn combines Newton's and secant iterations.

[BSS+16] extends the method of [BSSY18] for root clustering, *i.e.* it solves the LCP and is robust in the case of multiple roots; its implementation ([IPY18]) is a little more efficient than MPsolve for ROI's containing only several roots; when all the roots are sought, MPsolve remains the user's choice. The algorithms of [BSS+16] and [BSSY18] are direct successors of the previous subdivision algorithms of [Ren87] and [Pan00], presented under the name of Quad-tree algorithms (inherited from the earlier works by Henrici and Gargantini).

Besides Pellet's theorem, counting test in ROI can rely on Eq. (1) and winding numbers algorithms (see, e.g., [HG69,Ren87] and [ZZ19]).

1.2   Solving the LCP

C^0 and C^* Tests. The two tests C^0 and C^* discard boxes with no roots of p and count the number of roots in a box, respectively. For a given complex disc Δ, $C^0(\Delta, p)$ returns either -1 or 0, and returns 0 only if p has no root in Δ, while $C^*(\Delta, p)$ returns an integer $k \geq -1$ such that $k \geq 0$ only if p has k roots in Δ. Below, we may write $C^0(\Delta)$ for $C^0(\Delta, p)$ and $C^*(\Delta)$ for $C^*(\Delta, p)$.

In [BSS+16, BSSY18, IPY18], both C^0 and C^* are based on the so called "soft Pellet test" denoted $T^*(\Delta, p)$ or $T^*(\Delta)$ which returns an integer $k \geq -1$ such that $k \geq 0$ only if p has k roots in Δ:

$$C^0(\Delta) := \begin{cases} 0 & \text{if } T^*(\Delta) = 0 \\ -1 & \text{otherwise} \end{cases}$$

$$C^*(\Delta) := T^*(\Delta).$$

(4)

Boxes, Quadri-Section and Connected Components. The box B centered in $c = a + \mathbf{i}b$ with width w is defined as $[a - w/2, a + w/2] + \mathbf{i}[b - w/2, b + w/2]$. We denote by $w(B)$ the width of B. We call *containing disc* of B the disc $\Delta(B)$ defined as $D(c, \frac{3}{4}w(B))$. We define the four children of B as the four boxes centered in $(a \pm \frac{w}{4}) + \mathbf{i}(b \pm \frac{w}{4})$ with width $\frac{w}{2}$.

Recursive subdivisions of a ROI B_0 falls back to the construction of a tree rooted in B_0. Hereafter we refer to boxes that are nodes (and possibly leafs) of this tree as the boxes of the subdivision tree of B_0.

A *component* $\mathcal{C}$ is a set of connected boxes. The component box $B_{\mathcal{C}}$ of a component $\mathcal{C}$ is a smallest square box subject to $\mathcal{C} \subseteq B_{\mathcal{C}} \subseteq B_0$, where B_0 is the initial ROI. We write $\Delta(\mathcal{C})$ for $\Delta(B_{\mathcal{C}})$ and $w(\mathcal{C})$ for $w(B_{\mathcal{C}})$. Below we consider components made up of boxes of the same width; such a component is *compact* if $w(\mathcal{C})$ is at most 3 times the width of its boxes. Finally, a component $\mathcal{C}$ is *separated* from a set S if $\forall \mathcal{C}' \in S, 4\Delta(\mathcal{C}) \cap \mathcal{C}' = \emptyset$ and $4\Delta(\mathcal{C}) \subseteq 2B_0$.

A Root Clustering Algorithm. We give in Algorithm 1 a simple root clustering algorithm based on subdivision of ROI B_0. For convenience we assume that p has no root in $2B_0 \setminus B_0$ but this limitation can easily be removed. The paper [BSS+16] proves that Algorithm 1 terminates and output correct solution provided that the C^0 and C^*-tests are as in Eq. (4).

Note that in the **while** loop of Algorithm 1, components with widest containing box are processed first; together with the definition of a separated component, this implies the following remark:

Remark 1. *Let $\mathcal{C}$ be a component in Algorithm 1 that passes the test in step 4. Then $\mathcal{C}$ satisfies $\#(\Delta(\mathcal{C})) = \#(4\Delta(\mathcal{C}))$.*

2   Counting the Number of Roots in a Well Isolated Disc

In this section we cover a new test for counting the number of roots with multiplicity of p in a disc Δ provided that the roots in Δ are well isolated from the other roots of p. Let us first formalize this notion:

Definition 2 (Isolation ratio). *A complex disc Δ has isolation ratio ρ for a polynomial p if $\rho > 1$ and $Zero(\frac{1}{\rho}\Delta) = Zero(\rho\Delta)$.*

Let $\mathrm{Zero}(\Delta) = \{\alpha_1, \ldots, \alpha_{d_\Delta}\}$ and let m_i be the multiplicity of α_i. The h-th power sum of the roots in Δ is the complex number

$$s_h = \sum_{i=1}^{d_\Delta} m_i \times \alpha_i^h \tag{5}$$

In our test, called hereafter P^*-test, we approximate the 0-th power sum s_0 of the roots of p in Δ equal to the number of roots of p in Δ (counted with multiplicity). We obtain precise s_0 from s_0^* where p and its derivative p' are evaluated on only a small number of points on the contour of Δ. For instance, if Δ has isolation ratio 2 and p has degree 500, our test amounts to evaluating p and p' on $q = 11$ points; s_0 is recovered from these values in $O(q)$ arithmetic operations.

Algorithm 1. Root Clustering Algorithm

Input: A polynomial $p \in \mathbb{C}[z]$, a ROI B_0, $\epsilon > 0$; suppose p has no roots in $2B_0 \setminus B_0$
Output: Set R of components solving the LCP.
1: $R \leftarrow \emptyset$, $Q \leftarrow \{B_0\}$ // *Initialization*
2: **while** Q is not empty **do** // *Main loop*
3:     $\mathcal{C} \leftarrow Q.pop()$ //$\mathcal{C}$ *has the widest containing box in* Q
        // *Validation*
4:     **if** $w(\mathcal{C}) \leq \epsilon$ **and** $\mathcal{C}$ is compact **and** $\mathcal{C}$ is separated from Q **then**
5:         $k \leftarrow C^*(\Delta(\mathcal{C}), p)$
6:         **if** $k > 0$ **then**
7:             $R.push((\mathcal{C}, k))$
8:             **break**
        // *Bisection*
9:     $S \leftarrow$ empty set of boxes
10:    **for** each box B of $\mathcal{C}$ **do**
11:        **for** each child B' of B **do**
12:            **if** $C^0(\Delta(B'), p)$ returns -1 **then**
13:                $S.push(B')$
14:    $Q.push($ connected components in S $)$
15: **return** R

If p and its derivative can be evaluated at a low computational cost, e.g. when p is sparse or p is defined by a recurrence as the Mandelbrot polynomial (see [BF00][Eq. (16)] or Eq. (3) above), our P^*-test can be substantially cheaper to apply than the T^*-test presented above. Notice however that it requires the isolation ratio of Δ (or at least a lower bound) to be known.

2.1   Approximation of the 0-th Power Sum of the Roots in a Disc

[Sch82] and [Pan18] give formulas for approximating the powers sums s_h of the roots in the unit disc. Here we compute s_0 in any complex disc $\Delta = D(c, r)$.

For a positive integer q, define

$$s_0^* = \frac{r}{q} \sum_{g=0}^{q-1} \omega^g \frac{p'(c + r\omega^g)}{p(c + r\omega^g)} \qquad (6)$$

where $\omega = e^{\frac{2\pi i}{q}}$ denotes a primitive q-th root of unity.

The theorem below shows that the latter number approximates the 0-th power sum with an error that can be made as tight as desired by increasing q, providing that Δ has isolation ratio noticeably exceeding 1.

Theorem 3. *Let Δ have isolation ratio ρ for p, let $\theta = 1/\rho$, let s_0 be the 0-th power sum of the roots of p in Δ, and let s_0^* be defined as in Eq. 6. Then*

(i) $|s_0^* - s_0| \le \dfrac{d\theta^q}{1 - \theta^q}$.

(ii) *Fix $e > 0$. If $q = \lceil \log_\theta(\frac{e}{d+e}) \rceil$ then $|s_0^* - s_0| \le e$.*

Proof of Theorem 3. Let $p_\Delta(z)$ be the polynomial $p(c + rz)$. Thus $p_\Delta'(z) = rp'(c + rz)$ and Eq. (6) rewrites $s_0^* = \frac{1}{q} \sum_{g=0}^{q-1} \omega^g \frac{p_\Delta'(\omega^g)}{p_\Delta(\omega^g)}$. In addition, the unit disc $D(0,1)$ has isolation ratio ρ for p_Δ and contains s_0 roots of p_Δ. Then apply equation (12.10) in [Sch82] (with $e^{-\delta} = \theta, e^\delta = \rho$) to $p_\Delta(z)$ to obtain (i). (ii) is a direct consequence of (i).

□

For example, if Δ has isolation ratio 2, p has degree 500 and one wants to approximate s_0 with an error less than $1/4$, it suffices to apply formula in Eq. (6) for $q = 11$, that is to evaluate p and its derivative p' at 11 points.

Remark that in (ii), the required number q of evaluation points increases as the logarithm of ρ: if Δ has isolation ratio $\sqrt{\rho}$ (resp. ρ^2) instead of ρ, $\frac{1}{2}q$ (resp. $2q$) evaluation points are required. Thus doubling the number of evaluation points has the same effect as root squaring operations.

2.2   Black Box for Evaluating a Polynomial on an Oracle Number

Our goal is to give an effective description of our P^*-test; to this end, let us introduce the notion of *oracle numbers* that correspond to black boxes giving arbitrary precision approximations of any complex number. Such oracle numbers can be implemented through arbitrary precision interval arithmetic or ball arithmetic. Let $\Box\mathbb{C}$ be the set of complex intervals. If $\Box a \in \Box\mathbb{C}$, then $w(\Box a)$ is the maximum width of real and imaginary parts of $\Box a$.

For a number $a \in \mathbb{C}$, we call *oracle* for a a function $\mathcal{O}_a : \mathbb{N} \to \Box\mathbb{C}$ such that $a \in \mathcal{O}_a(L)$ and $w(\mathcal{O}_a(L)) \le 2^{-L}$ for any L. Let $\mathcal{O}_\mathbb{C}$ be the set of oracle numbers.

For a polynomial $p \in \mathbb{C}[z]$, we call *evaluation oracle* for p a function $\mathcal{I}_p : (\mathcal{O}_\mathbb{C}, \mathbb{N}) \to \Box\mathbb{C}$, such that if $\mathcal{O}_a$ is an oracle for a and $L \in \mathbb{N}$, then $p(a) \in \mathcal{I}_p(\mathcal{O}_a, L)$ and $w(\mathcal{I}_p(\mathcal{O}_a, L)) \le 2^{-L}$.

We consider evaluation oracles $\mathcal{I}_p$ and $\mathcal{I}_{p'}$ for p and its derivative p'. If p is given by $d + 1$ oracles for its coefficients, one can easily construct $\mathcal{I}_p$ and

$\mathcal{I}_{p'}$ by using for instance Horner's rule. However for some polynomials defined by a procedure, for instance the Mandelbrot polynomial (see Eq. (3)), one can construct fast evaluation oracles $\mathcal{I}_p$ and $\mathcal{I}_{p'}$ from the procedurial definition.

2.3   The P^*-test

Algorithm 2 counts the number of roots of p in a disc $\Delta = D(c, r)$ having isolation ratio at least ρ. For such a disc, any positive integer q and any integer $0 \leq g < q$, one has $p(c+r\omega^g) \neq 0$. As a consequence, there exist an L' s.t $\forall L \geq L', \forall 0 \leq g \leq q-1, 0 \notin \mathcal{I}_p(\mathcal{O}_{c+r\omega^g}, L)$ and the intervals $\square s_0^*$ computed in step 4 of Algorithm 2 have strictly decreasing width as of $L \geq L'$. This shows the termination of Algorithm 2. Its correctness is stated in the following proposition:

Algorithm 2. $P^*(\mathcal{I}_p, \mathcal{I}_{p'}, \Delta, \rho)$

Input: $\mathcal{I}_p$, $\mathcal{I}_{p'}$ evaluation oracles for p and p', $\Delta = D(c, r)$, $\rho > 1$. p has degree d.
Output: an integer in $\{0, \dots, d\}$
  1: $L \leftarrow 53$, $w \leftarrow 1$, $e \leftarrow 1/4$, $\theta \leftarrow 1/\rho$
  2: $q \leftarrow \lceil \log_\theta(\frac{e}{d+e}) \rceil$
  3: **while** $w \geq 1/2$ **do**

  4:       Compute interval $\square s_0^*$ as $\frac{r}{q} \sum_{g=0}^{q-1} \mathcal{O}_{\omega^g}(L) \frac{\mathcal{I}_{p'}(\mathcal{O}_{c+r\omega^g}, L)}{\mathcal{I}_p(\mathcal{O}_{c+r\omega^g}, L)}$

  5:       $w \leftarrow w(\square s_0^*)$
  6:       $L \leftarrow 2*L$
  7: $\square s_0 \leftarrow \square s_0^* + [-1/4, 1/4] + \mathbf{i}[-1/4, 1/4]$
  8: **return** the unique integer in $\square s_0$

Proposition 4. *Let k be the result of the call $P^*(\mathcal{I}_p, \mathcal{I}_{p'}, \Delta, \rho)$. If Δ has isolation ratio at least ρ for p, then p has k roots in Δ counted with multiplicity.*

Proof of Proposition 4. Once the **while** loop in Algorithm 2 terminates, the interval $\square s_0^*$ contains s_0^* and $w(\square s_0^*) < 1/2$. In addition, by virtue of statement *(ii)* of Theorem 3, one has $|s_0^* - s_0| \leq 1/4$, thus $\square s_0$ defined in step 7 satisfies: $w(\square s_0) < 1$ and $s_0 \in \square s_0$. Since $\square s_0$ contains at most one integer, s_0 is the unique integer in $\square s_0$, and is equal to the number of roots in Δ.

$\square$

3   Using the P^*-test in a Subdivision Framework

Let us discuss the use of the P^*-test as C^0 and C^*-tests in order to speed up Algorithm 1. Table 2 covers runs of Algorithm 1 on Mignotte and Mandelbrot's polynomials. t is the running time when C^0 and C^* tests are defined by Eq. (4). Columns nb show the respective numbers of C^0 and C^*-tests performed, column t_0 and t_0/t (resp. t_* and t_*/t) show time and ratio of times spent in C^0 (resp. C^*) tests when it is defined by Eq. (4).

One can readily use the P^*-test to implement the C^*-test by defining

$$C^*(\Delta) := P^*(\mathcal{I}_p, \mathcal{I}_{p'}, 2\Delta, 2) \tag{7}$$

Following Remark 1, the C^*-test is called in Algorithm 1 for components $\mathcal{C}$ satisfying $\#(\Delta(\mathcal{C})) = \#(4\Delta(\mathcal{C}))$. Hence $2\Delta(\mathcal{C})$ has isolation ratio 2 and by virtue of Proposition 4, $C^*(\Delta(\mathcal{C}))$ returns $r \geq 0$ only if $\Delta(\mathcal{C})$ contains r roots.

However this would not imply much improvements in itself. Column t'_* in Table 2 shows the time that would be spent in C^*-tests if it was defined by Eq. (7): it is far less than t_*, but the ratio of time spent in C^*-tests (see column t_*/t) is very small. In contrast, about 90% of the running time of Algorithm 1 is spent in C^0-tests (see column t_0/t). We propose to use a modified version of the P^*-test as a filter in the C^0-test to decrease its running time.

Table 2. Details on runs of Algorithm 1 on Mignotte and Mandelbrot's polynomials.

	C^0-tests							C^*-tests			
		T^*-tests		$\widetilde{P^*}$-tests					T^*-tests		P^*-tests
	nb	t_0	t_0/t (%)	t'_0	n_{-1}	n_{-2}	n_{err}	nb	t_*	t_*/t (%)	t'_*
Mignotte$_{128}$	4508	4.73	90.9	0.25	276	0	12	128	0.07	1.46	0.01
Mignotte$_{256}$	8452	27.8	91.2	0.60	544	0	20	256	0.58	1.92	0.02
Mandelbrot$_7$	4548	6.34	88.1	0.28	168	0	28	131	0.11	1.51	0.01
Mandelbrot$_8$	8892	35.6	88.4	0.67	318	0	57	256	0.69	1.71	0.03

Algorithm 3. $\widetilde{P^*}(\mathcal{I}_p, \mathcal{I}_{p'}, \Delta, \rho)$

Input: $\mathcal{I}_p$, $\mathcal{I}_{p'}$ evaluation oracles for p and p', $\Delta = D(c, r)$, $\rho > 1$. p has degree d.
Output: an integer in $\{-2, -1, 0, \ldots, d\}$

3.1 An Approximate P^*-test

The approximate version of the P^*-test is aimed at being applied to a disc $\Delta = D(c, r)$ with unknown isolation ratio. Unless Δ has isolation ratio $\rho > 1$, the very unlikely case where for some $0 \leq g < q$, $p(c + r\omega^g) = 0$, leads to a non-terminating call of $P^*(\mathcal{I}_p, \mathcal{I}_{p'}, \Delta, \rho)$. Also, $\square s_0$ computed in step 7 of Algorithm 2 could contain no integer or an integer that is not s_0. We define the $\widetilde{P^*}$-test specified in Algorithm 3 by modifying Algorithm 2 as follows:

1. after step 3, if an $\mathcal{I}_p(\mathcal{O}_{c+r\omega^g}, L)$ contains 0, the result -2 is returned;
2. step 7 is replaced with: $\square s_0 \leftarrow \square s_0^* + [-1/2, 1/2] + \mathbf{i}[-1/2, 1/2]$,
3. after step 7, unless $\square s_0$ contains a unique integer, the result -1 is returned.

Modification 1 ensures termination when Δ does not have isolation ratio $\rho > 1$. With modification 2, $\square s_0$ can have width greater than 1 and contain more than one integer. With modification 3, the $\widetilde{P}^*$-test can return -1 which means that no conclusion can be made. If $\widetilde{P}^*(\mathcal{I}_p, \mathcal{I}_{p'}, \Delta, \rho)$ returns a positive integer, this result has still to be checked, for instance, with the T^*-test.

In Table 2, column n_{-2} (resp. n_{-1}) shows the number of times $\widetilde{P}^*(\mathcal{I}_p, \mathcal{I}_{p'}, \Delta, 2)$ returns -2 (resp. -1) when applied in place of $T^*(\Delta)$ in the C^0-test. Column n_{err} shows the number of times the conclusion of $\widetilde{P}^*$ was wrong, and t'_0 shows the total time spent in $\widetilde{P}^*$-tests.

3.2   Using the P^* and $\widetilde{P}^*$-test in a Subdivision Framework

Our improvement of Algorithm 1 is based on the following heuristic remarks. First, it is very unlikely that $\widetilde{P}^*(\mathcal{I}_p, \mathcal{I}_{p'}, \Delta, 2)$ returns -2 (see column n_{-2} in Table 2). Second, when $P^*(\mathcal{I}_p, \mathcal{I}_{p'}, \Delta, 2)$ returns $k \geq 0$, it is very likely that Δ contains k roots counted with multiplicity (see column n_{err} in Table 2).

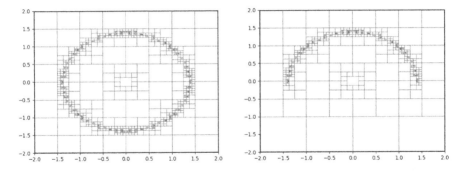

Fig. 2. Computing clusters for `Mignotte`$_{64}$ in the ROI $[-2,2] + \mathbf{i}[-2,2]$. Left: The subdivision tree for Algorithm 1. Right: The subdivision tree for Algorithm 5.

We define the C^0-test as follows:

$$C^0(\Delta) := \begin{cases} -1 & \text{if } \widetilde{P}^*(\mathcal{I}_p, \mathcal{I}_{p'}, \Delta, 2) \notin \{-2, 0\}, \\ -1 & \text{if } \widetilde{P}^*(\mathcal{I}_p, \mathcal{I}_{p'}, \Delta, 2) \in \{-2, 0\} \text{ and } T^*(\Delta) \neq 0, \\ 0 & \text{if } \widetilde{P}^*(\mathcal{I}_p, \mathcal{I}_{p'}, \Delta, 2) \in \{-2, 0\} \text{ and } T^*(\Delta) = 0. \end{cases} \quad (8)$$

If $C^0(\Delta)$ is defined in Eq. (8), it returns 0 only if Δ contains no root. Thus Algorithm 1 with C^0 and C^*-tests defined by Eqs. (8) and (7) is correct.

Remark now that if a square complex box B of width w does not contain root and is at a distance at least $\frac{3}{2}w$ from a root, then $\Delta(B)$ has isolation ratio 2, and $\widetilde{P}^*(\mathcal{I}_p, \mathcal{I}_{p'}, \Delta(B), 2)$ returns 0 or -2. As a consequence, the termination of Algorithm 1 with C^0 and C^*-tests defined in Eqs. (8) and (7) amounts to the termination of Algorithm 1 with C^0 and C^* defined in Eq. (4).

4   Clustering Roots of Polynomials with Real Coefficients

We consider here the special case where $p \in \mathbb{R}[z]$, and show how to improve a subdivision algorithm for solving the LCP. We propose to leverage on the geometric structure of the roots of p, that are either real, or imaginary and come in complex conjugated pairs: if $\alpha \in \mathbb{C}$ is a root of p so is $\overline{\alpha}$ where $\overline{\alpha}$ is the complex conjugate of α. The modified subdivision algorithm we propose deals only with the boxes of the subdivision tree of the ROI B_0 that have a positive imaginary part; the roots with positive imaginary parts are in the latter boxes. The roots with negative imaginary parts are implicitly represented by the former ones. In Fig. 2 are shown two subdivision trees constructed for clustering roots of a Mignotte polynomial of degree 64; the left-most one is obtained when applying Algorithm 1; the right-most one results of our improvement.

Below, we suppose that B_0 is symmetric with respect to the real axis and that p has no root in $2B_0 \setminus B_0$. These two limitations can easily be removed.

Algorithm 4. *Quadrisect($\mathcal{C}$)*

Input: A polynomial $p \in \mathbb{R}[z]$ and a component $\mathcal{C}$
Output: A list R of disjoint and not imaginary negative components
1: $S \leftarrow$ empty list of boxes
2: **for** each constituent box B of $\mathcal{C}$ **do**
3:     **for** each child B' of B **do**
4:         **if** B is **not** imaginary negative **then**
5:             **if** $C^0(\Delta(B'), p)$ returns -1 **then**
6:                 $S.push(B')$
7: $R \leftarrow$ group boxes of S in components
8: **return** R

Notations. Let B be a box centered in c. We define its conjugate $\overline{B}$ as the box centered in $\overline{c}$ with width $w(B)$. We say that B is *imaginary positive* (resp. *imaginary negative*) if $\forall b \in B$, $Im(b) > 0$ (resp. $Im(b) < 0$).

Let $\mathcal{C}$ be a component of boxes of the subdivision tree of B_0. We define $\overline{\mathcal{C}}$ as the component which boxes are the conjugate of the boxes of $\mathcal{C}$. We call *conjugate closure* of $\mathcal{C}$, and we denote it by $\mathcal{C}_{\overline{\cup}}$ the set of boxes $\mathcal{C} \cup (\overline{\mathcal{C}} \setminus \mathcal{C})$. If $\mathcal{C}$ intersects $\mathbb{R}$, $\mathcal{C}_{\overline{\cup}}$ is a component. We say that $\mathcal{C}$ is *imaginary positive* (resp. *imaginary negative*) if each box in $\mathcal{C}$ is imaginary positive (resp. imaginary negative).

Solving the LCP for Polynomials with Real Coefficients. We describe in Algorithm 4 a procedure to bisect a component, that discards boxes that are imaginary negative in addition to those that contain no root.

Our algorithm for solving the LCP for polynomials with real coefficients is presented in Algorithm 5. It maintains in the queue Q only components of boxes that are imaginary positive or that intersect the real line. Components with only imaginary negative boxes are implicitly represented by the imaginary positive ones. Components that intersect the real line are replaced by their conjugate closure. Components in Q are ordered by decreasing width of their containing boxes. The termination of Algorithm 5 is a consequence of the termination of Algorithm 1 that is proved in [BSS+16].

Let $\{(\mathcal{C}^1, m^1), \ldots, (\mathcal{C}, m^\ell)\}$ be the list returned by Algorithm 5 called for arguments p, B_0, ϵ. Then $\{(\Delta(\mathcal{C}^1), m^1), \ldots, (\Delta(\mathcal{C}^\ell), m^\ell)\}$ is a solution of the LCP problem for p, B_0, ϵ, $i.e.$:

(i) the $\Delta(\mathcal{C}^i)$'s are pairwise disjoint with radius less that ϵ,
(ii) $\forall 1 \leq i \leq \ell$, $(\mathcal{C}^i, m^i)$ satisfies $\#(\Delta(\mathcal{C}^i)) = \#(3\Delta(\mathcal{C}^i)) = m^i$,
(iii) $\texttt{Zero}(B_0, p) \subseteq \bigcup_{i=1}^\ell \texttt{Zero}(\Delta(\mathcal{C}^i), p) \subseteq \texttt{Zero}(2B_0, p)$.

In what follow we may write R for the list of connected components in R. $(i), (ii)$ and (iii) are direct consequences of the following proposition:

Proposition 5. *Consider Q and R after any execution of the* **while** *loop in Algorithm 5. Decompose Q in two lists Q^1 and Q^2 containing respectively the imaginary positive components of Q and the non imaginary components of Q. Note $\overline{Q^1}$ the list of the conjugates of the components in Q^1 and $Q^2_{\overline{\sqcup}}$ the list of the conjugate closures of the components in Q^2, and let $Q_{\overline{\sqcup}}$ be $\overline{Q^1} \cup Q^2_{\overline{\sqcup}}$. One has:*

Algorithm 5. Local root clustering for polynomials with real coefficients

---

Input: A polynomial $p \in \mathbb{R}[z]$, a ROI B_0, $\epsilon > 0$; assume p has no roots in $2B_0 \setminus B_0$, and B_0 is symmetric with respect to the real axis.
Output: A set R of components solving the LCP.
1: $R \leftarrow \emptyset$, $Q \leftarrow \{\{B_0\}\}$ // *Initialization*
2: **while** Q is not empty **do** // *Main loop*
3:     $\mathcal{C} \leftarrow Q.pop()$ //$\mathcal{C}$ *has the widest containing box in Q*
4:     $sFlag \leftarrow$ **false**
5:     **if** $\mathcal{C}$ is **not** imaginary positive **then** //*Note: $\mathcal{C} \cap \mathbb{R} \neq \emptyset$*
6:         $\mathcal{C} \leftarrow \mathcal{C}_{\overline{\sqcup}}$
7:         $sFlag \leftarrow \mathcal{C}$ is separated from Q
8:     **else**
9:         $sFlag \leftarrow (\mathcal{C}$ is separated from $Q)$ **and** $(4\Delta(\mathcal{C}) \cap \overline{\mathcal{C}} = \emptyset)$
10:    **if** $w(\mathcal{C}) \leq \epsilon$ **and** $\mathcal{C}$ is compact **and** $sFlag$ **then** // *Validation*
11:        $m \leftarrow C^*(\Delta(\mathcal{C}), p)$
12:        **if** $m > 0$ **then**
13:            $R.push((\mathcal{C}, m))$
14:            **if** $\mathcal{C}$ is imaginary positive **then**
15:                $R.push((\overline{\mathcal{C}}, m))$
16:            **break**
17:    $Q.push(Quadrisect(\mathcal{C}))$ // *Bisection*
18: **return** R

---

(1) any $\alpha \in \mathbf{Zero}(B_0)$ is in $R \cup Q \cup Q_{\overline{U}}$,
(2) any $\mathcal{C} \in R$ is separated from $(R \setminus \{\mathcal{C}\}) \cup Q \cup Q_{\overline{U}}$,
(3) any $(\mathcal{C}, m)$ in R is such that $m = \#(\Delta(\mathcal{C})) = \#(3\Delta(\mathcal{C}))$.

Proposition 5 is a consequence of Remark 1 and the following remark.

Remark 6. *Let $p \in \mathbb{R}[z]$ and $\mathcal{C}$ be a component. If $\mathcal{C}$ is imaginary negative or imaginary positive and if there exists m such that $m = \#(\Delta(\mathcal{C})) = \#(3\Delta(\mathcal{C}))$, then $m = \#(\Delta(\overline{\mathcal{C}})) = \#(3\Delta(\overline{\mathcal{C}}))$.*

5   Numerical Results

We implemented the two improvements of Sects. 3 and 4 in Ccluster. Ccluster0 refers to the original version of Ccluster. Both CclusterR and CclusterPs implement Algorithm 5. In CclusterPs, C^0 and C^* are defined by Eqs. (8) and (7).

Testing Suite. We tested our improvements on Mignotte and Mandelbrot's polynomials and on Bernoulli and Runnel's polynomials: the Bernoulli polynomial of degree d is $\mathbf{Bernoulli}_d(z) = \sum_{k=0}^{d} \binom{d}{k} b_{d-k} z^k$ where the b_i's are the Bernoulli numbers. It has about $d/2$ non-zero coefficients and, as far as we know, cannot be evaluated substantially faster than with Horner's scheme. It has real coefficients, and about 2/3 of its roots are real or imaginary positive (see left part of Fig. 3). Let $r = 2$, $q_0(z) = 1$, $q_1(z) = z$ and $q_{k+1}(z) = q_k(z)^r + z q_{k-1}(z)^{r^2}$. We define the Runnel's polynomial of parameter k as $\mathbf{Runnels}_k = q_k$. It has real coefficients, a multiple root (zero), and can be evaluated fast. The 107 distinct roots of $\mathbf{Runnels}_8$ are drawn on right part of Fig. 3.

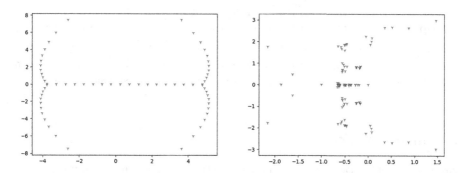

Fig. 3. Left: 64 clusters of roots for the Bernoulli polynomial of degree 64. Right: 107 clusters of roots for the Runnel's polynomial of degree 170.

Results. Table 3 gives details concerning the execution of `CclusterO`, `CclusterR` and `CclusterPs` for polynomials with increasing degrees. We used $\epsilon = 2^{-53}$ and the ROI $B_0 = [-500, 500] + \mathbf{i}[-500, 500]$ that contains all the roots of all the considered polynomials. Column (#Clus,#Sols) shows the number of clusters and the total multiplicity found. Columns (depth, size) show the depth and the size (*i.e.* number of nodes) of the subdivision tree for each version. t_1, t_2 and t_3 stand respectively for the running time in second of `CclusterO`, `CclusterR` and `CclusterPs`.

Algorithm 5 achieves speed up t_1/t_2. It is almost 2 for Mignotte polynomials, since about half of its roots are above the real axis. This speed up is less important for the three other families of polynomials, which have a non-negligible ratio of real roots. The speed up achieved by using the P^*-test is t_2/t_3. It is significant for Mignotte's polynomial, which is sparse, and Mandelbrot and Runnel's polynomials for which one can construct fast evaluation procedures.

Table 3. Details on runs of `CclusterO`, `CclusterR` and `CclusterPs` for polynomials in $\mathbb{R}[z]$ with increasing degree.

	CclusterO			CclusterR		CclusterPs			
	(#Clus, #Sols)	(depth, size)	t_1	(depth, size)	t_1/t_2	(depth, size)	t_3	t_2/t_3	t_1/t_3
Bernoulli$_{128}$	(128, 128)	(100, 4732)	6.30	(100, 3708)	1.72	(100, 4104)	3.30	1.10	1.90
Bernoulli$_{191}$	(191, 191)	(92, 7220)	20.2	(92, 5636)	1.74	(92, 6236)	10.7	1.08	1.88
Bernoulli$_{256}$	(256, 256)	(93, 9980)	41.8	(93, 7520)	1.67	(91, 8128)	21.9	1.14	1.90
Bernoulli$_{383}$	(383, 383)	(93, 14504)	120	(93, 11136)	1.82	(93, 11764)	53.5	1.23	2.25
Mignotte$_{128}$	(127, 128)	(96, 4508)	5.00	(92, 3212)	1.92	(92, 3484)	1.81	1.43	2.75
Mignotte$_{191}$	(190, 191)	(97, 6260)	15.5	(97, 4296)	2.01	(97, 4688)	4.34	1.77	3.58
Mignotte$_{256}$	(255, 256)	(94, 8452)	31.8	(94, 5484)	2.04	(94, 6648)	10.7	1.44	2.95
Mignotte$_{383}$	(382, 383)	(97, 12564)	79.7	(97, 8352)	1.98	(97, 9100)	26.8	1.49	2.97
Mandelbrot$_7$	(127, 127)	(96, 4548)	7.17	(96, 2996)	1.62	(96, 3200)	2.88	1.52	2.48
Mandelbrot$_8$	(255, 255)	(96, 8892)	40.6	(96, 5576)	1.71	(96, 6208)	15.1	1.56	2.69
Mandelbrot$_9$	(511, 511)	(100, 17956)	266	(100, 11016)	1.89	(100, 11868)	97.1	1.44	2.74
Runnels$_8$	(107, 170)	(96, 4652)	13.3	(96, 3252)	1.61	(96, 3624)	6.51	1.26	2.04
Runnels$_9$	(214, 341)	(99, 9592)	76.2	(99, 6260)	1.70	(99, 6624)	32.2	1.38	2.36
Runnels$_{10}$	(427, 682)	(100, 19084)	479	(100, 12288)	1.69	(100, 12904)	211	1.34	2.26

6 Future Works

Our main contribution is a significant practical progress in subdivision root-finding based on a new test for counting roots in a well-isolated disc. If the latter assumption does not hold, the test result is not guaranteed but is very likely to be correct. In a subdivision framework, we have proposed to use a test based on Pellet's theorem to verify its result. We aim to do so by using only evaluations of p and p'. This would imply a very significant improvement of the root clustering algorithm when p and p' can be evaluated very efficiently.

References

[BF00] Bini, D.A., Fiorentino, G.: Design, analysis, and implementation of a multi-precision polynomial rootfinder. Numer. Algorithms **23**(2), 127–173 (2000)

[BR14] Bini, D.A., Robol, L.: Solving secular and polynomial equations: a multi-precision algorithm. J. Comput. Appl. Math. **272**, 276–292 (2014)

[BSS+16] Becker, R., Sagraloff, M., Sharma, V., Xu, J., Yap, C.: Complexity analysis of root clustering for a complex polynomial. In: Proceedings of the ACM on International Symposium on Symbolic and Algebraic Computation, ISSAC 2016, pp. 71–78. ACM, New York (2016)

[BSSY18] Becker, R., Sagraloff, M., Sharma, V., Yap, C.: A near-optimal subdivision algorithm for complex root isolation based on Pellet test and Newton iteration. J. Symb. Comput. **86**, 51–96 (2018)

[HG69] Henrici, P., Gargantini, I.: Uniformly convergent algorithms for the simultaneous approximation of all zeros of a polynomial. In: Constructive Aspects of the Fundamental Theorem of Algebra, pp. 77–113. Wiley-Interscience, New York (1969)

[IPY18] Imbach, R., Pan, V.Y., Yap, C.: Implementation of a near-optimal complex root clustering algorithm. Math. Soft. - ICMS **2018**, 235–244 (2018)

[KRS16] Kobel, A., Rouillier, F., Sagraloff, M.: Computing real roots of real polynomials ... and now for real! In: Proceedings of the ACM on International Symposium on Symbolic and Algebraic Computation, ISSAC 2016, pp. 303–310. ACM, New York (2016)

[Pan00] Pan, V.Y.: Approximating complex polynomial zeros: modified Weyl's quadtree construction and improved newton's iteration. J. Complex. **16**(1), 213–264 (2000)

[Pan02] Pan, V.Y.: Univariate polynomials: nearly optimal algorithms for numerical factorization and root-finding. J. Symb. Comput. **33**(5), 701–733 (2002)

[Pan18] Pan, V.Y.: Old and new nearly optimal polynomial root-finders. arXiv preprint arXiv:1805.12042 (2018)

[PT13] Pan, V.Y., Tsigaridas, E.P.: On the Boolean complexity of real root refinement. In: Proceedings of the 38th International Symposium on Symbolic and Algebraic Computation, ISSAC 2013, pp. 299–306. ACM, New York (2013)

[PT16] Pan, V.Y., Tsigaridas, E.P.: Nearly optimal refinement of real roots of a univariate polynomial. J. Symb. Comput **74**, 181–204 (2016)

[Ren87] Renegar, J.: On the worst-case arithmetic complexity of approximating zeros of polynomials. J. Complex. **3**(2), 90–113 (1987)

[Sch82] Schönhage, A.: The fundamental theorem of algebra in terms of computational complexity. Manuscript. University of Tübingen, Germany (1982)

[SM16] Sagraloff, M., Mehlhorn, K.: Computing real roots of real polynomials. J. Symb. Comput. **73**, 46–86 (2016)

[XY19] Xu, J., Yap, C.: Effective subdivision algorithm for isolating zeros of real systems of equations, with complexity analysis. arXiv preprint (2019). arXiv:1905.03505

[ZZ19] Zaderman, V., Zhao, L.: Counting roots of a polynomial in a convex compact region by means of winding number calculation via sampling. arXiv preprint arXiv:1906.10805 (2019)

On the Chordality of Simple Decomposition in Top-Down Style

Chenqi Mou[1,2(✉)] and Jiahua Lai[1]

[1] LMIB–School of Mathematical Sciences, Beihang University, Beijing 100191, China
[2] Beijing Advanced Innovation Center for Big Data and Brain Computing,
Beijing University, Beijing 100191, China
{chenqi.mou,jiahualai}@buaa.edu.cn

Abstract. Simple decomposition of polynomial sets computes conditionally squarefree triangular sets or systems with certain zero or ideal relationships with the polynomial sets. In this paper we study the chordality of polynomial sets occurring in the process of simple decomposition in top-down style. We first reformulate Wang's algorithm for simple decomposition in top-down style so that the decomposition process can be described in an inductive way. Then we prove that for a polynomial set whose associated graph is chordal, all the polynomial sets in the process of Wang's algorithm for computing simple decomposition of this polynomial set have associated graphs which are subgraphs of the input chordal graph.

Keywords: Chordal graph · Simple decomposition · Top-down style · Triangular system

1 Introduction

Triangular decomposition is the process to decompose an arbitrary multivariate polynomial set into finitely many polynomial sets in triangular shape, called triangular sets, with associated zero or ideal relationships between the polynomial set and triangular sets. Here the triangular shape means that the greatest variables of the polynomials in the triangular sets increase strictly according to a given variable ordering. This special shape makes triangular sets particularly suitable for polynomial elimination and polynomial system solving. With extensive study on their properties and computation [1,6,11,16,24,27,28], triangular sets have become an indispensable tool for handling polynomials and polynomial ideals symbolically like Gröbner bases [4,8,9], with diverse applications in, e.g., automatic geometric theorem proving [28,29] and cryptanalysis [5,14].

This paper focuses on how to apply the chordal graphs to study and analyze the behaviors of simple decomposition in top-down style. The study in this paper

This work was partially supported by the National Natural Science Foundation of China (NSFC 11971050 and 11771034) and the Fundamental Research Funds for the Central Universities in China (YWF-19-BJ-J-324).

ⓒ Springer Nature Switzerland AG 2020
D. Slamanig et al. (Eds.): MACIS 2019, LNCS 11989, pp. 138–152, 2020.
https://doi.org/10.1007/978-3-030-43120-4_12

is directly inspired by the pioneering work of Cifuentes and Parrilo [7], where the connections between chordal graphs and triangular decomposition were established for the first time. After that the properties and behaviors of algorithms for triangular decomposition in top-down style were analyzed via the changes of associated graphs of polynomial sets in the process of decomposition [19, 20]. In particular, several algorithms for triangular decomposition in top-down style from [24, 26, 27] were proved to preserve chordality of polynomial sets in the decomposition. This fine property explains one experimental observation in [7] that algorithms due to Dongming Wang become more efficient when the polynomial sets to decompose are associated with chordal graphs from a theoretical point of view, and gives birth to efficient sparse algorithms for triangular decomposition in top-down style which make full use of the sparsity and chordality of the polynomial sets. It is worth mentioning that the results obtained in [19, 20] can be viewed as multivariate generalization of the existing role of chordal graphs in sparse Gaussian elimination in linear algebra [13, 21, 22]. In particular, experimentally the sparse algorithms proposed in [20] for triangular decomposition are more efficient when the chordal polynomial set becomes sparser.

One key idea for the aforementioned sparse triangular decomposition is to use the perfect elimination ordering from the chordal graph as the variable ordering for triangular decomposition. In fact, other concepts and tools from graph theory have also been applied to find "good" variable orderings to speedup the computation of triangular decomposition, for example the Dulmage-Mendelssohn decomposition of a bipartite graph associated to the polynomial set [10] and the computation of strongly connected components of a digraph associated to the polynomial set by using Tarjan's algorithm [18]. The authors feel that there should be more potential in the applications of graph theory in studying triangular decomposition.

One specific kind of algorithms not covered in [19, 20] are those for simple decomposition in top-down style. Simple sets are special triangular sets which are squarefree conditionally (and thus they are also called squarefree regular sets) [17, 25]. Due to this property, simple sets are useful for counting the numbers of solutions of polynomial systems and have been successfully applied to study differential systems [2, 3, 12]. This paper aims at proving similar theoretical results on one typical algorithm in top-down style for decomposing polynomial sets into simple sets or systems, that is the one due to Wang based on subresultant regular subchains [25], to those proved in [19, 20]. As one may find later in this paper, structurally this algorithm is much more complicated than those already studied in [19, 20].

The main contributions of this paper include the following: (1) Reformulation of Wang's algorithm for simple decomposition in [25], which essentially handles the nodes in the decomposition tree in a depth-first way, into a form which focuses on how child nodes are spawned from its parent node in the decomposition tree and thus suitable for the inductive proof (See Sect. 3 below). (2) Proof that this algorithm also preserves chordality of the polynomial set in the process of decomposition (see Sect. 5, in particular Theorem 1, below). These theoretical results provide better understanding on algorithms in top-down style for simple decomposition and add Wang's algorithm for simple decomposition to the list of potential algorithms to use as a subroutine in sparse algorithms for triangular decomposition [20].

2   Preliminaries

Let $\mathbb{K}[x_1, \ldots, x_n]$ be the multivariate polynomial ring over a field $\mathbb{K}$ in the variables $x_1, \ldots, x_n$. We fix a variable ordering $x_1 < \cdots < x_n$ throughout this paper. For simplicity we write $\mathbb{K}[x_1, \ldots, x_n]$ as $\mathbb{K}[\boldsymbol{x}]$.

2.1   Triangular and Simple Systems

Let F be a polynomial in $\mathbb{K}[\boldsymbol{x}]$. Then the greatest variable appearing in F is called the *leading variable* of F and denoted by $\mathrm{lv}(F)$. Suppose that $\mathrm{lv}(F) = x_k$. Then F can be written as $F = I x_k^d + R$ such that $I \in \mathbb{K}[x_1, \ldots, x_{k-1}]$, $R \in \mathbb{K}[x_1, \ldots, x_k]$, and $\deg(R, x_k) < d$. The polynomials I and R here are called the *initial* and *tail* of F and denoted by $\mathrm{ini}(F)$ and $\mathrm{tail}(F)$ respectively.

Denote the algebraic closure of $\mathbb{K}$ by $\overline{\mathbb{K}}$. For two polynomial sets $\mathcal{F}, \mathcal{G} \subset \mathbb{K}[\boldsymbol{x}]$, the set of common zeros in $\overline{\mathbb{K}}^n$ of the polynomials in $\mathcal{F}$ is denoted by $\mathsf{Z}(\mathcal{F})$, and $\mathsf{Z}(\mathcal{F}/\mathcal{G}) := \mathsf{Z}(\mathcal{F}) \setminus \mathsf{Z}(\prod_{G \in \mathcal{G}} G)$.

Definition 1. An ordered set of non-constant polynomials $\mathcal{T} = [T_1, \ldots, T_r] \subset \mathbb{K}[\boldsymbol{x}]$ is called a *triangular set* if $\mathrm{lv}(T_1) < \cdots < \mathrm{lv}(T_r)$. A pair $(\mathcal{T}, \mathcal{U})$ with $\mathcal{T}, \mathcal{U} \subset \mathbb{K}[\boldsymbol{x}]$ is called a *triangular system* if $\mathcal{T}$ is a triangular set, and for each $i = 2, \ldots, r$ and any $\overline{\boldsymbol{x}}_{i-1} \in \mathsf{Z}([T_1, \ldots, T_{i-1}]/\mathcal{U})$, we have $\mathrm{ini}(T_i)(\overline{\boldsymbol{x}}_{i-1}) \neq 0$.

Definition 2. Let T be a polynomial in $\mathbb{K}[x_1, \ldots, x_k]$ and $\overline{\boldsymbol{x}}_{k-1}$ be an element in $\tilde{\mathbb{K}}^{k-1}$, where $\tilde{\mathbb{K}}$ is some field extension of $\mathbb{K}$. Then the univariate polynomial $T(\overline{\boldsymbol{x}}_{k-1}, x_k)$ is said to be *squarefree* with respect to (written as w.r.t. hereafter) x_k if

$$\gcd(T(\overline{\boldsymbol{x}}_{k-1}, x_k), \frac{\partial T}{\partial x_k}(\overline{\boldsymbol{x}}_{k-1}, x_k)) \in \tilde{\mathbb{K}},$$

where $\frac{\partial T}{\partial x_k}$ is the formal derivative of T w.r.t. x_k.

For a polynomial set $\mathcal{F} \subset \mathbb{K}[\boldsymbol{x}]$, we denote its subset $\{F \in \mathcal{F} : \mathrm{lv}(F) = x_i\}$ by $\mathcal{F}^{(i)}$ for an integer i ($1 \leq i \leq n$).

Definition 3. ([27, Def. 3.3.1]) For two polynomial sets $\mathcal{T}, \mathcal{U} \subset \mathbb{K}[\boldsymbol{x}]$, the pair $(\mathcal{T}, \mathcal{U})$ is called a *simple system* if the following conditions hold.

(a) $\mathcal{T}$ and $\mathcal{U}$ are either triangular sets in $\mathbb{K}[\boldsymbol{x}]$ or emptysets;
(b) for each $i = 1, \ldots, n$, $\#\mathcal{T}^{(i)} + \#\mathcal{U}^{(i)} \leq 1$;
(c) for each $i = 1, \ldots, n$, if $\mathcal{T}^{(i)} \cup \mathcal{U}^{(i)} \neq \emptyset$, then for any $P \in \mathcal{T}^{(i)} \cup \mathcal{U}^{(i)}$ and $\overline{\boldsymbol{x}}_{i-1} \in \mathsf{Z}(\mathcal{T}^{(<i)}/\mathcal{U}^{(<i)})$, $\mathrm{ini}(P)(\overline{\boldsymbol{x}}_{i-1}) \neq 0$ and $P(\overline{\boldsymbol{x}}_{i-1}, x_i)$ is squarefree w.r.t. x_i, where $\mathcal{T}^{(<k)}$ denotes the truncated triangular set $[\mathcal{T}^{(1)}, \ldots, \mathcal{T}^{(k-1)}]$ and $\mathcal{U}^{(<k)}$ is similarly defined.

A triangular set T is called a *simple set* if $(T, \emptyset)$ forms a simple system or there exists another triangular set U such that (T, U) forms a simple system.

Simple systems are also referred to as Thomas systems [3, 12] and simple sets as squarefree regular sets [15]. Compared to the regular sets or regular chains (see. e.g., [1, 16, 30]), simple systems or sets impose stronger requirements on the polynomials within so that they are conditionally squarefree. This property of squarefreeness is particularly useful for counting the number of solutions of polynomial systems [2].

Definition 4. For an arbitrary non-empty polynomial set $\mathcal{F} \subset \mathbb{K}[\boldsymbol{x}]$, a finite number of triangular sets $T_1, \ldots, T_s \subset \mathbb{K}[\boldsymbol{x}]$ (or triangular systems $(T_1, U_1), \ldots,$ (T_s, U_s) respectively) are said to form a *triangular decomposition* of $\mathcal{F}$ if the zero relationship $\mathsf{Z}(\mathcal{F}) = \bigcup_{i=1}^{s} \mathsf{Z}(T_i / \operatorname{ini}(T_i))$ holds, where $\operatorname{ini}(T_i) := \{\operatorname{ini}(T) : T \in T_i\}$ (or $\mathsf{Z}(\mathcal{F}) = \bigcup_{i=1}^{s} \mathsf{Z}(T_i / U_i)$ holds respectively). A triangular decomposition is called a *simple decomposition* if each of its triangular sets or systems is simple.

The process for computing the triangular decomposition of a polynomial set $\mathcal{F}$ is also called *triangular decomposition* of $\mathcal{F}$.

There exist many algorithms for decomposing polynomial sets into triangular sets or systems with different properties. One of the main strategies for these algorithms for triangular decomposition is to perform reduction on polynomials which contain the greatest unprocessed variable until there is only one polynomial left whose greatest variable is this variable, at the same time producing new polynomials whose leading variables are strictly smaller than this variable. Algorithms for triangular decomposition with this strategy are said to be *in top-down style* and the readers are referred to [20] for more details on such algorithms.

2.2   Subresultant Regular Subchains

Let F and G be two polynomials in $\mathbb{K}[\boldsymbol{x}]$ such that $m = \deg(F, x_k) \geq \deg(G, x_k) = \ell$, and $\mathbf{M}$ be the Sylvester matrix of F and G w.r.t. x_k. For two integers i, j ($0 \leq i \leq j < \ell$), define $\mathbf{M}_{ij}$ to be the submatrix of $\mathbf{M}$ obtained by deleting the last j rows of F's coefficients, the last j rows of G's coefficients, and the last $2j + 1$ columns except the $(m + \ell - i - j)$-th one. Then the polynomial $H_j = \sum_{i=0}^{j} |\mathbf{M}_{ij}| x_k^i$ is called the jth *subresultant* of F and G w.r.t. x_k. In particular, the jth subresultant H_j is said to be *regular* if $\deg(H_j, x_k) = j$.

Definition 5. Let F, G, H_j ($j = 0, \ldots, \mu - 1$) $\in \mathbb{K}[\boldsymbol{x}]$ be defined as above, where $\mu := m - 1$ when $m > \ell$ and $\mu := \ell$ otherwise. Then the sequence $F, G, H_{\mu-1}, H_{\mu-2}, \ldots, H_0$ is called the *subresultant chain* of F and G w.r.t. x_k. Furthermore, let $H_{d_1}, \ldots, H_{d_r}$ be the regular ones in $H_{\mu-1}, \ldots, H_0$ with $d_1 > \cdots > d_r$. Then the sequence $F, G, H_{d_1}, \ldots, H_{d_r}$ is called the *subresultant regular subchain* of F and G w.r.t. x_k.

Lemma 1. ([27, Lem. 3.3.3]) *Let P be a polynomial in $\mathbb{K}[\boldsymbol{x}]$ with $\mathrm{lv}(P) = x_k$ and $H_2, \ldots, H_r$ be the subresultant regular subchain of P and its formal derivative $\frac{\partial P}{\partial x_k}$ w.r.t. x_k. Let*

$$H_2^* = H_2, \quad H_i^* = H_i/I \qquad (i = 3, \ldots, r), \tag{1}$$

where $I = \mathrm{lc}(P, x_k)$. Then

$$Z(P/I) = \bigcup_{i=2}^{r} Z(\{Q_i, I_{i+1}, \ldots, I_r\}/II_i), \quad Z(\emptyset/PI) = \bigcup_{i=2}^{r} Z(\{I_{i+1}, \ldots, I_r\}/Q_i II_i),$$

where $Q_i = \mathrm{pquo}(P, H_i^*, x_k)$ and $I_i = \mathrm{lc}(H_i^*, x_k)$ for $i = 2, \ldots, r$. In particular, for any $i = 2, \ldots, r$ and $\bar{\boldsymbol{x}}_{k-1} \in Z(\{I_{i+1}, \ldots, I_r\}/II_i)$, $Q_i(\bar{\boldsymbol{x}}_{k-1}, x_k)$ is square-free.

2.3   Chordal Graphs and Polynomial Sets

Let F be a polynomial in $\mathbb{K}[\boldsymbol{x}]$. The set of variables which effectively appear in F is called the *variable support* of F, denoted by $\mathrm{supp}(F)$. For a polynomial set $\mathcal{F} \subset \mathbb{K}[\boldsymbol{x}]$, define $\mathrm{supp}(\mathcal{F}) := \bigcup_{F \in \mathcal{F}} \mathrm{supp}(F)$. We associate an undirected graph (V, E) to $\mathcal{F}$ with the vertex set $V = \mathrm{supp}(\mathcal{F})$ and the edge set $E = \{(x_i, x_j) : 1 \leq i \neq j \leq n \text{ and } \exists F \in \mathcal{F} \text{ such that } x_i, x_j \in \mathrm{supp}(F)\}$ and denote the associated graph by $G(\mathcal{F})$.

Definition 6. Let $G = (V, E)$ be a graph with $V = \{x_1, \ldots, x_n\}$. Then an ordering $x_{i_1} < x_{i_2} < \cdots < x_{i_n}$ of the vertices is called a *perfect elimination ordering* of G if for each $j = i_1, \ldots, i_n$, any two distinct vertices in $\{x_j\} \cup \{x_k : x_k < x_j \text{ and } (x_k, x_j) \in E\}$ are connected with an edge of G. A graph G is said to be *chordal* if there exists a perfect elimination ordering of it.

Chordality of a graph G implies that whenever $(x_i, x_k) \in G$ and $(x_j, x_k) \in G$ with $x_i < x_k$ and $x_j < x_k$, we have $(x_i, x_j) \in G$. We will frequently use this property in the proofs in the sequel. Whether an arbitrary graph is chordal or not can be effectively tested with existing algorithms [23] (a perfect elimination ordering will also be returned if chordality is verified).

Definition 7. A polynomial set $\mathcal{F} \subset \mathbb{K}[\boldsymbol{x}]$ is said to be *chordal* if its associated graph $G(\mathcal{F})$ is chordal.

The associated graph of the chordal polynomial set $\mathcal{P} = \{x_2 + x_1, x_3 + x_1, x_4^2 + x_2, x_4^3 + x_3, x_5 + x_2, x_5 + x_3 + x_2\}$ is illustrated in Fig. 1 below.

3   Reformulation of Wang's Algorithm for Simple Decomposition in Top-Down Style

In this section we reformulate Wang's top-down algorithm for simple decomposition, which essentially handles all the nodes in the decomposition tree in

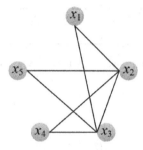

Fig. 1. Associated chordal graph $G(\mathcal{P})$ ($x_1 < x_2 < x_3 < x_4 < x_5$ is one perfect elimination ordering)

a depth-first way, into the following Algorithm 1, which focuses on how all the child nodes are spawned from each node and is thus convenient for the study on the chordality in an inductive way in Sect. 5.

Since in the decomposition process into simple systems, the squarefreeness of the polynomial in the equation or inequation part needs to be recorded, and thus the underlying data structure we use in Algorithm 1 is $(\mathcal{P}, \mathcal{Q}, k, \mathrm{sqf})$, which is slightly different from the one used in the previous related papers of the authors [19,20]. In this data structure, $\mathcal{P}$ is the polynomial set representing equations, $\mathcal{Q}$ is the polynomial set representing inequations, k is the index of the variable x_k under process, and sqf is a flag for recording whether the polynomial in $\mathcal{P}$ is squarefree or not: sqf $= 1$ means that $\#\mathcal{P}^{(k)} = 1$ and the polynomial in $\mathcal{P}^{(k)}$ is squarefree w.r.t. x_k. For the set Φ consisting of such 4-tuples and for an integer i ($0 \leq i \leq n$), $\Phi^{(i)}$ denotes $\{(\mathcal{P}, \mathcal{Q}, k, \mathrm{sqf}) \in \Phi : k = i\}$.

In Algorithm 1 below, the function pop($\mathcal{S}$) returns an element from S and then remove it from $\mathcal{S}$, the function SubRegSubchain(T_1, T_2) returns the subresultant regular subchain $(H_2, \ldots, H_r)$ of the two polynomials T_1 and T_2 w.r.t. $\mathrm{lv}(T_2)$, and the function SubRegSubchain*$(T, \frac{\partial T}{\partial x_k})$ returns $(H_2^*, \ldots, H_r^*)$, which are as defined in (1).

Next we justify the correctness of the reformulation as Algorithm 1, focusing on where two important changes happen: Case 1) When the parameter "k" in $(\mathcal{P}, \mathcal{Q}, k, \mathrm{sqf})$ becomes $k - 1$ (lines 6, 10, 14, and 33 in Algorithm 1, meaning that the process on the variable x_k finishes); Case 2) When the parameter "sqf" becomes 1 from 0 (lines 27, 31, and 38 in Algorithm 1, meaning that the polynomial in the equation part becomes squarefree).

Case 1. In this case we need to show that the process for handling the current variable x_k finishes, namely polynomial sets $\mathcal{P}$ and $\mathcal{Q}$ in the node $(\mathcal{P}, \mathcal{Q}, k, \mathrm{sqf})$ satisfy the conditions (b) and (c) of Definition 3 for $i = k$.

Line 6: now $\mathcal{P}^{(k)} = \mathcal{Q}^{(k)} = \emptyset$ and the conditions are satisfied trivially.

Line 10: now $\mathcal{P}^{(k)} = \emptyset$ and $T = \prod_{Q \in \mathcal{Q}^{(k)}} Q$ is of degree 1, which means $\mathcal{Q}^{(k)}$ contains only one polynomial which is T itself. Clearly $T(\bar{x}_{k-1}, x_k)$ is squarefree w.r.t. x_k for any $\bar{x}_{k-1}$, and the conditions are satisfied (note that $\mathrm{lv}(\mathrm{ini}(T)) < x_k$).

Algorithm 1: Wang's algorithm for simple decomposition $\Psi :=$ SimDec($\mathcal{F}$)

Input: $\mathcal{F}$, a polynomial set in $\mathbb{K}[\boldsymbol{x}]$
Output: Ψ, a set of finite simple systems which form a simple decomposition of $\mathcal{F}$

1   $\Phi := \{(\mathcal{F}, \emptyset, n, 0)\}; \quad \Psi := \emptyset;$
2   **for** $k = n, \ldots, 1$ **do**
3   **while** $\Phi^{(k)} \neq \emptyset$ **do**
4       $(\mathcal{P}, \mathcal{Q}, k, \mathrm{sqf}) := \mathrm{pop}(\Phi^{(k)});$
5   **if** $\mathcal{P}^{(k)} = \emptyset$ **then**
6   **if** $\mathcal{Q}^{(k)} = \emptyset$ **then** $\Phi := \Phi \cup \{(\mathcal{P}, \mathcal{Q}, k-1, 0)\};$
7   **else**
8           $T := \prod_{Q \in \mathcal{Q}^{(k)}} Q;$
9           $\Phi := \Phi \cup \{(\mathcal{P} \cup \mathrm{ini}(T), \mathcal{Q} \setminus \mathcal{Q}^{(k)} \cup \{\mathrm{tail}(T)\}, k, 0)\};$
10   **if** $\deg(T, x_k) = 1$ **then** $\Phi := \Phi \cup \{(\mathcal{P}, \mathcal{Q} \setminus \mathcal{Q}^{(k)} \cup \{T, \mathrm{ini}(T)\}, k-1, 0)\};$
11   **else**
12            $(H_2^*, \ldots, H_r^*) := \mathsf{SubRegSubchain}^*(T, \frac{\partial T}{\partial x_k});$
13   **for** $i = 2, \ldots, r$ **do**
14              $\Phi := \Phi \cup \{(\mathcal{P} \cup \{\mathrm{lc}(H_{i+1}^*, x_k)\}, \ldots, \mathrm{lc}(H_r^*, x_k)\}, \mathcal{Q} \setminus \mathcal{Q}^{(k)} \cup$
                $\{\mathrm{pquo}(T, H_i^*, x_k), \mathrm{lc}(H_i^*, x_k), \mathrm{ini}(T)\}, k-1, 0)\};$

15   **else**
16         $T_2 :=$ a polynomial in $\mathcal{P}^{(k)}$ of least degree in x_k;
17         $\Phi := \Phi \cup \{(\mathcal{P} \setminus \{T_2\} \cup \{\mathrm{ini}(T_2), \mathrm{tail}(T_2)\}, \mathcal{Q}, k, 0)\};$
18   **if** $\#\mathcal{P}^{(k)} > 1$ **then**
19           $T_1 :=$ a polynomial in $\mathcal{P}^{(k)} \setminus \{T_2\}$;
20           $(H_2, \ldots, H_r) := \mathsf{SubRegSubchain}(T_1, T_2);$
21   **if** $\mathrm{lv}(H_r) = x_k$ **then** $\overline{r} := r$; **else** $\overline{r} := r - 1$;
22   **for** $i = 2, \ldots, \overline{r} - 1$ **do**
23             $\Phi := \Phi \cup \{(\mathcal{P} \setminus \{T_1, T_2\} \cup \{H_i, \mathrm{lc}(H_{i+1}, x_k)\}, \ldots, \mathrm{lc}(H_{\overline{r}}, x_k)\}, \mathcal{Q} \cup$
            $\{\mathrm{ini}(T_2), \mathrm{lc}(H_i, x_k)\}, k, 0)\};$
24           $\Phi := \Phi \cup \{(\mathcal{P} \setminus \{T_1, T_2\} \cup \{H_r, H_{\overline{r}}\}, \mathcal{Q} \cup \{\mathrm{ini}(T_2), \mathrm{lc}(H_{\overline{r}}, x_k)\}, k, 0)\};$
25   **else**
26   **if** $\mathrm{sqf} = 0$ **then**
27   **if** $\deg(T_2) = 1$ **then** $\Phi := \Phi \cup \{(\mathcal{P}, \mathcal{Q} \cup \{\mathrm{ini}(T_2)\}, k, 1)\};$
28   **else**
29               $(H_2^*, \ldots, H_r^*) := \mathsf{SubRegSubchain}^*(T_2, \frac{\partial T_2}{\partial x_k});$
30   **for** $i = 2, \ldots, r$ **do**
31                 $\Phi := \Phi \cup \{(\mathcal{P} \setminus \{T_2\} \cup \{\mathrm{pquo}(T_2, H_i^*, x_k), \mathrm{lc}(H_{i+1}^*, x_k), \ldots,$
                 $\mathrm{lc}(H_r^*, x_k)\}, \mathcal{Q} \cup \{\mathrm{ini}(T_2), \mathrm{lc}(H_i^*, x_k)\}, k, 1)\};$

32   **else**
33   **if** $\mathcal{Q}^{(k)} = \emptyset$ **then** $\Phi := \Phi \cup \{(\mathcal{P}, \mathcal{Q}, k-1, 0)\};$
34   **else**
35               $T_1 :=$ a polynomial in $\mathcal{Q}^{(k)}$;
36               $(H_2, \ldots, H_r) := \mathsf{SubRegSubchain}(T_1, T_2);$
37   **for** $i = 2, \ldots, r$ **do**
38                 $\Phi := \Phi \cup \{(\mathcal{P} \setminus \{T_2\} \cup \{\mathrm{pquo}(T_2, H_i, x_k), \mathrm{lc}(H_{i+1}, x_k), \ldots,$
                 $\mathrm{lc}(H_r, x_k)\}, \mathcal{Q} \setminus \{T_1\} \cup \{\mathrm{lc}(H_i, x_k)\}, k, 1)\};$

39   **for** $(\mathcal{P}, \mathcal{Q}, 0) \in \Phi^{(0)}$ **do**
40   **if** $\mathcal{P}^{(0)} \cup \mathcal{Q}^{(0)} \setminus \{0\} = \emptyset$ **then**
41       $\Psi := \Psi \cup \{(\mathcal{P} \setminus \{0\}, \mathcal{Q} \setminus \{0\})\};$

42   **return** Ψ;

Line 14: now $\mathcal{P}^{(k)} = \emptyset$ and $\deg(T) > 1$. Note that for each i, the first two polynomial sets of the node adjoined to Φ are

$$\tilde{\mathcal{P}} := \mathcal{P} \cup \{\mathrm{lc}(H_{i+1}^*, x_k), \ldots, \mathrm{lc}(H_r^*, x_k)\},$$

$$\tilde{\mathcal{Q}} := \mathcal{Q} \backslash \mathcal{Q}^{(k)} \cup \{\mathrm{pquo}(T, H_i^*, x_k), \mathrm{lc}(H_i^*, x_k), \mathrm{ini}(T)\}.$$

First in the second polynomial set above, we see that $\mathrm{lv}(\mathrm{pquo}(T, H_i^*, x_k)) = x_k$, $\mathrm{lv}(\mathrm{lc}(H_i^*, x_k)) < x_k$, and $\mathrm{lv}(\mathrm{ini}(T)) < x_k$, and thus there is only one polynomial $\mathrm{pquo}(T, H_i^*, x_k)$ in the updated $\tilde{\mathcal{Q}}^{(k)}$. Then by Lemma 1 we know that $\mathrm{pquo}(P, H_i^*, x_k)(\overline{\boldsymbol{x}}_{k-1}, x_k)$ is squarefree for any $\overline{\boldsymbol{x}}_{k-1} \in Z(\{\mathrm{lc}(H_{i+1}^*, x_k), \ldots, \mathrm{lc}(H_r^*, x_k)\}/\mathrm{lc}(H_i^*, x_k)\,\mathrm{ini}(T))$. By the definition of $\tilde{\mathcal{P}}$ and $\tilde{\mathcal{Q}}$ above, we have

$$Z(\tilde{\mathcal{P}}^{(<k)}/\tilde{\mathcal{Q}}^{(<k)}) \subset Z(\{\mathrm{lc}(H_{i+1}^*, x_k), \ldots, \mathrm{lc}(H_r^*, x_k)\}/\mathrm{lc}(H_i^*, x_k)\,\mathrm{ini}(T)),$$

and thus the conditions are satisfied.

Line 33: now $\mathrm{sqf} = 1$, meaning that for the only one polynomial $T \in \mathcal{P}^{(k)}$, $T(\overline{\boldsymbol{x}}_{k-1}, x_k)$ is squarefree w.r.t. x_k and $\mathcal{Q} = \emptyset$, and thus the conditions are satisfied.

Case 2. In this case we need to show that in lines 27, 31, and 38, the polynomial set $\mathcal{P}$ of the node $(\mathcal{P}, \mathcal{Q}, k, \mathrm{sqf})$ to adjoin to Φ has only one polynomial whose leading variable is x_k and it is squarefree w.r.t. x_k.

The first argument is straightforward by viewing that all the three lines are governed by a "else" statement in line 15 with which the condition is "$\mathcal{P}^{(k)} = \emptyset$" and also governed by a "else" statement in line 25 with which the condition is "$\#\mathcal{P}^{(k)} > 1$".

The second argument is for the squarefreeness of the unique polynomial T in the polynomial set $\mathcal{P}$ of the node: in line 27 the squarefreeness of T is trivial since $\deg(T, x_k) = 1$, in line 31 the squarefreeness of T can be proved in a similar way to that in line 14 of Case 1 above, and in line 38 the polynomial $T = \mathrm{pquo}(T_2, H_i, x_k)$ is squarefree w.r.t. x_k because here T_2 is already squarefree w.r.t. x_k.

4   Decomposition Tree of Wang's Algorithm for Simple Decomposition in Top-Down Style

Based on the descriptions of Wang's algorithm for simple decomposition as $\mathsf{SimDec}()$ (Algorithm 1), now we can construct the decomposition tree of the algorithm in the following way. Let $\mathcal{F}$ be the input polynomial set of $\mathsf{SimDec}()$. The decomposition tree of $\mathsf{SimDec}()$ is rooted at the node $(\mathcal{F}, \emptyset, n, 0)$. Then any node $(\mathcal{P}, \mathcal{Q}, k, \mathrm{sqf})$ in the tree spawns its child nodes according the number of polynomials in $\mathcal{P}^{(k)}$, whether the polynomial in $\mathcal{P}^{(k)}$ is squarefree w.r.t. x_k or not, etc. This process is called splitting in the terminologies of triangular decomposition.

Next we identify all the four possible cases of splitting from a node $(\mathcal{P}, \mathcal{Q}, k, \mathrm{sqf})$ in the decomposition tree.

(I) When $\#\mathcal{P}^{(k)} > 1$, algorithm SimDec() picks a polynomial $T_2 \in \mathcal{P}^{(k)}$ of least degree in x_k and another polynomial $T_1 \in \mathcal{P}^{(k)}$ and applies elimination of x_k to these two polynomials by computing their subresultant regular subchain to spawn the child nodes $(\mathcal{P}', \mathcal{Q}', k, 0)$ (in line 17 of Algorithm 1) and $(\mathcal{P}_i, \mathcal{Q}_i, k, 0)$ for $i = 2, \ldots, \bar{r}$ (in lines 23 and 24) of $(\mathcal{P}, \mathcal{Q}, k, \mathrm{sqf})$, where

$$
\begin{aligned}
\mathcal{P}' &:= \mathcal{P} \backslash \{T_2\} \cup \{\mathrm{ini}(T_2), \mathrm{tail}(T_2)\}, \\
\mathcal{Q}' &:= \mathcal{Q}, \\
\mathcal{P}_i &:= \mathcal{P} \backslash \{T_1, T_2\} \cup \{H_i, \mathrm{lc}(H_{i+1}, x_k), \ldots, \mathrm{lc}(H_{\bar{r}}, x_k)\}, \quad i = 2, \ldots, \bar{r}-1, \quad (2) \\
\mathcal{P}_{\bar{r}} &:= \mathcal{P} \backslash \{T_1, T_2\} \cup \{H_r, H_{\bar{r}}\}, \\
\mathcal{Q}_i &:= \mathcal{Q} \cup \{\mathrm{ini}(T_2), \mathrm{lc}(H_i, x_k)\}, \qquad\qquad\qquad i = 2, \ldots, \bar{r}.
\end{aligned}
$$

(II) When $\#\mathcal{P}^{(k)} = 1$ and $\mathrm{sqf} = 0$, algorithm SimDec() computes the subresultant regular subchain of the unique polynomial $T_2 \in \mathcal{P}^{(k)}$ and its derivative to spawn the child nodes $(\mathcal{P}', \mathcal{Q}', k, 0)$ (in line 17) as in case (I) above and $(\mathcal{P}_i^*, \mathcal{Q}_i^*, k, 1)$ for $i = 2, \ldots, r$ (in line 31) of $(\mathcal{P}, \mathcal{Q}, k, \mathrm{sqf})$, where

$$
\begin{aligned}
\mathcal{P}_i^* &:= \mathcal{P} \backslash \{T_2\} \cup \{\mathrm{pquo}(T_2, H_i^*, x_k), \mathrm{lc}(H_{i+1}^*, x_k), \ldots, \mathrm{lc}(H_r^*, x_k)\}, \ i = 2, \ldots, r, \\
\mathcal{Q}_i^* &:= \mathcal{Q} \cup \{\mathrm{ini}(T_2), \mathrm{lc}(H_i^*, x_k)\}, \qquad\qquad\qquad\qquad\quad i = 2, \ldots, r.
\end{aligned} \quad (3)
$$

(III) When $\#\mathcal{P}^{(k)} = 1$ and $\mathrm{sqf} = 1$, algorithm SimDec() computes the subresultant regular subchain of the unique polynomial $T_2 \in \mathcal{P}^{(k)}$ and some polynomial T_1 in $\mathcal{Q}^{(k)}$ to spawn the child nodes $(\mathcal{P}', \mathcal{Q}', k, 0)$ (in line 17) as in case (I) above and $(\mathcal{P}_i', \mathcal{Q}_i', k, 1)$ for $i = 2, \ldots, r$ (in line 38) of $(\mathcal{P}, \mathcal{Q}, k, 1)$, where

$$
\begin{aligned}
\mathcal{P}_i' &:= \mathcal{P} \backslash \{T_2\} \cup \{\mathrm{pquo}(T_2, H_i, x_k), \mathrm{lc}(H_{i+1}, x_k), \ldots, \mathrm{lc}(H_r, x_k)\}, \ i = 2, \ldots, r, \\
\mathcal{Q}_i' &:= \mathcal{Q} \backslash \{T_1\} \cup \{\mathrm{lc}(H_i, x_k)\}, \qquad\qquad\qquad\qquad\quad i = 2, \ldots, r.
\end{aligned} \quad (4)
$$

(IV) When $\mathcal{P}^{(k)} = \emptyset$, algorithm SimDec() computes the subresultant regular subchain of the product T of all the polynomial in $\mathcal{Q}^{(k)}$ and its derivative to spawn the child nodes $(\mathcal{P}^\#, \mathcal{Q}^\#, k, 0)$ (in line 9) and $(\mathcal{P}_i^\#, \mathcal{Q}_i^\#, k-1, 0)$ for $i = 2, \ldots, r$ (in line 14) of $(\mathcal{P}, \mathcal{Q}, k, \mathrm{sqf})$, where

$$
\begin{aligned}
\mathcal{P}^\# &:= \mathcal{P} \cup \{\mathrm{ini}(T)\}, \\
\mathcal{Q}^\# &:= \mathcal{Q} \backslash \mathcal{Q}^{(k)} \cup \{\mathrm{tail}(T)\}, \\
\mathcal{P}_i^\# &:= \mathcal{P} \cup \{\mathrm{lc}(H_{i+1}^*, x_k), \ldots, \mathrm{lc}(H_r^*, x_k)\}, \qquad\qquad i = 2, \ldots, r, \\
\mathcal{Q}_i^\# &:= \mathcal{Q} \backslash \mathcal{Q}^{(k)} \cup \{\mathrm{pquo}(T, H_i^*, x_k), \mathrm{lc}(H_i^*, x_k), \mathrm{ini}(T)\}, \quad i = 2, \ldots, r.
\end{aligned} \quad (5)
$$

Based on the analysis above, the decomposition tree of the algorithm SimDec() for simple decomposition is illustrated in Fig. 2 below. We would like to remark that due to the heavy use of computation of subresultant regular subchains, the number of child nodes at one node in this tree is usually more than 2 and not fixed and thus dynamic.

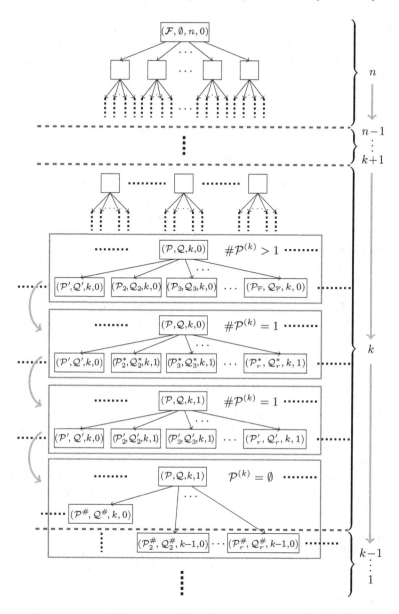

Fig. 2. Dynamic multi-branch decomposition tree of SimDec() for simple decomposition

5   Chordality of Polynomial Sets in Wang's Algorithm for Simple Decomposition in Top-Down Style

In this section we prove case by case that when the polynomial set $\mathcal{F}$ of SimDec() is chordal and a perfect elimination ordering is used as the variable ordering for

SimDec(), the two polynomial sets $\mathcal{P}$ and $\mathcal{Q}$ of an arbitrary node $(\mathcal{P}, \mathcal{Q}, k, \mathrm{sqf})$ in the decomposition tree have associated graphs $G(\mathcal{P})$ and $G(\mathcal{Q})$ which are subgraphs of $G(\mathcal{F})$, namely algorithm SimDec() preserves the chordality of the input polynomial set.

Proposition 1. *Let* $(\mathcal{P}, \mathcal{Q}, k, \mathrm{sqf})$ *be any node in the decomposition tree of* SimDec$(\mathcal{F})$ *such that* $\#\mathcal{P}^{(k)} \geq 1$, T_2 *be a polynomial in* $\mathcal{P}^{(k)}$ *with least degree in* x_k, *and* $\mathcal{P}'$ *be as defined in* (2). *Then we have* $G(\mathcal{P}') \subset G(\mathcal{P})$.

Proof. See the proof of [20, Prop. 20].   □

Proposition 2. *Let* $\mathcal{F} \subset \mathbb{K}[\boldsymbol{x}]$ *be a chordal polynomial set with* $x_1 < \cdots < x_n$ *as one perfect elimination ordering of* $G(\mathcal{F})$, *and* $(\mathcal{P}, \mathcal{Q}, k, \mathrm{sqf})$ *be an arbitrary node in the decomposition tree of* SimDec$(\mathcal{F})$ *such that* $\#\mathcal{P}^{(k)} > 1$, $G(\mathcal{P}) \subset G(\mathcal{F})$, *and* $G(\mathcal{Q}) \subset G(\mathcal{F})$. *Let* T_2 *be a polynomial in* $\mathcal{P}^{(k)}$ *with least degree in* x_k, T_1 *be another polynomial in* $\mathcal{P}^{(k)}$, *and* $\mathcal{P}_i, \mathcal{Q}_i$ $(i = 2, \ldots, \bar{r})$ *be as defined in* (2). *Then* $G(\mathcal{P}_i) \subset G(\mathcal{F})$ *and* $G(\mathcal{Q}_i) \subset G(\mathcal{F})$ *for* $i = 2, \ldots, \bar{r}$.

Proof. See the proof of [20, Prop. 24] for the inclusion $G(\mathcal{P}_i) \subset G(\mathcal{F})$. To prove $G(\mathcal{Q}_i) \subset G(\mathcal{F})$ for $i = 2, \ldots, \bar{r}$, it suffices to show that any edge $(x_p, x_q) \in G(\mathcal{Q}_i)$ is also an edge of $G(\mathcal{F})$. Since $\mathcal{Q}_i = \mathcal{Q} \cup \{\mathrm{ini}(T_2), \mathrm{lc}(H_i, x_k)\}$, if there exists a polynomial $T \in \mathcal{Q}$ such that $x_p, x_q \in \mathrm{supp}(T)$, then by the assumption $G(\mathcal{Q}) \subset G(\mathcal{F})$ clearly we have $(x_p, x_q) \in G(\mathcal{F})$; otherwise there exists a polynomial $T \in \{\mathrm{ini}(T_2), \mathrm{lc}(H_i, x_k)\}$ such that $x_p, x_q \in \mathrm{supp}(T)$, and by similar arguments in the proof of [20, Prop. 24] we can show that $(x_p, x_q) \in G(\mathcal{F})$.   □

Proposition 3. *Let* $\mathcal{F} \subset \mathbb{K}[\boldsymbol{x}]$ *be a chordal polynomial set with* $x_1 < \cdots < x_n$ *as one perfect elimination ordering of* $G(\mathcal{F})$, *and* $(\mathcal{P}, \mathcal{Q}, k, \mathrm{sqf})$ *be an arbitrary node in the decomposition tree of* SimDec$(\mathcal{F})$ *such that* $\#\mathcal{P}^{(k)} = 1$ *and* $\mathrm{sqf} = 0$. *Let* T_2 *be the unique polynomial in* $\mathcal{P}^{(k)}$, $H_2, \ldots, H_r$ *be the subresultant regular subchain of* T_2 *and* $\frac{\partial T_2}{\partial x_k}$, *and* $\mathcal{P}_i^*$ $(i = 2, \ldots, r)$ *be as defined in* (3). *Then* $G(\mathcal{P}_i^*) \subset G(\mathcal{P})$ *for* $i = 2, \ldots, r$. *In particular, if* $G(\mathcal{P}) \subset G(\mathcal{F})$ *and* $G(\mathcal{Q}) \subset G(\mathcal{F})$, *we have* $G(\mathcal{Q}_i^*) \subset G(\mathcal{F})$ *for* $i = 2, \ldots, r$.

Proof. Note that for each $i = 2, \ldots, r$, $\mathcal{P}_i^*$ is constructed from $\mathcal{P}$ by removing T_2 and adding pquo(T_2, H_i^*, x_k), $\mathrm{lc}(H_{i+1}^*, x_k), \ldots$, and $\mathrm{lc}(H_r^*, x_k)$. First all the polynomials $H_2, \ldots, H_r$ in the subresultant regular subchain, and thus $H_2^*, \ldots, H_r^*$ as defined in (1), are constructed from the polynomial T_2 only, and so are pquo(T_2, H_i^*, x_k), $\mathrm{lc}(H_{i+1}^*, x_k), \ldots$, and $\mathrm{lc}(H_r^*, x_k)$. This leads to the inclusions $\mathrm{supp}(\mathrm{pquo}(T_2, H_i^*, x_k)) \subset \mathrm{supp}(T_2)$ and $\mathrm{supp}(\mathrm{lc}(H_i^*, x_k)) \subset \mathrm{supp}(T_2)$ for $i = 2, \ldots, r$.

For any edge $(x_p, x_q) \in G(\mathcal{P}_i^*)$, if there exists a polynomial $T \in \mathcal{P} \setminus \{T_2\}$ such that $x_p, x_q \in \mathrm{supp}(T)$, then clearly $(x_p, x_q) \in G(\mathcal{P})$. Otherwise there exists a polynomial $T \in \{\mathrm{pquo}(T_2, H_i^*, x_k), \mathrm{lc}(H_{i+1}^*, x_k), \ldots, \mathrm{lc}(H_r^*, x_k)\}$ such that $x_p, x_q \in \mathrm{supp}(T)$. By the argument above we know that $x_p, x_q \in \mathrm{supp}(T_2)$, and thus $(x_p, x_q) \in G(\mathcal{P})$. This proves the first part of the conclusion.

By (3) we know that $\mathcal{Q}_i^* := \mathcal{Q} \cup \{\mathrm{ini}(T_2), \mathrm{lc}(H_i^*, x_k)\}$. Clearly $\mathrm{supp}(\mathcal{Q}_i^*) \subset \mathrm{supp}(\mathcal{P}) \cup \mathrm{supp}(\mathcal{Q}) \subset \mathrm{supp}(\mathcal{F})$. To prove $G(\mathcal{Q}_i^*) \subset G(\mathcal{F})$, it suffices to show that any edge $(x_p, x_q) \in G(\mathcal{Q}_i^*)$ is also an edge of $G(\mathcal{F})$. If there exists a polynomial $T \in \mathcal{Q}$ such that $x_p, x_q \in \mathrm{supp}(T)$, then by the assumption $G(\mathcal{Q}) \subset G(\mathcal{F})$ we have $(x_p, x_q) \in G(\mathcal{F})$; otherwise there exists a polynomial $T \in \{\mathrm{ini}(T_2), \mathrm{lc}(H_i^*, x_k)\}$ such that $x_p, x_q \in \mathrm{supp}(T)$, and in a similar way as in the proof above we know that $(x_p, x_q) \in G(\mathcal{P}) \subset G(\mathcal{F})$.   □

Proposition 4. *Let $\mathcal{F} \subset \mathbb{K}[\boldsymbol{x}]$ be a chordal polynomial set with $x_1 < \cdots < x_n$ as one perfect elimination ordering of $G(\mathcal{F})$ and $(\mathcal{P}, \mathcal{Q}, k, \mathrm{sqf})$ be an arbitrary node in the decomposition tree of $\mathsf{SimDec}(\mathcal{F})$ such that $\mathrm{sqf} = 1$, $G(\mathcal{P}) \subset G(\mathcal{F})$, and $G(\mathcal{Q}) \subset G(\mathcal{F})$. Let T_2 be the unique polynomial in $\mathcal{P}^{(k)}$, T_1 be some polynomial in $\mathcal{Q}^{(k)}$, $H_2, \ldots, H_r$ be the subresultant regular subchain of T_1 and T_2, and $\mathcal{P}_i', \mathcal{Q}_i'$ $(i = 2, \ldots, r)$ be as defined in (4). Then $G(\mathcal{P}_i') \subset G(\mathcal{F})$ and $G(\mathcal{Q}_i') \subset G(\mathcal{F})$ for $i = 2, \ldots, r$.*

Proof. For each $i = 2, \ldots, r$, clearly we have $\mathrm{supp}(\mathcal{P}_i') \subset \mathrm{supp}(\mathcal{P}) \cup \mathrm{supp}(\mathcal{Q}) \subset \mathrm{supp}(\mathcal{F})$. For any edge $(x_p, x_q) \in G(\mathcal{P}_i')$, if there exists a polynomial $T \in \mathcal{P} \setminus \{T_2\}$ such that $x_p, x_q \in \mathrm{supp}(T)$, then $(x_p, x_q) \in G(\mathcal{P}) \subset G(\mathcal{F})$. Otherwise there exists a polynomial $T \in \{\mathrm{pquo}(T_2, H_i, x_k), \mathrm{lc}(H_{i+1}, x_k), \ldots, \mathrm{lc}(H_r, x_k)\}$ such that $x_p, x_q \in \mathrm{supp}(T)$, and thus $x_p, x_q \in \mathrm{supp}(T) \subset \mathrm{supp}(T_1) \cup \mathrm{supp}(T_2)$. We consider the following three cases.

(a) When $x_p, x_q \in \mathrm{supp}(T_1)$: by the assumption that $G(\mathcal{Q}) \subset G(\mathcal{F})$ we have $(x_p, x_q) \in G(\mathcal{Q}) \subset G(\mathcal{F})$.

(b) When $x_p, x_q \in \mathrm{supp}(T_2)$: similarly by $G(\mathcal{P}) \subset G(\mathcal{F})$ we have $(x_p, x_q) \in G(\mathcal{P}) \subset G(\mathcal{F})$.

(c) When $x_p \in \mathrm{supp}(T_1)$ and $x_q \in \mathrm{supp}(T_2)$ (without loss of generality, we can assume that): by $T_1 \in \mathcal{Q}^{(k)}$ we know that $x_k \in \mathrm{supp}(T_1)$ and thus $(x_p, x_k) \in G(\mathcal{Q}) \subset G(\mathcal{F})$. Similarly we also have $(x_q, x_k) \in G(\mathcal{F})$. Then by the chordality of $G(\mathcal{F})$ we know that $(x_p, x_q) \in G(\mathcal{F})$.

To summarize, we have $G(\mathcal{P}_i') \subset G(\mathcal{F})$ for $i = 2, \ldots, r$.

Next we prove the inclusions $G(\mathcal{Q}_i') \subset G(\mathcal{F})$. For each $i = 2, \ldots, r$, we know that $\mathcal{Q}_i' := \mathcal{Q} \setminus \{T_1\} \cup \{\mathrm{lc}(H_i, x_k)\}$ and it is easy to see that $\mathrm{supp}(\mathcal{Q}_i') \subset \mathrm{supp}(\mathcal{P}) \cup \mathrm{supp}(\mathcal{Q}) \subset \mathrm{supp}(\mathcal{F})$. For any edge $(x_p, x_q) \in G(\mathcal{Q}_i')$, if there exists a polynomial $T \in \mathcal{Q} \setminus \{T_1\}$, then clearly $(x_p, x_q) \in G(\mathcal{P}) \subset G(\mathcal{F})$; otherwise $x_p, x_q \in \mathrm{lc}(H_i, x_k)$, and by the same arguments above we know that $(x_p, x_q) \in G(\mathcal{F})$.   □

Proposition 5. *Let $\mathcal{F} \subset \mathbb{K}[\boldsymbol{x}]$ be a chordal polynomial set with $x_1 < \cdots < x_n$ as one perfect elimination ordering of $G(\mathcal{F})$ and $(\mathcal{P}, \mathcal{Q}, k, \mathrm{sqf})$ be an arbitrary node in the decomposition tree of $\mathsf{SimDec}(\mathcal{F})$ such that $\mathcal{P}^{(k)} = \emptyset$, $G(\mathcal{P}) \subset G(\mathcal{F})$, and $G(\mathcal{Q}) \subset G(\mathcal{F})$. Let $T = \prod_{Q \in \mathcal{Q}^{(k)}} Q$, $H_2, \ldots, H_r$ be the subresultant regular subchain of T and $\frac{\partial T}{\partial x_k}$, and $\mathcal{P}^\#$, $\mathcal{Q}^\#$, and $\mathcal{P}_i^\#, \mathcal{Q}_i^\#$ $(i = 2, \ldots, r)$ be as defined in (5). Then $G(\mathcal{P}^\#), G(\mathcal{Q}^\#), G(\mathcal{P}_i^\#), G(\mathcal{Q}_i^\#)$ $(i = 2, \ldots, r)$ are all subgraphs of $G(\mathcal{F})$.*

Proof. The inclusions $G(\mathcal{P}^\#) \subset G(\mathcal{F})$ and $G(\mathcal{Q}^\#) \subset G(\mathcal{F})$ are easy to derive with the assumptions $G(\mathcal{P}) \subset G(\mathcal{F})$ and $G(\mathcal{Q}) \subset G(\mathcal{F})$.

As in the first part of the proof for Proposition 3 above, we can show that $\mathrm{supp}(\mathrm{pquo}(T_1, H_i^*, x_k))$, $\mathrm{supp}(\mathrm{ini}(T))$, and $\mathrm{supp}(\mathrm{lc}(H_i^*, x_k))$ $(i = 2, \ldots, r)$ are all subsets of $\mathrm{supp}(T)$. For any edge $(x_p, x_q) \in G(\mathcal{Q}_i^\#)$, if there exists a polynomial $\tilde{T} \in \mathcal{Q} \setminus \mathcal{Q}^{(k)}$ such that $x_p, x_q \in \mathrm{supp}(\tilde{T})$, then clearly $(x_p, x_q) \in G(\mathcal{Q}) \subset G(\mathcal{F})$. Otherwise there exists a polynomial $\tilde{T} \in \{\mathrm{pquo}(T, H_i^*, x_k), \mathrm{lc}(H_i^*, x_k), \mathrm{ini}(T)\}$ such that $x_p, x_q \in \mathrm{supp}(\tilde{T})$. By the above arguments we know that $x_p, x_q \in \mathrm{supp}(T)$. Since $T = \prod_{Q \in \mathcal{Q}^{(k)}} Q$, there exist polynomials Q_1 and Q_2 in $\mathcal{Q}^{(k)}$ such that $x_p, x_k \in \mathrm{supp}(Q_1)$ and $x_q, x_k \in \mathrm{supp}(Q_2)$. Then $(x_p, x_k) \in G(\mathcal{Q}) \subset G(\mathcal{F})$ and $(x_q, x_k) \in G(\mathcal{Q}) \subset G(\mathcal{F})$, and the chordality of $G(\mathcal{F})$ implies $(x_p, x_q) \in G(\mathcal{F})$.

To prove the inclusions $G(\mathcal{P}_i^\#) \subset G(\mathcal{F})$, it suffices to show that each edge $(x_p, x_q) \in G(\mathcal{P}_i^\#)$ is also an edge of $G(\mathcal{F})$. If there exists a polynomial $\tilde{T} \in \mathcal{P}$ such that $x_p, x_q \in \mathrm{supp}(\tilde{T})$, then clearly $(x_p, x_q) \in G(\mathcal{P}) \subset G(\mathcal{F})$; otherwise there exists a polynomial $\tilde{T} \in \{\mathrm{lc}(H_{i+1}^*, x_k), \ldots, \mathrm{lc}(H_r^*, x_k)\}$ such that $x_p, x_q \in \mathrm{supp}(\tilde{T})$, and by the same arguments as above, we know that $(x_p, x_q) \in G(\mathcal{Q}) \subset G(\mathcal{F})$. This ends the proof. $\quad\square$

Theorem 1. *Let $\mathcal{F} \subset \mathbb{K}[\boldsymbol{x}]$ be a chordal polynomial set with $x_1 < \cdots < x_n$ as one perfect elimination ordering of $G(\mathcal{F})$. Then for any node $(\mathcal{P}, \mathcal{Q}, k, \mathrm{sqf})$ in the decomposition tree of $\mathsf{SimDec}(\mathcal{F})$, we have $G(\mathcal{P}) \subset G(\mathcal{F})$ and $G(\mathcal{Q}) \subset G(\mathcal{F})$.*

Proof. We prove this theorem by induction on the depth of the node in the decomposition tree of $\mathsf{SimDec}(\mathcal{F})$. When $d = 0$, clearly for the root $(\mathcal{F}, \emptyset, n, 0)$ the conclusions hold. Now suppose that for any node $(\tilde{\mathcal{P}}, \tilde{\mathcal{Q}}, \tilde{k}, \mathrm{sqf})$ of depth d in the decomposition tree, we have $G(\tilde{\mathcal{P}}) \subset G(\mathcal{F})$ and $G(\tilde{\mathcal{Q}}) \subset G(\mathcal{F})$. Let $(\mathcal{P}, \mathcal{Q}, k, \mathrm{sqf})$ be an arbitrary node of depth $d + 1$ in the decomposition tree, with its parent node $(\tilde{\mathcal{P}}, \tilde{\mathcal{Q}}, \tilde{k}, \mathrm{sqf})$ of depth d. Next we prove $G(\mathcal{P}) \subset G(\mathcal{F})$ and $G(\mathcal{Q}) \subset G(\mathcal{F})$.

In Algorithm 1 there are the following lines where new nodes are spawned, and we prove the conclusions case by case.

(1) Lines 6 and 33: trivially by the inductive assumption.
(2) Lines 9 and 14: the inclusions can be derived directly by Proposition 5.
(3) Line 10: it suffices to prove that for any $(x_p, x_q) \in G(\mathcal{Q})$, we have $(x_p, x_q) \in G(\tilde{\mathcal{Q}})$. If there exists a polynomial $\tilde{T} \in \tilde{\mathcal{Q}} \setminus \tilde{\mathcal{Q}}^{(k)}$ such that $x_p, x_q \in \mathrm{supp}(\tilde{T})$, then clearly $(x_p, x_q) \in G(\tilde{\mathcal{Q}})$; otherwise there exists a polynomial $\tilde{T} = T$ or $\mathrm{ini}(T)$ such that $x_p, x_q \in \mathrm{supp}(\tilde{T})$, where T is the unique polynomial in $\tilde{\mathcal{Q}}^{(k)}$ (which is true under the condition $\deg(T, x_k) = 1$ in line 10), then we know that $x_p, x_q \in \mathrm{supp}(T)$ and thus $(x_p, x_q) \in G(\tilde{\mathcal{Q}}) \subset G(\mathcal{F})$.
(4) Line 17: by Proposition 1 we have $G(\mathcal{P}) \subset G(\mathcal{F})$, and the inclusion $G(\mathcal{Q}) \subset G(\mathcal{F})$ holds trivially.
(5) Lines 23 and 24: the inclusions can be derived directly by Proposition 2.
(6) Line 27: easy to derive by viewing that $\mathrm{supp}(\mathrm{ini}(T_2)) \subset \mathrm{supp}(T_2)$.
(7) Line 31: the inclusions can be derived directly by Proposition 3.
(8) Line 38: the inclusions can be derived directly by Proposition 4.

This ends the inductive proof of the theorem. □

Corollary 1. *Let $\mathcal{F} \subset \mathbb{K}[\boldsymbol{x}]$ be a chordal polynomial set with $x_1 < \cdots < x_n$ as one perfect elimination ordering of $G(\mathcal{F})$ and $(\mathcal{T}_1, \mathcal{U}_1), \ldots, (\mathcal{T}_s, \mathcal{U}_s)$ be the simple systems computed by* SimDec($\mathcal{F}$). *Then $G(\mathcal{T}_i) \subset G(\mathcal{F})$ for $i = 1, \ldots, s$.*

Proof. Straightforward from Theorem 1. □

6 Concluding Remarks

In this paper we first reformulate Wang's algorithm for simple decomposition in top-down style into Algorithm 1 which is suitable for studying it in an inductive way. Then we prove in Theorem 1 that under the conditions that the polynomial set to decompose is chordal and one perfect elimination ordering is used, all the polynomial sets representing equations and inequations in the decomposition process of Wang's algorithm for simple decomposition have associated graphs which are subgraphs of the input chordal graph. This naturally leads to Corollary 1 that all the simple systems computed by this algorithm have associated graphs which are subgraphs of the chordal graph. In other words, we prove that Wang's algorithm for simple decomposition in top-down style preserves chordality of the input polynomial set.

References

1. Aubry, P., Lazard, D., Moreno Maza, M.: On the theories of triangular sets. J. Symbolic Comput. **28**(1–2), 105–124 (1999)
2. Bächler, T.: Counting solutions of algebraic systems via triangular decomposition. Ph.D. thesis, RWTH Aachen University (2014)
3. Bächler, T., Gerdt, V., Lange-Hegermann, M., Robertz, D.: Algorithmic Thomas decomposition of algebraic and differential systems. J. Symbolic Comput. **47**(10), 1233–1266 (2012)
4. Buchberger, B.: Ein Algorithmus zum Auffinden der Basiselemente des Restklassenrings nach einem nulldimensionalen Polynomideal. Ph.D. thesis, Universität Innsbruck, Austria (1965)
5. Chai, F., Gao, X.S., Yuan, C.: A characteristic set method for solving Boolean equations and applications in cryptanalysis of stream ciphers. J. Syst. Sci. Complex. **21**(2), 191–208 (2008)
6. Chen, C., Moreno Maza, M.: Algorithms for computing triangular decompositions of polynomial systems. J. Symbolic Comput. **47**(6), 610–642 (2012)
7. Cifuentes, D., Parrilo, P.A.: Chordal networks of polynomial ideals. SIAM J. Appl. Algebra Geom. **1**(1), 73–110 (2017)
8. Cox, D.A., Little, J.B., O'Shea, D.: Using Algebraic Geometry. Springer, Heidelberg (1998). https://doi.org/10.1007/978-1-4757-6911-1
9. Faugère, J.C.: A new efficient algorithm for computing Gröbner bases (F_4). J. Pure Appl. Algebra **139**(1–3), 61–88 (1999)
10. Gao, X.S., Jiang, K.: Order in solving polynomial equations. In: Gao, X.S., Wang, D. (eds.) Computer Mathematics, Proceedings of ASCM 2000, pp. 308–318. World Scientific (2000)

11. Gao, X.S., Chou, S.C.: Solving parametric algebraic systems. In: Wang, P. (ed.) Proceedings of ISSAC 1992, pp. 335–341. ACM (1992)
12. Gerdt, V., Robertz, D.: Lagrangian constraints and differential Thomas decomposition. Adv. Appl. Math. **72**, 113–138 (2016)
13. Gilbert, J.R.: Predicting structure in sparse matrix computations. SIAM J. Matrix Anal. Appl. **15**(1), 62–79 (1994)
14. Huang, Z., Lin, D.: Attacking Bivium and Trivium with the characteristic set method. In: Nitaj, A., Pointcheval, D. (eds.) AFRICACRYPT 2011. LNCS, vol. 6737, pp. 77–91. Springer, Heidelberg (2011). https://doi.org/10.1007/978-3-642-21969-6_5
15. Hubert, E.: Notes on triangular sets and triangulation-decomposition algorithms i: polynomial systems. In: Winkler, F., Langer, U. (eds.) SNSC 2001. LNCS, vol. 2630, pp. 1–39. Springer, Heidelberg (2003). https://doi.org/10.1007/3-540-45084-X_1
16. Kalkbrener, M.: A generalized Euclidean algorithm for computing triangular representations of algebraic varieties. J. Symbolic Comput. **15**(2), 143–167 (1993)
17. Mou, C., Wang, D., Li, X.: Decomposing polynomial sets into simple sets over finite fields: The positive-dimensional case. Theoret. Comput. Sci. **468**, 102–113 (2013)
18. Mou, C.: Symbolic detection of steady states of autonomous differential biological systems by transformation into block triangular form. In: Jansson, J., Martín-Vide, C., Vega-Rodríguez, M.A. (eds.) AlCoB 2018. LNCS, vol. 10849, pp. 115–127. Springer, Cham (2018). https://doi.org/10.1007/978-3-319-91938-6_10
19. Mou, C., Bai, Y.: On the chordality of polynomial sets in triangular decomposition in top-down style. In: Arreche, C. (ed.) Proceedings of ISSAC 2018, pp. 287–294. ACM (2018)
20. Mou, C., Bai, Y., Lai, J.: Chordal graphs in triangular decomposition in top-down style. J. Symbolic Comput. (2019, to appear)
21. Parter, S.: The use of linear graphs in Gauss elimination. SIAM Rev. **3**(2), 119–130 (1961)
22. Rose, D.J.: Triangulated graphs and the elimination process. J. Math. Anal. Appl. **32**(3), 597–609 (1970)
23. Rose, D.J., Tarjan, R.E., Lueker, G.S.: Algorithmic aspects of vertex elimination on graphs. SIAM J. Comput. **5**(2), 266–283 (1976)
24. Wang, D.: An elimination method for polynomial systems. J. Symbolic Comput. **16**(2), 83–114 (1993)
25. Wang, D.: Decomposing polynomial systems into simple systems. J. Symbolic Comput. **25**(3), 295–314 (1998)
26. Wang, D.: Computing triangular systems and regular systems. J. Symbolic Comput. **30**(2), 221–236 (2000)
27. Wang, D.: Elimination Methods. Springer, Wien (2001). https://doi.org/10.1007/978-3-7091-6202-6
28. Wu, W.T.: On zeros of algebraic equations: An application of Ritt principle. Kexue Tongbao **31**(1), 1–5 (1986)
29. Wu, W.T.: Mechanical Theorem Proving in Geometries: Basic Principles. Springer, Heidelberg (1994). https://doi.org/10.1007/978-3-7091-6639-0
30. Yang, L., Zhang, J.Z.: Searching dependency between algebraic equations: An algorithm applied to automated reasoning. In: Johnson, J., McKee, S., Vella, A. (eds.) Artificial Intelligence in Mathematics, pp. 147–156. Oxford University Press, Oxford (1994)

Automatic Synthesis of Merging and Inserting Algorithms on Binary Trees Using Multisets in *Theorema*

Isabela Drămnesc[1(✉)] and Tudor Jebelean[2]

[1] West University, Timişoara, Romania
isabela.dramnesc@e-uvt.ro
[2] Johannes Kepler University, Linz, Austria
Tudor.Jebelean@jku.at

Abstract. We demonstrate the automatic proof–based synthesis of merging and inserting algorithms for [sorted] binary trees, using the notion of multisets, in the *Theorema* system. Each algorithm is extracted from the proof of the conjecture based on the specification of the desired function, in the form of a list of [conditional] equalities, which can be directly executed. The proofs are performed in natural style, using general techniques, but most importantly efficient inference rules and strategies specific for the domains involved. In particular we present specific techniques for the construction of arbitrarily nested recursive algorithms by general Noetherian induction, as well as a systematic method for the generation of the conjectures and consequently of the algorithms for the auxiliary functions needed in the main function.

Keywords: Algorithm synthesis · Binary trees · Multisets · Theorema

1 Introduction

Automated synthesis of algorithms based on logical principles is an interesting alternative to algorithm verification, because it focuses on the study of the properties of the involved domains, from which correct algorithms are obtained automatically, instead of creating them by human ingenuity. The case studies presented in this paper are part of our research on systematic theory construction (*theory exploration* [2]) and automated synthesis in the domain of finite *binary trees* for which we also use finite *multisets*. In two related papers [9,10] we already investigated algorithms for deletion from lists and binary trees, as well as sorting algorithms for lists. Multisets allow to express in a natural way the fact that two trees have the same elements, but more importantly (as are revealed by our experiments) it leads to powerful proof techniques. For space reasons, in this presentation we focus on *one argument induction*[1] and also on *compositional construction*[2] and do not approach yet algorithms which use both

---

[1] For binary functions one may use simultaneous induction on both arguments.
[2] The construction of the object desired for the synthesis uses only the objects which are already present in the proof, and does not try to decompose some of them.

© Springer Nature Switzerland AG 2020
D. Slamanig et al. (Eds.): MACIS 2019, LNCS 11989, pp. 153–168, 2020.
https://doi.org/10.1007/978-3-030-43120-4_13

lists and trees. We approach automated synthesis as described in our previous work – see e.g. [8,15]. First one proves automatically a *synthesis conjecture* which is based on the *specification* (input and output conditions) of the desired function, then the algorithm is extracted automatically from the proof. We use the *Theorema* system [6], in which the inference rules and the logical formulae are presented in *natural style* – similar to the one used by humans. Since *Theorema* also allows the execution of algorithms, we can test them immediately in the system. The theoretical basis and the correctness of the proof based synthesis scheme is well–known, see [7,18].

Each algorithm is produced as a list of clauses, each clause being a (possibly conditional) universally quantified equality which is to be applied as a rewrite rule from left to right. The LHS[3] of each equality consists of the function symbol (of the desired algorithm) applied to a term which identifies a certain class of possible inputs (this is sometimes called *pattern matching* programming). The clauses are such that all possible inputs are considered (*covering*), and no two clauses may apply to the same input (*mutual exclusion*) – these properties are automatically insured by the synthesis method.

Related Work and Originality. [18] introduces deductive techniques for algorithm synthesis, in particular for constructing recursive algorithms. These techniques are applied in [23] to manually derive several sorting algorithms in the theories of integers and strings. They present also a rule for generating auxiliary algorithms, see also [20]. Later implementations using some of these principles are in [17,22]. We presented a more detailed survey of synthesis methods in [8]. In the current paper we follow some of the principles from [18,23], but we develop different proof–based techniques for algorithm synthesis.

The theory of *multisets* is well studied in the literature, including computational formalizations (see e.g. [19], where finite multisets are called *bags*). A presentation of the theory of multisets and a good survey of the literature related to multisets and their usage is [1] and some interesting practical developments are in [21]. In previous work on synthesis, multisets are not explicitly used in the process of proof–based algorithm synthesis. They are just mentioned in the problem specification (e.g., in expressing the permutation of two objects), but their definition and properties are not involved in the process of proof–based algorithm synthesis. In this paper we explicitly use multisets, their definition and properties in the entire process of algorithm synthesis.

In our previous work we study proof–based algorithm synthesis in the theories of lists, sets and binary trees [12] separately ([8,14,15]), but without using multisets.

A systematic formalization of the theory of lists using multisets for the proofs of correctness of various sorting algorithms is mechanized in Isabelle/HOL[4], but this does not address the problem of algorithm synthesis. An interesting formalization in a previous version of *Theorema* [5], which includes the theory exploration and the synthesis of a sorting algorithm is presented in [4], which also

---

[3] We use LHS for left hand side and RHS for right hand side.
[4] https://isabelle.in.tum.de/library/HOL/HOL-Library/Sorting_Algorithms.html.

constituted the inspiration of our previous research on proof–based synthesis. However, in that pioneering work, the starting point of the synthesis (besides the specification of the desired function) is a specific *algorithm scheme*, while in our approach we use induction principles and dynamic induction.

In contrast to other investigations and to our previous research, the current study *uses multisets* in the synthesis problem and in the entire process of algorithm synthesis, combined properties which are necessary in the process of algorithm synthesis, the automatically generated proofs are performed in the new version of the *Theorema* system [6,24], and the investigation is performed in parallel on the two domains. We already investigated the proof–based synthesis of auxiliary algorithms on binary trees: *Merge* [11], and *Insert* [13], see also [15], but we did not use multisets and we applied different proof techniques. In our current approach using multisets we investigate in companion papers the synthesis of *Delete* on lists and trees [9] and the synthesis of sorting algorithms on lists [10].

Moreover, this paper describes more precisely the practical techniques for *cascading* and for *general Noetherian induction*, and illustrates them in more detail on several examples. For the purposes above, **three novel inference rules** are introduced, and **seven inference rules** and **seven strategies** are extended for these case studies on binary trees using multisets.

2   Proof–Based Synthesis

2.1   Context

Notations. We use square brackets for function and for predicate application, for instance: $f[x]$ instead of $f(x)$ and $P[a]$ instead of $P(a)$. Quantified variables are placed under the quantifier, as in $\forall_X$ and $\exists_X$.

The objects occurring in the formulae are: *elements*—objects from a totally ordered domain (denoted a, b, c) which are members of composite objects; *multisets* denoted A, B, C; and *binary trees* denoted L, R, S, T, X, Y, Z. (Multisets and binary trees are also addressed as *composite objects*).

Knowledge Base. For space reasons, we list explicitly only the formulae which are used in the proofs presented in this paper, the theory exploration includes more statements.

Elements of various composite structures are any objects whose domain is totally ordered (notation $\leq$ and $<$). The ordering on elements is extended to orderings between an element and a composite object (denoted $\preceq, \prec$) and between composite objects (denoted $\ll$), by requiring that all elements of the composite object observe the ordering relation[5].

Finite multisets are composite objects which may contain the same elements several times, that is each element has a certain *multiplicity*. $\emptyset$ denotes the empty

---

[5] Note that this introduces exceptions to antisymmetry and transitivity when the empty composite object is involved.

multiset, $\{\{a\}\}$ denotes the multiset having only the element a with multiplicity 1. The union (additive) is denoted by $\uplus$: multiplicity is the sum of multiplicities— like in [16]. Union is commutative and associative with unit $\emptyset$, these properties are used implicitly by the prover. We use $\mathcal{M}$ for the multiset of elements of a tree. When two trees have the same elements (that is, their multisets are equal), we call them *equivalent*.

A finite binary **tree** is either ε (empty) or a triplet $\langle L, a, R \rangle$, where L and R are trees. The multiset of a tree has the following property:

Property 1. $\underset{a,L,R}{\forall} \left(\begin{array}{c} \mathcal{M}[\varepsilon] = \emptyset \\ \mathcal{M}[\langle L, a, R \rangle] = \mathcal{M}[L] \uplus \{\{a\}\} \uplus \mathcal{M}[R] \end{array} \right)$

Sorted trees are defined in the following way:

Definition 1.
$$\underset{a,L,R}{\forall} \left(\begin{array}{c} IsSorted[\varepsilon] \\ IsSorted[\langle L, a, R \rangle] \iff IsSorted[L] \wedge IsSorted[R] \wedge L \preceq a \preceq R \end{array} \right)$$

Problem and Approach. Given two trees X, Y, *merge* them into a tree Z. Moreover, if X, Y are sorted, then Z should be also sorted. The synthesis conjecture has the general structure $\underset{X,Y}{\forall}(I[X,Y] \implies \underset{Z}{\exists}O[X,Y,Z])$, where I is the input condition and O is the output condition.

In the general case X, Y, Z are not required to be sorted, there is no input condition[6] and the output condition $O[X,Y,Z]$ is $(\mathcal{M}[Z] = \mathcal{M}[X] \uplus \mathcal{M}[Y])$, thus we have:

Conjecture 1. $\underset{XYZ}{\forall\forall\exists}\left(\mathcal{M}[Z] = \mathcal{M}[X] \uplus \mathcal{M}[Y] \right)$

For sorted trees we also consider the input condition $I[X,Y]$: $(IsSorted[X] \wedge IsSorted[Y])$ and we add $IsSorted[Z]$ to the output condition, thus we have:

Conjecture 2. $\underset{XY}{\forall\forall}(IsSorted[X] \wedge IsSorted[Y]) \implies \underset{Z}{\exists}\left(\mathcal{M}[Z] = \mathcal{M}[X] \uplus \mathcal{M}[Y] \wedge IsSorted[Z] \right)$

One may try to prove the conjectures by various induction principles, using one argument or both. For space reasons we focus in the present case study on *domain definition based induction* and on *induction on first argument*: $\underset{X}{\forall}P[X]$ is proven by the induction principle established by the inductive definition of the domain. When necessary we refine this induction to a *dynamic induction* method which is applicable to any Noetherian domain: in the induction step we start to prove the induction conclusion $P[t]$ (t ground term) by assuming some induction hypotheses $P[X_0], \ldots, P[X_n]$ according to the inductive definition of the domain ($X_0, \ldots, X_n$ are Skolem constants). If during the proof we need

---

[6] This means that the input condition is the logical constant *True* and the implication from the synthesis conjecture reduces to $O[X,Y,Z]$.

some assumption $P[t']$ where t' (also ground term) represents an object which is strictly smaller than the object represented by t in the Noetherian ordering, then we may assume $P[t']$ holds, that is we can add it to the induction hypotheses. The soundness of this technique is presented in detail in [15], and it allows to discover concrete induction principles based on the general Noetherian induction. The principle of well–founded induction is described as a deduction rule in [18]. Similarly, we use the Noetherian ordering induced by the strict inclusion of the corresponding multisets, which conveniently extends to a meta–ordering between terms, induced by the strict inclusion of the constants occurring in the respective terms. The practical technique for this dynamic induction is described as proof strategy **ST-6** and is illustrated on several examples below.

Moreover we use the *cascading* method pioneered in [3]: when the proof fails, from the failed goal the prover constructs a conjecture synthesis statement which can be used to obtain the auxiliary function which is necessary for the current synthesis. We have been using this for the case of lists in [8,10], and in this paper we describe it in a more systematic manner as proof strategy **ST-7** and we illustrate it on several examples: all insertion algorithms are generated by cascading starting from failed merging–synthesis proofs. [23] presents a method as a generalization of [18] (an "eureka step" is presented as a rule) for generation auxiliary procedures. Their method seems to be similar to *cascading*, but they use different deductive steps to generate the new statement to be proven and the development of the corresponding auxiliary functions. Moreover, in this paper we present the *cascading* method as an automatic proof technique in the *Theorema* system.

Induction Principle for Binary Trees. We use the induction principle established by the domain definition. In order to prove $\underset{X}{\forall} P[X]$ (*base case*) prove $P[\varepsilon]$; (*induction step*) for Skolem constants a, L_0, R_0 assume *induction hypoteses* $P[L_0], P[R_0]$ and prove *induction conclusion* $P[\langle L_0, a, R_0 \rangle]$, where $\langle L_0, a, R_0 \rangle$ is the *subject* of the induction conclusion.

In order to synthesize a merging algorithm as a function $F[X, S]$ we prove Conjecture 1 (take S for Y and T for Z) by transforming S into a Skolem constant S_0 and performing induction on X:

Base Case: We prove $\underset{T}{\exists} O[\varepsilon, S_0, T]$. If the proof succeeds to find for T a ground witness $\Im_1[S_0]$ then we know that $F[\varepsilon, S] = \Im_1[S]$.

Step Case: For arbitrary but fixed a, L_0 and R_0 (new constants), assume: $\underset{T}{\exists} O[L_0, S_0, T]$ and $\underset{T}{\exists} O[R_0, S_0, T]$, which are Skolemized by introducing two new constants T_1 and T_2. We prove: $\underset{T}{\exists} O[\langle L_0, a, R_0 \rangle, S_0, T]$. If the proof succeeds to find a witness $\Im_2[a, L_0, R_0, S_0, T_1, T_2]$, then we know that $F[\langle L, a, R \rangle, S] = \Im_2[a, L, R, S, F[L, S], F[R, S]]$. T_1 and T_2 are replaced by $F[L, S]$ and $F[R, S]$, respectively. Multiple witnesses generate several conditional equalities. Additional arguments to $\Im_2$ may be introduced by dynamic induction as described above and also below at strategy **ST-6**.

In the case of sorted trees, the proof schema is the same, only the given trees $(L_0, R_0, S_0, T_1, T_2)$ are assumed to be sorted, and the witness obtained has to be also sorted.

2.2 Special Inference Rules and Strategies

Following natural style proving, we use *Skolem constants* (denoted with numerical underscore like V_1) introduced for existential assumptions and universal goals, as well as *metavariables* (denoted with star power like T^*) introduced for existential goals. The prover uses classical inference rules (split ground conjunctions, rewrite by equality, etc.) as well as special rules appropriate for trees and multisets.

The strategies are similar to the ones in [8,15]. The first four strategies are briefly described in [9] and the last three strategies extend the ones in [10] on binary trees. The inference rules: **IR-1**, **IR-2**, **IR-3**, **IR-4**, **IR-5**, **IR-6**, and **IR-8** are adapted for these current case studies of synthesis (they extend the inference rules for lists in [9] and [10]) and all the others presented in this section are novel.

These inference rules and strategies are not specific to the problem of tree merging, but are developed in general for the automation of proof based synthesis of algorithms on lists and trees.

Special Inference Rules

Each rule is illustrated with an example from the experiments presented in Sect. 3.

IR-1: *Eliminate assumed formulae from goal.* In a conjunctive goal, delete the part which is already an assumption, or an instance of it. For example goal (34) becomes (35).

IR-2: *Rewrite by equality.* Example: goal (5) is transformed into (6).

IR-3: *Transform to multiple atoms.* This rule transforms parts of the goals or of the assumptions (like e.g. *IsSorted*) into simpler atoms (e.g. by definition). Example: goal (33) becomes (34).

IR-4: *Transform union of M in goal.* Example: goal (7) becomes (8).

IR-5: *Solve metavariables.* Example: goal (32) to (33).

IR-6: *Reduce the goal using assumptions.* Example: transforms goal (35) using the assumption (24) into (36).

IR-7: *Generate branches for trees.* This rule extracts the symbols from a multiset, arranges the symbols and generates branches with new goals. Example: when the goal is $\mathcal{M}[T^*] = \mathcal{M}[L_0] \uplus \{\{a\}\} \uplus \mathcal{M}[R_0] \uplus \mathcal{M}[S_0]$ extracts the symbols: L_0, a, R_0, S_0 and generates the permutations of (L_0, R_0, S_0). The element a is considered to be the root of the obtained trees. From all permutations only those are considered which correspond to the current assumptions about ordering.

IR-7-a: *Generate branches for binary non-sorted trees.* Example: if the assumptions are (3) and (4), then the prover generates an OR node with four branches, having goals: (7), (9), (11), and (13).

IR-7-b: *Generate branches for binary sorted trees.* Example: the assumptions are the ones above in **IR-7-a** and also (17), (18), and the goal (19), then the prover generates an OR node with two branches, having in the goal *IsSorted*$[T^*]$ and also: on one branch goal (20) and on another branch goal (23).

IR-8: *Simple goal conditional assumption.* When the proof fails and the current goal is ground and contains only simple elements (not composite objects), then the proof stops and its result is considered to be this goal (as opposite to *True* when the proof succeeds, or *False* when it fails). Typically this happens in branches generated by the rule **IR-7**, and the unproved goal will become a condition in the synthesized algorithm, as explained below at strategy **ST-4**. Example: goal (37).

Strategies

ST-1: *Quantifier reduction.* This strategy organizes the inference rules for quantifiers (e. g. when applying an induction principle), and it is more effective on goals. For the soundness of the prover it is necessary to keep track of the order in which Skolem constants and metavariables have been introduced, because a Skolem constant which cannot be generated before a certain meta-variable cannot be used in a solution for that meta–variable.

ST-2: *Priority of local assumptions.* We consider as local assumptions ground formulae which are generated during the current proof and as global assumptions definitions and properties in the knowledge base. The strategy consists in using first the local assumptions. Example: when the goal is $\mathcal{M}[W^*] = \{\{a\}\} \uplus \mathcal{M}[U_0] \uplus \mathcal{M}[V_0]$ and the assumption is $\mathcal{M}[W_1] = \mathcal{M}[U_0] \uplus \mathcal{M}[V_0]$, the new goal will be $\mathcal{M}[W^*] = \{\{a\}\} \uplus \mathcal{M}[W_1]$ because we give priority to terms containing the Skolem constants generated by the induction hypothesis (they correspond to recursive calls).

ST-3: *Generate more local assumptions.* Example: apply Modus Ponens on local assumptions.

ST-4: *Conditional branches.* Alternative branches generated by the rule **IR-7** may finish with success (proof value is *True*), failure (*False*), or some "simple" goal (proof value is this goal) as explained at **IR-8**. One may see the corresponding OR node of the proof as constituting the logical operation "or" applied to the proof values. If the result is *True* – that is, the disjunction is a logical consequence of the current theory (we can just say "it holds"), then the proof can be considered successful, and in fact it can be transformed by eliminating the false proof values, and by considering the remaining disjunction as a basis for proof by cases – which will now be an AND node, having on each branch the previous proof value as assumption. This approach in fact discovers automatically the basis for the case distinction proof. Moreover, if there are subsets of the disjunction which already hold disjunctively (we can say they are "covering"), then each such subset can be a basis for the case distinction, thus we can have several successful proof alternatives.

The strategy we employ does not actually transform the proof, because we are only interested in the algorithm. Instead, the respective proof values (simple failed goals) on the branches are taken as conditions for the logical equalities which compose the synthesized algorithm.

ST-5: *Pair multisets.* Often the goal contains an equality like $\mathcal{M}[Y^*] = \mathcal{M}[t_1] \uplus \mathcal{M}[t_2] \uplus \ldots$, where Y^* is the metavariable we need to solve, and $t_1, t_2, \ldots$ are ground terms. The main flow of the proof consists in transforming the union on the LHS of the equality into a single $\mathcal{M}[t]$, because this gives the solution $Y^* \to t$. Therefore the prover groups pairs of operands of $\uplus$ together (no matter whether they are contingent or not, because commutativity), creating alternatives for different groupings. (Consequently the pair will be transformed into an single multiset term by equality rewriting, or it will be treated by strategy **ST-6** or **ST-7**).

ST-6: *Dynamic Induction.* As mentioned in Subsect. 2.1, we use Noetherian induction based on the well–founded ordering between composite objects determined by the strict inclusion of the corresponding multisets. This is checked syntactically by the meta-relation between terms induced by the strict inclusion of the multisets of constants occuring in the terms. When a ground term t' occurring in the goal is smaller than the subject t of the current induction conclusion $P[t]$, then $P[t']$ is used as: $\underset{Y}{\forall}(I[t', Y] \implies \underset{Z}{\exists} O[t', Y, Z])$. Then the prover chooses a ground instantiation s (also part of the goal) for Y, it checks whether $I[t', s]$ holds, it creates a new Skolem constant like for instance Z_1 and it assumes $O[t', s, Z_1]$ holds. In the synthesized algorithm Z_1 will be replaced by $F[t', s]$ (where F is the name of the currently synthesized function). Typically the terms t' and s come from a pair of multiset terms by application of the strategy **ST-5**. Example: goals (40) and (42) are obtained using $P[X]$ (15).

ST-7: *Cascading.* When a pair of multiset terms $t_1[x], t_2[y]$ (x, y constants) is chosen by applying strategy **ST-5**, it may be that there exists no equalities among the current assumptions for reducing it to a single multiset term, or the reduction does not lead to a successful proof. In this case the prover constructs the conjecture: $\underset{X,Y}{\forall}(I[X, Y] \implies \underset{Z}{\exists}(\mathcal{M}[Z] = t_1[X] \uplus t_2[Y] \wedge Q[X, Y, Z]))$, whose proof results in the synthesis of a new function $F[X, Y]$ having the properties required by the current proof situation: $I[X, Y]$ is composed conjunctively from the assumptions which contain *only* the constants x, y (which are replaced by X, Y) and $Q[X, Y, Z]$ is inferred from the current goal.

3   Experiments

3.1   Synthesis of Merging on Non–sorted Binary Trees

The proof of Conjecture 1 by **Induction** on X proceeds as described in Subsect. 2.1 for the formula $P[X]$: $\underset{S}{\forall} \underset{T}{\exists}\left(\mathcal{M}[T] = \mathcal{M}[X] \uplus \mathcal{M}[S]\right)$. On both branches (base case and induction step), the universal S is Skolemized to S_0 ("arbitrary but fixed") and the existential T is replaced by the metavariable T^* (unknown witness), according to **ST-1**.

Proof. Base case: Prove

$$\mathcal{M}[T^*] = \mathcal{M}[\varepsilon] \uplus \mathcal{M}[S_0]. \tag{1}$$

Apply **IR-4** using Property 1 and the goal becomes:

$$\mathcal{M}[T^*] = \mathcal{M}[S_0]. \tag{2}$$

Apply **IR-5**, the obtained substitution is $\{T^* \to S_0\}$.

Induction step: Assume

$$\mathcal{M}[T_1] = \mathcal{M}[L_0] \uplus \mathcal{M}[S_0], \tag{3}$$

$$\mathcal{M}[T_2] = \mathcal{M}[R_0] \uplus \mathcal{M}[S_0] \tag{4}$$

and prove:

$$\mathcal{M}[T^*] = \mathcal{M}[\langle L_0, a, R_0 \rangle] \uplus \mathcal{M}[S_0]. \tag{5}$$

Apply **IR-2** using Property 1 and the goal becomes:

$$\mathcal{M}[T^*] = \mathcal{M}[L_0] \uplus \{\{a\}\} \uplus \mathcal{M}[R_0] \uplus \mathcal{M}[S_0]. \tag{6}$$

Apply **IR-7-a:** using the assumptions (3), (4) and generate and OR node with four branches:

Branch-1: The new goal is:

$$\mathcal{M}[T^*] = \mathcal{M}[L_0] \uplus \{\{a\}\} \uplus \mathcal{M}[T_2]. \tag{7}$$

Apply **IR-4** and the goal becomes:

$$\mathcal{M}[T^*] = \mathcal{M}[\langle L_0, a, T_2 \rangle]. \tag{8}$$

Apply **IR-5** and the obtained substitution on this branch is $\{T^* \to \langle L_0, a, T_2 \rangle\}$.

Branch-2: The new goal is:

$$\mathcal{M}[T^*] = \mathcal{M}[T_1] \uplus \{\{a\}\} \uplus \mathcal{M}[R_0]. \tag{9}$$

Apply **IR-4** and the goal becomes:

$$\mathcal{M}[T^*] = \mathcal{M}[\langle T_1, a, R_0 \rangle] \tag{10}$$

and the substitution is $\{T^* \to \langle T_1, a, R_0 \rangle\}$.

Branch-3: The new goal is:

$$\mathcal{M}[T^*] = \mathcal{M}[R_0] \uplus \{\{a\}\} \uplus \mathcal{M}[T_1]. \tag{11}$$

Apply **IR-4**, the goal becomes:

$$\mathcal{M}[T^*] = \mathcal{M}[\langle R_0, a, T_1 \rangle] \tag{12}$$

and the substitution is $\{T^* \to \langle R_0, a, T_1 \rangle\}$.

Branch-4: The new goal is:

$$\mathcal{M}[T^*] = \mathcal{M}[T_2] \uplus \{\!\{a\}\!\} \uplus \mathcal{M}[L_0]. \tag{13}$$

Apply **IR-4**, the goal becomes:

$$\mathcal{M}[T^*] = \mathcal{M}[\langle T_2, a, L_0 \rangle] \tag{14}$$

and the substitution is $\{T^* \to \langle T_2, a, L_0 \rangle\}$.

Since all branches succeed, each of them generates an alternative algorithm, thus we have:

Algorithm 1. Concatenation of trees.

$$\forall_{a,L,R,S} \left(\begin{array}{c} Conc[\varepsilon, S] = S \\ Conc[\langle L, a, R \rangle, S] = \langle L, a, Conc[R, S] \rangle \end{array} \right)$$

as well as three other concatenation algorithms where the RHS of the second equality is: $\langle F[L, S], a, R \rangle, \langle R, a, F[L, S] \rangle$, or $\langle F[R, S], a, L \rangle$.

3.2 Synthesis of Merging on Sorted Binary Trees

The proof of Conjecture 2 by **Induction** on X proceeds as described in Subsect. 2.1 for the formula $P[X]$:

$$\forall_{S} \Big((IsSorted[X] \wedge IsSorted[S]) \implies \exists_{T} (\mathcal{M}[T] = \mathcal{M}[X] \uplus \mathcal{M}[S] \wedge IsSorted[T] \Big) \tag{15}$$

On both branches (base case and induction step), the universal S is Skolemized to S_0 ("arbitrary but fixed") and the existential T is replaced by the metavariable T^* (unknown witness), according to **ST-1**. The proof is similar with the previous one, with the difference that at the induction step in addition to the induction hypothesis (3), (4) one obtains more assumptions regarding the ordering.

Proof.

$$IsSorted[T_1] \wedge IsSorted[T_2], \tag{16}$$

$$IsSorted[\langle L_0, a, R_0 \rangle]. \tag{17}$$

By **IR-3** from (17) obtain:

$$IsSorted[L_0] \wedge L_0 \preceq a \wedge a \preceq R_0 \wedge IsSorted[R_0]. \tag{18}$$

The goal is similar to (6), in addition T^* has to be sorted:

$$\mathcal{M}[T^*] = \mathcal{M}[L_0] \uplus \{\!\{a\}\!\} \uplus \mathcal{M}[R_0] \uplus \mathcal{M}[S_0] \wedge IsSorted[T^*]. \tag{19}$$

Apply **IR-7-b** using the induction hypothesis (3), (4), and also (17), (18) and generate two branches:

Branch-1: The new goal is

$$\mathcal{M}[T^*] = \mathcal{M}[T_1] \uplus \{\{a\}\} \uplus \mathcal{M}[R_0] \wedge \textit{IsSorted}[T^*]. \tag{20}$$

Apply **IR-4** and the goal is:

$$\mathcal{M}[T^*] = \mathcal{M}[\langle T_1, a, R_0 \rangle] \wedge \textit{IsSorted}[T^*]. \tag{21}$$

Apply **IR-5**, the obtained substitution is $\{T^* \rightarrow \langle T_1, a, R_0 \rangle\}$ and the remaining goal is:

$$\textit{IsSorted}[\langle T_1, a, R_0 \rangle]. \tag{22}$$

Apply **IR-3** using Property 1, **IR-2** using (16), (18), **IR-6** using (3), (18) and the remaining goal is $S_0 \preceq a$. The proof fails on this branch.

Branch-2: The new goal is

$$\mathcal{M}[T^*] = \mathcal{M}[L_0] \uplus \{\{a\}\} \uplus \mathcal{M}[T_2] \wedge \textit{IsSorted}[T^*]. \tag{23}$$

Similarly, the obtained substitution is $\{T^* \rightarrow \langle L_0, a, T_2 \rangle\}$ and the remaining goal is $a \preceq S_0$. The proof fails.

However synthesis is still possible by the technique described below.

Cascading–Synthesis of Insertion on Binary Trees: The prover applies strategy **ST-5** (pair multisets) by grouping $\{\{a\}\}$ and $\mathcal{M}[R_0]$ – for which we already know $a \preceq R_0$ – and then strategy **ST-7** (cascading), producing the conjecture:

Conjecture 3. $\underset{a,R}{\forall} \Big(\big(\textit{IsSorted}[R] \wedge a \preceq R \big) \implies \underset{S}{\exists} \big(\mathcal{M}[S] = \{\{a\}\} \uplus \mathcal{M}[R] \wedge \textit{IsSorted}[S] \big) \Big)$

By proving this conjecture we obtain the algorithm *Prepend* which places a given element as the leftmost node of a given tree:

Algorithm 2. Prepend an element to a tree.
$$\underset{a,b,L,R}{\forall} \left(\begin{array}{c} \textit{Prepend}[a, \varepsilon] = \langle \varepsilon, a, \varepsilon \rangle \\ \textit{Prepend}[a, \langle L, b, R \rangle] = \langle \textit{Prepend}[a, L], b, R \rangle \end{array} \right)$$

However, by using this auxiliary function the main proof still does not succeed, therefore a merging algorithm cannot be found.

Similarly, for the goal (23), by grouping $\mathcal{M}[L_0]$ and $\{\{a\}\}$ – for which we already know $L_0 \preceq a$, we obtain the conjecture for the synthesis of the auxiliary function *Append*, which places a given element at the rightmost node of a given tree, but in this case the synthesis of the merging algorithm still fails.

If for proving (19) we group $\mathcal{M}[S_0]$ and $\{\{a\}\}$, then there is no more ordering between them, and the conjecture is:

Conjecture 4. $\underset{a,X}{\forall} \left(IsSorted[X] \implies \underset{S}{\exists} \left(\mathcal{M}[S] = \{\{a\}\} \uplus \mathcal{M}[X] \wedge IsSorted[S] \right) \right)$

By proving this conjecture we obtain the function *Insert* which places a given element as the appropriate position in a sorted tree.

Prove Conjecture 4 by applying **Induction** on X.

Proof. Base Case: The obtained substitution is $\{T^* \rightarrow \langle \varepsilon, a, \varepsilon \rangle\}$.

Induction Step: Assume

$$\mathcal{M}[S_1] = \{\{a\}\} \uplus \mathcal{M}[L_0], \tag{24}$$

$$\mathcal{M}[S_2] = \{\{a\}\} \uplus \mathcal{M}[R_0], \tag{25}$$

$$IsSorted[S_1] \wedge IsSorted[S_2], \tag{26}$$

$$IsSorted[\langle L_0, b, R_0 \rangle], \tag{27}$$

$$IsSorted[L_0] \wedge L_0 \preceq b \wedge b \preceq R_0 \wedge IsSorted[R_0] \tag{28}$$

and prove:

$$\mathcal{M}[S^*] = \{\{a\}\} \uplus \mathcal{M}[\langle L_0, b, R_0 \rangle] \wedge IsSorted[S^*]. \tag{29}$$

Apply **IR-2** using Property 1 and the new goal is:

$$\mathcal{M}[S^*] = \{\{a\}\} \uplus \mathcal{M}[L_0] \uplus \{\{b\}\} \uplus \mathcal{M}[R_0] \wedge IsSorted[S^*]. \tag{30}$$

Apply **IR-7-b** considering b to be the root of the obtained tree, using the assumptions (24), (25), (27) and generate two branches:

Branch-1: The new goal is:

$$\mathcal{M}[S^*] = \mathcal{M}[S_1] \uplus \{\{b\}\} \uplus \mathcal{M}[R_0] \wedge IsSorted[S^*]. \tag{31}$$

Apply **IR-4** using Property 1 and prove:

$$\mathcal{M}[S^*] = \mathcal{M}[\langle S_1, b, R_0 \rangle] \wedge IsSorted[S^*]. \tag{32}$$

Apply **IR-5**, the substitution is $\{T^* \rightarrow \langle S_1, b, R_0 \rangle\}$ and the new goal is:

$$IsSorted[\langle S_1, b, R_0 \rangle]. \tag{33}$$

Apply **IR-3** using Definition 1 and the goal becomes:

$$IsSorted[S_1] \wedge S_1 \preceq b \wedge b \preceq R_0 \wedge IsSorted[R_0]. \tag{34}$$

Apply IR − 1 using (26), (28) and the remaining goal is: $S_1 \preceq b.$ (35)

Apply IR − 6 using (24) and the new goal is: $a \leq b \wedge L_0 \preceq b.$ (36)

Apply IR − 1 using (28) and the remaining goal is: $a \leq b.$ (37)

By **IR-8**, (37) becomes the conditional assumption on this branch.

Branch-2: The new goal is:

$$\mathcal{M}[S^*] = \mathcal{M}[L_0] \uplus \{\{b\}\} \uplus \mathcal{M}[S_2] \wedge IsSorted[S^*]. \tag{38}$$

Similar as in the previous branch, the obtained substitution is $\{T^* \rightarrow \langle L_0, b, S_2 \rangle\}$ and the conditional assumption on this branch is $b \leq a$. By **ST-4** we obtain:

Algorithm 3. Insertion in a sorted tree.

$$\underset{a,b,L,R}{\forall} \left(\begin{array}{c} Ins[a, \varepsilon] = \langle \varepsilon, a, \varepsilon \rangle \\ Ins[a, \langle L, b, R \rangle] = \left\{ \begin{array}{l} \langle Ins[a, L], b, R \rangle, \text{ if } a \leq b \\ \langle L, b, Ins[a, R] \rangle, \text{ if } b < a \end{array} \right. \end{array} \right)$$

By the cascading strategy **ST-7**, we continue the proof of the merging conjecture by replacing in the goal (19) the subterm $\{\{a\}\} \uplus \mathcal{M}[S_0]$ (which generated the conjecture for synthesizing Ins) by the corresponding instance $Ins[a, S_0]$:

Proof.

$$\mathcal{M}[T^*] = \mathcal{M}[L_0] \uplus \mathcal{M}[R_0] \uplus \mathcal{M}[Ins[a, S_0]] \wedge IsSorted[T^*]. \tag{39}$$

Apply strategy **ST-5** (pair multisets) and **ST-6** (dynamic induction) to $\mathcal{M}[R_0]$ and $\mathcal{M}[Ins[a, S_0]]$. The object represented by R_0 is smaller in the well founded ordering than the object $\langle R_0, a, L_0 \rangle$, which is the subject of the current induction conclusion (formula (15) with substitution $X \longrightarrow \langle R_0, a, L_0 \rangle$). Therefore we may assume $P[R_0]$ holds, and use $Ins[a, S_0]$ for the instantiation of the second argument, thus by Skolemization we obtain an object R_1 observing:

$$\mathcal{M}[R_1] = \mathcal{M}[R_0] \uplus \mathcal{M}[Ins[a, S_0]] \wedge IsSorted[R_1]. \tag{40}$$

Apply equality rewriting using this to transform goal (39) into:

$$\mathcal{M}[T^*] = \mathcal{M}[L_0] \uplus \mathcal{M}[R_1] \wedge IsSorted[T^*]. \tag{41}$$

Since the object represented by L_0 is smaller in the well founded ordering than $\langle R_0, a, L_0 \rangle$, we can again apply Noetherian induction to obtain L_1 with:

$$\mathcal{M}[L_1] = \mathcal{M}[L_0] \uplus \mathcal{M}[R_0] \wedge IsSorted[L_1]. \tag{42}$$

Apply equality rewriting using this to transform goal (41) into:

$$\mathcal{M}[T^*] = \mathcal{M}[L_1] \wedge IsSorted[T^*] \tag{43}$$

which gives the solution $T^* = L_1$ and the proof succeeds, giving the algorithm:

Algorithm 4. Merge sorted trees, version 1.

$$\underset{a,L,R,S}{\forall} \left(\begin{array}{c} Merge[\varepsilon, S] = S \\ Merge[\langle L, a, R \rangle, S] = Merge[L, Merge[R, Ins[a, S]]] \end{array} \right)$$

Note how a *nested recursion*—for which a concrete induction principle would be difficult to guess—is produced automatically by our method. This algorithm is interesting because it is probably optimal: essentially it inserts one by one the elements of the first tree into the (sorted) second tree. Note also that the assumptions (17) and (18) are not necessary for the success of the proof, and indeed the algorithm produces a sorted tree even if the first argument is not sorted. Similarly to the situation with lists [10], since the first argument does not need to be sorted, both this algorithm and the next one can be used for sorting as $Merge[T, \varepsilon]$. Sorting is performed by traversing the tree and inserting the elements one by one in a new sorted tree, which appears to be optimal.

There are many ways in which the subterms of the RHS of the equality in (19) can be grouped pairwise and then be used in a similar manner to cascade new auxiliary functions and to produce new merging algorithms. We present here only one other alternative, which is interesting because it is tail recursive, and only slightly less efficient than the previous one.

The proof is modified as follows:

Proof. Strategy **ST-5** (pair multisets) on the goal (39) groups the subterms $\mathcal{M}[L_0]$ and $\mathcal{M}[R_0]$, and then strategy **ST-7** (cascade) generates the conjecture:

$$\underset{L,R}{\forall} \underset{X}{\exists} \mathcal{M}[X] = \mathcal{M}[L] \uplus \mathcal{M}[R] \wedge IsSorted[X] \tag{44}$$

The proof of this is very similar to the proof of Conjecture 1 (for synthesis of merging on non-sorted trees) presented at the beginning of Sect. 3.1, with the difference that the proof starts with the additional assumptions $IsSorted[L_0, a, R_0]$, $IsSorted[T_1]$, $IsSorted[T_2]$, while the goal has also $IsSorted[T^*]$. Therefore the proof succeeds on the first branch with the same witness $\langle L_0, a, T_2 \rangle$, which is proven sorted by applying the definition and the properties of ordering to the assumptions—so the same Algorithm 1 $Conc$ also concatenates sorted trees into a sorted tree.

Strategy **ST-7** (cascading) replaces in goal (19) the pair $\mathcal{M}[L_0] \uplus \mathcal{M}[R_0]$ by $\mathcal{M}[Conc[L_0, R_0]]$ to get:

$$\mathcal{M}[T^*] = \mathcal{M}[Conc[L_0, R_0]] \uplus \mathcal{M}[Ins[a, S_0]] \wedge IsSorted[T^*]. \tag{45}$$

Since $Conc[L_0, R_0]$ is smaller in the well–founded ordering than $\langle L_0, a, R_0 \rangle$, strategy **ST-6** (dynamic induction) uses it together with the instantiation $\mathcal{M}[Ins[a, S_0]]$ for the second argument, and obtains L_2 with the property:

$$\mathcal{M}[L_2] = \mathcal{M}[R_0] \uplus \mathcal{M}[Ins[a, S_0]] \wedge IsSorted[L_2]. \tag{46}$$

By equality rewriting this transforms the goal (45) into:

$$\mathcal{M}[T^*] = \mathcal{M}[L_2] \wedge IsSorted[T^*] \tag{47}$$

which gives the solution $T^* = L_2$ and the proof succeeds, giving the algorithm:

Algorithm 5. Merge sorted trees, version 2.

$$\underset{a,L,R,S}{\forall} \left(\begin{array}{c} Merge[\varepsilon, S] = S \\ Merge[\langle L, a, R \rangle, S] = Merge[Conc[L, R], Ins[a, S]]] \end{array} \right)$$

Similarly to the other version, since the first argument does not need to be sorted, this can also be used for sorting as $Merge[T, \varepsilon]$. This algorithm is interesting because it is tail recursive, even as it is slightly less efficient than the previous one.

4   Conclusions and Further Work

Our experiments demonstrate the possibility of automatic synthesis of complex algorithms on (possibly sorted) binary trees, using the notion of multiset. In certain cases, depending on the proof strategy, several algorithms are produced for the same function or from different proofs the same algorithm is produced.

Even as some of the synthesized algorithms are relatively straightforward and sometimes not optimal, this case study helps in at least three ways. First, the study develops the underlying theory and helps understand better the principles of theory exploration, for instance by a parallel development one has hints about interesting functions on trees suggested by the classical operations on multisets (insertion corresponds to union with one element, merging corresponds to union, etc.). Second, the study helps to develop efficient proof methods for these domains, in particular by using specific inference rules and strategies which are also taylored for synthesis proofs, notably for discovering induction principles for nested recursion. Finally, the various algorithms which are produced can constitute a test field for methods of automatic evaluation of efficiency, time and space consumption, etc.

A distinctive feature of our approach is the use of natural–style proofs, which is facilitated by the *Theorema* system. The natural style of proving (as formula notation, as proof text, and as inference steps) has the advantage of allowing human inspection in an intuitive way, and this facilitates the development of intuitive inference rules which embed the knowledge about the underlying domains.

The experiments presented here continue our previous work on synthesis of deletion algorithms and sorting algorithms on lists using multisets and is prerequisite for further work on synthesis of more complex algorithms for sorting and searching, including algorithms which combine operations on several domains.

References

1. Blizard, W.D.: Multiset theory. Notre Dame J. Formal Logic **30**(1), 36–66 (1989). https://doi.org/10.1305/ndjfl/1093634995
2. Buchberger, B.: Theory exploration with theorema. Analele Universitatii Din Timisoara, Seria Matematica-Informatica **XXXVII**(2), 9–32 (2000)
3. Buchberger, B.: Algorithm invention and verification by lazy thinking. Analele Universitatii din Timisoara, Seria Matematica - Informatica **XLI**, 41–70 (2003)
4. Buchberger, B., Craciun, A.: Algorithm synthesis by lazy thinking: using problem schemes. In: Proceedings of SYNASC, pp. 90–106 (2004)
5. Buchberger, B., et al.: The theorema project: a progress report. In: Calculemus 2000, pp. 98–113. A.K. Peters, Natick (2000)
6. Buchberger, B., Jebelean, T., Kutsia, T., Maletzky, A., Windsteiger, W.: Theorema 2.0: computer-assisted natural-style mathematics. J. Formal. Reason. **9**(1), 149–185 (2016). https://doi.org/10.6092/issn.1972-5787/4568
7. Bundy, A., Dixon, L., Gow, J., Fleuriot, J.: Constructing induction rules for deductive synthesis proofs. Electron. Notes Theor. Comput. Sci. **153**, 3–21 (2006). https://doi.org/10.1016/j.entcs.2005.08.003

8. Dramnesc, I., Jebelean, T.: Synthesis of list algorithms by mechanical proving. J. Symb. Comput. **68**, 61–92 (2015). https://doi.org/10.1016/j.jsc.2014.09.030
9. Dramnesc, I., Jebelean, T.: Case studies on algorithm discovery from proofs: the delete function on lists and binary trees using multisets. In: SISY 2019, pp. 213–220. IEEE Xplore (2019)
10. Dramnesc, I., Jebelean, T.: Proof-based synthesis of sorting algorithms using multisets in theorema. In: FROM 2019, EPTCS 303, pp. 76–91 (2019). https://doi.org/10.4204/EPTCS.303.6
11. Dramnesc, I., Jebelean, T., Stratulat, S.: Combinatorial techniques for proof-based synthesis of sorting algorithms. In: SYNASC 2015, pp. 137–144 (2015). https://doi.org/10.1109/SYNASC.2015.30
12. Dramnesc, I., Jebelean, T., Stratulat, S.: Theory exploration of binary trees. In: SISY 2015, pp. 139–144. IEEE (2015). https://doi.org/10.1109/SISY.2015.7325367
13. Dramnesc, I., Jebelean, T., Stratulat, S.: A case study on algorithm discovery from proofs: the insert function on binary trees. In: SACI 2016, pp. 231–236. IEEE (2016). https://doi.org/10.1109/SACI.2016.7507376
14. Drămnesc, I., Jebelean, T., Stratulat, S.: Proof–based synthesis of sorting algorithms for trees. In: Dediu, A.-H., Janoušek, J., Martín-Vide, C., Truthe, B. (eds.) LATA 2016. LNCS, vol. 9618, pp. 562–575. Springer, Cham (2016). https://doi.org/10.1007/978-3-319-30000-9_43
15. Dramnesc, I., Jebelean, T., Stratulat, S.: Mechanical synthesis of sorting algorithms for binary trees by logic and combinatorial techniques. J. Symb. Comput. **90**, 3–41 (2019). https://doi.org/10.1016/j.jsc.2018.04.002
16. Knuth, D.E.: The Art of Computer Programming, Volume 2: Seminumerical Algorithms, 3rd edn. Addison-Wesley, Boston (1998). https://doi.org/10.1137/1012065
17. Korukhova, Y.: Automatic deductive synthesis of lisp programs in the system ALISA. In: Fisher, M., van der Hoek, W., Konev, B., Lisitsa, A. (eds.) JELIA 2006. LNCS (LNAI), vol. 4160, pp. 242–252. Springer, Heidelberg (2006). https://doi.org/10.1007/11853886_21
18. Manna, Z., Waldinger, R.: A deductive approach to program synthesis. ACM Trans. Program. Lang. Syst. **2**(1), 90–121 (1980). https://doi.org/10.1145/357084.357090
19. Manna, Z., Waldinger, R.: The Logical Basis for Computer Programming, vol. 1: Deductive Reasoning. Addison-Wesley, Boston (1985). https://doi.org/10.2307/2275898
20. Manna, Z., Waldinger, R.: Fundamentals of deductive program synthesis. IEEE Trans. Softw. Eng. **18**(8), 674–704 (1992). https://doi.org/10.1109/32.153379
21. Radoaca, A.: Properties of multisets compared to sets. In: SYNASC 2015, pp. 187–188 (2015). https://doi.org/10.1109/SYNASC.2015.37
22. Smith, D.R.: Kids: a semiautomatic program development system. IEEE Trans. Softw. Eng. **16**(9), 1024–1043 (1990). https://doi.org/10.1109/32.578788
23. Traugott, J.: Deductive synthesis of sorting programs. J. Symb. Comput. **7**(6), 533–572 (1989). https://doi.org/10.1016/S0747-7171(89)80040-9
24. Windsteiger, W.: Theorema 2.0: a system for mathematical theory exploration. In: Hong, H., Yap, C. (eds.) ICMS 2014. LNCS, vol. 8592, pp. 49–52. Springer, Heidelberg (2014). https://doi.org/10.1007/978-3-662-44199-2_9

Algebraic Analysis of Bifurcations and Chaos for Discrete Dynamical Systems

Bo Huang[1,2] and Wei Niu[3,4(✉)]

[1] LMIB-School of Mathematical Sciences, Beihang University, Beijing 100191, China
bohuang0407@buaa.edu.cn
[2] Courant Institute of Mathematical Sciences, New York University,
New York 10012, USA
[3] Ecole Centrale de Pékin, Beihang University, Beijing 100191, China
wei.niu@buaa.edu.cn
[4] Beijing Advanced Innovation Center for Big Data and Brain Computing,
Beihang University, Beijing 100191, China

Abstract. This paper deals with the stability, bifurcations and chaotic behaviors of discrete dynamical systems by using methods of symbolic computation. We explain how to reduce the problems of analyzing the stability, bifurcations and chaos induced by snapback repellers to algebraic problems, and solve them by using an algorithmic approach based on methods for solving semi-algebraic systems. The feasibility of the symbolic approach is demonstrated by analyses of the dynamical behaviors for several discrete models.

Keywords: Bifurcations · Chaos · Discrete systems · Symbolic computation · Snapback repeller

1 Introduction

Many biological phenomena, control and economic problems may be modeled mathematically by dynamical systems (see [14,17,26]). Most of such systems are nonlinear, and it is difficult to find their analytical solutions in general, so studying the qualitative behaviors of their solutions becomes an important issue. The most concerned behaviors of such systems are stability of fixed points, bifurcations, chaos and so on.

This work was done while Bo Huang was visiting NYU Courant. The first author wishes to thank Professor Chee Yap for his profound concern. Both authors thank Professor Dongming Wang for his valuable suggestions and the anonymous referees for their helpful comments on improving the presentation. The work was partially supported by China Scholarship Council (No. 201806020128), by the Academic Excellent Foundation of BUAA for PhD Students, by the NSF grant #CCF-1708884, and by the NSFC project 11601023.

© Springer Nature Switzerland AG 2020
D. Slamanig et al. (Eds.): MACIS 2019, LNCS 11989, pp. 169–184, 2020.
https://doi.org/10.1007/978-3-030-43120-4_14

Consider the following first-order autonomous discrete difference equations (high-order systems can be transformed into first-order ones)

$$
\begin{cases}
x_1(t+1) = \phi_1(\mu_1, \ldots, \mu_m, x_1(t), \ldots, x_n(t)), \\
\quad\vdots \\
x_n(t+1) = \phi_n(\mu_1, \ldots, \mu_m, x_1(t), \ldots, x_n(t)),
\end{cases}
\tag{1}
$$

where $\mu_1, \ldots, \mu_m$ are parameters independent of t, $x_1, \ldots, x_n$ are variables, and $\phi_i : \mathcal{K}^{m+n} \to \mathcal{K}$ is a map for $i = 1, \ldots, n$ with $\mathcal{K}$ a field. For discrete biological models in the form of (1), a general algebraic approach has been proposed in [20] for the detection and analysis of stability and bifurcations of real fixed points. More recently, this approach has also been applied to the analysis of chaos induced by snapback repeller for discrete dynamical systems [15]. Discrete dynamical systems in the form of (1) could serve as the underlying mathematical model for many practical problems. The dynamical behaviors (e.g., bifurcations and chaos) of such systems are of significance in these practical problems (see [11,17]). Therefore it is of importance and our interest to adapt, extend, and apply the algebraic approach to analyze the bifurcations and chaos for interesting and challenging discrete models.

In this paper we focus on a class of n-dimensional discrete dynamical systems in the form (1) with

$$
\boldsymbol{x}(t+1) = \boldsymbol{f}(\boldsymbol{\mu}, \boldsymbol{x}(t)),
\tag{2}
$$

where $\boldsymbol{f}$ is a $\mathcal{C}^1$ nonlinear map with parameters $\boldsymbol{\mu}$ from $\mathbb{R}$, the real number field. Our objective in this paper is to study algebraically and symbolically the conditions on the parameters $\boldsymbol{\mu}$ for the discrete system (2) to have a prescribed number of (stable) fixed points, certain types of bifurcations and chaotic behaviors. More concretely, we are interested in the following problem.

Problem. Let $\bar{x}$ be a fixed point of system (2) (i.e., $\boldsymbol{f}(\boldsymbol{\mu}, \bar{x}) = \bar{x}$). Determine the explicit conditions on the parameters $\boldsymbol{\mu}$ such that

1. the fixed point $\bar{x}$ is stable for system (2);
2. system (2) may undergo some important bifurcations;
3. $\bar{x}$ is a snapback repeller of system (2).

Remark 1. In this paper we consider the Neimark-Sacher bifurcation, the period doubling bifurcation and the stationary bifurcation for the discrete system (2). These important bifurcations will be introduced in Sect. 2.

This paper reports our current study on the use of algebraic methods based on Gröbner bases [3,9], triangular decomposition [27,28], quantifier elimination [5,13], real solution classification [30] and discriminant varieties [19] for stability, bifurcations and chaos analysis of discrete systems in the form of (1). The outline of our work is as follows. In Sect. 2, we first explain how to reduce the stability, bifurcations and chaos problems of discrete systems over $\mathbb{R}$ to purely algebraic problems and then solve these problems by using algebraic methods. An illustrative example and some experimental results together with remarks are provided in Sect. 3. This paper ends with a discussion in Sect. 4.

2   Algebraic Criteria for Stability, Bifurcations and Chaos

In this section, we recall some important bifurcations and the snapback repeller for a discrete dynamical system together with some theorems on the algebraic criteria of zero distribution with respect to the unit circle, then we explain how to reduce the problems of stability, bifurcations and chaos analyses to algebraic problems.

2.1   Stability Analysis of Fixed Points

We first describe some notions for system (2). Let f^k denote the k times of compositions of f with itself. A point x, is said to be a p-periodic point of f if $f^p(x) = x$ but $f^k(x) \neq x$ for $p > k \geq 1$. If $p = 1$, i.e., $f(x) = x$, then x is called a fixed point. Let $f'(x)$ be the Jacobian matrix of f with determinant $|f'(x)|$.

Clearly we can use the equation: $[x - f(\mu, x) = 0]$ to detect the fixed points of system (2). After computing the fixed points, we want to analyze the stability of (2) at each fixed point or to determine conditions on the parameters for the fixed point to be stable. To this end, we let

$$A(\lambda) = \lambda^n + a_{n-1}\lambda^{n-1} + \ldots + a_0 \tag{3}$$

be the characteristic polynomial of the Jacobian matrix $f'(x)$ of system (2), where $a_i = a_i(\mu, x)$, $i = 0, \ldots, n - 1$. The following theorem tells us how to determine the stability of a fixed point for a discrete system.

Theorem 1 *(see, e.g., [10]). Let $\bar{x}$ be a fixed point of system (2). If all the eigenvalues λ_i of the Jacobian matrix $f'(x)$ are inside the unit circle, i.e. $|\lambda_i| < 1$ for all i, then $\bar{x}$ is asymptotically stable.*

In this paper we will use a sequence of symmetric polynomials of descending degrees for the characteristic polynomial $A(\lambda)$ to determine the stability of a fixed point (see Theorem 4 in Sect. 2.3), instead of using the Schur-Cohn criterion. Since the Schur-Cohn criterion is expressed in terms of certain determinants formed by the coefficients of $A(\lambda)$, and the computational complexity of these determinants may grow very fast with the dimension n of system (2). We remark that for the bifurcation analysis we will still use the generalized forms of the Schur-Cohn criterion (see Sect. 2.2). Exploring how to extend Theorem 4 to analyze the bifurcations of discrete systems is a question for further study.

2.2   Bifurcation Analysis

Similar to the continuous case, there may be many different situations for discrete systems. In this subsection, we recall some important bifurcations for discrete systems and explain how to reduce the problems of bifurcation analysis to algebraic problems.

(1) Neimark-Sacher bifurcation. The Neimark-Sacher bifurcation for discrete dynamical systems corresponds to the Hopf bifurcation in the continuous case. For this bifurcation, the Jacobian matrix $f'(x)$ has a pair of complex conjugate eigenvalues on the unit cycle and all other eigenvalues inside the circle. The following theorem is generalized from the Schur-Cohn criterion.

Theorem 2 *[29]. A pair of complex conjugate roots of $A(\lambda)$ lie on the unit circle and the other roots of $A(\lambda)$ all lie inside the unit circle if and only if*

(a) $A(1) > 0$ *and* $(-1)^n A(-1) > 0$,
(b) $D_1^{\pm} > 0$, $D_3^{\pm} > 0,\dots$, $D_{n-3}^{\pm} > 0$, $D_{n-1}^{+} > 0$, $D_{n-1}^{-} = 0$ *(when n is even), or*
$D_2^{\pm} > 0$, $D_4^{\pm} > 0,\dots$, $D_{n-3}^{\pm} > 0$, $D_{n-1}^{+} > 0$, $D_{n-1}^{-} = 0$ *(when n is odd),*

where

$$
D_i^{\pm} = \left| \begin{pmatrix} 1 & a_{n-1} & a_{n-2} & \cdots & a_{n-i+1} \\ 0 & 1 & a_{n-1} & \cdots & a_{n-i+2} \\ 0 & 0 & 1 & \cdots & a_{n-i+3} \\ \vdots & \vdots & \vdots & \ddots & \vdots \\ 0 & 0 & 0 & \cdots & 1 \end{pmatrix} \pm \begin{pmatrix} a_{i-1} & a_{i-2} & \cdots & a_1 & a_0 \\ a_{i-2} & a_{i-3} & \cdots & a_0 & 0 \\ \vdots & \vdots & \ddots & \vdots & \vdots \\ a_1 & a_0 & \cdots & 0 & 0 \\ a_0 & 0 & \cdots & 0 & 0 \end{pmatrix} \right|
$$

are the same as in the Schur-Cohn criterion (see, e.g., [20]).

In this case, system (2) may undergo a Neimark-Sacher bifurcation.

(2) Period Doubling Bifurcation. A period doubling bifurcation (or flip bifurcation) can arise only in a discrete dynamical system. At this bifurcation, the system switches to a new behavior with twice the period of the original system. A series of period doubling bifurcations may lead the system from order to chaos. In this situation, the Jacobian matrix $f'(x)$ has one real eigenvalue which equals -1, and the other eigenvalues of $f'(x)$ are all inside the unit circle. Obviously, a necessary and sufficient condition for the characteristic polynomial $A(\lambda)$ to have one real root -1 and all other roots inside the unit circle is

(a) $A(1) > 0$ and $A(-1) = 0$,
(b) $D_1^{\pm} > 0$, $D_3^{\pm} > 0,\dots$, $D_{n-3}^{\pm} > 0$, $D_{n-1}^{\pm} > 0$ (when n is even), or
$D_2^{\pm} > 0$, $D_4^{\pm} > 0,\dots$, $D_{n-3}^{\pm} > 0$, $D_{n-1}^{\pm} > 0$ (when n is odd).

(3) Stationary Bifurcation. If the Jacobian matrix $f'(x)$ has one real eigenvalue which equals 1, then the system (2) may undergo a *saddle-node* (also called *fold bifurcation*), *transcritical* or *pitchfork bifurcation*. These bifurcations are all called stationary bifurcations, see [20] for more details on these types of bifurcations. Replacing condition (a) for period doubling bifurcation by $A(1) = 0$ and $(-1)^n A(-1) > 0$, one can obtain the condition under which system (2) may undergo a stationary bifurcation, but the determination of the type of stationary bifurcation for a concrete system requires further analysis.

Therefore, the problem of determining the conditions on parameters under which a bifurcation of certain type may occur can be reduced to an algebraic problem, see ([20], Sect. 3) for more details.

Note that the conditions for the bifurcations analyzed above are the critical conditions for one of the eigenvalues to reach the unit circle. This eigenvalue should not be stationary on the unit circle, but continue to go outside of the circle as the parameters vary. So whether a bifurcation indeed occurs or not depends on the critical conditions together with the so-called *transversality* (or *crossing*) condition and the *non-resonance* condition. In this paper, we focus our study on the critical conditions, which provides possibilities for the occurrence of bifurcations. The derivation of necessary and sufficient bifurcation conditions and further analysis of the types and stability of bifurcations are our ongoing research.

2.3   Snapback Repeller and Marotto's Theorem

We now describe the notion of snapback repeller and Marotto's theorem. For the C^1 nonlinear map f of (2), we let $B_r(x)$ be the closed ball of radius r centered at x under a given norm $|| \cdot ||$ in $\mathbb{R}^n$. We say that a fixed $\bar{x}$ is a *repelling fixed point* of f with respect to the norm $|| \cdot ||$ if there exists a constant $s > 1$ such that $||f(x) - f(y)|| > s \cdot ||x - y||$ for any $x, y \in B_r(\bar{x})$ with $x \neq y$, where $B_r(\bar{x})$ is defined on this norm $|| \cdot ||$, called a repelling neighborhood of $\bar{x}$.

Definition 1. *Let $\bar{x}$ be a repelling fixed point of f in $B_r(\bar{x})$ for some $r > 0$. We say that $\bar{x}$ is a snapback repeller of f if there exist a point $x_0 \in B_r(\bar{x})$ with $x_0 \neq \bar{x}$ and an integer $m > 1$, such that $x_m = \bar{x}$ and $|f'(x_k)| \neq 0$ for $1 \leq k \leq m$, where $x_k = f^k(x_0)$.*

The point x_0 in this definition is called a *snapback point* of f. Under this definition, the following theorem by Marotto holds [22, 23].

Theorem 3. *If f possesses a snapback repeller, then f is chaotic in the following sense: There exist (i) a positive integer N such that for each integer $p \geq N$, f has a periodic point of period p; (ii) a "scrambled set" of f, i.e., an uncountable set S containing no periodic points of f such that*

(a) $f(S) \subset S$,
(b) $\limsup\limits_{k \to \infty} || f^k(u) - f^k(v) || > 0$, for all $u, v \in S$ with $u \neq v$,
(c) $\limsup\limits_{k \to \infty} || f^k(u) - f^k(v_p) || > 0$, for all $u \in S$ and any periodic point v_p of f;

(iii) an uncountable subset S_0 of S such that $\liminf\limits_{k \to \infty} || f^k(u) - f^k(v) || = 0$, for every $u, v \in S_0$.

In this work we study the snapback repeller of system (2) by quoting the following lemma from [18] which can be used to determine a repelling fixed point of f under the Euclidean norm.

Lemma 1. *Let $\bar{x}$ be a fixed point of f which is continuously differentiable in $B_r(\bar{x})$. If*

$$\lambda > 1, \quad \text{for all eigenvalues } \lambda \text{ of } \left(f'(\bar{x})\right)^T f'(\bar{x}), \tag{4}$$

then there exist $s > 1$ and $r' \in (0, r]$ such that $\|f(x) - f(y)\|_2 > s \cdot \|x - y\|_2$, for all $x, y \in B_{r'}(\bar{x})$ with $x \neq y$, and all the eigenvalues of $\left(f'(x)\right)^T f'(x)$ exceed one for all $x \in B_{r'}(\bar{x})$.

Our objective is to present a symbolic computation approach to detect the chaotic behavior of system (2) by using Marotto's theorem. In the following, we will describe the algebraic criterion for Marotto's theorem.

Algebraic Criterion for Zeros Distribution with Respect to the Unit Circle. Our aim is to derive the algebraic criterion for all zeros of a given polynomial to be inside the unit circle (IUC) or outside the unit circle (OUC). To this end, we will use a sequence of symmetric polynomials of descending degrees for the characteristic polynomial, see [4] (or [15]).
    Let

$$D(\lambda) = d_0 + d_1\lambda + \ldots + d_n\lambda^n \tag{5}$$

be this characteristic polynomial, where $d_i = d_i(\mu, x)$, $i = 0, \ldots, n$. Then denote by $D^*(\lambda)$ the reciprocated polynomial of $D(\lambda)$, namely, $D^*(\lambda) = \lambda^n D(\lambda^{-1}) = d_n + d_{n-1}\lambda + \cdots + d_0\lambda^n$.
    Given the polynomial $D(\lambda)$, we assign to it a sequence of $n+1$ polynomials $T_n(\lambda), T_{n-1}(\lambda), \cdots, T_0(\lambda)$ according to the following formal definition:

$$T_n(\lambda) = D(\lambda) + D^*(\lambda), \quad T_{n-1}(\lambda) = [D(\lambda) - D^*(\lambda)]/(\lambda - 1),$$
$$T_{k-2}(\lambda) = \lambda^{-1}[\delta_k(\lambda + 1)T_{k-1}(\lambda) - T_k(\lambda)], \quad k = n, n-1, \ldots, 2, \tag{6}$$

where $\delta_k = T_k(0)/T_{k-1}(0)$. The recursion requires the normal conditions $T_{n-i}(0) \neq 0$, $i = 0, 1, \ldots, n$. The construction is interrupted when a $T_k(0) = 0$ occurs, and in [4], such singular cases can be classified into two types. The following theorem will be used to analyze the stability and chaotic behaviors of a discrete system, showing that there is no need to consider such singular cases. The proofs of Theorem 4 can be found in [4,15].

Theorem 4. *All zeros of $D(\lambda)$ are IUC (or OUC) if and only if the normal conditions $T_{n-i}(0) \neq 0$, $i = 0, 1, \ldots, n$ hold and $\nu_n = \text{Var}\{T_n(1), \ldots, T_0(1)\} = 0$ (or $\nu_n = n$).*

Semi-algebraic Systems for Marotto's Theorem. In the previous subsection, we have explained how to formulate critical algebraic criterion for all zeros of a given polynomial to be OUC. Based on this work, we will deduce the critical algebraic conditions for analysing Marotto's theorem.
    Let

$$\bar{A}(\lambda) = \lambda^n + \ldots + \bar{a}_1\lambda + \bar{a}_0 \tag{7}$$

be the characteristic polynomial of $\left(\boldsymbol{f}'(\bar{\boldsymbol{x}})\right)^T \boldsymbol{f}'(\bar{\boldsymbol{x}})$, where $\bar{a}_i = \bar{a}_i(\boldsymbol{\mu}, \bar{\boldsymbol{x}})$, $i = 0, \ldots, n-1$. According to Eq. (6) we can associate with $\bar{A}(\lambda)$ a sequence $\{\bar{T}_i(\lambda)\}_{i=0}^n$ which can be used to detect eigenvalue assignment. More precisely, we can use the following theorem to analyze the chaotic behavior of system (2).

Theorem 5. *[15] The general n-dimensional discrete system (2) is chaotic in the sense of Marotto if one of the following semi-algebraic systems has at least one real solution:*

$$\Psi_j : \begin{cases} \bar{\boldsymbol{x}} - \boldsymbol{f}(\boldsymbol{\mu}, \bar{\boldsymbol{x}}) = 0, \quad \boldsymbol{f}^m(\boldsymbol{x}_0) - \bar{\boldsymbol{x}} = 0, \\ (-1)^{i+j-1}\bar{T}_{n-i}(1)|_{\bar{\boldsymbol{x}}} > 0, \quad i = 0, \ldots, n, \\ (-1)^{i+j-1}\bar{T}_{n-i}(1)|_{\boldsymbol{x}_0} > 0, \quad i = 0, \ldots, n, \\ \bar{T}_{n-i}(0)|_{\boldsymbol{x}_0} \neq 0, \quad \bar{T}_{n-i}(0)|_{\bar{\boldsymbol{x}}} \neq 0, \quad i = 0, \ldots, n, \\ \boldsymbol{x}_0 \neq \bar{\boldsymbol{x}}, \quad |\boldsymbol{f}'(\boldsymbol{x}_k)| \neq 0, \quad k = 1, \ldots, m, \end{cases} \qquad (8)$$

where $j = 1, 2$, $\boldsymbol{\mu}$ and $\bar{\boldsymbol{x}}$ are respectively the parameters and fixed point of system (2) and m (≥ 2) is a given positive integer number.

In Theorem 5, what we want to find are the conditions on the parameters $\boldsymbol{\mu}$ for each of the semi-algebraic system (8) to have at least one real solution. There exist algebraic methods based on Gröbner bases, triangular decomposition, quantifier elimination, real solution classification, and discriminant varieties which can be used to solve such semi-algebraic systems.

3   Experiments

In this section, we first present the analysis of stability, bifurcations and chaos for a generalized Mira 2 map as an illustration of our algebraic approach explained above, and then study the stability, bifurcations and chaos of a 3D Hénon-like map of degree 2 and report some results on Tinkerbell map. Finally, we formulate a problem about planar quadratic maps based on our experiments. All the experiments were made in Maple 17, running under Windows 10 Professional Edition on a Inter(R) Core(TM) i7-7500U CPU @ 2.70 GHz 2.90 Ghz with 8G RAM.

3.1   Illustrative Example

In this subsection, we consider the following generalized Mira 2 map which takes the form

$$\begin{cases} x_{n+1} = Ax_n + Cy_n, \\ y_{n+1} = x_n^2 + By_n, \end{cases} \qquad (9)$$

where $A > 1$, $B \neq 1$ and C are nonzero real numbers. Mira 2 map [$C = 1$, $y_{n+1} = x_n^2 + B$] was first introduced in [21], and many details on the basin bifurcations are shown in [16].

We first detect the fixed points of (9) by using the following algebraic system

$$
\begin{cases}
P_1 = Ax + Cy - x = 0, \quad P_2 = x^2 + By - y = 0, \\
A > 1, \ B \neq 0, \ C \neq 0, \ B - 1 \neq 0.
\end{cases}
\tag{10}
$$

Note that the Jacobian matrix $\boldsymbol{f}'(\bar{\boldsymbol{x}})$ of map (9) evaluated at the fixed point $\bar{\boldsymbol{x}} = (x, y)$ is given by $\boldsymbol{f}'(\bar{\boldsymbol{x}}) = \begin{pmatrix} A & C \\ 2x & B \end{pmatrix}$, and the characteristic polynomial of the matrix $\boldsymbol{f}'(\bar{\boldsymbol{x}})$ can be written as $A(\lambda) = \lambda^2 - (A + B)\lambda + AB - 2Cx$.

To analyze the stability of each fixed point, we first use Theorem 4 together with Eq. (6) for the polynomial $A(\lambda)$ to obtain inequality polynomials $T_2(1), \ldots, T_0(1)$ and $T_2(0), \ldots, T_0(0)$, and then reduce the problem to that of solving the following semi-algebraic system

$$
\begin{cases}
P_1 = Ax + Cy - x = 0, \quad P_2 = x^2 + By - y = 0, \\
T_2(1) = 2AB - 4Cx - 2A - 2B + 2 > 0, (\text{or} - T_2(1) > 0), \\
T_1(1) = -2AB + 4Cx + 2 > 0, (\text{or} - T_1(1) > 0), \\
T_0(1) = 2AB - 4Cx + 2A + 2B + 2 > 0, (\text{or} - T_0(1) > 0), \\
T_2(0) = AB - 2Cx + 1 \neq 0, \quad T_1(0) = -AB + 2Cx + 1 \neq 0, \\
T_0(0) = 2AB - 4Cx + 2A + 2B + 2 \neq 0, \\
A - 1 > 0, \quad B \neq 0, \quad C \neq 0, \quad B - 1 \neq 0.
\end{cases}
\tag{11}
$$

The above semi-algebraic systems (10) and (11) may be solved by using the method of Yang and Xia [30] for real solution classification (implemented as a Maple package DISCOVERER by Xia), or the method of discriminant varieties of Lazard and Rouillier [19] (implemented as a Maple package DV by Moroz and Rouillier). Firstly, we list the results on the classification for the number of (stable) fixed points.

(a) System (9) always has two fixed points when $[A > 1, BC \neq 0, B - 1 \neq 0]$.
(b) When $[B - 1 < 0, 0 < AB - 2A - 2B + 3, AB - 3A - 3B + 1 < 0]$ system (9) has one stable fixed point; system (9) cannot have two stable fixed points.

Next we determine necessary bifurcation conditions on parameters A, B and C for system (9) to have certain type of bifurcations. For the Neimark-Sacher bifurcation, the problem may be formulated as that of detecting the conditions for the following semi-algebraic system to have at least one real solution:

$$
\begin{cases}
P_1 = Ax + Cy - x = 0, \quad P_2 = x^2 + By - y = 0, \quad -AB + 2Cx + 1 = 0, \\
AB - 2Cx - A - B + 1 > 0, \quad AB - 2Cx + A + B + 1 > 0, \\
AB - 2Cx + 1 > 0, \quad A - 1 > 0, \quad BC \neq 0, \quad B - 1 \neq 0.
\end{cases}
$$

Solving the above system by using DISCOVERER, we find that system (9) may undergo a Neimark-Sacher bifurcation if the condition $[B^2 + 2B - 7 < 0, AB - 2A - 2B + 3 = 0]$ holds. In a similar way, we obtain the necessary condition $[B^2 + 2B - 7 < 0, AB - 3A - 3B + 1 = 0]$ for the period doubling bifurcation to occur, and no stationary bifurcation occurs for this system.

Finally, we determine the conditions on A, B and C under which the fixed point $\bar{x}$ of system (9) is a snapback repeller.

Due to the Marotto's Theorem, we need to find one point $\boldsymbol{x}_0 = (x_0, y_0)$ in a repelling neighbourhood $B_{r'}(\bar{\boldsymbol{x}})$ such that $\boldsymbol{x}_0 \neq \bar{\boldsymbol{x}}$, $\boldsymbol{f}^m(\boldsymbol{x}_0) = \bar{\boldsymbol{x}}$ and $|\boldsymbol{f}'(\boldsymbol{x}_k)| \neq 0$ $(1 \leq k \leq m)$ hold for some positive integer m. Here, we consider the map $\boldsymbol{f}^2$ especially. Note the characteristic polynomial of the matrix $\left(\boldsymbol{f}'(\bar{x})\right)^T \boldsymbol{f}'(\bar{x})$ can be written as $\bar{A}(\lambda) = \lambda^2 - (A^2 + B^2 + C^2 + 4x^2)\lambda + A^2B^2 - 4ABCx + 4C^2x^2$.

Then for this polynomial $\bar{A}(\lambda)$ we can obtain the inequality polynomials $\bar{T}_2(1), \ldots, \bar{T}_0(1)$ and $\bar{T}_2(0), \ldots, \bar{T}_0(0)$ in Theorem 5 by using Eq. (6).

More precisely, using Theorem 5 we have the following semi-algebraic system for $j = 1$, which can be used for analyzing whether system (9) is chaotic or not in the sense of Marotto:

$$\Psi_1 : \begin{cases} P_1 = Ax + Cy - x = 0, P_2 = x^2 + By - y = 0, \\ P_3 = A(Ax_0 + Cy_0) + C(x_0^2 + By_0) - x = 0, \\ P_4 = (Ax_0 + Cy_0)^2 + B(x_0^2 + By_0) - y = 0, \\ \bar{T}_2(1)|_{\bar{x},x_0} > 0, \quad -\bar{T}_1(1)|_{\bar{x},x_0} > 0, \quad \bar{T}_0(1)|_{\bar{x},x_0} > 0, \\ \bar{T}_2(0)|_{\bar{x},x_0} \neq 0, \quad \bar{T}_1(0)|_{\bar{x},x_0} \neq 0, \quad \bar{T}_0(0)|_{\bar{x},x_0} \neq 0, \\ x_0 \neq \bar{x}, \quad |\boldsymbol{f}'(\boldsymbol{x}_1)| = AB - 2C(Ax_0 + Cy_0) \neq 0, \\ |\boldsymbol{f}'(\boldsymbol{x}_2)| = AB - 2Cx \neq 0, \quad A - 1 > 0, \quad BC(B-1) \neq 0, \end{cases}$$

where

$$\bar{T}_2(1)|_{\bar{x}} = 2A^2B^2 - 8ABCx + 8C^2x^2 - 2A^2 - 2B^2 - 2C^2 - 8x^2 + 2,$$
$$\bar{T}_1(1)|_{\bar{x}} = -2A^2B^2 + 8ABCx - 8C^2x^2 + 2,$$
$$\bar{T}_0(1)|_{\bar{x}} = 2A^2B^2 - 8ABCx + 8C^2x^2 + 2A^2 + 2B^2 + 2C^2 + 8x^2 + 2,$$
$$\bar{T}_2(0)|_{\bar{x}} = A^2B^2 - 4ABCx + 4C^2x^2 + 1,$$
$$\bar{T}_1(0)|_{\bar{x}} = -A^2B^2 + 4ABCx - 4C^2x^2 + 1,$$
$$\bar{T}_0(0)|_{\bar{x}} = 2A^2B^2 - 8ABCx + 8C^2x^2 + 2A^2 + 2B^2 + 2C^2 + 8x^2 + 2$$

with x, y, x_0, y_0 are the variables and A, B, C are the real parameters.

Solving the semi-algebraic system Ψ_1 by using DISCOVERER, we obtain that the semi-algebraic system Ψ_1 when $A = 5$ has at least one real solution if and only if one of the following conditions holds:

$$C_{1,1} = [0 < R_{1,1}, 0 < R_{1,2}],$$
$$C_{1,2} = [0 < R_{1,1}, 0 \leq R_{1,4}, R_{1,5} < 0],$$
$$C_{1,3} = [0 < R_{1,1}, 0 \leq R_{1,3}, 0 \leq R_{1,4}, 0 < R_{1,5}],$$

where

$$R_{1,1} = 8B^2C^2 - C^4 - 48BC^2 - 64B^2 + 40C^2 + 128B - 64,$$
$$R_{1,2} = 8B^2C^2 - C^4 - 48BC^2 - 4B^2 + 40C^2 - 32B - 64,$$
$$R_{1,3} = 20B^2C^2 - C^4 - 20BC^2 - 46B^2 - 120C^2 + 20B + 96,$$
$$R_{1,4} = 22B^2C^2 + C^4 - 20BC^2 + 46B^2 - 70C^2 - 20B - 96,$$
$$R_{1,5} = 400B^4C^4 - 40B^2C^6 + C^8 - 800B^3C^4 + 40BC^6 - 1840B^4C^2$$
$$- 4308B^2C^4 + 240C^6 + 2640B^3C^2 + 4760BC^4 + 16B^4 + 14080B^2C^2$$
$$+ 14208C^4 + 160B^3 - 8640BC^2 + 1168B^2 - 23040C^2 + 3840B + 9216.$$

Similar to the above steps, we can solve and obtain that there is no given number of real solution(s) for the semi-algebraic system Ψ_2 when $A = 5$. We remark that the polynomial expressions involved in the analysis are huge, and Maple was unable to reclaim sufficient memory during a calculation for the three free parameters A, B and C. From the above analyses, we have the following theorem.

Theorem 6. *The fixed point $\bar{x}$ is a snapback repeller of system* (9) *if one of the conditions: $\mathcal{C}_{1,1}$ or $\mathcal{C}_{1,2}$ or $\mathcal{C}_{1,3}$ is satisfied, and hence system* (9) *is chaotic in the sense of Marotto.*

Now we take the condition $\mathcal{C}_{1,3}$ to illustrate the parametric region where the fixed point $\bar{x}$ is a snapback repeller for a visual understanding on the conditions above. The dotted lines in Fig. 1 are the critical boundaries determined

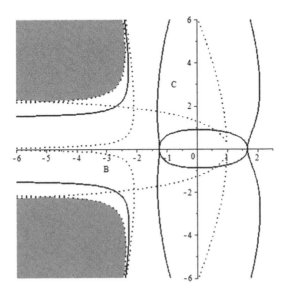

Fig. 1. Parameter space determined by $\mathcal{C}_{1,3}$ for system (9) is chaotic.

by the polynomials appearing in condition $\mathcal{C}_{1,3}$, and the shadowed region is the parameter domain where all the inequalities hold.

3.2   Other Models and Remarks

A 3D Hénon-Like Map. The Hénon map $[x_{n+1} = 1 + y_n - Ex_n^2, y_{n+1} = Dx_n]$, first introduced in Hénon [12], is a two dimensional and invertible map and is one of the most known and studied examples of a dynamical system with a strange attractor. Here we study the following 3D Hénon-like map:

$$\begin{cases} x_{n+1} = 1 + y_n - Ex_n^2, \\ y_{n+1} = Dx_n + Ez_n, \\ z_{n+1} = -Ey_n, \end{cases} \tag{12}$$

where $E \neq 0$ and $D > 0$ are real parameters. We have the following results for this discrete system.

(a) When $[R_{2,1} < 0]$, system (12) has no fixed point; when $[R_{2,1} = 0]$, system (12) has a unique fixed point; when $[0 < R_{2,1}]$, system (12) has two fixed points.

(b) When $[0 < R_{2,1}, 0 \leq R_{2,2}, R_{2,3} < 0]$, system (12) has two fixed points, of which one is stable; system (12) cannot have two stable fixed points.

(c) When $[-D < 0, D - 1 < 0, R_{2,3} = 0]$, system (12) may undergo a period doubling bifurcation.

(d) When $[-D < 0, D - 1 < 0, R_{2,1} = 0]$, system (12) may undergo a stationary bifurcation.

(e) When $E = 2$, system (12) has a snapback repeller if one of the following conditions holds:

$$\mathcal{C}_{2,1} = [-D < 0, D - 4 < 0],$$
$$\mathcal{C}_{2,2} = [0 < D - 4, 8D^4 + 45D^3 - 501D^2 + 135D + 2475 < 0].$$

The explicit expressions of $R_{2,i}$ for $i = 1, \ldots, 3$ are as follows:

$$R_{2,1} = 4E^5 + E^4 + 8E^3 - 2E^2 D + 2E^2 + D^2 + 4E - 2D + 1,$$
$$R_{2,2} = E^2 - D + 1,$$
$$R_{2,3} = 4E^5 - 3E^4 + 8E^3 + 6E^2 D - 6E^2 - 3D^2 + 4E + 6D - 3.$$

We remark that when $D > 0$ there is no Neimark-Sacher bifurcation occurring. Note that the conditions for system (12) to have a snapback repeller at $\bar{x} = (x, y, z)$ are obtained by taking $E = 2$. Because Maple was consuming too much of the CPU during the calculation for the two free parameters D and E.

The Tinkerbell Map. The Tinkerbell map [1,7] is a discrete dynamical system given by the equations

$$\begin{cases} x_{n+1} = x_n^2 - y_n^2 + ax_n + by_n, \\ y_{n+1} = 2x_ny_n + cx_n + dy_n, \end{cases} \tag{13}$$

where a, b, c and d are real parameters. This map has been extensively analyzed numerically, and has found its place in various applications, for example in an optical ring phase conjugated resonator [2], in PID controller design [8] and in pseudo-random number generators [25]. Applying our algebraic approach, we have the following results for this system.

(a) When $[R_{3,1} > 0]$, system (13) has two fixed points; when $[R_{3,1} = 0]$, system (13) has three fixed points; when $[R_{3,1} < 0]$, system (13) has four fixed points. Here

$$\begin{aligned} R_{3,1} = &\ 8a^3d - 4a^2b^2 - 4a^2bc - a^2c^2 - 12a^2d^2 + 4ab^2d + 22abcd + 10ac^2d + 6ad^3 \\ &- 8b^3c - 12b^2c^2 - b^2d^2 - 6bc^3 - 10bcd^2 - c^4 + 2c^2d^2 - d^4 - 8a^3 + 4ab^2 \\ &- 14abc - 8ac^2 + 6ad^2 - 2b^2d - 2bcd - 14c^2d - 2d^3 + 12a^2 - 6ad - b^2 \\ &+ 8bc + 11c^2 - 6a + 2d + 1. \end{aligned}$$

(b) When $b = d = 0$, system (13) has one stable fixed point if one of the following conditions holds

$$C_{3,1} = [0 < R_{3,2}, R_{3,3} < 0, R_{3,4} < 0, 0 < R_{3,5}, 0 < R_{3,6}, R_{3,8} \leq 0],$$
$$C_{3,2} = [0 < R_{3,2}, R_{3,3} < 0, R_{3,4} < 0, 0 < R_{3,5}, 0 < R_{3,6}, 0 \leq R_{3,8}, 0 \leq R_{3,9}],$$

$$\vdots$$

$$C_{3,11} = [0 < R_{3,2}, 0 < R_{3,3}, 0 < R_{3,4}, 0 < R_{3,5}, R_{3,6} < 0, R_{3,7} \leq 0, R_{3,8} \leq 0],$$

and has two stable fixed points if one of the following conditions holds

$$C_{3,12} = [0 < R_{3,2}, R_{3,3} < 0, 0 < R_{3,5}, 0 < R_{3,6}, 0 < R_{3,4}, R_{3,8} \leq 0, R_{3,9} \leq 0],$$
$$C_{3,13} = [0 < R_{3,2}, 0 < R_{3,3}, 0 < R_{3,5}, 0 < R_{3,6}, R_{3,10} \leq 0],$$

where $R_{3,2} = a + 1$, $R_{3,3} = a - 1$, $R_{3,4} = a^2c^2 + c^4 + 8a^3 + 8ac^2 - 12a^2 - 11c^2 + 6a - 1$. The expressions of $R_{3,5}, \ldots, R_{3,10}$ are too long, so we omit them here.

(c) When $b = d = 0$, system (13) may undergo a Neimark-Sacher bifurcation if one of the following conditions holds

$$C_{3,14} = [0 < R_{3,11}, R_{3,12} < 0, 0 < R_{3,13}, 0 < R_{3,14}, R_{3,15} = 0],$$
$$C_{3,15} = [0 < R_{3,11}, 0 < R_{3,12}, 0 < R_{3,13}, 0 < R_{3,14}, R_{3,15} = 0],$$

where

$$R_{3,11} = c^4 + 29c^2 - 1, \quad R_{3,12} = 4c^4 + 113c^2 - 1210,$$
$$R_{3,13} = 7c^4 + 75c^2 + 1521, \quad R_{3,14} = c^8 + 20c^6 + 350c^4 + 2052c^2 + 9,$$
$$\begin{aligned} R_{3,15} = &\ 2a^3c^2 + 2ac^4 + 18a^4 + 15a^2c^2 - 3c^4 - 66a^3 - 66ac^2 + 65a^2 + 61c^2 \\ &- 24a + 3. \end{aligned}$$

(d) When $b = d = 0$ and $[a + 1 = 0, R_{3,16} := 3a^3c^2 + 3ac^4 + 32a^4 + 31a^2c^2 - c^4 - 96a^3 - 105ac^2 - 54a^2 + 27c^2 + 135a + 81 = 0]$, system (13) may undergo a period doubling bifurcation.

(e) When $b = 0$, system (13) may undergo a stationary bifurcation if one of the following conditions holds

$$C_{3,16} = [0 < d + 1, d - 1 < 0, a - 1 = 0],$$
$$C_{3,17} = [R_{3,17} < 0, 0 < R_{3,18}, R_{3,19} = 0],$$
$$C_{3,18} = [0 < R_{3,17}, R_{3,18} < 0, R_{3,19} = 0],$$

where

$$R_{3,17} = c^4 + 29c^2d^2 - d^4 - 58c^2d + 4d^3 + 29c^2 - 6d^2 + 4d - 1,$$
$$R_{3,18} = c^4 + 29c^2d^2 - d^4 - 50c^2d + 16d^3 + 25c^2 - 90d^2 + 200d - 125,$$
$$R_{3,19} = 8a^3d - a^2c^2 - 12a^2d^2 + 10ac^2d + 6ad^3 - c^4 + 2c^2d^2 - d^4 - 8a^3 - 8ac^2$$
$$+ 6ad^2 - 14c^2d - 2d^3 + 12a^2 - 6ad + 11c^2 - 6a + 2d + 1.$$

(f) System (13) has a snapback repeller if one of the following conditions $C_{3,i}$ for $i = 19, \ldots, 25$ holds.

Case 1: When $a = b = c = 0$, we have

$$C_{3,19} = [0 < d + 1, 0 < d^2 - 2d - 1, 0 < R_{3,20}],$$
$$C_{3,20} = [d + 1 < 0, 0 < d^2 - 2d - 1, R_{3,20} < 0],$$
$$C_{3,21} = [d + 1 < 0, 0 < d^2 - 2d - 1, R_{3,21} < 0, 0 < R_{3,22}],$$
$$C_{3,22} = [0 < d + 1, 0 < d^2 - 2d - 1, 0 < R_{3,21}, 0 < R_{3,22}].$$

Here again we omit the expressions of $R_{3,20}, \ldots, R_{3,22}$ for brevity.

Case 2: When $a = c = d = 0$, we have

$$C_{3,23} = [2b^3 - 2b - 1 < 0, 2b^3 - 2b + 1 < 0],$$
$$C_{3,24} = [0 < 2b^3 - 2b - 1, 0 < 2b^3 - 2b + 1].$$

Case 3: When $b = c = d = 0$, we have $C_{3,25} = [-1 + 2a > 0]$.

The computations for these three cases in DISCOVERER took about 154.360, 17141.657 and 60.813 s respectively. We also tried to analyze the case $a = b = d = 0$ for system (13) to have a snapback repeller, but Maple was consuming too much of the CPU during the calculation.

Our experiments demonstrate the feasibility of our algebraic approach for stability, bifurcations and chaos analysis of discrete dynamical systems. However, the polynomial expressions involved in the analysis are huge, which makes the computation very difficult. Some explicit conditions on the parameters for the bifurcations and chaos can only be obtained under some constrains. So far we can only deal with systems of low (may less than five) dimensions. Even for the low dimensional systems, the determination of the conditions on all the parameters involved for the bifurcations and chaos is highly nontrivial. Below we list an interesting subproblem of our problem in Sect. 1 on the planar quadratic maps for researchers on related fields.

Problem on Planar Quadratic Maps. Consider the following planar quadratic map

$$
\begin{aligned}
x_{n+1} &= a_1 x_n^2 + a_2 x_n y_n + a_3 y_n^2 + a_4 x_n + a_5 y_n + a_6, \\
y_{n+1} &= b_1 x_n^2 + b_2 x_n y_n + b_3 y_n^2 + b_4 x_n + b_5 y_n + b_6.
\end{aligned}
\tag{14}
$$

Let $\bar{x}$ be a fixed point of this system (if any). Then what are the explicit conditions on the parameters a_i and b_i for $i = 1, \ldots, 6$ such that

1. $\bar{x}$ is stable for system (14)?
2. system (14) undergos certain types of bifurcations?
3. $\bar{x}$ is a snapback repeller of system (14)?

Remark that since many famous and well-studied maps are subclasses of system (14), such as (generalized) Mira 2 map, (generalized) Hénon map and Tinkerbell map, now we have obtained partial results for this problem.

4 Discussion

This paper presents an algebraic approach to detect fixed points and to analyze their stability, bifurcations and chaos for discrete dynamical systems. Illustrative examples and experiments are provided, showing the feasibility of the proposed approach. This work extends previous investigations on stability and bifurcations [20] to more complex dynamics (chaotic behaviors) of discrete systems.

The bifurcation conditions we can derive for discrete dynamical systems are only necessary ones. How to check the sufficiency of the conditions and how to determine the type of each stationary bifurcation are questions that remain for further study. It is of great interest to employ our algebraic approach to analyze the stability, bifurcations and chaotic behaviors in many different fields (biology, physics, engineering, etc.). On the other hand, we note that sometimes it might be very hard to determine the explicit conditions on the parameters such that the considered discrete systems have stable fixed points (or may have a certain type of bifurcations or have a snapback repeller). In practical terms, there exist some values of the parameters that satisfy the required properties. Advanced techniques for analysing the bistability and oscillations in biological networks (see [6]) will be studied and integrated into the current approach to tackle this kind of problems. How to simplify and optimize the steps of symbolic computations in the current approach is another question worthy of study.

References

1. Alligood, K.T., Sauer, T.D., Yorke, J.A.: Chaos: An Introduction to Dynamical Systems. Springer, Berlin (1996). https://doi.org/10.1007/b97589
2. Aboites, V., Wilson, M., Bosque, L., del Campestre, L.: Tinkerbell chaos in a ring phase-conjugated resonator. Int. J. Pure Appl. Math. **54**(3), 429–435 (2009)
3. Buchberger, B.: Gröbner bases: an algorithmic method in polynomial ideal theory. In: Multidimensional Systems Theory, pp. 184–232. Reidel, Dordrecht (1985)
4. Bistritz, Y.: Zero location with respect to the unit circle of directe-time linear system polynomials. Proc. IEEE **72**(9), 1131–1142 (1984)
5. Collins, G.E., Hong, H.: Partial cylindrical algebraic decomposition for quantifier elimination. J. Symb. Comput. **12**(3), 299–328 (1991)
6. Dickenstein, A., Millán, M., Shiu, A., Tang, X.: Multistationarity in structured reaction networks. Bull. Math. Biol. **81**(5), 1527–1581 (2019)
7. Davidchack, R.L., Lai, Y.C., Klebanoff, A., Bollt, E.M.: Towards complete detection of unstable periodic orbits in chaotic systems. Phys. Lett. A **287**(1–2), 99–104 (2001)
8. Coelho, L.S., Mariani, V.C.: Firefly algorithm approach based on chaotic Tinkerbell map applied to multivariable PID controller tuning. Comput. Math. Appl. **64**(8), 2371–2382 (2012)
9. Faugère, J.-C.: A new efficient algorithm for computing Gröbner bases (F4). J. Pure Appl. Algebra. **139**(1–3), 61–88 (1999)
10. Galor, O.: Discrete Dynamical Systems. Springer, Berlin (2007). https://doi.org/10.1007/3-540-36776-4
11. Glendinning, P.: Bifurcations of snap-back repellers with application to border-collision bifurcations. Int. J. Bifurcat. Chaos **20**(2), 479–489 (2010)
12. Hénon, M.: A two-dimensional mapping with a strange attractor. Commun. Math. Phys. **50**(1), 69–76 (1976)
13. Hong, H., Liska, R., Steinberg, S.: Testing stability by quantifier elimination. J. Symb. Comput. **24**(2), 161–187 (1997)
14. Hong, H., Tang, X., Xia, B.: Special algorithm for stability analysis of multistable biological regulatory systems. J. Symb. Comput. **70**(1), 112–135 (2015)
15. Huang, B., Niu, W.: Analysis of snapback repellers using methods of symbolic computation. Int. J. Bifurcat. Chaos **29**(4), 1950054-1-13 (2019)
16. Kitajima, H., Kawakami, H., Mira, C.: A method to calculate basin bifurcation sets for a two-dimensional nonivertible map. Int. J. Bifurcat. Chaos **10**(8), 2001–2014 (2000)
17. Kaslik, E., Balint, S.: Complex and chaotic dynamics in a discrete-time-delayed Hopfield neural network with ring architecture. Neural Networks **22**(10), 1411–1418 (2009)
18. Li, C., Chen, G.: An improved version of the Marotto theorem. Chaos Solit. Fract. **18**(1), 69–77 (2003)
19. Lazard, D., Rouillier, F.: Solving parametric polynomial systems. J. Symb. Comput. **42**(6), 636–667 (2007)
20. Li, X., Mou, C., Niu, W., Wang, D.: Stability analysis for discrete biological models using algebraic methods. Math. Comput. Sci. **5**(3), 247–262 (2011)
21. Mira, C., Barugola, A., Gardini, L.: Chaotic Dynamics in Two-Dimensional Nonvertible Map. World Scientific, Singapore (1996)
22. Marotto, F.: Snap-back repellers imply chaos in $\mathbb{R}^n$. J. Math. Anal. Appl. **63**(1), 199–223 (1978)

23. Marotto, F.: On redefining a snap-back repeller. Chaos Solit. Fract. **25**(1), 25–28 (2005)
24. Niu, W., Shi, J., Mou, C.: Analysis of codimension 2 bifurcations for high-dimensional discrete systems using symbolic computation methods. Appl. Math. Comput. **273**, 934–947 (2016)
25. Stoyanov, B., Kordov, K.: Novel secure pseudo-random number generation scheme based on two Tinkerbell maps. Adv. Stud. Theor. Phys. **9**(9), 411–421 (2015)
26. Sang, B., Huang, B.: Bautin bifurcations of a financial system. Electron. J. Qual. Theory Differ. Equ. **2017**(95), 1–22 (2017)
27. Wu, W.-T.: Mathematics Mechanization. Science Press/Kluwer Academic, Beijing (2000)
28. Wang, D.: Elimination Methods. Springer, New York (2001). https://doi.org/10.1007/978-3-7091-6202-6
29. Wen, G.: Criterion to identify Hopf bifurcations in maps of arbitrary dimension. Phys. Rev. E. **72**(2), 026201-1-4 (2005)
30. Yang, L., Xia, B.: Real solution classifications of parametric semi-algebraic systems. In: Algorithmic Algebra and Logic-Proceedings of the A3L, pp. 281–289. Herstellung und Verlag, Norderstedt (2005)

Security and Cryptography

Acceleration of Spatial Correlation Based Hardware Trojan Detection Using Shared Grids Ratio

Fatma Nur Esirci$^{(\boxtimes)}$ and Alp Arslan Bayrakci

Gebze Technical University, Gebze, Turkey
{fesirci,abayrakci}@gtu.edu.tr

Abstract. Due to mostly economic reasons almost all countries including the developed ones have to handle integrated circuit designs to a foreign fab for manufacturing, which raises the security issues like intentional malicious circuit (hardware Trojan) insertion by an adversary. A previously proposed method to address these security issues detects hardware Trojan using the spatial correlations in accordance with delay based side channel analysis. However, it is never applied to full circuits and it requires too many path delay computations to select correlated path pairs. In this paper, we first apply the method and present the results for full circuits and then, the method is accelerated by proposing a novel path selection criterion which avoids the computation of path delays. In terms of detection success, the resultant method performs similar to the previous one, but in a much faster fashion.

Keywords: Hardware security · Hardware Trojan · Side channel analysis · Spatial correlations · Process variations

1 Introduction

The fabrication of chips is a sophisticated process that can only be performed in state-of-art fabrication facilities. Given this increasingly expanding cost and complexity of foundries, the semiconductor business model has largely shifted to a contract foundry business model over the past two decades. For example, Texas Instruments and Advanced Micro Devices, two chip making giants that have traditionally used their in-house facilities for fabricating their chips, have both in 2010 announced outsourcing most of their sub-45 nm fabrication to major contract foundries worldwide [15]. One of the most crucial effects of this compulsory shift is on hardware security. Handling the design to manufacturing fab and the difficulty of detecting any malicious alteration on the manufactured chip make the system vulnerable to attacks especially during the manufacturing. Any such malicious alteration on the circuit is called hardware Trojan (HT). Several papers [14] and the IEEE Spectrum magazine articles [1] comprehensively describe the hardware Trojan threat on security and the difficulty of detection.

© Springer Nature Switzerland AG 2020
D. Slamanig et al. (Eds.): MACIS 2019, LNCS 11989, pp. 187–201, 2020.
https://doi.org/10.1007/978-3-030-43120-4_15

Especially, the mission critical circuits such as the ones used for cryptography are main targets for such hardware Trojan attacks [5,11]. There are many different types of proposed hardware Trojans as well as many detection methods until now [15]. Destructive methods can provide exact results but only for the investigated chip, by also making the chip useless after the destructive analysis. It cannot guarantee the authenticity of the remaining chips either. Therefore, non-destructive methods like side channel analysis are worked on to detect HT in the chips. Side channel analysis based Trojan detection methods investigates the measurable side channel signals like delay [13], power [3] and temperature [12] to reveal any HT existence. Yet, the unavoidable process variations can easily hide the effect of the inserted Trojan. This makes especially the detection of small Trojans very hard.

The spatial correlation based HT detection method in [8] claims to detect even the smallest type of Trojan composed of only one XOR gate under realistic process variations using delay based HT detection. The method takes advantage of spatial correlations that are inherently present in manufactured chips. However, it is not tested on full circuit and also it is computationally complex. In this paper, we first adapt the method in [8] to full circuits and report the results on full circuits. Then, we propose using a new criterion, called shared grids ratio, for the selection of correlated path among numerous candidates. Theoretical cost analysis of both methods as well as the experimental results are presented in the paper. The results show that the proposed improvement can speed up the method about 10 times in correlated path selection on the average, which in turn accelerates the whole method more than 2 times on the average over benchmark circuits. And this enhancement comes with almost no cost on the HT detection capability of the method.

This paper is organized as follows: Sect. 2 gives some background on circuit representation as graph, delay based HT detection, the effect of variations on detection and summarizes the spatial correlation based HT detection method in [8] by separating it into four stages. Section 3 adapts this method to full circuits and presents the results. A cost analysis for this adaptation is performed in Sect. 4. Section 5 introduces a new criterion to accelerate the method. The results and comparisons of both the previous and the new method are explained in Sect. 6.

2   Preliminaries

2.1   Representation of Circuits

We use graph structure to express digital circuits, where each gate in the circuit corresponds to a node and each interconnect between two gates corresponds to an edge of the graph. A path in the circuit starts from a primary input, traverses through gates (nodes) and ends at a primary output. Any edge of the circuit is assumed to have the potential of a Trojan circuit insertion.

2.2   Delay Based Trojan Detection

One of the most effective methods in the literature is delay based HT detection, which is a sub-branch of side channel analysis (SCA). Normally, a smart Trojan is designed to stay at passive state so that it cannot be detected through conventional functional tests. Yet, at least a tiny part (payload) of the Trojan must be inserted on a wire in the circuit in order to be able to alter the signal at that wire when it gets active. Thus, the payload part brings some delay add-on to the original circuit. The power of delay based detection is due to the fact that they do not require to make HT active for detection in contrast to functional test based methods. Also they can be applied by widely used feasible delay tests without destroying the chip in contrast to destructive detection methods. The success of the Trojan detection based on delay is dependent on Trojan size because the bigger the Trojan is, the more delay add-on it has. And its main drawback is the process variations that can easily hide the delay add-on of the Trojan circuitry.

2.3   Variation Effect and Difficulties

Process variation is due to the nature of the chip manufacturing process. It is undesirable but inevitable. The circuits are designed according to specific constraints such as functionality, speed and power consumption. At the post manufacturing stage, the chip set obtained by manufacturing are examined to see if they meet these constraints. Yet, due to manufacturing process variations on circuit components like gate length and threshold voltage, chips do not exactly meet the same specification but instead each manufactured chip comes with different properties.

Any realistic variation model must include both inter-die (between chips) and intra-die (within the chip) variation components. As the integrated circuits scale down in feature size with developing technology, the effect of intra-die variation increases. The intra-die variation component inherently exhibits spatial correlations. As a result of the spatial correlations, the random parameters of the transistors closer to each other are affected more similar from the variations when compared with the ones residing far from each other. If the results of a method are justified by circuit simulations, it is very important to use variation models that can consider all variation components as well as accurate variation amounts corresponding to current technology [9].

The main challenge of SCA based detection is to distinguish the HT effect from the effect of process variations. SCA based detection methods either fail to use accurate variation models or fail to detect very small Trojans. To overcome this challenge, we require a method that enables us to get rid of the variation effect even under the accurate variation model.

The spatial correlation based HT detection method proposed in [8] uses such a variation model and precise transistor level Spice simulations to justify the proposed method. It also claims to detect even the smallest type of Trojans. As opposed to most SCA based methods [15], it can even work in the absence of a golden model when only a fraction of the manufactured chips have an inserted Trojan. Such selective insertion is preferred by the adversary because otherwise destructive analysis of any chip can easily reveal the Trojan existence. However, the scalability of the method is unknown as it is not executed on full circuits. Also, it is computationally very complex to detect a correlated path pair for each edge (interconnect) in the circuit requiring numerous path delay and correlation coefficient computations. Next section summarizes this method.

2.4   Review of Spatial Correlation Based HT Detection Method [8]

Getting rid of variation effect is a hard task as the variations neither can be avoided nor can be exactly measured due their random nature. One technique is to divide components that are affected from the variations very similar so that the effect of variations is canceled out [11, 16].

Spatial Correlation Based HT Detection [8] extracts correlated paths by taking advantage of spatial correlations. As any Trojan circuit must be connected to at least one edge in the circuit, the method traverses all edges in the circuit to detect whether there is a connected Trojan on that edge. It is composed of the following stages executed for each edge in the circuit: (i) extraction of one suspected path for each edge, (ii) extraction of correlated path candidates for each suspected path, (iii) selection of one correlated path among the candidates, (iv) measurement and division of path delays of suspected and correlated paths. The first three stages are pre-manufacturing but the last stage is post manufacturing and must be applied to manufactured chips.

(i) It is easier to detect Trojan using short (small delay) paths in the circuit for its increased relative effect on delay. Therefore, for each edge e, the shortest path passing through that edge is selected as the suspected path (P_{susp}^e). The cost of suspected path extraction is not high as the shortest path is detected according to the nominal delay values of nodes (logic gates).

(ii) The second stage is the extraction of all possible path candidates which may be correlated with the suspected path. For that, spatial correlation information is used. Due to the spatial correlations, a path which has logic gates residing at very close locations with another path must have correlated path delays. This means that if one can find a very closely located path for a suspected path, the ratio of path delays of these two paths can cancel out the variation component. In this case, any alteration like Trojan insertion can be easily revealed by detecting the deviation in path delay ratio. In order to find path candidates correlated with a suspected path, the circuit is divided into grids (Fig. 3) and then, all paths whose gates are located either at the same grid or at the adjacent grids of the suspected path are extracted and collected in correlated path candidates set. At the end, each suspected path has a corresponding correlated path candidates set.

(iii) At the third stage, first of all, the path delays of all correlated path candidates are computed for all samples (chips). Then, for each path pair consisting of the corresponding suspected path and a correlated path candidate, the correlation coefficient is computed based on these path delays. The correlated path candidate resulting in the best correlation coefficient is nominated as the correlated path (P^e_{corr}) for the corresponding suspected path of edge e.

(iv) This is the post manufacturing test stage. The path delays of the suspected path and the nominated correlated path are measured and divided to cancel out the variation component, which reveals any HT existence. The computation of delay ratio for a sample edge e is shown by (1), where d denotes the path delay. The algorithm is successful in Trojan detection without requiring golden model if the resultant delay ratios for Trojan-free and Trojan-inserted samples can be separated from each other.

$$R_e = \frac{d(P^e_{susp})}{d(P^e_{corr})} \tag{1}$$

For delay computations above, a delay model called Stochastic Logical Effort (SLE) and constructed by precise transistor level Spice simulations [4] is employed. Path delay computation is performed by summing up the individual delays of gates on the path. Delay of each gate is computed by a fast and accurate gate delay model called Stochastic Logical Effort (SLE) as shown in (2). In this equation $d_r(S)$ is the delay of a logic gate r for sample S, $\tau(S)$ is the reference inverter delay, $p_r(S)$ and $g_r(S)$ is the parasitic component and logical effort for the same chip and h_r is the electrical fan-out for gate r. The further details of the model can be found in [4].

$$d_r(S) = \tau(S)(p_r(S) + g_r(S)h_r) \tag{2}$$

As it is quantified in this paper, one of the most time consuming part in the algorithm is stage (iii). Considering current deeply integrated circuits with even millions of gates, a numerous number of path delay and correlation coefficient computations are required as there may be a plenty of correlated path candidates considering all edges in the circuit.

3   CCM: Adaptation of the Method in Sect. 2.4 to Full Circuit

Spatial correlation based HT detection method in [8] is applied only to randomly selected three edges from each benchmark circuit. The paper also does not devise any method to discriminate Trojan inserted samples from the Trojan-free ones. It only reports the number of misclassified samples when the best separating line between the ratios of Trojan inserted chips and Trojan free ones is assumed. In the actual case the separation line is unknown.

To compensate these shortcomings, we adapt the algorithm to full circuit. Instead of randomly selecting just three edges in [8], we assume all edges in a circuit are suspected for Trojan existence and thus, the algorithm is executed for all edges in the circuit. Throughout the paper, this method is referred as Correlation Coefficient based Method (CCM). CCM is a direct adaptation of the stages explained in Sect. 2.4. Therefore, it uses correlation coefficients to eliminate correlated path candidates as stage (iii) of Sect. 2.4 explains. CCM requires correlation coefficients, as it searches for the path with the highest correlation coefficient to select the path correlated most with the corresponding suspected path among the candidates. Then, this path constitutes the path pair with the suspected path. This pair is used to compute delay ratio shown by (1).

We call an edge to be *covered* if the samples with a Trojan inserted on that edge can be detected by the method after post-manufacturing tests (stage (iv) tests). A Trojan inserted sample is said to be *detected* if, for that edge, the resultant delay ratio distributions of all Trojan-free and Trojan-inserted samples are separate from each other. The two distributions are separate if their 1.5σ have positive difference. The computation of 1.5σ difference between delay ratios of Trojan-free and Trojan-inserted samples for an edge e is shown by (3).

$$\Delta_{1.5\sigma_e} = (\mu_{\hat{R}_e} - 1.5\hat{\sigma}_{\hat{R}_e}) - (\mu_{R_e} - 1.5\sigma_{R_e}) \tag{3}$$

where $\Delta_{1.5\sigma_e}$ is the 1.5σ difference for edge e, μ_{R_e} and σ_{R_e} are mean and standard deviation of the delay ratios for the Trojan-free samples, $\mu_{\hat{R}_e}$ and $\hat{\sigma}_{\hat{R}_e}$ are mean and standard deviation of the delay ratios for the Trojan-inserted samples. Please remind that the delay ratio is computed by dividing the delay of the suspected path to the delay of the corresponding correlated path in the pair (1). Therefore, if the edge can be covered, this means that the CCM has picked the right path pair for that edge to detect any Trojan insertion on it.

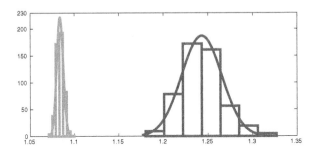

Fig. 1. Histogram of ratios without HT (green) and with HT (red) (Color figure online)

For instance, delay ratio distributions for a covered edge from c1908 benchmark circuit are plotted in Fig. 1. For that edge, the Trojan-inserted samples can be easily separated without a need for a golden model. Therefore, this edge is said to be secured or covered by the method. However, Fig. 2 shows another edge from c1908, which cannot be covered by the method as the delay ratios of Trojan-free and Trojan-inserted samples are intermixed into each other resulting in negative $\Delta_{1.5\sigma_e}$ and not possible to be separated if they were not colored.

Table 1 shows the results for CCM. Number of edges in the circuit, the resultant edge coverage and the number of total correlated path candidates are the

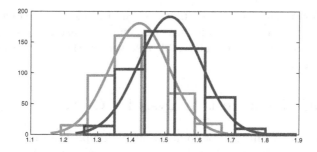

Fig. 2. Histogram of ratios without HT (green) and with HT (red) (Color figure online)

Table 1. Full circuit experimental results for CCM

Benchmark	# of edges	Edge coverage	# of candidates for CCM
432	255	95.7%	36,647
499	296	64.5%	29,320
880	507	94.7%	39,176
1355	856	96.6%	17,200,357
1908	1420	93.4%	5,353,429
2670	1850	95.2%	7,619,334

respective columns of Table 1. The number of <suspected path, correlated path> pairs is equal to the number of edges, therefore not reported in the table. The full coverage means that any Trojan inserted on any edge can be detected by the method. Results show that the coverage is just about 90% on the average over benchmark circuits in the table. It also shows the number of correlated path candidates required for CCM. This makes the stage (iii) of the algorithm explained in Sect. 2.4 the most unbearable part of the algorithm. Because the path delay and correlation coefficient for each correlated path candidate are computed in stage (iii).

Table 2. Time consumption for stages for CCM

Benchmark	Run time stage (i)	Run time stage (ii)	Run time stage (iii)
432	4 s	54 s	110 s
499	7 s	38 s	102 s
880	23 s	180 s	169 s
1355	1.2 m	13.21 h	21.76 h
1908	1.38 m	4.96 h	7.63 h
2670	7.2 m	6.39 h	9.18 h

Table 2 shows the amount of time consumption for each stage of CCM, explained in Sect. 2.4 – except for the post-manufacturing stage (stage (iv)) – when applied to full circuits. This table also verifies that the main bottleneck is stage (iii) for the method.

4   Computational Cost Analysis for CCM

The CCM has the main flaw of computational complexity due to mainly the stage (iii) computations. Because this stage computes the path delay using (2) for each sample chip and for each of the extracted correlated path candidates. Then, using these path delays, the correlation coefficient is computed again for each candidate. However, the number of correlated path candidates as shown in Table 1 can get very large with the increasing circuit size or complexity. Besides, the number of samples must be a big enough number to get accurate results, which also complicates stage (iii) computations.

The resultant computational cost of stage (iii) is represented by (4). In this equation, $\texttt{Nsamples}$ represents the number of chips, $\texttt{Ngates}_{\texttt{full}}$ is the number of logic gates in the full circuit, $\texttt{Cost}_{\texttt{SLE}}$ is the cost of computing SLE in (2) (two multiplications and one addition), $\texttt{Ncand}$ is the number of correlated path candidates, $\texttt{Ngates}_{\texttt{avg}}$ is the average number of gates over all correlated path candidates, $\texttt{Cost}_{\texttt{add}}$ is the cost of one addition used in path delay computation. $\texttt{Cost}_{\texttt{coeff}}$ is the unit cost for correlation coefficient computation. It utilizes arithmetic operations like addition, division and square root. Lastly $\texttt{Cost}_{\texttt{comp}}$ is the cost of comparing floating point numbers to find the max.

$$\begin{aligned}\texttt{Cost}^{\texttt{CCM}}_{\texttt{stage(iii)}} = &\ \texttt{Nsamples} \times \texttt{Ngates}_{\texttt{full}} \times \texttt{Cost}_{\texttt{SLE}}+\\ &\ \texttt{Ncand} \times \texttt{Nsamples} \times \texttt{Ngates}_{\texttt{avg}} \times \texttt{Cost}_{\texttt{add}}+\\ &\ \texttt{Ncand} \times \texttt{Nsamples} \times \log{(\texttt{Nsamples})} \times \texttt{Cost}_{\texttt{coeff}}+\\ &\ \texttt{Ncand} \times \texttt{Cost}_{\texttt{comp}}\end{aligned} \tag{4}$$

The first row of the equation shows the cost of computing SLE delays for each gate and for each sample chip, the second row shows the cost of path delay computations using SLE delays computed in the first row and performed for each chip and each correlated path candidate. The third row in the equation shows the correlation coefficient computation using path delays computed in the second row. The second row and especially the third row constitute the main source of complexity. The last row is for finding the candidate with the maximum correlation coefficient.

5   Shared Grids Method (SGM) for Accelerating CCM

When Table 2 is investigated, stage (i), i.e. the suspected path extraction has a negligible cost. However, stage (iii) is about 1.7 times slower on the average than even stage (ii), which makes it the most problematic stage of the method.

The cost analysis for stage (iii) is shown by (4). The rows in that equation that have a factor of (Ncand × Nsamples) are the main source of the cost. Number of correlated path candidates for each benchmark circuit is shown in Table 1. For instance c2670 having 1850 edges resulted in more than 7.5 million correlated path candidates. Considering a thousand samples as we do in this paper, the factor above becomes about 7.5 billion which is a huge number. In actual case, the number of samples can be much larger resulting in much higher costs for stage (iii).

In this section, we propose a much faster method to select the best correlated path candidate. Due to the spatial correlations, the correlation between two paths depends on the spatial distance between them. But first of all, let us detail the actual problem with CCM.

CCM takes advantage of spatial correlation to find the correlated path pairs. It finds such a pair for each edge in the circuit so that the path delay ratio of the pair cancels the variation component which reveals any HT existence for the corresponding circuit edge. For that purpose, at stage (ii) explained in Sect. 2.4, the correlated path candidates of each suspected path are extracted so that all of them have their logic gates located very close to the corresponding suspected path. This closeness is guaranteed by first dividing the circuit into grids as shown in Fig. 3 and then selecting the paths residing at the same or adjacent grids of the gates of suspected path. Without loss of generalization let us assume that a suspected path has all its logic gates located at the dark shaded grids in Fig. 3. Then, the stage (ii) of CCM collects all correlated path candidates, whose logic gates are located at either the dark shaded grids or their adjacent grids that are shaded lightly on the figure.

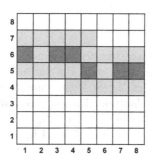

Fig. 3. The division of circuit layout to grids.

Due to the spatial correlation, one expects that all candidates (especially the ones residing at only dark shaded grids for our example case) must have a good correlation and hence a good correlation coefficient. In such a case picking just one correlated path candidate would be fairly enough to have a correlated pair instead of enumerating all of the candidates for each edge (or suspected path), and then computing the path delays and correlation coefficients for all of them.

But when we investigate the candidates, we see that this is not the case. The different correlation coefficients for all correlated path candidates corresponding to just one suspected path are shown as an example in Fig. 4. The candidates are sorted with the ascending coefficient values. For this sample case, some of the candidates may have very bad correlation coefficients down to 0.75. Please note that, empirical results show us that the correlation coefficient must have a value very close to 1 like 0.95 and above in order to be able to cancel the effect of variations and reveal the existence of Trojans. If the Trojan is as small as only one logic gate even a correlation coefficient of 0.95 may not be enough for detection. This necessitates the computation of correlation coefficient for each correlated path candidate to nominate the one with the largest coefficient as the correlated path of the pair.

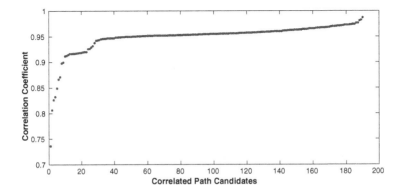

Fig. 4. CCM values for candidates of correlated path

With a further investigation, the actual reason behind that reveals the fact that being at even the same grids with the suspected path does not mean to be highly correlated with it just because the number of shared (common) grids can be fractionally very low. Without loss of generality, let us assume that the suspected path has gates distributed to n_s different grids and one correlated path candidate for that suspected path has all its gates located at n_c different grids, where the number of union and intersection of n_s and n_c grids are denoted by $n_\cup$ and $n_\cap$ respectively. The resultant correlation between these two paths would not be good enough to cancel variation effect if $n_\cap$ is much smaller than $n_\cup$. We name the $n_\cap / n_\cup$ ratio as *shared grids ratio* (SG). Shared grids ratio for a path pair $< P_{susp}, P_{corr}^i >$ can be computed as shown in (5). i denotes the index of the correlated path candidate for the suspected path. The correlation between two paths tends to increase by the increasing shared grids ratio.

$$SG^i = \frac{\text{number of shared grids for } < P_{susp}, P_{corr}^i > \text{ pair}}{\text{number of all grids in } P_{susp} \cup P_{corr}^i} \qquad (5)$$

For the acceleration of stage (iii), we propose selecting the correlated path candidate with largest SG ratio computed by (5) instead of the one with the largest correlation coefficient. For a sample suspected path, Fig. 5 demonstrates how correlation coefficient has a rise trend despite some fluctuations while SG increases. Usage of SG is based on the fact that the more grids the two paths share, the more spatial correlation they would have, which means better detection.

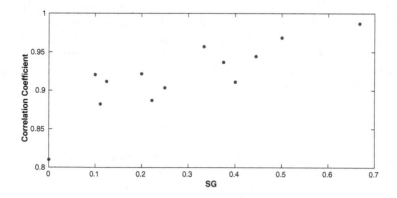

Fig. 5. CCM vs SGM for candidates of correlated path

SG only requires the detection of the number of total grids that both paths reside at (Cost_{DNM}) as well as the grids that are shared by both paths (Cost_{NM}). Then, one division is enough to get SG (Cost_{div}). No path delays and no costly correlation coefficient computation are required. Moreover, it is not performed for each sample as SG does not change from chip to chip. As a result, the new cost of stage (iii) can be written as shown by (6). $\text{Ngrids}_{\text{avg}}$ shows the average number of grids that are occupied by a path pair. It is certain that $\text{Ngrids}_{\text{avg}}$ is much smaller than the number of samples. $\text{Ncand} \times \text{Cost}_{\text{comp}}$ is for finding the correlated path candidate i with the maximum SG^i similar to (4).

$$\text{Cost}^{\text{SGM}}_{\text{stage(iii)}} = \text{Ncand} \times \text{Ngrids}_{\text{avg}} \times (\text{Cost}_{\text{DNM}} + \text{Cost}_{\text{NM}} + \text{Cost}_{\text{div}}) + \text{Ncand} \times \text{Cost}_{\text{comp}} \tag{6}$$

Especially when the denominator of SG equation is a small number, more than one correlated path candidate can have the largest SG value. In such a case, the best candidate can be detected by computing path delays for only the candidates having that largest shared grid ratio. It should be noted that this cost must be added to (6). But it is difficult to theoretically represent it because the number of such candidates having the same largest shared grids ratio is unknown a priori. Yet, we take into account this additional path delay cost for all experimental results in Sect. 6. Also, Table 3 reports the total number of such candidate paths as the last column.

6   Results: Comparison of CCM and SGM

For all experiments in this paper realistic variation model considering both inter-die and intra-die variations with spatial correlations [2] is employed. The benchmark circuits are synthesized for 45 nm open cell library of Nangate [10]. The most significant random parameters are taken to be transistor channel length (L_{eff}) and threshold voltage (V_t) as devised in [8]. The $3\sigma/\mu$ ratio of 12% and 20% are assumed for L_{eff} and V_t respectively according to the International Technology Roadmap for Semiconductors (ITRS) report [9]. Well known ISCAS'85 benchmark test circuits [6] are used for the experiments. All computations and simulations are performed on HP z620 workstation with Xeon E5-2620, six-core, 2-GHz processors and 24 GB of RAM. A very small Trojan of one XOR gate is employed to test the limits of the proposed method and see their detection performance.

Table 3. Comparison of CCM with SGM results

Benchmark	Edge coverage for CCM	Edge coverage for SGM	# of candidates for CCM	# of candidates for SGM
432	95.7%	92.5%	36,647	580
499	64.5%	56.7%	29,320	521
880	94.7%	92.5%	39,176	1782
1355	96.6%	95.8%	17,200,357	102,781
1908	93.4%	91.5%	5,353,429	19,583
2670	95.2%	93.5%	7,619,334	77,390

We compare the correlation coefficient based method (CCM) in Sect. 3 with shared grid ratio based method (SGM) proposed in Sect. 5. Please remind that a covered edge means that any Trojan bigger or equal to one XOR gate inserted to that edge can be detected by the method without requiring golden model. Table 3 shows the comparison results. The first deduction from the table is that both CCM and SGM can almost cover or secure the whole circuit resulting in about 90% edge coverage. This means that the methods can detect any Trojan inserted on any place in 90% of the circuit. Excluding c499, which is an obvious outlier, the edge coverage of SGM even becomes about 93% on the average. In January 2008, Dean Collins, deputy director of DARPA's Microsystems Technology Office and manager for the Trust in IC initiative initiates a hardware Trojan detection contest among three companies: Raytheon, Luna Innovations and Xradia. The Trojan circuit is inserted by MIT Lincoln Labs. Collins states to IEEE Spectrum magazine that the goal is a 90% detection rate [1], which confirms the sufficiency of 90% coverage.

The last two columns shows the number of correlated path candidates that must be examined for CCM and SGM respectively. The number of candidates is more than 100 times less for SGM because it eliminates all candidates except the ones having the largest shared grids (SG) ratio for each suspected path. This table shows that SGM does not lose accuracy although it performs path delay computations for a much smaller set of correlated path candidates.

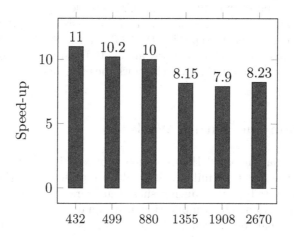

Fig. 6. Stage (iii) speed-up of SGM over CCM

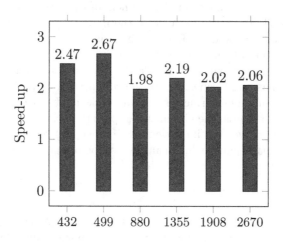

Fig. 7. Complete speed-up by SGM over CCM

As Table 2 suggests the most time consuming part of the spatial correlation based HT detection by CCM is stage (iii). This is why SGM is proposed to speed up that stage. To quantify the speed-up by SGM over CCM at stage (iii) computations, we have recorded the time required for the computation of stage (iii) by both methods. Figure 6 plots the resultant stage (iii) speed-up for each benchmark circuit as a bar graph. SGM accelerates stage (iii) of CCM about 9 times on the average over test circuits, which is a serious speed improvement. Please note that all additional path delay computations due to the candidates shown at the last column of Table 3 are taken into account at the speed-up values of Figs. 6 and 7.

The resultant speed-up of SGM over CCM considering the total time for all three stages (from (i) to (iii)) is shown in Fig. 7. When executed on full circuit the proposed SGM can double the speed of the CCM on the average over benchmark circuits. More precisely, SGM achieves about 100% speed improvement with only 3% edge coverage reduction, which shows the efficiency and accuracy of the proposed method.

7   Discussion and Future Work

The spatial correlation based HT detection proposed in [8] is adapted to full circuit and for the first time full circuit results are presented in this paper. The method is accelerated by introducing shared grids ratio instead of correlation coefficient computation. The computational cost analysis of both methods shows the efficiency comparison as well as the empirical results, which show that both methods can secure more than the 90% of the circuit. But usage of shared grids can increase the speed of the whole method more than twice on the average.

Although the method is applied and tested on combinational circuits, it can be generalized to sequential circuits by the help of the techniques like enhanced-scan delay tests [7]. To further accelerate the method, primarily parallelization by GPU utilization can be used. Because, especially stage (ii) and stage (iii) are suitable for distributed computation.

The method in this paper is developed with a focus on improving pre-manufacturing phase and especially to speed up stage (iii). However, due to stage (iv) i.e. post-manufacturing tests, it may take a lot of time to obtain path-delay tests. In other words, the aim of this paper is to decrease the required time to extract path pairs, yet the improvement of stage (iv) requires the extraction of less number of path pairs, which can be a scope of another paper.

References

1. Adee, S.: The hunt for the kill switch. IEEE Spectr. **45**(5), 34–39 (2008)
2. Agarwal, A., Blaauw, D., Zolotov, V.: Statistical timing analysis for intra-die process variations with spatial correlations. In: Proceedings of the 2003 IEEE/ACM International Conference on Computer-Aided Design, p. 900. IEEE Computer Society (2003)
3. Banga, M., Hsiao, M.S.: A region based approach for the identification of hardware trojans. In: 2008 IEEE International Workshop on Hardware-Oriented Security and Trust, pp. 40–47. IEEE (2008)
4. Bayrakci, A.A.: Stochastic logical effort as a variation aware delay model to estimate timing yield. Integr. VLSI J. **48**, 101–108 (2015)

5. Bhasin, S., Danger, J.L., Guilley, S., Ngo, X.T., Sauvage, L.: Hardware trojan horses in cryptographic IP cores. In: 2013 Workshop on Fault Diagnosis and Tolerance in Cryptography, pp. 15–29. IEEE (2013)
6. Brglez, F.: A neural netlist of 10 combinational benchmark circuits. In: Proceedings of the IEEE ISCAS: Special Session on ATPG and Fault Simulation, pp. 151–158 (1985)
7. Bushnell, M., Agrawal, V.: Essentials of Electronic Testing for Digital, Memory and Mixed-Signal VLSI Circuits, vol. 17. Springer, Boston (2004). https://doi.org/10.1007/b117406
8. Esirci, F.N., Bayrakci, A.A.: Hardware trojan detection based on correlated path delays in defiance of variations with spatial correlations. In: Proceedings of the Conference on Design, Automation & Test in Europe, pp. 163–168. European Design and Automation Association (2017)
9. ITRS Commitee: International technology roadmap for semiconductors (ITRS) 2011 report. http://www.itrs2.net/2011-itrs.html
10. Nangate: 45nm open cell library. http://www.nangate.com/
11. Narasimhan, S., et al.: Hardware trojan detection by multiple-parameter side-channel analysis. IEEE Trans. Comput. 62(11), 2183–2195 (2012)
12. Nowroz, A.N., Hu, K., Koushanfar, F., Reda, S.: Novel techniques for high-sensitivity hardware trojan detection using thermal and power maps. IEEE Trans. Comput. Aided Des. Integr. Circuits Syst. 33(12), 1792–1805 (2014)
13. Rai, D., Lach, J.: Performance of delay-based trojan detection techniques under parameter variations. In: 2009 IEEE International Workshop on Hardware-Oriented Security and Trust, pp. 58–65. IEEE (2009)
14. Tehranipoor, M., Koushanfar, F.: A survey of hardware trojan taxonomy and detection. IEEE Des. Test Comput. 27(1), 10–25 (2010)
15. Tehranipoor, M., Wang, C.: Introduction to Hardware Security and Trust. Springer, New York (2011). https://doi.org/10.1007/978-1-4419-8080-9
16. Yoshimizu, N.: Hardware trojan detection by symmetry breaking in path delays. In: 2014 IEEE International Symposium on Hardware-Oriented Security and Trust (HOST), pp. 107–111. IEEE (2014)

A Parallel GPU Implementation of SWIFFTX

Metin Evrim Ulu$^{(\boxtimes)}$ and Murat Cenk$^{(\boxtimes)}$

Middle East Technical University, Ankara, Turkey
evrimulu@gmail.com, mcenk@metu.edu.tr
http://iam.metu.edu.tr

Abstract. The SWIFFTX algorithm is one of the candidates of SHA-3 Hash Competition that uses the number theoretic transform (NTT). It has 256-byte input blocks and 65-byte output blocks. In this paper, a parallel implementation of the algorithm and particular techniques to make it faster on GPU are proposed. We target version 6.1 of NVIDIA$^{®}$CUDA^{TM} compute architecture that employs an ISA (Instruction Set Architecture) called Parallel Thread Execution (PTX) which possesses special instrinsics, hence we modify the reference implementation for better results. Experimental results indicate almost 10x improvement in speed and 5 W decrease in power consumption per 2^{16} hashes.

Keywords: Hash function · SWIFFTX · SHA-3 · NTT · GPU · CUDA

1 Introduction

SWIFFT is a collection of compression functions [5,6,12]. The security of it is based on the computationally hard lattice problems that provides this function with the property of being provably collision resistant. Therefore, it may be used in digital signatures and authentication protocols. However, the SWIFFT compression function has some undesirable properties such as linearity and lack of pseudorandomness. In order to remedy this situation and remove these undesirable properties, a new compression function called SWIFFTX, one of the candidates of SHA-3 competition, was proposed in [1].

SWIFFTX has 256-byte input blocks and 65-byte output blocks. In the default configuration, input byte string is shaped as a 32 column matrix where each column comprises 8 bytes. The initial round first executes a Number Theoretic Transform (NTT) on each column and the result is a 64 by 32 matrix. This matrix is then multiplied by three different constant matrices $A_i, \forall i \in \mathbb{Z}, 0 \leq i < 3$ separately. Next, the diagonals of these three matrices are extracted to form three vectors of dimension 64. These vectors are then translated to byte strings by a translation algorithm and results are concatenated to form a single byte string. To provide non-linearity, this byte string is passed through a SBox before fed into the second round. The second round is similar to the first one

© Springer Nature Switzerland AG 2020
D. Slamanig et al. (Eds.): MACIS 2019, LNCS 11989, pp. 202–217, 2020.
https://doi.org/10.1007/978-3-030-43120-4_16

yet only a single matrix multiplication is done where constants are provided by A_0. Only 25 columns of A_0 are used in matrix multiplication. Finally, there is a carry propagation operation at the end of the round that assembles the final byte of the output. In SWIFFTX, arithmetics are carried out in the finite field of characteristic $p = 2^8 + 1 = 257$. The total number of constants in the matrices $A_i, \forall i \in \mathbb{Z}, 0 \leq i < 3$ is $3NM$ where $N = 64$ is the number of rows and $M = 32$ is the number of columns. These constants are designed to be random and derived from the expansion of the transcendental number π via a certain algorithm (see [6]) in order values to fit into the given field.

Compute Unified Device Architecture (CUDA) is very different from general purpose architectures such as x86 and AMD64. It has a great number of threads. A group of 32 threads is called a warp. This is the minimal number of threads that can be spawned simultaneously. Warps can be arranged to form a block. Therefore, the number of threads in a block is a multiple of 32. Blocks can also be arranged to form larger blocks called grids. Blocks and grids can be 1, 2 or 3 dimensional to fit into the requirements of the implementation. In this paper, we target the GP104 (GP104-400-A1) chip manufactured by NVIDIA®. This chip is a member of the sixth generation NVIDIA Pascal™ microarchitecture. In this particular chip, the number of threads in a block is limited to $2^{10} = 1024$ threads. A Streaming Multiprocessor (SM) in this chip can run two 1024-thread blocks simultaneously and there are 20 of them.

In terms of cache, GP104 has 48 KiB Unified Cache and 2 MiB L2 Cache. In CUDA, Unified Cache can be used for local/global loads/stores. L2 is a little bit larger and can be employed for caching global loads/stores. There is also a Texture Cache in Unified Cache which is used for loading constants. Among others, GP104 has another 96 KiB local fast memory per SM. This memory is called Shared Memory and it is particularly useful in terms of optimizing an implementation. Basically, the name *shared* comes from the fact that this memory area can be divided into smaller chunks and moreover, can be shared among a block. Although this region is declared to be fast as registers (2 clocks), it has a limited size and most of the time, determines the maximum number of warps that can be spawned simultaneously along with other factors such as number of registers per block, number of threads per block and number of threads per multiprocessor.

In this paper, we present an efficient parallel implementation of SWIFFTX on GPU. In order to obtain high performance, we have optimized memory access according to memory transaction coalescing rules and optimized arithmetic operations using intrinsics. These are essential for realization of a fast implementation. Furthermore, shared memory is used to hold all intermediate values. Representing elements of $\mathbb{F}_{257}$ in *signed char* posed a certain challenge however, this is resolved by a map and another additional small routine. Moreover, the serial base 257 to base 256 translation algorithm is parallelized by using a binomial matrix. Experimental results (Sect. 6) show that our implementation is approximately 1000 times faster than the single-threaded x86 reference implementation and 10 times faster than the ported reference implementation. In terms of power consumption, our implementation performs $5\,\mathrm{W}$ better per 2^{16} hashes and $13\,\mathrm{W}$ better per 2^{19} hashes.

The rest of the paper is organized as follows: In Sect. 2, definition of SWIFFT function and description of SWIFFTX algorithm are given. In Sect. 3, we discuss the reference x86 implementation. We port this implementation to CUDA without applying a particular optimization, evaluate its characteristics and determine its performance bottlenecks. In Sect. 4, we implement a parallel version of the SWIFFTX algorithm. In order to achieve our goal, we investigate further possible optimizations specific to the given hardware and propose solutions to discords between hardware and software. In Sect. 5, we discuss further improvements such as improving cache hits rates and fixing memory bank conflicts. In Sect. 6, we develop necessary methodology to evaluate two implementation and obtain our results. These results are mostly GPU specific. The conclusion is presented in Sect. 7.

2   SWIFFT and SWIFFTX

In this section, we provide a description for the SWIFFT and the SWIFFTX. Let $p = 2^8 + 1 = 257$, $N = 64$ and $M = 32$ with $2N \mid p - 1$. These are the concrete parameters given in [6]. First, we start with a few definitions.

Definition 1. *Let $n \in \mathbb{Z}^+, \mathbf{X} \in \mathbb{F}_p^{n \times n}, \mathbf{Y} \in \mathbb{F}_p^n$. Define the column operator $C_j : \mathbb{F}_p^{n \times n} \to \mathbb{F}_p^n$ as $C_j(\mathbf{X}) = \mathbf{Y}$ where $y_i = x_{ij}$, $\forall i \in \mathbb{Z}$, $0 \leq i < n$.*

Definition 2. *Let $n \in \mathbb{Z}^+, \mathbf{X} \in \mathbb{F}_p^{n \times n}, \mathbf{Y} \in \mathbb{F}_p^n$ be a square matrix. Define the main diagonal operator $D : \mathbb{F}_p^{n \times n} \to \mathbb{F}_p^n$ as $D(\mathbf{X}) = \mathbf{Y}$ where $y_i = x_{ii}$, $\forall i \in \mathbb{Z}$, $0 \leq i < n$.*

Definition 3. *The Number Theoretic Transform employed in SWIFFT is defined as $NTT_N : \mathbb{F}_2^N \to \mathbb{F}_p^N$ where $NTT_N(u_0, \ldots, u_{n-1}) = (v_0, \ldots, v_{n-1})$, $v_j = \sum_{i=0}^{N-1} u_i \omega^{2ij}$, $\forall j \in \mathbb{Z}$, $0 \leq j < N$, and $\omega \in \mathbb{F}_p$ is the 2N-th root of unity such that $\omega^{2N} = 1$.*

Definition 4. *Define the unary operator $E_M : \mathbb{F}_2^{N \times M} \to \mathbb{F}_p^{N \times M}$ as $E(\mathbf{X}) = \mathbf{YX}$, where $\mathbf{Y} \in \mathbb{F}_p^{N \times N}$ with $y_{ij} = \omega^{2j} \in \mathbb{F}_p$, $\forall i, j \in \mathbb{Z}$, $0 \leq i, j < N$ and ω is the 2N-th root of unity such that $\omega^{2N} = 1$.*

It is now possible to define the first part of the SWIFFT compression function in terms of the primitives above.

Definition 5. *Let $\mathbf{U}, \mathbf{A} \in \mathbb{F}_p^{N \times M}$. Define $SWIFFT'$ as follows:*

$$SWIFFT'_M : \mathbb{F}_2^{N \times M} \times \mathbb{F}_p^{N \times M} \to \mathbb{F}_p^N$$
$$\mathbf{U} \times \mathbf{A} \mapsto D(\mathbf{VA}^\mathbf{T})$$

where $C(\mathbf{V})_j = NTT_N \circ C_j \circ E_M(\mathbf{U})$, $\forall j \in \mathbb{Z}$, $0 \leq j < M$.

The above definition shows how to calculate j-th column of the matrix $\mathbf{V}$ denoted by $C(\mathbf{V})_j$. Finally, $\mathbf{V}$ is multiplied by $\mathbf{A}^{\mathbf{T}}$ and D is applied to obtain the result.

$SWIFFT'$ basically captures the crucial part of the $SWIFFT$. The rest deals with the translation of vectors with elements in $\mathbb{F}_p$ to vectors with elements in $\mathbb{F}_2$.

Definition 6. *Let* $\mathbf{X} \in \mathbb{F}_p^N$ *be a matrix and let* $N' = N/8$. *Define the map* G *as:*

$$G : \mathbb{F}_p^N \to \mathbb{F}_p^{N' \times N'}$$
$$\mathbf{X} \mapsto \mathbf{Y}$$

where $y_{ij} = x_{i+8*j}$, $\forall i, j \in \mathbb{Z}$, $0 \le i, j < 8$.

Definition 7. *Let* $N' = N/8$. *Define the translation map* T *as:*

$$T : \mathbb{Z}_p^{N'} \to \mathbb{Z}_{256}^{N'} \times \mathbb{Z}_{256}$$
$$\mathbf{a} = \sum_{i=0}^{N'-1} a_i p^i \mapsto \mathbf{b} = \sum_{i=0}^{N'-1} b_i 256^i \times ((\mathbf{a} - (\mathbf{a} \bmod 256^{N'})) \gg N).$$

The above function basically translates vectors from base 256 to base 257 with the rightmost component being the carry. It is now possible to define SWIFFT.

Definition 8. *Let* $\mathbf{U} \in \mathbb{F}_p^{N \times M}$ *be an input matrix and* $\mathbf{A} \in \mathbb{F}_p^{N \times M}$ *be a constant matrix. Then,* $SWIFFT_M$ *is defined as:*

$$SWIFFT_M : \mathbb{F}_2^{N \times M} \times \mathbb{F}_p^{N \times M} \to \mathbb{Z}_{256}^N \times \mathbb{Z}_{256}$$
$$\mathbf{U} \times \mathbf{A} \mapsto \sum_{i=0}^{N'-1} \pi_1(a_i) 256^{iN'} \times \bigvee_{i=0}^{N'-1} \pi_2(a_i) 2^i$$

where $N' = N/8$, $\mathbf{U}' = SWIFFT'_M(\mathbf{U}, \mathbf{A})$, $a_i = T \circ C_i \circ G(\mathbf{U}')$ *and* π_j *is the projection operator onto the j-th component.*

SWIFFTX employs SWIFFT as a building block. However, there are two variations. The first round employs $SWIFFT_M$ with parameter $M = 32$ and the second round sets $M = 25$ denoted by M' in the following definition.

Definition 9. *Let* $\mathbf{X} \in \mathbb{F}_2^{N \times M}$ *be an input matrix and* $\mathbf{A_i} \in \mathbb{F}_p^{N \times M}$, $\forall i \in \mathbb{Z}$, $0 \le i < 3$ *be three constant matrices. Then,* $SWIFFTX_M$ *is defined as follows:*

$$SWIFFTX_M : \mathbb{F}_2^{N \times M} \to \mathbb{Z}_{256}^{N+1}$$
$$\mathbf{X} \mapsto \pi_1(\mathbf{Z}) \,\|\, \pi_2(\mathbf{Z})$$

where $\mathbf{Y_i} = SWIFFT_M(\mathbf{X}, \mathbf{A_i})$, $\forall i \in \mathbb{Z}$, $0 \le i < 3$, $\mathbf{U} = \pi_1(\mathbf{Y_1}) \,\|\, \pi_1(\mathbf{Y_2}) \,\|\, \pi_1(\mathbf{Y_3})$,
$\mathbf{V} = \pi_2(\mathbf{Y_1}) \,\|\, \pi_2(\mathbf{Y_2}) \,\|\, \pi_2(\mathbf{Y_3})$, $\mathbf{P} = 0 \in \mathbb{Z}_{256}^5$ *is a five byte padding,*
$\mathbf{Z} = SWIFFT_{M'}(SBox(\mathbf{U} \,\|\, \mathbf{V} \,\|\, \mathbf{P}), \mathbf{A_0})$, *SBox is a lookup operation and*
$\|$ *is the concatenation operator.*

3   The Reference Implementation

Next we continue with the x86 reference implementation included in Crypto-Streams [3]. The outline of this implementation is given in Fig. 1. We have kept the variable names unaltered so that the reader can trace them back to the source code. Apart from the definition of SWIFFTX in the previous section, this implementation re-uses the common NTT output for the sake of performance. Assuming a word is 16-bits, elements of $\mathbb{F}_p$ are kept in words. Powers of 64-th root of unity are centered toward zero in the initialization stage. Moreover, NTT is performed via a lookup table. Similarly, SBox lookup is done on byte basis. Translation to base 256 from $\mathbb{F}_p$ is done in 6 iterations in a very efficient manner. Although this implementation is very efficient on x86, it still runs on a single-thread. We ported this implementation to CUDA without applying any further optimizations other than migrating constants to the device memory.

```
ALGORITHM: SWIFFTX
INPUT:   uint8_t input[256]; // Input
         int16_t A_0[N*M], A_1[N*M], A_2[N*M]; // Constants
OUTPUT:  uint8_t output[65]; // Output

int32_t fftOut[N*M]; // NTT Output
int32_t sum[3*N]; // Three vectors of dimension N
uint8_t intermediate[3*N+8]; // Output of the first round

doNTT_32(input, fftOut); // 32 Column NTT
doMultiply_and_Diag_3(fftOut, A_0, A_1, A_2, sum); // Multipl. and Diagonal
doTranslate_3(sum, intermediate); // Translate to base 256
doSBox(intermediate); // Apply SBox
doNTT_25(intermediate, fftOut); // 25 Column NTT
doMultiply_and_Diag_1(fftOut, A_0, sum); // Multipl. and Diagonal
doTranslate(sum, output); // Translate to base 256
```

Fig. 1. SWIFFTX algorithm

Unlike x86, CUDA architecture provides a high number of registers. The maximum number of registers per block a CUDA kernel can employ is 255. In cases where more registers are required, spills occur and load/stores are served by Unified Cache. The reference implementation is register rich. With a block size of a warp (32 threads), the compiler decides to use 228 registers (Fig. 2) for the default optimization level 3 (-O3) although we haven't forced any loop to unroll. Since the implementation is for x86, it does not employ any shared memory.

According to GP104 specification [7], each SM has a 256 KiB register file. Assuming all 32 bits, a SM can hold up to 65536 registers simultaneously. This kernel has a block size of 32 (a single warp) and employs 228 registers. A calculation shows, each block requires 7296 registers. Therefore, each SM can run

$65536/7296 \approx 8.9$ warps. NVIDIA®Visual Profiler ($nvvp$, [10]) tells that the actual value is 8 warps. Moreover, each SM can run 2048 threads or 64 warps simultaneously, so the utilization is only 12.5%. Contrary to its high register usage, the kernel still requires an additional 8656 bytes stack frame which further slows the execution down.

```
ptxas info : 16786 bytes gmem
ptxas info : Compiling entry function '_Z14swifftx_kernelPhS_i' for 'sm_61'
ptxas info : Function properties for _Z14swifftx_kernelPhS_i
            8656 bytes stack frame, 0 bytes spill stores, 0 bytes spill loads
ptxas info    : Used 228 registers, 340 bytes cmem[0]
```

Fig. 2. Reference implementation compiler stats

4   A Parallel Implementation

In this section, we present a parallel CUDA implementation of the algorithm. The outline of our implementation is given in Fig. 3.

We select 64 threads, 2 warps per block. This number matches the number of rows of constant matrices A_i, $\forall i \in \mathbb{Z}$, $0 \leq i < 3$. The details of the proposed parallel implementation are as follows.

First, we reserve enough shared memory per block for fast access to intermediate values. These intermediate values are NM words for the NTT output ($fftOut[]$), $3N$ words for the diagonals ($sum[]$) and $3N+8$ bytes for the output of the first round ($intermediate[]$). In total, $(NM)2+(3N)2+(3N+8) = 4680$ bytes. To enable fast access to 256 bytes hash input, at the begining of the kernel, we copy the input to a shared memory location, specifically, to the space reserved for the diagonal output. This space will not be used until NTT is completed therefore we can use it temporarily. Copy is implemented by pointer dereferencing. A 64-thread block can copy the input in a single step (256 bytes = 64×4 bytes). For the hash output, only 17 threads are running, the others are idle (17×4 bytes = 68 bytes > 65 bytes). The input is fetched from memory via 128-byte transactions, obeying the memory coalescing rules. Similarly, the hash output is written to the device memory via 64-byte transactions almost all the time except the last 8 bytes.

Now, NTT for a column can be done in 8 steps or strides. Therefore, we divide the NTT into 8 strides where each stride i is responsible for output row $8k + i, \forall k \in \mathbb{Z}$, $0 \leq k < 8$. A 64-thread block can process 8 columns at once and 32 columns in 4 steps. Furthermore, the multiplication at the begining of the NTT is transformed into a lookup. The size of the lookup table is $256 \times 8 \times 8$ words with values in $\mathbb{F}_{257}$. These values are constant and served by the Texture Cache. At the end of the NTT, we transpose the output to help the inner product operation in the next stage. Reduction of the field elements is accomplished by the following macro:

$$\#define \quad Q\_REDUCE(a) \quad (((a) \ \& \ 0xFF) - ((a) \gg 8))$$

```
ALGORITHM: SWIFFTX
INPUT:  uint8_t input[256]; // Input
        int8_t A_0[N*M], A_1[N*M], A_2[N*M]; // Constants
OUTPUT: uint8_t output[65]; // Output
NUMBER OF THREADS PER BLOCK: 64

__shared__ int16_t S_fftOut[N*(M+2)]; // Adjusted NTT Output
__shared__ int16_t S_sum[3*N+12]; // Adjusted sum output
__shared__ uint8_t S_intermediate[4*N]; // Enlarged intermediate
uint32_t *_input = S_sum;
// Copy input to shared (4-bytes per thread)
doParallel_Copy(input, _input); __syncThreads();
// 32 Column NTT
doParallel_NTT_32((int8_t *)_input, S_fftOut); __syncThreads();
// Multiply and take the diagonal
doParallel_Multiply_and_Diag_3(S_fftOut, A_0, A_1, A_2, S_sum);
__syncThreads();
// Fix leaps
doParallel_Adjust(S_fftOut, S_sum); __syncThreads();
// Translate to base 256
doParallel_Translate_3(S_sum, S_intermediate); __syncThreads();
// Apply SBox
doParallel_SBox(S_intermediate); __syncThreads();
// 25 Column NTT
doParallel_NTT_25(S_intermediate, S_fftOut); __syncThreads();
// Multipl. and Diagonal
doParallel_Multiply_and_Diag_1(S_fftOut, A_0, S_sum); __syncThreads();
// Translate to base 256
doParallel_Translate(S_sum, S_intermediate); __syncThreads();
// Copy results back to device memory
doParallel_Copy(intermediate, output);
```

Fig. 3. Our proposed parallel SWIFFTX algorithm

Basically, this macro subtracts the 8–15th bits from the 0–7th bits. This reduction is an instrinsic property of the nega-cyclic field.

We have the NTT output in shared memory. We need to calculate three diagonals for products $A_i V^T, \forall i \in \mathbb{Z}, \ 0 \le i < 3$. Although this seems to be a straightforward calculation, Pascal tuning guide [9] informs that the multiplication is a multi-clock operation in GP104 and the compiler can compile a single multiplication upto 20 instructions. To remedy this situation, we employ an intrinsic called $\_dp\_2a$, a two-way dot product. The definition of this operator is given in Fig. 4 (PTX Manual [8], Section 9).

Syntax:
```
dp2a.mode.atype.btype d, a, b, c;
.atype = .btype = { .u32, .s32 };
.mode = { .lo, .hi };
```

Description:
Two-way 16-bit to 8-bit dot product which is accumulated
in 32-bit result. Operand a and b are 32-bit inputs. Operand a holds
two 16-bits inputs in packed form and operand b holds 4 byte inputs
in packed form for dot product. Depending on the .mode specified,
either lower half or upper half of operand b will be used for dot
product. Operand c has type .u32 if both .atype and .btype are .u32
else operand c has type .s32 .

Semantics:
```
d = c;
// Extract two 16-bit values from a 32-bit input and sign or zero extend
// based on input type.
Va = extractAndSignOrZeroExt_2(a, .atype);
// Extract four 8-bit values from a 32-bit input and sign or zero extend
// based on input type.
Vb = extractAndSignOrZeroExt_4(b, .btype);
b_select = (.mode == .lo) ? 0 : 2;
for (i = 0; i < 2; ++i) {
    d += Va[i] * Vb[b_select + i];
}
```

Fig. 4. dp2a two-way dot product-accumulate operator

Now we face the problem that the entries in matrices A_i do not fit into $int8\_t$'s
(*signed char*). According to C++11 standard, *signed char* can hold values from
-128 to 127 if the compiler employs *Two's complement* representation. Fortunately, the CUDA compiler *nvcc* employs *Two's complement* representation;
therefore, we can map our field according to the following function.

$$f : \mathbb{F}_{257} \to Int8$$
$$f(a) = \begin{cases} a & if \ 0 \le a \le 127, \\ 127 & if \ a = 128, \\ a - 257 & otherwise. \end{cases}$$

This representation is different than the *diminished-one number system*
employed in [4]. In [4], $\mathbb{F}_{257}$ is considered to be the integers in the range 0 to 256
inclusive and the field is mapped to 8-bits by subtracting 1 from each element
while excluding the zero. The zero case is detected by an additional signal and
handled exclusively. However, in GPU, we have no way of knowing whether a
value is zero or not unless a predicate is executed. Unfortunately, predicates are
sources of divergence therefore very expensive especially in loop bodies, hence
we propose a slightly altered approach by defining the above function f. The
codomain of f is selected to be 8-bit *signed char* just to make it compatible with

the $\_\_dp\_2a$ operator. We have determined 23 $+128$'s in $A_i, \forall i \in \mathbb{Z}, 0 \leq i < 3$. There are nine in A_0, seven in A_1, and seven in A_2. We first replace those values with $+127$'s and do the multiplication. After computing the diagonal, we add missing values by a small routine called *doParallel_Adjust* which only employs 23 of the 64 threads in a block. This calculation is done as follows:

$$sum_j = \sum_i a_{ji} \times fftOut_{ij},$$

$$sum_j = \sum_i b_{ji} \times fftOut_{ij} + \sum_i 128 \times fftOut_{ij}, \qquad b_{ij} \neq 128,$$

$$sum_j = \sum_i b_{ji} \times fftOut_{ij} + \sum_i 127 \times fftOut_{ij} + \sum_i fftOut_{ij},$$

$$sum_j = \sum_i c_{ji} \times fftOut_{ij} + \sum_i fftOut_{ij}, \qquad c_{ij} \neq 128.$$

with $\forall j \in \mathbb{Z}, 0 \leq j < N$ and $\forall i \in \mathbb{Z}, 0 \leq i < M$. Note that, the rightmost summation on the last line is the residue that needs to be added to its respective row j. To keep *doParallel_Adjust* procedure simple, we represent a particular adjustment in a dword. The first byte is the matrix the entry is in, the second is the row, the third is the column and the fourth is always zero. The last byte is kept for the sake of memory alignment. Totally, it consumes $23 \times 4 = 92$ bytes.

Since $\_\_dp\_2a$ is a two-way dot product, 32 columns can be processed totally in 8 calls to variants $\_\_dp\_2a\_lo$ and $\_\_dp\_2a\_hi$. Specifically, this means fetching 2 dwords from *fftOut*, a dword from A_i and computing the dot product twice using each variant once. This is a prominent improvement over the original iteration count of 32 and now, the loop can be unrolled without overloading the intruction fetch queue. Moreover, this approach also halves the memory transaction size required for fetching the entries in the matrices A_i.

Next step is to translate the diagonal entries in $\mathbb{F}_{257}$ to $\mathbb{F}_{256}$. In the reference implementation this is done efficiently in 6 iterations. However, each column has to be processed by a single thread. To make it parallel, we employ the following binomial matrix:

```
typedef int8_t swift_int8_t;
swift_int8_t binom1[8*8] = {
    1, 1, 1, 1,   1,  1,  1,  1,
    0, 1, 2, 3,   4,  5,  6,  7,
    0, 0, 1, 3,   6, 10, 15, 21,
    0, 0, 0, 1,   4, 10, 20, 35,
    0, 0, 0, 0,   1,  5, 15, 35,
    0, 0, 0, 0,   0,  1,  6, 21,
    0, 0, 0, 0,   0,  0,  1,  7,
    0, 0, 0, 0,   0,  0,  0,  1,
};
```

This matrix is based on the fact that the equation $257^n = (256 + 1)^n = \sum_i \binom{n}{i} 256^i$ holds. Therefore, the elements b_{ij} of binom1 are defined as follows:

$$b_{ij} = \begin{cases} \binom{j}{i} & if \ i \leq j, \\ 0 & otherwise, \end{cases} \quad \forall i, j \in \mathbb{Z}, \ 0 \leq i, j \leq 7.$$

We compute the product of this *binom1* matrix and the vector *sum* again using *_dp_2a* operator. Then for 24 columns, we serially propagate carry bits using only 24 threads. Finally using three threads we propagate three final carry bytes and write them to the 25-th column.

Next, SBox lookup is done. In SWIFFTX, an 8 by 8 bits SBox is employed to provide nonlinearity and this table is accessed byte by byte. Inputs and outputs are $3N + 8 = 204$ bytes long. In GP104, shared memory banks are 4-bytes wide. Processing the input byte by byte therefore creates 4 times more shared memory write transactions than necessary. Instead, we lookup 4 values, combine them using logical shifts and write them at once to comply with the physical shared memory structure. This concludes the first stage of the algorithm.

Second stage starts with a NTT executed on 25 columns. This does not fit well into our 64-thread per block implementation. We execute three and a half NTT iterations to process 25 columns. Adjustment is done only on eight values since the ninth +128 is in column 30. Translation to $\mathbb{F}_{256}$ is applied using the same technique but this time the output is only $8 + 1 = 9$ columns. Finally, the 65-byte output in the shared memory is written to device memory dword by dword to comply with the memory transaction coalesing rules. This finalizes major optimizations done on the algorithm.

5   Further Improvements and Occupancy Analysis

In Pascal microarchitecture, shared memory is divided into 16 banks where each bank is 4 bytes. This physical constraint leads to conflicts while writing the NTT output. To overcome this situation, we add two unused rows to the transposed NTT output and adjust pointer arithmetic accordingly. Now, the NTT output is $(M + 2)N = 34 \times 64$ words. In pointer arithmetic, multiplication by 34 is implemented as shift by 5 plus shift by 1. However, this introduces latency when compared to a single shift. Similarly, we add four rows to *sum* area, it is $3N + 12$ words now. Finally, to prevent any other alignment issues, we set the size of the *intermediate* area to $4N$ bytes. Since we copy the input to a shared memory location at begining of the kernel, we do not face any global memory access inefficiencies related to the input. Similarly, the output is written directly from shared memory to the global memory at the end of the kernel therefore it is efficient. However, the size of the output is 65 bytes. If several thousands of hashes are calculated in bulk, this leads to an alignment issue for the output of the consecutive hashes. Therefore, we modify the output size and set it to 72 bytes to prevent any issues of this kind.

Now, NTT lookup table is of size $256 \times 8 \times 8$ words. Each lookup returns a word. *nvvp* shows an inefficiency in global load L2 transaction, specifically, the ideal is 2 but the current is 4 transactions per access. This can be remedied only if the lookup returns a dword. However, the table size in that case becomes too large to fit into L2 Cache therefore left unoptimized. Similarly, *nvvp* shows an inefficiency of 7.5 to 1 L2 transactions per access while looking up the table SBox. We have tried to implement the same routine by an 16×16 bits SBox yet the speed is reduced, therefore we left it as is.

Finally, we calculate the occupancy. In one hand, *nvcc* compiler tells that our kernel employs 48 registers. GP104, register file is 256 KiB, assuming all 4 bytes, there are 65536 registers in total per SM. A thread employs 48 and a block employs $48 \times 64 = 3072$ registers. Dividing 65536 by this number leads to ≈ 21.3 block limit. On the other hand, we have the following definitions:

```
__shared__ int16_t S_fftOut[(M+2)*N];
__shared__ int16_t S_sum[3*N+12];
__shared__ unsigned char S_intermediate[4*N];
```

Now, *S_fftOut* is $(32+2) \times 64 \times 2 = 4352$ bytes, *S_sum* is $[(3 \times 64) + 12] \times 2 = 408$ bytes and *S_intermediate* is $4 \times 64 = 256$ bytes. Therefore, Shared Memory usage is 5016 bytes in total. Dividing the Shared Memory size 96 Kib by this number leads to ≈ 19.6 blocks. Therefore, our kernel has a block limit of $min(21.3, 19.6) = 19.6$. This leads to $19 \times 64 = 1216$ threads per SM or in other words we have $1216/2048 = 0.594$, 59.4% occupancy per SM.

6   Methodology and Results

All results are obtained on an Intel®E5410 CPU system where Linux version is 4.14.104, GNU glibc version is 2.27, CUDA version is 10.0 and NVIDIA®driver version is 415.27. The method the results are obtained is as follows. Each test round consists of 2^{14} hashes. First, there is a step to warm the CPU and the GPU up for 10 rounds. Then, we generate test data for each set of input and sequencially run the algorithm on CPU, then on the GPU and collect the results. Generated input data is classified as follows: *(i) All zeroes:* weight 0/byte, *(ii) All ones:* weight 8/byte, *(iii) All random:* non-constant random weight/byte, *(iv) All random:* weight 4/byte.

This classification allows us to see whether or not the Hamming weight of the input does effect the performance of our kernel. All random data is generated by glibc *random(3)*. For *all random weight 4/byte* test, we employ *Fisher-Yates* shuffling algorithm [2]. We have generated enough random data for 2^{14} input blocks (2^{22} bytes) per test round. Furthermore, we set the affinity of the process via *sched_affinity(2)* to CPU 0 to get consistent results, avoid kernel rescheduling and cache invalidation.

The results are given in Table 1. These results are acquired using GNU gprof v2.31.1 and nvprof v10.0.130 profilers. Table 1 shows execution times of different implementations. First one is the x86 reference implementation, the second, GPU ported reference implementation and the last one is our parallel implementation. These results strongly indicate that the Hamming weight of the input is irrelevant, hence all data in the following tables are collected using non-constant weight *All random* data set. Table 2 depicts cache hit rates. Table 3 depicts memory throughput metrics obtained by the profiler. Figure 5 depicts kernel stall reasons of implementations. For the reference, test device properties are also included in Table 6.

Table 1. Experimental results, test round: 2^{14} hashes

	Intel® Xeon^{TM} E5410 reference impl.	NVIDIA® GeForce GTX^{TM} 1080 ported reference impl.	NVIDIA® GeForce GTX^{TM} 1080 our parallel impl.	Unit
All zeroes	3.50×10^5	3.76×10^3	3.78×10^2	μsec
All ones	3.50×10^5	3.75×10^3	3.78×10^2	μsec
All random	3.50×10^5	3.77×10^3	3.80×10^2	μsec
All random weight 4	3.50×10^5	3.76×10^3	3.80×10^2	μsec

Table 1 shows almost 10x increase in speed compared to the ported reference implementation. Global memory accesses in our implementation are very efficient. For *all random* test, *nvvp* shows global store efficiency of 70.8% and global load efficiency of 74.7%. The kernel employs a total of 5016 bytes of shared memory per block. Unfortunately, shared memory efficiency is only 52.2%, nevertheless, it is compensated since it is very fast.

The use of shared memory makes data access very fast. However, it is limited only upto 96 KiB per SM therefore determines the number of warps spawned simultaneously. Parallel kernel requires only 48 registers. Decreasing this number via _launch_bounds_ leads to spill loads and stores degrading the performance. This number is sufficient for 19 warps to be spawned simultaneously. Our kernel performes L2 Cache hit rate of 33.7% and Unified Cache hit rate of 96.2% in the very same test (Table 2). These numbers indicate caches are efficiently utilized. Also, measured occupancy per SM is 57.4%. Since this is above 50%, it is enough to hide the arithmetic latency of the ALU inside GP104. This is discussed in detail in [13].

Table 2. Cache hit rates

Metric	Description	Reference impl.	Parallel impl.
tex_cache_hit_rate	Unified Cache Hit Rate	67.05%	96.22%
l2_tex_hit_rate	Hit rate at L2 cache for all requests from texture cache	9.92%	33.75%
global_hit_rate	Hit rate for global load and store in unified L1/Tex cache	92.77%	92.27%
local_hit_rate	Hit rate for local loads and stores	50.12%	0.00%

Table 3. Memory throughput metrics

Metric	Description	Reference impl.	Parallel impl.
sysmem_read_throughput	System memory read	0.00000 B/s	0.00000 B/s
sysmem_write_throughput	System memory write	41.824 KB/s	475.05 KB/s
dram_read_throughput	Device memory read	132.83 GB/s	11.888 GB/s
dram_write_throughput	Device memory write	70.031 GB/s	6.5270 GB/s
local_load_throughput	Local memory load	265.30 GB/s	0.00000 B/s
local_store_throughput	Local memory store	72.083 GB/s	0.00000 B/s
gld_throughput	Global load	117.50 GB/s	580.49 GB/s
gst_throughput	Global store	8.4957 GB/s	4.4536 GB/s
shared_load_throughput	Shared memory load	0.00000 B/s	1644.9 GB/s
shared_store_throughput	Shared memory store	0.00000 B/s	914.47 GB/s
tex_cache_throughput	Unified cache	327.27 GB/s	1101.5 GB/s
l2_tex_read_throughput	L2 (Texture Reads)	138.68 GB/s	71.209 GB/s
l2_tex_write_throughput	L2 (Texture Writes)	80.579 GB/s	4.4536 GB/s
l2_read_throughput	L2 (Reads)	139.02 GB/s	71.643 GB/s
l2_write_throughput	L2 (Writes)	80.579 GB/s	4.4548 GB/s

In Table 3, memory throughput metrics are given. First of all, system memory access is negligible in both kernels since data is copied to the device memory beforehand. Next, the device memory usage is reduced making it a bottleneck no more. Since our kernel employs a less number of registers, there is no local load or store. Global load throughput is increased by five times and stores are reduced by a half. Similary, it is possible to observe Shared Memory throughput which makes a big difference in our parallel implementation. Unified Cache throughput is increased since we instruct the assembler to cache everything via the flag *(-Xptxas -dlcm=ca)*. This is also the reason for the reductions in L2 throughputs.

Figure 5 depicts the percentage of stall reasons per kernel. In the reference implementation, kernel stalls 29% due the memory dependences. Also loading constants from Texture Cache generates a lot of traffic (24%). Moreover, it is not possible to learn what is included in *stall_other* (29%) so, it is better to keep it low under normal circumstances. In the parallel kernel, the largest percentage is owned by the execution dependency. In order to lower this value, it is possible to unroll loops so that compiler can move instructions around and optimize execution dependency. However, all of the loops in our kernel other than the one wrapping the NTT iterations are already unrolled so there is nothing that can be done. Unrolling that particular loop leads to register spills so we left it as is. The actual unrolling effect can also be observed by the 18% *stall_inst_fetch* metric. Too much unrolling is likely to overload the instruction fetch queue. In our case, the above configuration works well. Other metrics are around 10% and almost equally distributed. This is an indication of a balance between trade-offs.

Additional user-space benchmarking shows that we can calculate 2^{14} hashes in 420 ms on a single x86 thread. The same can be achieved only in 4 ms on the test device. This data also includes the duration of copying input to the device memory and getting it back to system memory over PCIe bus. According to these indicators, the throughput of x86 implementation is $2^{14} \times 2^8$ bytes/420 ms = 4 MiB/420 ms $\approx$ 10 MiB/s per thread while the throughput of our CUDA implementation is $2^{14} \times 2^8$ bytes/4 ms = 4 MiB/4 ms $\approx$ 1 GiB/s where 2^8 bytes is the hash input block size.

Power consumption metrics have also been collected using nvprof profiler (Tables 4 and 5). The program is run for only a single test round without a warmup stage. Collected data shows on *adaptive* power mode, our board consumes 40 W per test round on Reference Implementation and 35 W on Parallel Implementation. This suggests our implementation consumes almost 5 W less on *adaptive* mode per test round. These values become 56 W to 43 W when test round hash count is increased to 2^{19}. The difference is almost 13 W.

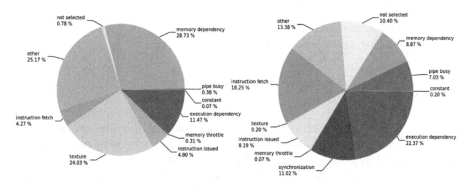

Fig. 5. Kernel stall reasons, reference impl. (left) vs our parallel impl. (right)

Table 4. Ported reference implementation power consumption data

Data/PowerMizer mode	Adaptive (Min/Avg/Max)	Max. Perf. (Min/Avg/Max)	Unit
SM clock	139.00/1313.40/1607.00	1607.00/1607.00/1607.00	MHz
Memory clock	405.00/3789.80/5005.00	4513.00/4709.80/5005.00	MHz
Temperature	51.00/51.56/52.00	53.00/53.00/53.00	C
Power	10039.00/39879.00/53144.00	46522.00/50178.67/53231.00	mW
Fan	0.00/0.00/0.00	0.00/0.00/0.00	%

Table 5. Our parallel implementation power consumption data

Data/PowerMizer mode	Adaptive (Min/Avg/Max)	Max. Perf. (Min/Avg/Max)	Unit
SM clock	139.00/1019.80/1607.00	1607.00/1607.00/1607.00	MHz
Memory clock	405.00/2968.20/5005.00	4513.00/4759.00/5005.00	MHz
Temperature	52.00/52.33/53.00	53.38/53.00/54.00	C
Power	9995.00/34898.33/53135.00	47012.00/49984.00/53718.00	mW
Fan	0.00/0.00/0.00	0.00/0.00/0.00	%

Table 6. Test device properites

Property	Value	Unit
Name, Brand	Asus® NVIDIA GTX^{TM} 1080	
Architecture	NVIDIA Pascal^{TM}	
Total amount of global memory	8120	Mbytes
Number of Stream Multiprocessors	20	
Number of cores per SM	128	
Total number of cores	2560	
GPU/Memory clock	1734 / 5005	MHz
L2	2097152	bytes
Total amount of constant memory	65536	bytes
Total amount of shared memory per block	49152	bytes
Total number of registers available per block	65536	
Warp size	32	
Maximum number of threads per multiprocessor	2048	
Maximum number of threads per block	1024	

7   Conclusion

SWIFFTX is one of the lattice based hash function that provides provable collision resistance and pseudo-randomness. In this paper, we have presented an efficient parallel implementation of SWIFFTX on GPU. Our tests have showed that the proposed implementation is approximately 1000 times faster than the single-thread x86 implementation and 10 times faster than the ported reference implementation. Moreover, the throughput is also increased by 100 times. In terms of power consumption, our implementation performs 5 W better per 2^{16} hashes and 13 W better per 2^{19} hashes.

It should be noted that there are newer architectures such as Volta and Turing than Pascal, a member of the sixth generation CUDA. These newer generations have higher memory bandwidth and more computation capabilities. Furthermore, the technology called Independent Thread Scheduling (ITS) is built into those new architectures. This technology will probably allow GPU's to utilize resources more efficiently in terms of scheduling and synchronization and deliver more speed and throughput if properly implemented. First idea basically aims to increase the occupancy. It might be possible to implement a version of the algorithm that does not employ any shared memory instead, passes data across threads via Warp Shuffling. Consequently, this new implementation and the one that uses shared memory can be run simultaneously in the presence of ITS and hence a higher occupancy will be achieved. On the other hand, this might not lead to a significant improvement since we are still facing the burden of field arithmetic assigned to ALU. The second idea might target the time lost during synchronization. Figure 5 indicates that our kernel is stalled by synchronization primitives by 11%. This situation might be improved by ITS and defining explicit memory reads and writes using volatile keyword.

References

1. Arbitman, Y., Dogon, G., Lyubashevsky, V., Micciancio, D., Peikert, C., Rosen, A.: SWIFFTX: a proposal for the SHA-3 standard. In: The First SHA-3 Candidate Conference (2008)
2. Durstenfeld, R.: Algorithm 235: random permutation. Commun. ACM **7**(7), 420 (1964)
3. Centre for Research on Cryptography and Brno Czech Republic Security, Masaryk University. Tool for generation of data from cryptoprimitives (block and stream ciphers, hash functions). https://github.com/crocs-muni/CryptoStreams. Accessed Dec 2018
4. Györfi, T., Cret, O., Hanrot, G., Brisebarre, N.: High-throughput hardware architecture for the swifft/swifftx hash functions. IACR Cryptology ePrint Archive, 2012:343 (2012)
5. Lyubashevsky, V., Micciancio, D.: Generalized compact knapsacks are collision resistant. In: 33rd International Colloquium Automata, Languages and Programming, ICALP 2006, Venice, Italy, 10–14 July 2006, Proceedings, Part II, pp. 144–155 (2006)
6. Lyubashevsky, V., Micciancio, D., Peikert, C., Rosen, A.: SWIFFT: a modest proposal for FFT hashing. In: Nyberg, K. (ed.) FSE 2008. LNCS, vol. 5086, pp. 54–72. Springer, Heidelberg (2008). https://doi.org/10.1007/978-3-540-71039-4_4
7. NVIDIA: GeForce GTX 1080 Whitepaper. https://international.download.nvidia.com/geforce-com/international/pdfs/GeForce. Accessed Dec 2018
8. NVIDIA: Parallel Thread Execution ISA. https://docs.nvidia.com/cuda/parallel-thread-execution/index.html. Accessed Apr 2018
9. NVIDIA: Pascal Tuning Guide. https://docs.nvidia.com/cuda/pascal-tuning-guide/index.html. Accessed Apr 2018
10. NVIDIA: Visual Profiler. https://docs.nvidia.com/cuda/profiler-users-guide/index.html. Accessed Apr 2018
11. CUDA NVIDIA: NVIDIA CUDA C programming guide. Nvidia Corporation **120**(18), 8 (2011)
12. Peikert, C., Rosen, A.: Efficient collision-resistant hashing from worst-case assumptions on cyclic lattices. In: Theory of Cryptography, Third Theory of Cryptography Conference, TCC 2006, New York, NY, USA, 4–7 March 2006, Proceedings, pp. 145–166 (2006)
13. Volkov, V.: Better performance at lower occupancy. Proc. GPU Technol. Conf. **10**, 16 (2010)

Computing an Invariant of a Linear Code

Mijail Borges-Quintana[1(✉)], Miguel Ángel Borges-Trenard[2],
Edgar Martínez-Moro[3(✉)], and Gustavo Torres-Guerrero[1]

[1] Department of Mathematics, Faculty of Natural and Exact Sciences,
University of Oriente, Santiago de Cuba, Cuba
mijail@uo.edu.cu, gtorresguerrero85@gmail.com
[2] Doctorate in Mathematics Education, University Antonio Nariño,
Bogotá, Colombia
borgestrenard2014@gmail.com
[3] Institute of Mathematics IMUVa, University of Valladolid,
Valladolid, Castilla, Spain
edgar.martinez@uva.es

Abstract. In this work we present an efficient algorithm that generates the leader codewords of a linear code in an incremental form. On the other hand, using the set of leader codewords we define a transformation that remains invariant only if the codes are equivalent which is used as a signature for checking the code equivalence problem. An upper bound on the weight of the codewords is imposed to this algorithm in order to get a smallest set that can be also used as a signature for the 'Code Equivalence Problem'.

Keywords: Leader codewords · Code equivalence · Coset leaders

1 Introduction

In this work we are interested in the mathematical aspects of the set of leader codewords of a linear code related with two main issues, its computation and getting a signature for the 'Code Equivalence Problem'. In [3] this set is defined for binary codes and it vis given an algorithm for its computation in an incremental form based on the Gröbner representation [9] of the code. The extension of those results for general linear codes is analyzed in [5].

We formulate a kind of Möller's algorithm for Gröbner representation techniques that generates the leader codewords in an incremental form. Nevertheless, we state and proof the correctness of the algorithm without the need of using Gröbner basis. An upper bound on the weight of the codewords is imposed the algorithm in order to get only those leader codewords bounded by a given weight. We also show how can be used a suitable subset of the leader codewords in the 'Code Equivalence Problem', i.e. the problem of determining whether two given

E. Martínez-Moro—Partially supported by the Spanish State Research Agency (AEI) under Grants MTM2015-65764-C3-1, PGC2018-096446-B-C21.

linear codes are permutation-equivalent. If they are, we also want to recover this permutation group. In [10] the authors proved that this problem is not NP-complete but also that it is at least as hard as the Graph Isomorphism Problem. On the other hand, the support splitting algorithm [11] solves the computational version of the problem in polynomial time for all but an exponentially small proportion of the instances. In that paper it is stated that the main difficulty in the implementation of the algorithm lies in the choice of the invariant since usually the computation rapidly becomes intractable when its size grows.

Note that the role played by the Gröbner representation in the equivalence of codes was introduced in [4]. The set of leader codewords proposed in this paper is a structure which is considerably smaller that the invariant proposed in [4]. Despite of this, this set grows fast as the size of the code increase; so we impose an upper bound on the weight of the codewords to be included and prove that is enough to consider this subset of leader codewords as invariant. We use this subset for finding the permutation between equivalent codes. Note also that it can be used in any algorithm based on partitions and refinements like those in [8,11]. In particular, we have adapted the support splitting algorithm by defining a specific signature corresponding to this subset as invariant.

The structure of the paper is as follows. In Sect. 2 we present some preliminary facts and notations. In Sect. 3 we define the set of leader codewords and describe the algorithm. Section 4 provides a formal proof that this subset is an invariant for the code and we show how it can be used for finding the permutation group between equivalent codes. The algorithm is described and formalized in Sect. 5. Finally in Sect. 6 we present some experimental results.

2 Preliminaries

2.1 Linear Codes

From now on we shall denote by $\mathbb{F}_q$ the finite field with $q = p^m$ elements, p a prime. A *linear code* $\mathcal{C}$ over $\mathbb{F}_q$ of length n and dimension k is a k-dimensional subspace of $\mathbb{F}_q^n$. We will call the vectors $\mathbf{v}$ in $\mathbb{F}_q^n$ words and those $\mathbf{v} \in \mathcal{C}$, codewords. For every word $\mathbf{v} \in \mathbb{F}_q^n$ its *support* is defined as $\text{supp}(\mathbf{v}) = \{i \mid v_i \neq 0\}$ and its *Hamming weight*, denoted by $w_H(\mathbf{v})$ as the cardinality of $\text{supp}(\mathbf{v})$ and the *Hamming distance* $d_H(\mathbf{x}, \mathbf{y})$ between two words $\mathbf{x}, \mathbf{y} \in \mathbb{F}_q^n$ is $d_H(\mathbf{x}, \mathbf{y}) = w_H(\mathbf{x} - \mathbf{y})$. The *minimum distance* $d(\mathcal{C})$ of a linear code $\mathcal{C}$ is defined as the minimum weight among all nonzero codewords.

The set of words of minimal Hamming weight in all the cosets of $\mathbb{F}_q^n/\mathcal{C}$ is the *set of coset leaders* of the code $\mathcal{C}$ in $\mathbb{F}_q^n$ and we will denote it by $\text{CL}(\mathcal{C})$. $\text{CL}(\mathbf{y})$ will denote the subset of coset leaders corresponding to the coset $\mathbf{y} + \mathcal{C}$. Given a coset $\mathbf{y} + \mathcal{C}$ we define the *weight of the coset* $w_H(\mathbf{y} + \mathcal{C})$ as the smallest Hamming weight among all vectors in the coset, or equivalently the weight of one of its leaders. It is well known that given $t = \lfloor \frac{d(\mathcal{C})-1}{2} \rfloor$ where $\lfloor \cdot \rfloor$ denotes the greatest integer function then every coset of weight at most t has a unique coset leader.

2.2   The Weak Order Ideal of the Coset Leaders

Let $f(X)$ be an irreducible polynomial over $\mathbb{F}_p$ of degree m and β be a root of $f(X)$, then any element $a \in \mathbb{F}_q$ can be represented as $a_1 + a_2\beta + \ldots + a_m\beta^{m-1}$ with $a_i \in \mathbb{F}_p$ for $i \in \{1, \ldots, m\}$.

Definition 1. *We define the* generalized support *of a vector* $\mathbf{v} = (\mathbf{v}_1, \ldots, \mathbf{v}_n) \in \mathbb{F}_q^n$ *as the support of the nm-tuple given by the concatenations of the p-adic expansion of each component* $\mathbf{v}_i = v_{i_1} + v_{i_2}\beta + \ldots + v_{i_m}\beta^{m-1}$ *of* $\mathbf{v}$. *That is* $\mathrm{supp}_{\mathrm{gen}}(\mathbf{v}) = (\mathrm{supp}((v_{i_1}, \ldots, v_{i_m})) : i = 1, \ldots, n)$, *and* $\mathrm{supp}_{\mathrm{gen}}(\mathbf{v})[i] = \mathrm{supp}((v_{i_1}, \ldots, v_{i_m}))$. *We will say that* $i_j \in \mathrm{supp}_{\mathrm{gen}}(\mathbf{v})$ *if the corresponding* v_{i_j} *is not zero.*

The set $\mathrm{Can}(\mathbb{F}_q, f) = \left\{ \mathbf{e}_{ij} = \beta^{j-1}\mathbf{e}_i : i = 1, \ldots, n; j = 1, \ldots, m \right\}$ represents the canonical basis of $(\mathbb{F}_q^n, +)$. We state the following connection between $\mathbb{F}_q^n$ and $\mathbb{N}^{nm}$:

$$\Delta : \mathbb{F}_q^n \to \mathbb{N}^{nm}$$

$$\mathbf{v} \mapsto (\psi(v_{i_j}) : i = 1, \ldots, n, j = 1, \ldots, m),$$

where the mapping $\psi : \mathbb{F}_p \to \mathbb{N}$ is defined as $k \cdot 1_{\mathbb{F}_p} \mapsto k \bmod p$. On the other hand we define the mapping $\nabla : \mathbb{N}^{nm} \to \mathbb{F}_q^n$ as $\mathbf{a} \mapsto (a_{m(i-1)+1} + a_{m(i-1)+2}\beta + \ldots + a_{m(i-1)+m}\beta^{m-1})$, $i = 1, \ldots, n$.

Definition 2. *Given* $\mathbf{x}, \mathbf{y} \in (\mathbb{F}_q^n, +)$, $\mathbf{x} = \sum_{i,j} x_{i_j} \mathbf{e}_{ij}$, $\mathbf{y} = \sum_{i,j} y_{i_j} \mathbf{e}_{ij}$, *we say* $\mathbf{x} \subset \mathbf{y}$ *if* $\psi(x_{i_j}) \le \psi(y_{i_j})$ *for all* $i \in \{1, \ldots, n\}$ *and* $j \in \{1, \ldots, m\}$.

The map Δ relates orders on $\mathbb{F}_q^n$ with orders on $\mathbb{N}^{nm}$, and vice versa. An *admissible order* on $(\mathbb{N}^{nm}, +)$ is a total order $<$ on $\mathbb{N}^{nm}$ satisfying the following two conditions

1. $\mathbf{0} < \mathbf{x}$, for all $\mathbf{x} \in \mathbb{N}^{nm}$, $\mathbf{x} \ne \mathbf{0}$.
2. If $\mathbf{x} < \mathbf{y}$, then $\mathbf{x} + \mathbf{z} < \mathbf{y} + \mathbf{z}$, for all $\mathbf{z} \in \mathbb{N}^{nm}$.

In particular, any admissible order on $(\mathbb{N}^{nm}, +)$, (lexicographical, degree lexicographical, degree reverse lexicographical ...) induces an order on $(\mathbb{F}_q^n, +)$. A representation of a word $\mathbf{v}$ as an nm-tuple over $\mathbb{N}$ is said to be in *standard form* if $\Delta(\nabla(\mathbf{v})) = \mathbf{v}$. We will denote the standard form of $\mathbf{v}$ as $\mathrm{SF}(\mathbf{v}, f)$ (note that $\nabla(\mathbf{v}) = \nabla(\mathrm{SF}(\mathbf{v}, f))$). Therefore, $\mathbf{v}$ is in standard form if $\mathbf{v} = \mathrm{SF}(\mathbf{v}, f)$ (we will also say $\mathbf{v} \in \mathrm{SF}(\mathbb{F}_q^n, f)$). In shake of brevity, from now on we will consider the polynomial f fixed and we will use $\mathrm{Can}(\mathbb{F}_q)$ and $\mathrm{SF}(\mathbb{F}_q^n)$ instead of $\mathrm{Can}(\mathbb{F}_q, f)$ and $\mathrm{SF}(\mathbb{F}_q^n, f)$ respectively.

Definition 3. *A subset* $\mathcal{O}$ *of* $\mathbb{N}^k$ *is an* order ideal *if for all* $\mathbf{w} \in \mathcal{O}$ *and* $\mathbf{v} \in \mathbb{N}^k$ *s.t.* $\mathbf{v}_i \le \mathbf{w}_i$, $i = 1, \ldots, k$, *then* $\mathbf{v} \in \mathcal{O}$.

In the same fashion as the previous definition, we say that a subset $\mathcal{S}$ of $\mathbb{F}_q^n$ is an order ideal if $\Delta(\mathcal{S})$ is an order ideal in $\mathbb{N}^{nm}$. It is easy to check that an equivalent definition for the order ideal would be that for all $\mathbf{w} \in \mathcal{S}$, and for all $i_j \in \mathrm{supp}_{\mathrm{gen}}(\mathbf{w})$, and $\mathbf{v} \in \mathbb{F}_q^n$ s.t. $\mathbf{w} = \mathbf{v} + \mathbf{e}_{ij}$ we have $\mathbf{v} \in \mathcal{S}$. If we change it slightly and instead of for all $i_j \in \mathrm{supp}_{\mathrm{gen}}(\mathbf{w})$ the condition is satisfied at least for one $i_j \in \mathrm{supp}_{\mathrm{gen}}(\mathbf{w})$ we say that the set $\mathcal{S}$ is a *weak order ideal*. More formally,

Definition 4. *A subset* S *of* $\mathbb{F}_q^n$ *is a weak order ideal if for all* $\mathbf{w} \in S \setminus \mathbf{0}$ *there exists a* $i_j \in \mathrm{supp}_{\mathrm{gen}}(\mathbf{w})$ *such that for* $\mathbf{v} \in \mathbb{F}_q^n$ *with* $\mathbf{w} = \mathbf{v} + \mathbf{e}_{ij}$ *then* $\mathbf{v} \in S$.

In the above situation we will say that the word $\mathbf{w}$ is an *ancestor* of the word $\mathbf{v}$, and that $\mathbf{v}$ is a *descendant* of $\mathbf{w}$. In non binary case a coset leader could be an ancestor of another coset leader or an ancestor of a word at Hamming distance 1 to a coset leader (this last case is not possible in the binary case).

The first idea that allows us to compute incrementally the set of all coset leaders for a linear code was introduced in [4] using the additive structure of $\mathbb{F}_q^n$ and the set of canonical generators $\mathrm{Can}(\mathbb{F}_q)$. Unfortunately in [4] most of the chosen coset representatives may not be coset leaders if the weight of the coset is greater than t. In order to incrementally generate all coset leaders starting from $\mathbf{0}$ adding elements in $\mathrm{Can}(\mathbb{F}_q)$, we must consider words with weight one more than the previous chosen coset leader (see [5]).

Definition 5. *Given* $\prec_1$ *an admissible order on* $(\mathbb{N}^{nm}, +)$ *we define the weight compatible order* $\prec$ *on* $(\mathbb{F}_q^n, +)$ *associated to* $\prec_1$ *as the ordering given by*

1. $\mathbf{x} \prec \mathbf{y}$ *if* $\mathrm{w_H}(\mathbf{x}) < \mathrm{w_H}(\mathbf{y})$ *or*
2. if $\mathrm{w_H}(\mathbf{x}) = \mathrm{w_H}(\mathbf{y})$ *then* $\Delta(\mathbf{x}) \prec_1 \Delta(\mathbf{y})$.

In other words, the words in $\mathbb{F}_q^n$ are ordered according their Hamming weights and the order $\prec_1$ break ties. These class of orders is a subset of the class of monotone α-orderings in [7]. In fact we will need a little more than monotonicity, we will also need the following condition: for every pair $\mathbf{v}, \mathbf{w} \in \mathrm{SF}(\mathbb{F}_q^n)$ such that $\mathbf{v} \subset \mathbf{w}$ one has that $\mathbf{v} \prec \mathbf{w}$. Note that this last condition is indeed true for a weight compatible order. In addition, for any weight compatible order $\prec$ every strictly decreasing sequence terminates (due to the finiteness of the set $\mathbb{F}_q^n$). In the binary case the behavior of the coset leaders can be translated to the fact that the set of coset leader is an order ideal of $\mathbb{F}_2^n$; whereas, for non binary linear codes this is no longer true even if we try to use the characterization of order ideals given in [6], where order ideals do not need to be associated with admissible orders.

Definition 6. *We define* the weak order ideal of the coset leaders *of a linear code* C *as the set* $\mathcal{O}(C)$ *of elements in* $\mathbb{F}_q^n$ *verifying the following items,*

1. $\mathbf{0} \in \mathcal{O}(C)$.
2. If $\mathbf{v} \in \mathcal{O}(C)$ *and* $\mathrm{w_H}(\mathbf{v}) = \mathrm{w_H}(\mathbf{v} + C)$ *then*

$$\left\{ \mathbf{v} + \mathbf{e}_{ij} \mid \Delta(\mathbf{v}) + \Delta(\mathbf{e}_{ij}) \in \mathrm{SF}(\mathbb{F}_q^n) \right\} \subset \mathcal{O}(C).$$

3. If $\mathbf{v} \in \mathcal{O}(C)$ *and* $\mathrm{w_H}(\mathbf{v}) = \mathrm{w_H}(\mathbf{v} + C) + 1$ *then*

$$\left\{ \mathbf{v} + \mathbf{e}_{ij} \mid i \in \mathrm{supp}(\mathbf{v}), \Delta(\mathbf{v}) + \Delta(\mathbf{e}_{ij}) \in \mathrm{SF}(\mathbb{F}_q^n), \mathbf{v} - \mathbf{v}_i \in \mathrm{CL}(C) \right\} \subset \mathcal{O}(C).$$

Note that it is clear by items 2 and 3 in the definition above that $\mathcal{O}(C)$ is a weak order ideal. Note also that the definition of the set $\mathcal{O}(C)$ also gives an algorithmic process to built this set, which result very important to construct the set $\mathrm{CL}(C)$ taking into account that $\mathrm{CL}(C) \subset \mathcal{O}(C)$. The following two theorems show the connections between the set of coset leaders and the weak order ideal of the coset leaders.

Theorem 1 (See [5]). *Let* $\mathbf{w} \in \mathbb{F}_q^n$. *If there exists* $i \in 1, \ldots, n$ *s.t.* $\mathbf{w} - \mathbf{w}_i \in$ CL$(\mathcal{C})$ *then* $\mathbf{w} \in \mathcal{O}(\mathcal{C})$.

Theorem 2 (See [5]). *Let* $\mathbf{w} \in \mathbb{F}_q^n$ *and* $\mathbf{w} \in$ CL$(\mathcal{C})$ *then* $\mathbf{w} \in \mathcal{O}(\mathcal{C})$.

3   Leader Codewords of Linear Codes

Definition 7. *The set of* leader codewords *of a linear code* $\mathcal{C}$ *is defined as*

$$
\mathrm{L}(\mathcal{C}) = \left\{ \begin{array}{c} \mathbf{v}_1 + \mathbf{e}_{ij} - \mathbf{v}_2 \in \mathcal{C} \setminus \{\mathbf{0}\} \mid \Delta(\mathbf{v}_1) + \Delta(\mathbf{e}_{ij}) \in \mathrm{SF}(\mathbb{F}_q^n), \\ \mathbf{v}_2 \in \mathrm{CL}(\mathcal{C}) \ and \ \mathbf{v}_1 - \mathbf{v}_{1i} \in \mathrm{CL}(\mathcal{C}) \end{array} \right\}.
$$

Note that the definition is a bit more elaborated that the one for binary codes in [3] due to the fact that in the general case not all coset leaders need to be ancestors of coset leaders. The name of leader codewords comes from the fact that one could compute all coset leaders of a corresponding word knowing the set L$(\mathcal{C})$ adapting [3, Algorithm 3]. Theorem 1 guarantees that $\mathbf{w} \in \mathcal{O}(\mathcal{C})$ provided that $\mathbf{w} - \mathbf{w}_i \in$ CL$(\mathcal{C})$ for some i, then the associated set of leader codewords may be computed as $\{\mathbf{w} - \mathbf{v} : \mathbf{w} \in \mathcal{O}(\mathcal{C}), \mathbf{w} - \mathbf{w}_i \in \mathrm{CL}(\mathcal{C}), \mathbf{v} \in \mathrm{CL}(\mathbf{w}) \text{ and } \mathbf{v} \neq \mathbf{w}\}$.

3.1   Computing Algorithm

In [3] it is presented a Möller's like algorithm for computing the leader codewords for binary linear codes. Given a weight compatible ordering $\prec$, it is introduced an incremental form of generating the set of leader codewords. The generation of these elements is based on the construction of an object List (a crucial object in a Möller-like algorithm). The object List for general linear codes is related exactly with the computation of the set $\mathcal{O}(\mathcal{C})$; i.e. List is the smallest ordered set of elements in $\mathbb{F}_q^n$ verifying the following properties:

1. $\mathbf{0} \in$ List.
2. Criterion 1: If $\mathbf{v} \in$ List and $\mathrm{w}_H(\mathbf{v}) = \mathrm{w}_H(\mathbf{v} + \mathcal{C})$ then
$$
\{\mathbf{v} + \mathbf{e}_{ij} \mid \Delta(\mathbf{v}) + \Delta(\mathbf{e}_{ij}) \in \mathrm{SF}(\mathbb{F}_q^n)\} \subset \text{List}.
$$
3. Criterion 2: If $\mathbf{v} \in$ List and $\mathrm{w}_H(\mathbf{v}) = \mathrm{w}_H(\mathbf{v} + \mathcal{C}) + 1$ then
$$
\{\mathbf{v} + \mathbf{e}_{ij} \mid i \in \mathrm{supp}(\mathbf{v}), \Delta(\mathbf{v}) + \Delta(\mathbf{e}_{ij}) \in \mathrm{SF}(\mathbb{F}_q^n),
$$
$$
\mathbf{v} - \mathbf{v}_i \in \mathrm{CL}(\mathcal{C})\} \subset \text{List}.
$$

Given a weight compatible order $\prec$ and a linear code $\mathcal{C}$, the algorithm will incrementally generate all elements in List and also all coset leaders, starting from the zero codeword in List. Then Theorem 1 guarantees that

$$
\mathbf{w} \in \text{List provided that } \mathbf{w} - \mathbf{w}_i \in \mathrm{CL}(\mathcal{C}) \text{ for some } i, \tag{1}
$$

and the associated set of leader codewords may be computed as $\{\mathbf{w} - \mathbf{v} : \mathbf{v} \in$ CL$(\mathbf{w})$ and $\mathbf{v} \neq \mathbf{w}\}$.

3.2   Computing up to a Given Level

Let $\mathcal{Q}$ be a set of elements in $\mathbb{F}_q^n$. We will call a *level of weight k* to the set $\mathcal{Q}'$ such that $\mathcal{Q}' = \{\mathbf{v} \in \mathcal{Q} \mid \mathrm{w}_H(\mathbf{v}) = k\}$. We can get a partition of the set $\mathcal{Q}$ ordered by the weight of each level $0 \le k_1 < k_2 < \ldots < k_s$. We will refer to the i-th set $\mathcal{Q}_i$ in this partition by the level of weight k_i of $\mathcal{Q}$ and we will denote as $\mathcal{Q}_{[i]}$ to the set of all words up to the level i.

As it was discussed in the previous section, the leader codewords of a linear code $\mathcal{C}$ are generated in an incremental form according to a weight compatible order, so we can set an upper bound if we only want the leader codewords up to a given level. The following proposition establishes a connection between the weight of the elements belonging to List and the weight of their corresponding leader codewords.

Proposition 1. *Let $\mathbf{c} \in \mathrm{L}(\mathcal{C})$ and $\mathbf{w} \in$ List the least element w.r.t to $\prec$ such that $\mathbf{c} = \mathbf{w} - \mathbf{v}$, $\mathbf{w} - \mathbf{w}_i \in \mathrm{CL}(\mathcal{C})$ for some $i \in 1, \ldots, n$, $\mathbf{v} \in \mathrm{CL}(\mathcal{C})$. Then $2\mathrm{w}_H(\mathbf{w}) - 1 \le \mathrm{w}_H(\mathbf{c})$.*

Proof. Since $\mathbf{c} = \mathbf{w} - \mathbf{v}$ and $\mathbf{v} \in \mathrm{CL}(\mathcal{C})$, we have $\mathbf{w} \in \mathbf{v} + \mathcal{C}$. Then $\mathrm{w}_H(\mathbf{v}) \le \mathrm{w}_H(\mathbf{w})$. If we suppose $\mathrm{w}_H(\mathbf{v}) = \mathrm{w}_H(\mathbf{w}) - 2$ then $\mathbf{c} = (\mathbf{w}_i - \mathbf{v}) - (-(\mathbf{w} - \mathbf{w}_i)) = \mathbf{a} - \mathbf{b}$, where $\mathbf{a} - \mathbf{a}_i = -\mathbf{v} \in \mathrm{CL}(\mathcal{C})$, $\mathbf{b} = -(\mathbf{w} - \mathbf{w}_i) \in \mathrm{CL}(\mathcal{C})$. Now, $\mathrm{w}_H(\mathbf{a}) \le \mathrm{w}_H(\mathbf{w}_i) + \mathrm{w}_H(\mathbf{v}) = \mathrm{w}_H(\mathbf{w}) - 1$. This is $\mathrm{w}_H(\mathbf{a}) < \mathrm{w}_H(\mathbf{w})$ and so $\mathbf{a} \prec \mathbf{w}$. Finally, by (1), $\mathbf{a} - \mathbf{a}_i \in \mathrm{CL}(\mathcal{C})$ implies $\mathbf{a} \in$ List, which is a contradiction because $\mathbf{w}$ is the least element in List to obtain $\mathbf{c}$.

Therefore, $\mathrm{w}_H(\mathbf{v}) \ge \mathrm{w}_H(\mathbf{w}) - 1$, from where it is obtained $2\mathrm{w}_H(\mathbf{w}) - 1 \le \mathrm{w}_H(\mathbf{c})$. $\qquad\qquad\square$

Remark 1. As a direct consequence of the previous result we have that, in order to compute all leader codewords up to a weight k, it is enough to stop the algorithm in the first element of List of weight t such that $2t - 1 > k$.

Algorithm 1 below summarizes the aspects discussed above. There are three functions needed to understand the algorithm:

- InsertNexts[t, List] inserts all sums $\mathbf{t} + \mathbf{e}_{ij}$ in List, where $\Delta(\mathbf{v}) + \Delta(\mathbf{e}_{ij}) \in \mathrm{SF}(\mathbb{F}_q^n)$, keeping the increasing order $\prec$ in List.
- NextTerm[List] returns the first element from List and deletes it from that set.
- Member[obj, G] returns the position j of obj in G, if $obj \in G$, and false otherwise.

Proposition 2. *Algorithm 1 computes the set of leader codewords of a linear code $\mathcal{C}$ up to a given level.*

Algorithm 1: Computation of the leader codewords up to a given level

---

input : A weight compatible ordering $\prec$, a parity check matrix H of a code $\mathcal{C}$ and the level k.

output: $\mathrm{L}(\mathcal{C})_{[k]}$.

1 List $\leftarrow [0]$; $r \leftarrow 0$; $\mathrm{CL}(\mathcal{C}) \leftarrow \emptyset$; $\mathcal{S} \leftarrow \emptyset$; $\mathrm{L}(\mathcal{C}) \leftarrow \emptyset$; $k' \leftarrow 0$; $w_{k'} \leftarrow 0$;
  $w_k \leftarrow \infty$; $Stop \leftarrow$ false;

2 **while** List $\neq \emptyset$ and $Stop \neq$ true **do**

3      $\mathbf{t} \leftarrow$ NextTerm[List];

4   **if** $2\mathrm{w}_H(\mathbf{t}) - 1 \leq w_k$ **then**

5          $s \leftarrow \mathbf{t}H^T$;

6          $j \leftarrow$ Member$[s, \mathcal{S}]$;

7   **if** $j \neq$ false **then**

8   **if** $\mathrm{w}_H(\mathbf{t}) = \mathrm{w}_H(\mathrm{CL}(\mathcal{C})[j][1])$ **then** // Criterion 1 in List

9                  $\mathrm{CL}(\mathcal{C})[j] \leftarrow \mathrm{CL}(\mathcal{C})[j] \cup \{\mathbf{t}\}$;

10   List $\leftarrow$ InsertNext$[\mathbf{t}, \text{List}]$;

11   **end if**

12   **if** $\mathrm{w}_H(\mathbf{t}) = \mathrm{w}_H(\mathrm{CL}(\mathcal{C})[j][1]) + 1$ **then** // Criterion 2 in List

13   **for** $i \in \mathrm{supp}(\mathbf{t})$: $\mathbf{t} - \mathbf{t}_i \in \mathrm{CL}(\mathcal{C})$ **do**

14   List $\leftarrow$ InsertNext$[\mathbf{t}, \text{List}]$;

15   **end for**

16   **end if**

17   **for** $i \in \mathrm{supp}(\mathbf{t})$: $\mathbf{t} - \mathbf{t}_i \in \mathrm{CL}(\mathcal{C})$ **do**

18   **for** $\mathbf{t}' \in \mathrm{CL}(\mathcal{C})[j]$ and $(\mathbf{t} \neq \mathbf{t}')$ **do**

19   **if** $\mathrm{w}_H(\mathbf{t} - \mathbf{t}') > w_{k'}$ **then**

20                          $k' \leftarrow k' + 1$; $w_{k'} \leftarrow \mathrm{w}_H(\mathbf{t} - \mathbf{t}')$;

21   **if** $k' = k$ **then** // $\mathrm{L}(\mathcal{C})$ has reached the level k

22                              $w_k \leftarrow w_{k'}$;

23   **end if**

24   **end if**

25   **if** $\mathrm{w}_H(\mathbf{t} - \mathbf{t}') \leq w_k$ **then**

26                          $\mathrm{L}(\mathcal{C}) \leftarrow \mathrm{L}(\mathcal{C}) \cup \{\mathbf{t} - \mathbf{t}'\}$;

27   **end if**

28   **end for**

29   **end for**

30   **else**

31              $r \leftarrow r + 1$; $\mathrm{CL}(\mathcal{C})[r] \leftarrow \{\mathbf{t}\}$; $\mathcal{S}[r] \leftarrow s$;

32   List $=$ InsertNext$[\mathbf{t}, \text{List}]$ // Criterion 1 in List;

33   **end if**

34   **else**

35          $Stop \leftarrow$ true;

36   **end if**

37 **end while**

38 **return** $\mathrm{L}(\mathcal{C})$

Proof (Of Proposition 2). Let us first prove that all the words inserted in List satisfy the desired properties pointed in Sect. 3.1. By Step 1, $\mathbf{0} \in$ List, verifying the first property, then in Step 5 the syndrome (an element of the coset) of $\mathbf{t} = \mathtt{NextTerm[List]}$ is computed and based on the outcome of Step 6 we have two possible cases,

1. If $j = \mathtt{false}$ then the coset $\mathcal{C} + \mathbf{t}$ has not yet been considered, therefore it is created taking $\mathbf{t}$ as a representative of minimal weight. Step 32 guarantees Criterion 1 in the second property.
2. On the other hand, if $j \neq \mathtt{false}$, the coset $\mathcal{C} + \mathbf{t}_j$ has been created and in case of $\mathrm{w}_H(\mathbf{t}) = \mathrm{w}_H(\mathbf{t}_j)$ Step 10 guarantees Criterion 1. If $\mathrm{w}_H(\mathbf{t}) = \mathrm{w}_H(\mathbf{t}_j) + 1$ then Step 13 and Step 14 verify Criterion 2 in the third property of List.

Therefore Algorithm 1 constructs List fulfilling the required properties. Furthermore, in List is included the set $\mathcal{O}(\mathcal{C})$, then by Theorem 2, List contains all coset leaders, thus Step 9 and Step 31 assure the computation of the whole set of coset leaders. From Step 19 to Step 24 the algorithm keeps track of the current level of $\mathrm{L}(\mathcal{C})$ and the weight associated with that level. Finally, Step 25 and Step 26 create the set $\mathrm{L}(\mathcal{C})$ of leader codewords according to Definition 7. Meanwhile, the second stop condition of the loop (Step 2) given by Proposition 1 prevents from continuing when the current weight is greater than the given weight for the desired level k. $\qquad\square$

Of course note that if no level k is specified then Algorithm 1 computes the whole set of leader codewords.

4 $\mathrm{L}(\mathcal{C})$ as an Invariant for Linear Codes

It is clear that if two codes $\mathcal{C}, \mathcal{C}'$ are permutation equivalent so that for a given $\sigma \in S_n$ we have that $\mathcal{C}' = \sigma(\mathcal{C})$, then $\mathrm{L}(\mathcal{C}') = \sigma(\mathrm{L}(\mathcal{C}))$. In [4, Theorem 3] it is shown that two linear codes are equivalent if their so called Matphi structure are equivalent. These Matphi structures depend also on the cosets determined by the codes, but the size of this object is bigger than the set of leader codewords. The following result establishes that the set of leader codewords is also an invariant.

Theorem 3. *Let $\mathcal{C}, \mathcal{C}'$ be linear codes and $\sigma \in S_n$. Then $\mathcal{C}' = \sigma(\mathcal{C})$ if and only if $\mathrm{L}(\mathcal{C}') = \sigma(\mathrm{L}(\mathcal{C}))$.*

Proof. Let $\mathcal{C}' = \sigma(\mathcal{C})$ for $\sigma \in S_n$, in order to prove $\mathrm{L}(\mathcal{C}') = \sigma(\mathrm{L}(\mathcal{C}))$ it is enough to prove that $\sigma(\mathrm{L}(\mathcal{C})) \subset \mathrm{L}(\mathcal{C}')$. Let $\mathbf{c} \in \mathrm{L}(\mathcal{C})$, then $\mathbf{c} = \mathbf{v}_1 + \mathbf{e}_{ij} - \mathbf{v}_2$, $\Delta(\mathbf{v}) + \Delta(\mathbf{e}_{ij}) \in \mathrm{SF}(\mathbb{F}_q^n)$, $\mathbf{v}_2 \in \mathrm{CL}(\mathcal{C})$ and $\mathbf{v}_1 - \mathbf{v}_{1i} \in \mathrm{CL}(\mathcal{C})$. Thus, $\mathcal{C}' = \sigma(\mathcal{C})$ implies $\sigma(\mathbf{v}_1) - \sigma(\mathbf{v}_{1i}) = \sigma(\mathbf{v}_1 - \mathbf{v}_{1i}) \in \mathrm{CL}(\mathcal{C}')$, $\sigma(\mathbf{v}_2) \in \mathrm{CL}(\mathcal{C}')$ and $\Delta(\sigma(\mathbf{v}_1)) + \Delta(\mathbf{e}_{\sigma(i)j}) \in \mathrm{SF}(\mathbb{F}_q^n)$. Then, $\mathbf{c}' = \sigma(\mathbf{c}) = \sigma(\mathbf{v}_1) + \sigma(\mathbf{e}_{ij}) - \sigma(\mathbf{v}_2) = \sigma(\mathbf{v}_1) + \mathbf{e}_{\sigma(i)j} - \sigma(\mathbf{v}_2)$. Therefore, $\mathbf{c}' \in \mathrm{L}(\mathcal{C}')$.

Now, let us suppose that $\mathrm{L}(\mathcal{C}') = \sigma(\mathrm{L}(\mathcal{C}))$ and let $\mathbf{c} \in \mathcal{C}$. In [5] it was proved that the set $\mathrm{L}(\mathcal{C})$ is a test set for $\mathcal{C}$. This means, there exist $\mathbf{c}_1, \dots, \mathbf{c}_k$, $\mathbf{c}_i \in LC$, $i = 1, \dots, k$ such that $\mathbf{c} = \mathbf{c}_1 + \dots + \mathbf{c}_k$. That is, we have

$$\sigma(\mathbf{c}) = \sigma(\mathbf{c}_1) + \dots + \sigma(\mathbf{c}_k). \qquad (2)$$

But $\sigma(\mathbf{c}_i) \in L(\mathcal{C}')$, $i = 1, \ldots, k$, then, taking into account (2) we obtain $\sigma(\mathbf{c}) \in \mathcal{C}'$.

$\square$

Remark 2. A mapping is an invariant for a code means that it remains invariant under a permutation. The previous theorem shows that the set of leader codewords $L(\mathcal{C})$ may give a very strong invariant in the sense that it is preserved if and only if the codes are equivalent. Due to its prohibitive size as the code length increases we take the subset $L(\mathcal{C})_{[2]}$ and for this we have $L(\mathcal{C}')_{[2]} = \sigma(L(\mathcal{C})_{[2]})$ provided that $\mathcal{C}' = \sigma(\mathcal{C})$.

The following lemma allow us to state Theorem 4 in order to use the set of leader codewords up to a given level as invariant.

Lemma 1. *Let* $\mathcal{C} = \langle \mathcal{B} \rangle$ *and* $\mathcal{C}' = \langle \mathcal{B}' \rangle$ *two codes over* $\mathbb{F}_q$ *with spanning sets* $\mathcal{B}$ *and* $\mathcal{B}'$. *If there exists* $\sigma \in S_n$ *such that* $\mathcal{B}' = \sigma(\mathcal{B})$ *then* $\mathcal{C}' = \sigma(\mathcal{C})$.

Proof. Let $c' \in \mathcal{C}'$. Then $c' = \sum_{\alpha_i \in \mathbb{F}_q} \alpha_i \beta_i' = \sum \alpha_i \sigma(\beta_i) = \sigma \left(\sum \alpha_i \beta_i \right)$ and hence $c' \in \sigma(\mathcal{C})$. On the other hand, let $c \in \mathcal{C}$. Then $c = \sum_{\alpha_i \in \mathbb{F}_q} \alpha_i \beta_i$ and $\sigma(c) = \sum \alpha_i \sigma(\beta_i) = \sum \alpha_i \beta_i'$. Therefore $\sigma(c) \in \mathcal{C}'$.

$\square$

Theorem 4. *Let* $\mathcal{C}$ *and* $\mathcal{C}'$ *linear codes of* $\mathbb{F}_q^n$ *such that* $dim(\mathcal{C}) = dim(\mathcal{C}')$ *and* $k = \min_s \{ s \in \mathbb{N} \mid \mathcal{C} = \langle L(\mathcal{C})_{[s]} \rangle \}$. *For* $m \geq k$, *for any* $\sigma \in S_n$ *such that* $\sigma \left(L(\mathcal{C})_{[m]} \right) = L(\mathcal{C}')_{[m]}$ *then* $\mathcal{C}' = \sigma(\mathcal{C})$.

Proof. It is a consequence of the fact that $L(\mathcal{C})_{[m]}$ is a spanning set of $\mathcal{C}$ for $m \geq k$, Lemma 1 and $dim(\mathcal{C}) = dim(\mathcal{C}')$. Note that, by applying the lemma, $\mathcal{C} \sim \sigma(\mathcal{C})$. On the other hand, $\sigma(\mathcal{C})$ is a subspace of $\mathcal{C}'$ of the same dimension of $\mathcal{C}'$, so $\sigma(\mathcal{C}) = \mathcal{C}'$.

$\square$

Note that all codewords of minimum weight are leader codewords. Moreover, $L(C)_{[1]}$ is exactly this set of codewords. In case of codes that are generated by this set, $k = 1$ in Theorem 4 and it is enough to use $L(C)_{[1]}$ to compute the candidates permutations.

5   Finding the Permutation

The idea of using the subset $L(\mathcal{C})_{[2]}$ of the set $L(\mathcal{C})$ as an invariant can be applied for finding the permutations between equivalent codes and it can be used in any algorithm based on partitions and refinements like [8,11]. In particular, we have specified the algorithm described in [11] by defining a specific signature corresponding to $L(\mathcal{C})_{[2]}$. We have changed a little the definition of signature but keeping the central idea. The construction of the partition and the refinement process based on the signature follow similar procedures. A description of related algorithms for code equivalence is done in [12].

5.1   The Proposed Signature

One way of defining signatures for codes is by using an invariant, we are going to introduce a signature based on the set $L(\mathcal{C})_{[2]}$.

Definition 8 ([11]). *A signature S over a set Ω maps a code $\mathcal{C}$ of length n and an element $i \in I_n = \{1, \ldots, n\}$ into an element of Ω and is such that for all permutations $\sigma \in S_n$, $S(\mathcal{C}, i) = S(\sigma(\mathcal{C}), \sigma(i))$.*

Let $\mathbb{Z}[y_0, \ldots, y_n]$ be the polynomial ring of the $n+1$ variables $y_0, \ldots, y_n$ over the integers. We define a signature over $\Omega = \mathbb{Z}[y_0, \ldots, y_n] \times \mathbb{Z}[y_0, \ldots, y_n]$ which depends on the numbers of assignments of positions already done. Note at the beginning no assignment has been done yet.

Let $J \subset I_n$, $J = \{j_1, \ldots, j_s\}$ be the assignments of positions we assumed have been done to the set $J' \subset I_n$, $J' = \{j'_1, \ldots, j'_s\}$, J may be equal to the empty set and $s = 0$. Then for all permutations $\sigma \in S_n$, such that $\sigma(j_i) = j'_i$, $i = 1, \ldots, s$, we define for $i \in I_n \setminus J$

$$\mathrm{SLC}_s(\mathcal{C}, i) = (a_{i0}y_0 + \ldots + a_{is}y_0y_1 \cdots y_s, b_{i0}y_0 + \ldots + b_{is}y_0y_1 \cdots y_s),$$

where the first component $a_{i0}y_0 + \ldots + a_{is}y_0y_1 \cdots y_s$ stands for the subset $L(\mathcal{C})_1$ of $L(\mathcal{C})_{[2]}$ and the second component $b_{i0}y_0 + \ldots + b_{is}y_0y_1 \cdots y_s$ stands for the subset $L(\mathcal{C})_2$ of $L(\mathcal{C})_{[2]}$. Specifically a_{ik}, $k \in 0, \ldots, s$ means that there are a_{ik} elements $\mathbf{c} \in L(\mathcal{C})_1$ with $\mathbf{c}_i \neq 0$ and others exactly k positions from J which are not zero. Similarly, b_{ik}, $k \in 0, \ldots, s$ means that there are b_{ik} elements $\mathbf{c} \in L(\mathcal{C})_2$ with $\mathbf{c}_i \neq 0$ and other exactly k positions from J which are not zero.

Note that $\mathrm{SLC}_s(\mathcal{C}, i)$ counts the interactions between the position i and the set of positions J already assigned in the subsets $L(\mathcal{C})_1$ and $L(\mathcal{C})_2$ of $L(\mathcal{C})_{[2]}$. As it is expected, for all permutations $\sigma \in S_n$, such that $\sigma(j_i) = j'_i$, $i = 1, \ldots, s$, $\mathrm{SLC}_s(\mathcal{C}, i) = \mathrm{SLC}_s(\sigma(\mathcal{C}), \sigma(i))$ which is guaranteed by Theorem 3.

The $(\mathcal{C}, \mathrm{SLC}_s)$-partition (see [11]) is $\mathcal{P}(\mathcal{C}, \mathrm{SLC}_s) = \{J_e : e \in \Omega\}$, where $J_e = \{i \in I_n \setminus J : \mathrm{SLC}_s(\mathcal{C}, i) = e\}$. Note that the partition corresponding to a permutation is such that $\mathcal{P}(\sigma(\mathcal{C}), \mathrm{SLC}_s) = \{\sigma(J_e) : e \in \Omega\}$, for all permutations $\sigma \in S_n$, such that $\sigma(j_i) = j'_i$, $i = 1, \ldots, s$.

5.2   Refining the Partition

Given the linear codes $\mathcal{C}$ and $\mathcal{C}'$ and a subset of s positions J already assigned to J', such that $\mathrm{SLC}_i(\mathcal{C}, j_{i+1}) = \mathrm{SLC}_i(\mathcal{C}', j'_{i+1})$, $i = 0, \ldots, s - 1$.

We compute the partitions $\mathcal{P}(\mathcal{C}, \mathrm{SLC}_s) = \{J_e : e \in \Omega\}$ and $\mathcal{P}(\mathcal{C}', \mathrm{SLC}_s) = \{J'_e : e \in \Omega\}$ and then we take into account that a position from J_e must be transformed into a position of J'_e, so the next assignment is decided. For example, we may take the J_{e_1} subset of minimal cardinal and then the position $i \in J_{e_1}$ of minimal absolute value. Once a new position is chosen we select its image j such that $\mathrm{SLC}_s(\mathcal{C}', j) = \mathrm{SLC}_s(\mathcal{C}, i)$ and then $J = J \cup \{i\}$, $J' = J' \cup \{j\}$, $s = s + 1$.

In this process it is possible to detect some contradictions which means no permutation will be found by this path. For example, it is clear that the cardinal of the partitions for $\mathcal{C}$ and $\mathcal{C}'$ must be the same, and also $| J_e | = | J'_e |$ for all $e \in \Omega$.

5.3   Computing Algorithm

Proposition 3. *Algorithm 2 computes a permutation $\sigma \in S_n$ between the codes $\mathcal{C}$ and $\mathcal{C}'$, that is $\mathcal{C}' = \sigma(\mathcal{C})$. If no permutation is found then these codes are not permutation equivalent.*

Algorithm 2: Computing the permutation

---

1 **function** PermutationEquivCodes
> **input** : $\mathcal{C}, \mathcal{C}'$ and a weight compatible ordering $\prec$
> **output**: A permutation $\sigma \in S_n$, such that $\mathcal{C}' = \sigma(\mathcal{C})$

2 | $J \leftarrow \emptyset; J' \leftarrow \emptyset; s \leftarrow 0$
3 | Compute $\mathrm{L}(\mathcal{C})_{[2]}$ and $\mathrm{L}(\mathcal{C}')_{[2]}$ using $\prec$ as described in Section 3
4 | FindPermutation($\mathrm{L}(\mathcal{C})_{[2]}, \mathrm{L}(\mathcal{C}')_{[2]}, J, J', s$)
5 | **if** no permutation found **then**
6 | | **return** $\mathcal{C}, \mathcal{C}'$ are not permutation equivalent codes
7 | **end if**
8 | **return** $\sigma \leftarrow \sigma(J_i) = J'_i$, $i = 1, \ldots, n$
9 **end func**

10 **function** FindPermutation
> **input** : $\mathrm{L}(\mathcal{C})_{[2]}, \mathrm{L}(\mathcal{C}')_{[2]}$ and $J, J' \subset I_n$ where $s = |J| = |J'|$
> **output**: J, J' such that $\sigma(J_i) = J'_i$, $i = 1, \ldots, n$ if a permutation is
> > found. No permutation found is returned otherwise

11 | **if** $|J| = n$ **then**
12 | | **return** J, J'
13 | **end if**
14 | $\mathcal{P} \leftarrow \{J_e : \mathrm{SLC}_s(\mathcal{C}, i) = e, i \in I_n \setminus J, e \in \Omega\}$ // use $\mathrm{L}(\mathcal{C})_{[2]}$ as invariant
15 | $\mathcal{P}' \leftarrow \{J'_e : \mathrm{SLC}_s(\mathcal{C}', i) = e, i \in I_n \setminus J', e \in \Omega\}$ // to compute $\mathrm{SLC}_s(\mathcal{C}, i)$
16 | **if** $|J_e| = |J'_e|$ for all $J_e \in \mathcal{P}, J'_e \in \mathcal{P}'$ **then**
17 | | $J \leftarrow J \cup \{i\}, i \in J_{e_1}$ // J_{e_1}, i chosen randomly or by an heuristic
18 | | $J^* \leftarrow J'_{e_1}$ such that $J'_{e_1} \in \mathcal{P}'$
19 | | **while** no permutation found **and** $J^* \neq \emptyset$ **do**
20 | | | $J' \leftarrow J' \cup \{j\}, j \in J^*$
21 | | | FindPermutation($\mathrm{L}(\mathcal{C})_{[2]}, \mathrm{L}(\mathcal{C}')_{[2]}, J, J', s+1$)
22 | | | **if** a permutation were found **then**
23 | | | | **return** J, J'
24 | | | **end if**
25 | | | $J' \leftarrow J' \setminus \{j\}$; $J^* \leftarrow J^* \setminus \{j\}$
26 | | **end while**
27 | | $J \leftarrow J \setminus \{i\}$
28 | **end if**
29 | **return** no permutation found
30 **end func**

Proof (Of Proposition 3). It is clear that if there exists a permutation between two codes $\mathcal{C}$ and $\mathcal{C}'$, by Theorem 3, this permutation transforms $L(\mathcal{C})$ into $L(\mathcal{C}')$ and then defines the same signatures and partitions. Thus one of those permutations will be found by Algorithm 2. The process is finite because there is a finite number of permutations and therefore the process of analyzing different assignments following the signatures and partitions is finite.   □

Note that in Algorithm 2 the function `PermutationEquivCodes` do the initializations. Then the sets $L(\mathcal{C})_{[2]}$ and $L(\mathcal{C}')_{[2]}$ are computed (also they can be loaded from a precomputed database) and then a call to the recursive function `FindPermutation` is made which follows a refinement process following an n-ary tree structure, where a permutation is found when a node of level n is reached.

Finding All the Permutations. Given a linear code $\mathcal{C}$ of length n, the subgroup of all elements σ of S_n such that $\sigma(\mathcal{C}) = \mathcal{C}$ is called the permutation automorphism group of $\mathcal{C}$. Note that if the permutation automorphism group is nontrivial and if $\mathcal{C}'$ is permutation equivalent to $\mathcal{C}$, then several permutations could satisfy $\mathcal{C}' = \sigma(\mathcal{C})$. Algorithm 2 can be easily modified to compute all those permutations. This can be achieved if a list of pairs J, J' is returned instead of a single pair and before each successful return statement those pairs are added to this list. An n-ary tree transversal is made, where at each node of level n one permutation is considered. After that step an expurgation process should be carried out since some invalid permutations could be introduced because $\sigma(L(\mathcal{C})_{[2]}) = L(\mathcal{C}')_{[2]}$ may not be sufficient to guarantee that $\sigma(\mathcal{C}) = \mathcal{C}'$.

Example 1 (Toy Example). Consider the binary codes $\mathcal{C} = \langle (0,1,0,0,1),$ $(1,1,0,1,0), (0,1,1,0,0) \rangle$ and $\mathcal{C}' = \langle (1,1,0,0,0), (1,0,1,0,1), (1,0,0,1,0) \rangle$ and

$$L(\mathcal{C})_1 = \{(0,1,1,0,0), (0,1,0,0,1), (0,0,1,0,1)\},$$
$$L(\mathcal{C})_2 = \{(1,1,0,1,0), (1,0,1,1,0), (1,0,0,1,1)\},$$
$$L(\mathcal{C}')_1 = \{(1,1,0,0,0), (1,0,0,1,0), (0,1,0,1,0)\},$$
$$L(\mathcal{C}')_2 = \{(1,0,1,0,1), (0,1,1,0,1), (0,0,1,1,1)\}.$$

Note that $\mathcal{C}' = \sigma(\mathcal{C})$ with $\sigma = (1,3,4,5,2)$ and thus $L(\mathcal{C}')_{[2]} = \sigma(L(\mathcal{C})_{[2]})$. On the first call to `FindPermutation` we get

$\mathrm{SLC}_0(\mathcal{C}, 1) = (0, 3y_0)$, $\mathrm{SLC}_0(\mathcal{C}, 2) = (2y_0, y_0)$,
$\mathrm{SLC}_0(\mathcal{C}, 3) = (2y_0, y_0)$, $\mathrm{SLC}_0(\mathcal{C}, 4) = (0, 3y_0)$, $\mathrm{SLC}_0(\mathcal{C}, 5) = (2y_0, y_0)$,
$\mathrm{SLC}_0(\mathcal{C}', 1) = (2y_0, y_0)$, $\mathrm{SLC}_0(\mathcal{C}', 2) = (2y_0, y_0)$,
$\mathrm{SLC}_0(\mathcal{C}', 3) = (0, 3y_0)$, $\mathrm{SLC}_0(\mathcal{C}', 4) = (2y_0, y_0)$, $\mathrm{SLC}_0(\mathcal{C}', 5) = (0, 3y_0)$.

Now with $e_1 = (0, 3y_0), e_2 = (2y_0, y_0)$ such that $e_1, e_2 \in \Omega$ we get

$$\mathcal{P}(\mathcal{C}, \mathrm{SLC}_0) = \{J_{e_1} = \{1,4\}, J_{e_2} = \{2,3,5\}\},$$
$$\mathcal{P}'(\mathcal{C}', \mathrm{SLC}_0) = \{J'_{e_1} = \{3,5\}, J'_{e_2} = \{1,2,4\}\}.$$

At this point is verified that $|J_{e_1}| = |J'_{e_1}|$ and $|J_{e_2}| = |J'_{e_2}|$ and a coordinate must be chosen, that is, $i \in J_e$ such that $J_e \in \mathcal{P}$. Recall that this can be done at random or following some heuristics. Let us take the one with minimal cardinal and the position with minimal absolute value. Then $J = J \cup \{1\} \Rightarrow J = \{1\}$ and we may try with each element in J'_{e_1}, since $\sigma(1) \in J'_{e_1}$ provided that $\mathrm{SLC}_0(\mathcal{C}, 1) = \mathrm{SLC}_0(\mathcal{C}', \sigma(1))$. Starting with the minimum value $J' = J' \cup \{3\} \Rightarrow J' = \{3\}$, then a recursive call is made, meaning that $\sigma(1) = 3$. Note that if no permutation is found through this path, a new selection must be made following a tree structure. In the new call we get

$$\begin{aligned}
\mathrm{SLC}_1(\mathcal{C}, 2) &= (0, 2y_0 + y_0 y_1), \quad \mathrm{SLC}_1(\mathcal{C}, 3) = (0, 2y_0 + y_0 y_1), \\
\mathrm{SLC}_1(\mathcal{C}, 4) &= (0, 3y_0 y_1), \qquad\qquad \mathrm{SLC}_1(\mathcal{C}, 5) = (0, 2y_0 + y_0 y_1), \\
\mathrm{SLC}_1(\mathcal{C}', 1) &= (0, 2y_0 + y_0 y_1), \quad \mathrm{SLC}_1(\mathcal{C}', 2) = (0, 2y_0 + y_0 y_1), \\
\mathrm{SLC}_1(\mathcal{C}', 4) &= (0, 2y_0 + y_0 y_1), \quad \mathrm{SLC}_1(\mathcal{C}', 5) = (0, y_0 y_1),
\end{aligned}$$

$$\mathcal{P}(\mathcal{C}, \mathrm{SLC}_1) = \left\{ J_{e_1} = \{4\}, J_{e_2} = \{2, 3, 5\} \right\},$$
$$\mathcal{P}'(\mathcal{C}', \mathrm{SLC}_1) = \left\{ J'_{e_1} = \{5\}, J'_{e_2} = \{1, 2, 4\} \right\},$$

where $J = J \cup \{4\} \Rightarrow J = \{1, 4\}$ and $J' = J' \cup \{5\} \Rightarrow J' = \{3, 5\}$, meaning that $\sigma(1) = 3$ and $\sigma(4) = 5$. Following this refining procedure is obtained J and J' such that $\sigma(j_i) = j'_i$, $i = 1, \ldots, 5$.

6   Experimental Results

The algorithms in this paper were implemented in C++ using the GNU operating system based gcc compiler and performed using the high performance computing capabilities provided at University of Oriente, Cuba (http://www.uo.edu.cu). In Table 1 we show the advantage of choosing the set of leader codewords only up to the second level. A significant difference can be noticed in the execution time, since for level 2 there is a relatively slight change as the number of cosets increase,

Table 1. Execution time and number of leader codewords

Codes	# Cosets	Level 2		All levels	
		Time (sec.)	Num.	Time (sec.)	Num.
$\mathbb{F}_2\,[12, 6, 4]$	64	0.0001	28	0.0310	50
$\mathbb{F}_2\,[15, 8, 3]$	128	0.0001	9	0.1400	127
$\mathbb{F}_2\,[18, 10, 3]$	256	0.0150	16	0.6560	328
$\mathbb{F}_3\,[12, 6, 3]$	729	0.0620	16	6.9690	568
$\mathbb{F}_3\,[12, 5, 3]$	2187	0.0620	12	42.2510	472
$\mathbb{F}_3\,[13, 6, 4]$	2187	0.1250	28	57.8660	520
$\mathbb{F}_4\,[11, 5, 4]$	4096	0.2970	42	271.0270	816
$\mathbb{F}_4\,[12, 6, 3]$	4096	0.4100	15	401.0000	2435
$\mathbb{F}_4\,[14, 7, 4]$	16384	1.1720	27	10942.7280	4564

compared with the fast growth in computing the whole set. On the other hand, the number of leader codewords up to the second level remains stable and much more smaller (see Fig. 1).

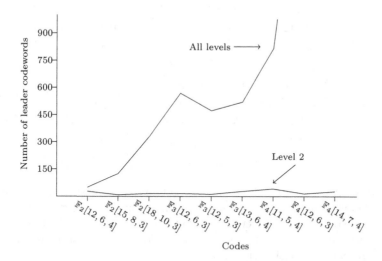

Fig. 1. Number of leader codewords up to the 2nd and all levels

In order to evaluate the performance of the algorithm for finding the permutation between two linear codes, a code is generated first at random and then is applied a permutation generated at random too. For these two codes we compute the set of the leader codewords up to the second level, and then, they are used as input for the algorithm that will give as output the first valid permutation.

Table 2 is shows execution times for the leader codewords up to the second level, only for the generated code, and in a different column is showed the time

Table 2. Execution time in seconds to find the first permutation between two linear codes randomly permuted

Codes ($\mathcal{C}$)	Cosets	$L(\mathcal{C})_{[2]}$	1st permutation
$\mathbb{F}_2\,[15, 7, 3]$	256	0.031	0.015
$\mathbb{F}_2\,[21, 12, 3]$	512	0.093	0.047
$\mathbb{F}_2\,[29, 18, 3]$	2048	0.328	0.078
$\mathbb{F}_2\,[34, 19, 4]$	32768	12.688	0.063
$\mathbb{F}_3\,[18, 8, 4]$	59049	22.891	0.078
$\mathbb{F}_3\,[23, 12, 5]$	177147	96.395	0.078
$\mathbb{F}_3\,[32, 20, 5]$	531441	567.770	0.240
$\mathbb{F}_4\,[20, 10, 5]$	1048576	647.020	0.060
$\mathbb{F}_4\,[26, 15, 5]$	4194304	3001.800	0.360
$\mathbb{F}_4\,[30, 18, 5]$	16777216	9402.400	0.140

Table 3. Execution time in seconds to find all permutations

Codes	Permutations	First (sec.)	All (sec.)
$\mathbb{F}_2\,[15,7,3]$	8	0.015	0.031
$\mathbb{F}_2\,[29,18,3]$	96	0.078	0.265
$\mathbb{F}_2\,[21,12,3]$	144	0.047	0.271
$\mathbb{F}_4\,[30,18,5]$	720	0.140	0.830
$\mathbb{F}_2\,[15,7,3]$	768	0.015	1.218
$\mathbb{F}_3\,[18,8,4]$	1536	0.015	2.828
$\mathbb{F}_3\,[23,12,4]$	3456	0.042	8.636
$\mathbb{F}_2\,[34,19,4]$	4608	0.063	9.000
$\mathbb{F}_4\,[30,18,5]$	10080	0.024	9.376

consumed by Algorithm 1. The codes are generated increasing the number of cosets as before, but this time these numbers are much more greater, showing the advantage of using the selected invariant $L(\mathcal{C})_{[2]}$. Note that the time used by the algorithm to get the first correct permutation is significantly smaller than the time spent in computing the invariant.

Finally Table 3 shows the execution times for the algorithm adapted to compute all the permutations and it is compared with the timing of the first permutation obtained. The time for finding the first permutation depends on how the elements are chosen in each refinement stage. This explain the fluctuating time according to the increasing number of elements in the permutation group.

Some Comments on Complexity Issues. Authors would like to emphasize that the main goal of this paper is the study the computation of leader codewords and the properties related to the permutation equivalent problem from the mathematical point of view. Some of the experiments are devoted to show the possibility of using part of the set of leader codewords instead of the whole set and to compare this two instance.

Algorithm 1 for computing the leader codewords is efficient because its computational complexity is linear on the size of the weak order ideal of the code, and because of the nature of the leader codewords this can not be improved much more. Also we adapted the algorithm to compute the set up to a given weight. Anyway the computation of this set becomes intractable when the code length increase, particularly, the redundancy of the code (the number of cosets).

On the other hand, we used the SSA Algorithm and we construct with the leader codewords a signature in order to use the scheme of this algorithm. The main limitation is the high complexity of computing the invariant, the set of leader codewords up to a given level. For complexity issues regarding the SSA Algorithm and other problems related with the code equivalence, we recommend [13], which examines also complexity issues of SSA, [1] it can be used as a comparison for works related to this problem, [2], useful reference and a standard in these works to compare against.

References

1. Babai, L., Codenotti, P., Grochow, J.A., Qiao, Y.: Code equivalence and group isomorphism. In: Proceedings of the Twenty-Second Annual ACM-SIAM Symposium on Discrete Algorithms, pp. 1395–1408. Society for Industrial and Applied Mathematics (2011)
2. Betten, A., Braun, M., Fripertinger, H., Kerber, A., Kohnert, A., Wassermann, A.: Error-Correcting Linear Codes: Classification by Isometry and Applications, vol. 18. Springer, Heidelberg (2006). https://doi.org/10.1007/3-540-31703-1
3. Borges-Quintana, M., Borges-Trenard, M., Márquez-Corbella, I., Martínez-Moro, E.: Computing coset leaders and leader codewords of binary codes. J. Algebra Appl. **14**(8), 19 (2015)
4. Borges-Quintana, M., Borges-Trenard, M., Martínez-Moro, E.: On a Gröbner bases structure associated to linear codes. J. Discrete Math. Sci. Cryptogr. **10**(2), 151–191 (2007)
5. Borges-Quintana, M., Borges-Trenard, M., Martínez-Moro, E.: On the weak order ideal associated to linear codes. Math. Comput. Sci. **12**(3), 339–347 (2018)
6. Braun, G., Pokutta, S.: A polyhedral characterization of border bases. SIAM J. Discrete Math. **30**(1), 239–265 (2016)
7. Helleseth, T., Kløve, T., Levenshtein, V.I.: Error-correction capability of binary linear codes. IEEE Trans. Inf. Theory **51**(4), 1408–1423 (2005)
8. Leon, J.S.: Computing automorphism groups of error-correcting codes. IEEE Trans. Inform. Theory **28**, 496–511 (1982)
9. Mora, T.: Solving Polynomial Equation Systems II: Macaulay's Paradigm and Gröbner Technology. Cambridge University Press, Cambridge (2005)
10. Petrank, E., Roth, R.: Is code equivalence easy to decide? IEEE Trans. Inform. Theory **43**(5), 1602–1604 (1997)
11. Sendier, N.: Finding the permutation between equivalent linear codes: the support splitting algorithm. IEEE Trans. Inform. Theory **46**(4), 1193–1203 (2000)
12. Sendier, N., Simos, D.: How easy is code equivalence over $\mathbb{F}_q$? In: International Workshop on Coding and Cryptography (2013). https://www.rocq.inria.fr/secret/PUBLICATIONS/codeq3.pdf
13. Sendrier, N., Simos, D.E.: The hardness of code equivalence over $\mathbb{F}_q$ and its application to code-based cryptography. In: Gaborit, P. (ed.) PQCrypto 2013. LNCS, vol. 7932, pp. 203–216. Springer, Heidelberg (2013). https://doi.org/10.1007/978-3-642-38616-9_14

Generalized Secret Sharing Schemes Using N$^\mu$MDS Codes

Sanyam Mehta[1] and Vishal Saraswat[2($\boxtimes$)] (iD)

[1] Goldman Sachs Services Pvt Ltd, Bangalore, India
`sanyam.mehta12@gmail.com`
[2] Robert Bosch Engineering & Business Solutions Pvt Ltd, Bangalore, India
`vishal.saraswat@gmail.com`

Abstract. Mehta et al. [11] recently proposed an NMDS code-based secret sharing scheme having a richer access structure than the traditional (t, n) threshold secret sharing schemes, and is based on two mutually nonmonotonic sets of user groups of sizes t and $t - 1$ respectively, where $n \geq t > 1$ corresponds to the total number of users. We give a full generalization of their scheme with complete security proofs. We propose an efficient generalized secret sharing scheme constructed using N$^\mu$MDS codes with time complexity of $O(n^3)$. The scheme accepts an access structure constructed using $\mu + 1$ mutually nonmonotonic sets of user groups with sizes, $t, t - 1, \ldots, t - \mu$, respectively, where $1 \leq \mu < t$, and the parameter t defines the threshold such that all user groups of size greater than t can recover the secret. The proposed secret sharing scheme is perfect and ideal and has robust cheating detection and cheater identification features.

Keywords: Secret sharing schemes · Generalized access structure · Near MDS codes · Almost MDS codes

1 Introduction

Secret sharing schemes allow a dealer, D, to split a secret s into n shares $s_1, \ldots, s_n$ and distribute these shares to a set $\mathcal{P}$ of n users, $P_1, \ldots, P_n$, according to an *access structure* $\Gamma \subset 2^{\mathcal{P}}$ such that a subset $\mathcal{A} \subseteq \mathcal{P}$ of users can form the secret using their shares if and only if $\mathcal{A} \in \Gamma$. Moreover the secret sharing scheme is called a (t, n) *threshold secret sharing scheme* if the access structure Γ is defined by

$$\mathcal{A} \in \Gamma \iff |\mathcal{A}| \geq t,$$

for some $t \in \{1, 2, \ldots, n\}$. Otherwise it is called a *generalized secret sharing scheme*.

Blakley [2] and Shamir [13] independently proposed secret sharing schemes in 1979. Shamir's scheme utilises the standard Lagrange interpolation and linear algebra whereas Blakley's scheme uses the concept of intersection of hyperplanes in finite geometries. Both of these schemes were threshold secret sharing schemes,

D. Slamanig et al. (Eds.): MACIS 2019, LNCS 11989, pp. 234–248, 2020.
https://doi.org/10.1007/978-3-030-43120-4_18

that is, they restricted users in such a way that if and only if the number of users exceeds the threshold, they could recover the secret. Ito et al. [8] introduced the notion of a secret sharing scheme with a generalized access structure. A generalized access structure consists of arbitrary subsets of users (irrespective of each subset's size), who could find the secret. They proposed a scheme in which the dealer assigned several copies of a (t, n)-threshold secret sharing scheme to every user. The dealer chooses two positive integers m and t and a prime power q satisfying $t \leq m < q$ and

- chooses $\alpha_{t-1} \in \mathrm{GF}(q) - \{0\}$ and $\alpha_1, \ldots, \alpha_{t-2}$ from $\mathrm{GF}(q)$ and computes $f(x) = s + \alpha_1 x + \alpha_2 x^2 + \cdots + \alpha_{t-1} x^{t-1}$, where $\mathrm{GF}(q)$ is the Galois Field of order q and $f(0) = s \in \mathrm{GF}(q)$ is the secret;
- chooses $x_1, \ldots, x_m \in \mathrm{GF}(q) - \{0\}$ and computes $s_j = f(x_j)$ $(1 \leq j \leq m)$;
- and finally, assigns a subset $S_i \subset \{(x_1, s_1), \ldots, (x_m, s_m)\}$ to the user P_i, $1 \leq i \leq n$.

The access structure of this scheme contains all those sets for which the size of the union of the users' shares $\geq t$. In the worst case, the share size is exponential in the size of the set of users. Benaloh and Leichter [1] proposed a secret sharing scheme with a generalized access structure which was simpler than that of the Ito et al.'s scheme [8]. Their construction utilizes the monotonicity property inherent in secret sharing schemes. They create a composition of multiple schemes with simple access structures and realize all access structures which can be defined using a small monotone formula. Although this scheme is simpler and more efficient than Ito et al.'s scheme [8], the share length is still exponential in the number of users.

Considering the secret sharing scheme proposed by Shamir once again, note that although a cheating user can not recover the secret by providing an incorrect share, but by getting a wrong key, he can misguide the honest users. Various ways of detecting and correcting the secret have been suggested by scholars. Some consider that there are only t shareholders for secret recovery and to check that the shares are not fake, the dealer gives an additional information such as using some check vectors to which will act like some kind of certificate for each user. Others have suggested to use error correcting codes where fake shares can be assumed to be errors and corrected like error correction of codes. Most of the initial schemes had concerns over cheater detection and identification and use of trusted third parties (combiners and dealers). Lein et al. [6] proposed a modification of Shamir's scheme [13] which allowed for cheater detection and identification. If $m > t$ users come together, where t is the threshold, then there are $\binom{m}{t}$ ways for the users to pool their shares and for each such way, a $t - 1$ degree recovery polynomial can be constructed through interpolation. The original polynomial can be then compared with the interpolated polynomial. Users who could not recover the original polynomial and are in the majority of groups are marked as possible cheaters and then the shares are corrected recursively until no cheater is left. This cheater detection and identification algorithm trades off space and time-complexities for secret recovery.

Researchers also observed that instead of using arbitrary matrices, using linear codes provided the following advantages

- A single generator matrix is sufficient to represent them.
- They enable easy transmission and easier error detection.
- Even though features for cheater detection, identification, and verification were added, schemes were still efficient.

McEliece and Sarwate [10] constructed a secret sharing scheme from Reed-Solomon codes and showed it to be essentially the same as the Shamir threshold scheme [13]. Later, Massey [9] gave a general construction of linear secret sharing schemes from linear codes (or linear matroids). Blakley and Kabatiansky [3] and Dijk [4] gave a generalization of Massey's scheme to multidimensional subspaces instead of vectors. Pieprzyk and Zhang [12] used Maximum Distance Separable (MDS) codes to construct a secret sharing scheme in which, an Maximum Distance Separable matrix G of dimension $(t \times n)$ along with a message vector $\mathbf{v}$ of dimension $1 \times t$ is chosen by the dealer. The dealer then finds the desired codeword by computing $\mathbf{v} \times G$. The secret is the first element of the codeword.

It was shown in [9] that the access structure of the resulting secret sharing schemes is determined by the minimal codewords in the dual code. However, determining the minimal codewords in a linear code and hence, the access structure, is hard. Dodunekov [5] proposed using NMDS codes instead of MDS codes to construct a secret sharing scheme while observing the following advantages:

- They are less space consuming and easier to implement.
- Their access structure is richer than MDS secret sharing.
- The generator matrix of the code is hard to identify by an adversary.
- Shares the same properties of cheating detection and cheater identification with MDS codes based schemes.

Mehta et al. [11] proposed an NMDS code-based secret sharing scheme having a richer access structure than the traditional (t, n) threshold secret sharing schemes and an access structure constructed using two mutually nonmonotonic sets of user groups having sizes, t and $t - 1$ respectively, where n corresponds to the total number of users.

1.1 Our Contribution

We have proposed an efficient generalized secret sharing scheme based on $\mathrm{N}^\mu\mathrm{MDS}$ codes. The use of the $\mathrm{N}^\mu\mathrm{MDS}$ matrices allows us to have *authorized* sets of varying sizes thus allowing the scheme to have a generalized and richer access structure. The proposed secret sharing scheme is perfect and ideal and has robust cheating detection and cheater identification features. The time complexity for the share distribution and share recovery phases is just $O(n^3)$, where n is the order of users. The proposed scheme has a finer access structure and provides a direction towards a fully generalized secret sharing scheme. The scheme constructs the access structure using $\mu + 1$ mutually nonmonotonic sets of user groups of sizes, $t, t-1, \ldots, t-\mu$, respectively, where $1 \leq \mu < t$, and the parameter t defines the threshold such that all user groups of size greater than t can recover the secret.

2   Preliminaries

We denote the Galois Field, $GF(q)$, of order q where $q = p^m$ is a prime power by $\mathbb{F}_q$. For $a_i \in \mathbb{F}_q$, $1 \leq i \leq n$, $(a_1, \ldots, a_n)$ denotes a vector in $\mathbb{F}_q^n$. We will also use the same notation, $(a_1, \ldots, a_n)$, to denote to denote a $n \times 1$ matrix (column) over $\mathbb{F}_q$. On the other hand, $[\, a_1 \;\; a_2 \;\; \cdots \;\; a_{n-1} \;\; a_n \,]$ denotes a $1 \times n$ matrix (row) over $\mathbb{F}_q$. For vectors $\mathbf{v}_i = (v_{i1}, \ldots, v_{it}) \in \mathbb{F}_q^t$, $1 \leq i \leq n$, $[\, \mathbf{v}_1 \;\; \mathbf{v}_2 \;\; \cdots \;\; \mathbf{v}_{n-1} \;\; \mathbf{v}_n \,]$ denotes the $t \times n$ matrix over $\mathbb{F}_q$ formed by considering $\mathbf{v}_i$ as columns. For a $t \times n$ matrix G over $\mathbb{F}_q$, the ith column of G is denoted $G[i] \in \mathbb{F}_q^t$, $0 \leq i \leq n$.

2.1   Coding Theory

Definition 1. *A non-empty subset* $\mathbf{C}$ *of* $\mathcal{A}^n$, *where* $\mathcal{A} = \{a_0, \ldots a_{q-1}\}$, *is called a q-ary block code of length n over $\mathcal{A}$, and a string in $\mathbf{C}$ is called a* codeword.

Definition 2. *The number of positions in which x and y differ is known as* Hamming distance $d(x,y)$ *between x and y. The* minimum distance *of a code $\mathbf{C}$ is defined as*

$$d(\mathbf{C}) = \min_{x \neq y \in \mathbf{C}} d(x,y)\,.$$

Definition 3. *A linear code, $\mathbf{L}$, of length n is a linear subspace of $\mathbb{F}_q^n$. If dimension of $\mathbf{L}$ is t then we call it an $[n,t]$-code (over $\mathbb{F}_q$). Further, if the minimum distance of $\mathbf{L}$ is d then we call it an $[n,t,d]$-code (over $\mathbb{F}_q$).*

Definition 4. *The set of non-zero coordinate positions of a codeword $c \in \mathbf{C}$ is called its* support, $\mathrm{Supp}(c)$. *The support of a code $\mathbf{C}$, $\mathrm{Supp}(\mathbf{C})$, is defined as*

$$\mathrm{Supp}(\mathbf{C}) = \cup_{c \in \mathbf{C}} \mathrm{Supp}(c)\,.$$

Definition 5. *The rth generalized Hamming distance, $d_r(\mathbf{C})$, is the cardinality of the minimum support of an $[n,r]$-subcode of $[n,t]$-code $\mathbf{C}$, where, $1 \leq r \leq t$.*

$$d_r(\mathbf{C}) = \min\{|\mathrm{Supp}\,\mathbf{D}| : \mathbf{D} \text{ is } [n,r]_q \text{ subcode of } \mathbf{C}\}\,.$$

Remark 1. The Hamming Distance of $\mathbf{C}$ $d(\mathbf{C}) = d_1(\mathbf{C})$.

Definition 6. *For an $[n,t,d]$-code $\mathbf{C}$, the* Singleton bound *states that the parameters of $\mathbf{C}$ must satisfy*

$$q^t \leq q^{n-d+1}\,.$$

In other words, $d \leq n - t + 1$.

Definition 7. *The rth generalized Singleton bound $d_r(\mathbf{C})$ states that*

$$d_r(\mathbf{C}) \le n - t + r \ where \ r = 1, 2, \ldots, t.$$

Definition 8. *A maximum distance separable (MDS) code is an $[n,t]$-linear code which achieves the Singleton bound, that is, it is an $[n, t, n-t+1]$-code.*

Proposition 1. *For an $[n, t, d]$ MDS code $\mathbf{L}$ over $\mathbb{F}_q$, let H be any of its parity check matrix of $\mathbf{L}$ and let $G = (I_t \mid A)$ be any of its generator matrix in standard form (ref. Remark 2). Then*

1. *Any $n - t$ columns of H are linearly independent.*
2. *Any t columns of G are linearly independent.*
3. *Any square submatrix of A is non singular.*

Definition 9. *The class of $[n, t]$-codes with*

$$d_1(\mathbf{C}) = n - t$$

are called almost-MDS (AMDS) *codes.*

Definition 10. *The class of $[n, t]$-codes with*

$$d_1(\mathbf{C}) = n - t,$$
$$and \ d_i(\mathbf{C}) = n - t + i, \qquad\qquad for \ i = 2, 3, \ldots, t,$$

are called near-MDS (NMDS) *codes.*

Definition 11. *The class of $[n, t]$-codes with*

$$d_i(\mathbf{C}) = n - t + 2i - \mu - 1, \qquad for \ i = 1, 2, \ldots, \mu$$
$$and \ d_i(\mathbf{C}) = n - t + i, \qquad for \ i = \mu + 1, \ldots, t,$$

are called $N^\mu MDS$ *codes.*

Remark 2. For the purposes of this work, we will assume that the generator matrices G are in their standard form, that is, $G = (I_t \mid A)$, where I_t is the identity matrix of size $t \times t$. Moreover, the MDS (or the N^μMDS) matrices correspond to the matrix A.

A detailed characterization of N^μMDS codes was provided in [14]. The relevant properties of N^μMDS matrices required for this paper are as follows.

Proposition 2 (Properties of N^μMDS Codes). *The matrix characterization of an $N^\mu MDS$ code with a generator matrix G is as follows:*

1. *For all $i = 1, 2, \ldots, \mu$,*
   (i) for $i < l \le min\{d_i - 1, t\}$, every $(l - 2i + 2 + \mu, l)$ submatrix has rank $\ge (l - i + 1)$.

(ii) there exists an l, $i < l \le min\{d_i, t\}$, and an $(l - 2i + 1 + \mu, l)$ submatrix with rank equal to $(l - i)$.

2. For all $i = \mu + 1, \ldots, t$,
   (i) for $1 < l \le min\{(n - t), (t - \mu)\}$, every $(l, l + \mu)$ submatrix has rank l.

Corollary 1 (Properties of N$^\mu$MDS Matrices.)   *The standard generator matrix for an $[n, t]$ N$^\mu$MDS code has the following properties:*

1. *Any $t - \mu + 2i$ columns of the generator matrix have rank $\ge t - \mu + i$, where $i = 0, 1, \ldots, \mu - 1$.*
2. *There exists a set of $t - \mu + 2i + 1$ columns with rank $t - \mu + i$, for $i = 0, 1, \ldots, \mu - 1$.*
3. *Any $t + \mu$ columns of the generator matrix have rank t and are linearly independent.*

2.2   Secret Sharing

Let $\mathcal{P} = P_1, \ldots, P_n$ be a set of n users. We call a subset $\mathcal{A}$ of $\mathcal{P}$ a *group* of users.

Definition 12. *A collection $\Gamma \subseteq 2^{\mathcal{P}}$ is called* monotone *if $\mathcal{A} \in \Gamma$ and $\mathcal{A} \subseteq \mathcal{B}$ then $\mathcal{B} \in \Gamma$.*

Definition 13. *We call two collections (sets) $\mathcal{G}^i, \mathcal{G}^j \subseteq 2^{\mathcal{P}}$ mutually nonmonotonic sets if for all $\mathcal{A} \in \mathcal{G}^i$, there is no $\mathcal{B} \in \mathcal{G}^j$, such that $\mathcal{B} \subset \mathcal{A}$ and vice versa.*

Definition 14. *$\Gamma \subseteq 2^{\mathcal{P}}$ is called an* access structure *if it is a monotone collection such that only the subsets of users in Γ are authorized to recover the secret. Subsets not in Γ are termed to be* unauthorized sets.

Definition 15. *A* distribution scheme *is denoted by Π with $\mathcal{S}$, the domain of secrets, and $\mathcal{R}$, a set of strings. For a secret $t \in \mathcal{S}$ and a string $r \in \mathcal{R}$ sampled randomly observing Δ, where Δ is the probability distribution on $\mathcal{R}$, a share vector $\Pi(t, r) = (s_1, s_2, \ldots, s_j)$ is computed and each share s_j is communicated to P_j via a secure channel.*

Definition 16. *A distribution scheme along with domain of secrets $\mathcal{S}$ realizing access structure Γ is called a* secret sharing scheme *$\Sigma = \langle \Pi, \Delta \rangle$.*

Definition 17. *A secret sharing scheme is* correct *if an authorized subset of users can always recover the secret. In other words, for any set $\mathcal{A} \in \Gamma$, there exists a recovery function or algorithm SRA such that for a key $k \in \mathcal{S}$,*

$$\Pr[SRA(\mathcal{A}) \text{ is } k] = 1.$$

Definition 18. *If $\mathcal{T}$ is the set of all possible shares and $\mathcal{S}$ is the set of all possible secrets, then the* information rate *ρ of the secret sharing scheme is defined to be*

$$\rho = \frac{\log(|\mathcal{S}|)}{\log(|\mathcal{T}|)}.$$

Definition 19. *A secret sharing scheme is* ideal *if the set of all secrets, $\mathcal{S}$, and the set of all shares, $\mathcal{T}$, are of same cardinality. That is, a secret sharing scheme is* ideal *if its information rate is one.*

Definition 20. *A secret sharing scheme is* perfect *if an unauthorized group of users, $\mathcal{C}$, cannot obtain any information about the secret from their pool of shares. That is, the probability of $\mathcal{C}$ recovering the secret using their pool of shares is equivalent to the probability of recovering the secret without using their pool of shares. In other words, for any subset $\mathcal{B} \notin \Gamma$, two secrets b and $c \in \mathcal{S}$ and every possible share vector $\langle s_j \rangle_{P_j \in \mathcal{B}}$,*

$$\Pr[\Pi(b,r)_{\mathcal{B}} = \langle s_j \rangle_{P_j \in \mathcal{B}}] = \Pr[\Pi(c,r)_{\mathcal{B}} = \langle s_j \rangle_{P_j \in \mathcal{B}}]$$

Definition 21. *A secret sharing scheme Σ is said to be* linear *over $\mathbb{F}_q$ if there exists a vector $\mathbf{v} = (v_0, v_1, \ldots, v_{t-1}) \in \mathbb{F}_q^t$ and a matrix $A \in \mathbb{F}_q^{t \times n}$, such that $\mathbf{v} \times A = (s_0, s_1, \ldots, s_{n-1})$ where s_0 is the secret and $(s_1, \ldots, s_{n-1})$ is the share vector.*

Definition 22. *During the secret recovery phase of a secret sharing scheme by an authorized subset of users $\mathcal{A}_c$, if a user P_i provides a wrong share, $\hat{s}_i$, instead of the correct one, s_i, it was assigned by the dealer during the share distribution phase, then the subset may fail to recover the secret, or worse, recover a wrong secret. Such a user is called a* cheater *and detection of occurrence of such an attack is called* cheating detection.

Definition 23. *Identification, with negligible error probability ϵ, of the user(s) providing wrong inputs while recovering the secret is called* cheater identification.

3   Proposed Secret Sharing Scheme

Though the scheme proposed in [11] has a richer access structure than the traditional (t, n) threshold secret sharing schemes, it only allows an access structure consisting of two mutually nonmonotonic sets of user groups of sizes, t and $t-1$, respectively. We propose a secret sharing scheme which admits a finer access structure based on $\mu + 1$, $1 \leq \mu \leq n - t$, mutually nonmonotonic sets of user groups of sizes, $t - \mu + 1 + i$, $1 \leq i \leq \mu + 1$, respectively. The proposed scheme is based on the properties of $N^\mu MDS$ matrices which allow us to have an access structure which is richer and independent of the field size.

3.1   Access Structure

The access structure of the proposed secret sharing scheme is definded using the properties of $N^\mu MDS$ matrices [14] and is a generalization of the one proposed in [11]. Let

$$G = [\ G[0]\ \ G[1]\ \ \ldots\ \ G[t-1]\ \ G[t]\ \ \ldots\ \ G[n]\ ]$$

be a standard generator matrix of an $[n+1, t, n-t-\mu+2]$ N$^\mu$MDS code over $\mathbb{F}_q$ where $G[i] \in \mathbb{F}_q^t$, $0 \le i \le n$.

Given a set $\mathcal{P}$ of n users, $P_1, \ldots, P_n$, we say that the column $G[i]$ corresponds to the user P_i and we define an *access structure* $\Gamma_\mu \subset 2^{\mathcal{P}}$ consisting of $\mu + 1$ mutually nonmonotonic sets, namely, $\mathcal{G}^0, \mathcal{G}^1, \ldots, \mathcal{G}^\mu$ defined as as follows:

1. $\mathcal{G}^i$, $i < \mu$, consists of all $(t - \mu + i)$ users whose corresponding columns in G, along with the first column, form $t - \mu + i + 1$ linearly dependent columns, and for all $\mathcal{A} \in \mathcal{G}^i$, there is no $\mathcal{B} \in \mathcal{G}^j$, $j < i$, such that $\mathcal{B} \subset \mathcal{A}$.
2. $\mathcal{G}^\mu$ consists of all (t) users whose corresponding columns in G are linearly independent, and for all $\mathcal{A} \in \mathcal{G}^\mu$, there is no $\mathcal{B} \in \mathcal{G}^j$, $j < \mu$, such that $\mathcal{B} \subset \mathcal{A}$.

Note that the access structure Γ_μ as defined above is a generalized access structure and satisfies the monotonicity property. Thus, the secret sharing scheme based on Γ_μ is a generalized secret sharing scheme.

3.2   Share Construction

To compute the n shares of a given secret $s_0 \in \mathbb{F}_q$, the dealer chooses $t-1$ random elements $\alpha_1, \ldots, \alpha_{t-1}$ from $\mathbb{F}_q$ and computes the codeword $(s_0, s_1, \ldots, s_n)$ by multiplying the generator matrix G by the t-length vector $(s_0, \alpha_1, \ldots, \alpha_{t-1})$. That is,

$$(s_0, s_1, \ldots, s_n) = (s_0, \alpha_1, \ldots, \alpha_{t-1}) \cdot G.$$

The elements $s_i \in \mathbb{F}_q$, $1 \le i \le n$, are the shares of the users $P_1, \ldots, P_n$ respectively. We say that the first column of G, $G[0]$, corresponds to the secret s_0 and the remaining columns $G[i]$, $1 \le i \le n$, correspond to the shares s_i of the users P_i.

3.3   Secret Recovery

The secret recovery algorithm SRA$_\mu$ is similar to the method proposed in [11] with modifications in the algorithm to allow for recovery of secret by user subsets of various sizes. Given a set of m users $\mathcal{B} = \{P_{j_1}, \ldots, P_{j_m}\} \in \Gamma_\mu$ and their respective shares $\{s_{j_1}, \ldots, s_{j_m}\}$, SRA$_\mu$ computes the secret as follows:

1. Construct the matrix

$$G' = \begin{bmatrix} G[j_1] & \cdots & G[j_m] & G[0] \end{bmatrix}$$

   formed by the columns which correspond to the shares of the users and the column which corresponds to the secret.
2. Row-reduce the matrix G' to make its first m (or t, whichever is minimum) rows and columns an identity matrix and denote the last column of this row-reduced matrix G' by $G[0]'$.
3. If $m < t$, add $t - m$ zeros to construct the pooled codeword

$$\mathsf{pool} = (s_{t_0}, s_{t_1}, \ldots, s_{t_{m-1}}, 0, \ldots, 0)$$

   and multiply pool to $G[0]'$ to obtain the secret.

4. Else multiply its sub-codeword $(s_{t_0}, s_{t_1}, \ldots, s_{t_{t-1}})$ to $G[0]'$ to obtain the secret.

Here, t_i's correspond to the t (or m) columns forming an identity matrix.

4   Analysis of the Proposed Scheme

Lemma 1. *For any $(t - \mu + 2i + 1)$ linearly dependent columns of an $[n, t, n - t - \mu + 1]$ N^{μ}MDS matrix, G, with rank $(t - \mu + i)$ where $0 \le i \le \mu - 1$, each of the remaining $n - (t - \mu + 2i + 1)$ columns is linearly independent of them.*

Proof. Without loss of generality, suppose the given $(t - \mu + 2i + 1)$ linearly dependent columns with rank $(t - \mu + 1)$ are $G[0], G[1], \ldots, G[t - \mu + 2i]$ and let $0 \le j \le (t - \mu + 2i)$ be such that

$$G[j] = \sum_{i=0, i \ne j}^{t-\mu+2i} a_i G[i], \text{ not all } a_i = 0.$$

Now, let $G[\ell]$ be a column from the remaining $n - (t - \mu + 2i + 1)$ columns of the matrix which is linearly dependent on the given $(t - \mu + 2i + 1)$ columns. That is,

$$G[\ell] = \sum_{i=0}^{t-\mu+2i} b_i G[i], \text{ not all } b_i = 0.$$

Substituting the value of $G[j]$, we get

$$G[\ell] = \sum_{i=0, i \ne j}^{t-\mu+2i} (a_i b_j + b_i) G[i],$$

where $0 \le j \le t - \mu + 2i$ and not all $a_i = 0$ and not all $b_i = 0$. Hence $G[\ell]$ is a linear combination of the remaining $(t - \mu + 2i)$ columns $G[i]$ $(0 \le i \le t - \mu + 2i, i \ne j)$.

Since both the columns $G[j]$ and $G[\ell]$ are a linear combination of remaining the $(t - \mu + 2i)$ columns, it makes the rank of these $(t - \mu + 2i + 2)$ columns less than or equal to $(t - \mu + i)$. But, from Property 1 of N^{μ}MDS codes, any $(t - \mu + 2i + 2)$ columns have rank $\ge (t - \mu + i + 1)$. Thus, our hypothesis is wrong and $G[\ell]$ must be linearly independent of the given $(t - \mu + 2i + 1)$ columns.

Proposition 3. *There exists a group of $(t - \mu + 2i + 1)$ users, $0 \le i \le \mu - 1$ which is unauthorized.*

Proof. By Lemma 1, for any $(t - \mu + 2i + 1)$ linearly dependent columns

$$\{G[j_1], G[j_2], \ldots, G[j_{t-\mu+2i+1}]\}$$

with rank $(t - \mu + i)$, the column $G[0]$ is linearly independent of them. Thus the secret s_0 cannot be recovered using just the shares

$$\{s_{j_1}, s_{j_2}, \ldots, s_{j_{t-\mu+2i+1}}\}.$$

Hence the users

$$\{P_{j_1}, \ldots, P_{j_{t-\mu+2i+1}}\}$$

form an unauthorized set.

Proposition 4. *There exists a group of $(t - \mu + 2i)$ users, $0 \le i \le \mu - 1$ which is unauthorized.*

Proof. If we take all columns except $G[j_\ell]$, $(0 \le \ell \le (t - \mu + 2i + 1))$, from the previous construction, we will get $(t - \mu + 2i)$ linearly dependent columns

$$\{G[j_1], \ldots, G[j_{\ell-1}], G[j_{\ell+1}], \ldots, G[j_{(t-\mu+2i+1)}]\}$$

with rank $(t - \mu + i)$, with the secret's column $G[0]$ being linearly independent from these $(t - \mu + 2i)$ columns. Thus, the $(t - \mu + 2i)$ users

$$\{P_{j_1}, \ldots, P_{j_{\ell-1}}, P_{j_{\ell+1}}, \ldots, P_{j_{(t-\mu+2i+1)}}\}$$

form an unauthorized set.

Theorem 1. *The proposed secret sharing scheme Σ_μ is correct.*

Proof. Let $\mathcal{B} \in \Gamma_\mu$. Then $\mathcal{B}$ is an authorized set and we show that $\mathcal{B}$ can correctly recover the secret. Let $s_{j_1}, \ldots, s_{j_m}$ be the shares of the users in $\mathcal{B}$, and s_0 be the secret.

Case 1: $\mathcal{B}$ *is from* $\mathcal{G}^i, i < \mu$: Note that, the column $G[0]$ which corresponds to the secret s_0 is linearly dependent on the columns which correspond to the users in $\mathcal{B}$. Therefore, the algorithm SRA_μ can find the coefficients a_i's (by row-reducing the matrix formed by these columns and the column $G[0]$) such that

$$s_0 = a_1 s_{j_1} + a_2 s_{j_2} + \ldots a_{t-\mu+i} s_{j_{t-\mu+i}}$$

and find the secret s_0.

Case 2: $\mathcal{B}$ *is from* $\mathcal{G}^\mu$: Since columns which correspond to the users in $\mathcal{B}$ are t linearly independent columns of G, any other column of G, including the column $G[0]$, must be linearly dependent on them. Thus, the algorithm SRA_μ can find the coefficients a_i's (by row-reducing the matrix formed by these columns and the column $G[0]$) such that

$$s_0 = a_1 s_{j_1} + a_2 s_{j_2} + \ldots a_t s_{j_t}$$

and find the secret s_0.

Case 3: $\mathcal{B}$ **is a superset of a group in** $\mathcal{G}^i$ **or** $\mathcal{G}^\mu$: If $\mathcal{B}$ is a superset of a group in $\mathcal{G}^i$, the users in $\mathcal{B}$ have at least $t - \mu + i$ linearly independent columns in G with the column $G[0]$ being linearly dependent on them by definition of $\mathcal{G}^i$. Therefore the algorithm SRA_μ, as in Case 1, can find the secret s_0. Otherwise, if $\mathcal{B}$ is a superset of a group in $\mathcal{G}^\mu$, then we already have t linearly independent columns in G which correspond to the group in $\mathcal{G}^\mu$ and the algorithm SRA_μ, as in Case 2, can find the secret s_0.

Hence, if $\mathcal{B}$ is an authorized set, then $\Pr[\mathsf{SRA}_\mu(\mathcal{B}) = s_0] = 1$ and hence the secret sharing scheme Σ_μ is correct.

Theorem 2. *The proposed secret sharing scheme Σ_μ has perfect privacy.*

Proof. Let $\mathcal{B}$ be an unauthorized set of m users which try to recover the secret. Since the secret $s_0 \xleftarrow{\$} \mathbb{F}_q$, the probability of randomly guessing the secret is $1/q$. Also, since $\mathrm{N}^\mu\mathrm{MDS}$ matrices have a high diffusion property, whenever a vector $\mathbf{v} \in \mathbb{F}_q^t$ is multiplied to its submatrix formed by its m columns, the output generated is uniformly distributed in $\mathbb{F}_q^m$. Hence, for any share $s_i, 1 \le i \le n$, the probability of randomly guessing s_i is $1/q$.

Case 1: $m \le t - \mu - 1$: Note that, by Property 1 of $\mathrm{N}^\mu\mathrm{MDS}$ matrices, the $m + 1 \le t - \mu$ columns in G which correspond to these m users along with the column $G[0]$ are linearly independent. Therefore the column $G[0]$ cannot be obtained as a linear combination of m columns which correspond to these users, that is, $\mathsf{SRA}_\mu(\mathcal{B}) \ne s_0$. Thus $\mathcal{B}$ will require at least one more correct share to compute the secret. But the probability of $\mathcal{B}$ guessing the correct secret (or another correct share) is $1/q$. Thus the probability of $\mathcal{B}$ obtaining the secret is less than or equal to $1/q$.

Case 2: $m = t - \mu + i, 0 \le i < \mu$: Since $\mathcal{B}$ is unauthorized, it neither belongs in $\mathcal{G}^i$ nor is a superset of a group in $\mathcal{G}^j, j < i$. This implies that the column $G[0]$ is linearly independent of the columns which correspond to the users in $\mathcal{B}$. Therefore the column $G[0]$ cannot be obtained as a linear combination of m columns which correspond to these users, that is, $\mathsf{SRA}_\mu(\mathcal{B}) \ne s_0$. Thus $\mathcal{B}$ will require at least one more correct share, or replace one of the pooled shares with a forged share, to compute the secret. But the probability of $\mathcal{B}$ guessing the correct secret (or another correct share) is $1/q$. Thus the probability of $\mathcal{B}$ obtaining the secret is less than or equal to $1/q$.

Case 3: $m = t + i, 0 \le i < \mu$: Since $\mathcal{B}$ is unauthorized, it neither belongs in $\mathcal{G}^\mu$ nor is a superset of a group in $\mathcal{G}^j, j \le \mu$. This implies that the columns which correspond to $\mathcal{B}$ are linearly dependent and the column $G[0]$ is independent of them (rendering any subset of $\mathcal{B}$ not a part of $\mathcal{G}^j$). Therefore the column $G[0]$ cannot be obtained as a linear combination of m columns which correspond to these users, that is, $\mathsf{SRA}_\mu(\mathcal{B}) \ne s_0$. Thus $\mathcal{B}$ will require at least one more correct share, or replace one of the pooled shares with a forged share, to compute the secret. But the probability of $\mathcal{B}$ guessing the correct secret (or another correct share) is $1/q$. Thus the probability of $\mathcal{B}$ obtaining the secret is less than or equal to $1/q$.

Note that, on an input of a random set of shares to SRA_μ, the probability of SRA_μ generating the correct secret s_0 is $1/q$. Therefore,

$$\Pr[\mathsf{SRA}_\mu(\mathcal{B}) = s_0] = \Pr[\mathsf{SRA}_\mu(\mathcal{B}) = \overline{s_0}]$$

and hence Σ_μ has perfect privacy.

Theorem 3. *The proposed secret sharing scheme Σ_μ is ideal.*

Proof. Since both the secret and the shares are elements of $\mathbb{F}_q$, the information rate ρ is

$$\rho = \frac{\log \mid \mathbb{F}_q \mid}{\log \mid \mathbb{F}_q \mid} = 1$$

and hence Σ_μ is ideal.

Theorem 4. *The proposed secret sharing scheme Σ_μ is a linear secret sharing scheme.*

Proof. By Definition 21 of a linear secret sharing scheme, and by the construction of the shares as in Subsect. 3.2, it is clear that the proposed secret sharing scheme is linear.

Proposition 5. *The time-complexity for the share construction and the secret recovery phase of the proposed scheme is $\mathcal{O}(n^3)$.*

Proof. That the complexity of the setup phase is $\mathcal{O}(n^3)$ is straight forward. We show that the complexity of the secret reconstruction phase is $\mathcal{O}(n^3)$.

The Step 2 of Algorithm SRA_μ computes the reduced row echelon form of the matrix G' constructed in Step 1. Since $m \leq n$, G' is at most a $(t \times n)$ matrix. Since row reduction of a $(t \times n)$ matrix can be done in $\mathcal{O}(t^2 n)$ operations and since $t \leq n$, the complexity of this step is $\mathcal{O}(n^3)$. That is the most complex step of the code because the remaining steps are linear in the size of the matrix. Hence, the complexity of the reconstruction phase is $\mathcal{O}(n^3)$.

4.1 Cheating Detection and Cheating Identification

The proofs in this section Σ_μ have been adapted from [11]. The following two lemmas, Lemmas 2 and 3, state standard properties of linear codes which we will use in this section. We refer the reader to [7] for the proof of Lemma 3.

Lemma 2. *Given an $[n, t, n - t - \mu + 1]$ N$^\mu$MDS code and its generator matrix G, if*

$$(s_0, s_1, \ldots, s_{n-1}) = (\alpha_0, \alpha_1, \ldots, \alpha_{t-1}) \cdot G$$

and

$$(\hat{s}_0, \hat{s}_1, \ldots, \hat{s}_{n-1}) = (\hat{\alpha}_0, \hat{\alpha}_1, \ldots, \hat{\alpha}_{t-1}) \cdot G$$

such that

$$(\alpha_0, \alpha_1, \ldots, \alpha_{t-1}) \neq (\hat{\alpha}_0, \hat{\alpha}_1, \ldots, \hat{\alpha}_{t-1}),$$

then

$$d((s_0, s_1, \ldots, s_{n-1}), (\hat{s}_0, \hat{s}_1, \ldots, \hat{s}_{n-1})) \geq n - t - \mu + 1.$$

Proof. Since $(\alpha_0, \alpha_1, \ldots, \alpha_{t-1})$ and $(\hat{\alpha}_0, \hat{\alpha}_1, \ldots, \hat{\alpha}_{t-1})$ are distinct, they generate different codewords of the $N^\mu MDS$ code. Hence, they generate different codewords $(s_0, s_1, \ldots, s_{n-1})$ and $(\hat{s}_0, \hat{s}_1, \ldots, \hat{s}_{n-1})$ are distinct. Thus, the Hamming distance between them must be greater than or equal to $n - t - \mu + 1$, the minimum distance of the code.

Lemma 3. *Let* $\mathbf{C}$ *be an* $[n, t, d]$ *linear code over* $GF(q)$. *Let* $\mathbf{C}^i$ *be the* punctured *code defined by dropping the* i^{th} *coordinate,* $1 \le i \le n$, *from the codewords of* $\mathbf{C}$. *Then,* $\mathbf{C}^i$ *is an* $[n - 1, \tilde{t}, \tilde{d}]$ *code where*

- $\tilde{t} = t$ *and* $\tilde{d} = d$ *if* $\mathbf{C}$ *does not have any codeword of weight* d *with a nonzero* i^{th} *coordinate;*
- $\tilde{t} = t$ *and* $\tilde{d} = d - 1$ *if* $d > 1$ *and* $\mathbf{C}$ *has a codeword of weight* d *with a nonzero* i^{th} *coordinate;*
- $\tilde{t} = t - 1$ *and* $\tilde{d} \ge d$ *if* $d = 1$, $t > 1$ *and* $\mathbf{C}$ *has a codeword of weight* d *with a nonzero* i^{th} *coordinate.*

Theorem 5. *The proposed scheme allows cheating detection if the number of cheaters in a group* m *users is less than* $m - t - 1$.

Proof. Suppose $P_{j_1}, \ldots, P_{j_m}$ submit the shares $\hat{s}_{j_1} = s_{j_1} + \delta_1, \ldots, \hat{s}_{j_m} = s_{j_m} + \delta_m$, $\delta_j \in GF(q)$, to the reconstruction algorithm. Then if $\delta_i = 0$, P_{j_i} is honest, and if $\delta_i \ne 0$, P_{j_i} is a cheater. Let G' be the $t \times m$ submatrix formed by the m columns of G indexed by $j_1, j_2, \ldots j_m$. Let

$$H_0 = \{(s_1, \ldots, s_m) \mid (s_1, \ldots, s_m) = (\alpha_0, \alpha_1, \ldots, \alpha_{t-1}) \cdot G', \ \alpha_i \in GF(q)\}.$$

Let $\mathbf{s} = (s_{j_1}, \ldots, s_{j_m})$, $\delta = (\delta_1, \ldots, \delta_m)$ and $\hat{\mathbf{s}} = \mathbf{s} + \delta = (\hat{s}_{j_1}, \ldots, \hat{s}_{j_m})$.

By Lemma 3, any two distinct codewords in H_0 have a Hamming distance of at least $m - t - 1$. Now, if the Hamming weight of δ is less than $m - t - 1$, then the Hamming distance between $\hat{\mathbf{s}}$ and $\mathbf{s}$ is less than $m - t - 1$. Thus by Lemma 2, $\hat{\mathbf{s}} \in H_0$ if and only if $\hat{\mathbf{s}} = \mathbf{s}$, that is, when $\delta = 0$. Hence, if the number of cheating users is less than $m - t - 1$, cheating by them can be detected.

Theorem 6. *The proposed scheme allows cheater identification if the number of cheaters in a group* m *users is less than* $\lfloor \frac{m-t-1}{2} \rfloor$.

Proof. Let P_{j_i}, $1 \le i \le m$, G', H_0, $\mathbf{s}$, δ and $\hat{\mathbf{s}}$ be as in Theorem 5. Let the Hamming weight of δ is less than $\lfloor \frac{m-t-1}{2} \rfloor$. Then the Hamming distance $d(\hat{\mathbf{s}}, \mathbf{s})$ is less than $\lfloor \frac{m-t-1}{2} \rfloor$. For any $\tilde{\mathbf{s}} \ne \mathbf{s} \in H_0$, by Lemma 3, $d(\mathbf{s}, \tilde{\mathbf{s}}) \ge m - t - 1$. Hence using the triangle inequality, we get

$$d(\hat{\mathbf{s}}, \tilde{\mathbf{s}}) \ge d(\mathbf{s}, \tilde{\mathbf{s}}) - d(\hat{\mathbf{s}}, \mathbf{s})$$

$$\ge (m - t - 1) - \left\lfloor \frac{m-t-1}{2} \right\rfloor = \left\lceil \frac{m-t-1}{2} \right\rceil \ge \left\lfloor \frac{m-t-1}{2} \right\rfloor = d(\hat{\mathbf{s}}, \mathbf{s}).$$

Hence, $d(\hat{\mathbf{s}}, \mathbf{s}) = \min\{d(\hat{\mathbf{s}}, \tilde{\mathbf{s}}) \mid \tilde{\mathbf{s}} \in H_0\}$. Thus standard error decoding techniques for linear codes can be used to decode $\hat{\mathbf{s}}$ to recover the secret $\mathbf{s}$. Then by computing $\delta = \hat{\mathbf{s}} - \mathbf{s}$, the user P_{j_i} is determined to be a cheater if $\delta_i \ne 0$.

Hence, if the number of cheating users is less than $\lfloor \frac{m-t-1}{2} \rfloor$, the secret can be reconstructed correctly and all the cheating users can be identified.

5   Conclusion and Future Work

We have proposed an efficient ideal and perfect generalized secret sharing scheme based on N^μMDS codes with desirable security features of cheating detection and cheater identification. The use of the N^μMDS matrices allows us to have authorized sets of varying sizes thus allowing the scheme to have a generalized and richer access structure. The proposed scheme allows an access structure consisting of $\mu + 1$ mutually nonmonotonic sets of user groups of sizes, $t, t - 1, \ldots, t - \mu$, respectively, where $1 \leq \mu < t$, where n is the number of users and the parameter t for the access structure is independent of the field size. The proposed scheme admits a finer access structure and provides a direction towards a fully generalized secret sharing scheme. We believe a fully generalized secret sharing scheme realizing arbitrary access structures should be possible with *almost* MDS codes. We are studying the properties of these codes and working on generating an almost MDS code for any given access structure.

Acknowledgments. The authors acknowledge the support of the Department of Mathematics, BITS Goa, Indian Institute of Technology, Jammu, and R. C. Bose Centre for Cryptology and Security, ISI Kolkata.

References

1. Benaloh, J., Leichter, J.: Generalized secret sharing and monotone functions. In: Goldwasser, S. (ed.) CRYPTO 1988. LNCS, vol. 403, pp. 27–35. Springer, New York (1990). https://doi.org/10.1007/0-387-34799-2_3
2. Blakley, G.: Safeguarding cryptographic keys. In: AFIPS, pp. 313–317. AFIPS Press (1979)
3. Blakley, G., Kabatiansky, G.: Generalized ideal secret-sharing schemes and matroids. Probl. Peredachi Informatsii **33**(3), 102–110 (1997)
4. Dijk, M.: A linear construction of perfect secret sharing schemes. In: De Santis, A. (ed.) EUROCRYPT 1994. LNCS, vol. 950, pp. 23–34. Springer, Heidelberg (1995). https://doi.org/10.1007/BFb0053421
5. Dodunekov, S.: Applications of near MDS codes in cryptography. In: Enhancing Cryptographic Primitives with Techniques from Error Correcting Codes, NATO Science for Peace and Security Series - D: Information and Communication Security, vol. 23, pp. 81–86. IOS Press (2009)
6. Harn, L., Lin, C.: Detection and identification of cheaters in (t, n) secret sharing scheme. Des. Codes Cryptograph. **52**(1), 15–24 (2009)
7. Huffman, W.C., Pless, V.: Fundamentals of Error-Correcting Codes. Cambridge University Press, Cambridge (2010)
8. Ito, M., Saito, A., Nishizeki, T.: Secret sharing scheme realizing general access structure. Electron. Commun. Jpn. (Part III: Fundam. Electron. Sci.) **72**(9), 56–64 (1989)
9. Massey, J.: Minimal codewords and secret sharing. In: Sixth Joint Swedish-Russian Workshop on Information Theory, Molle, Sweden, pp. 276–279 (1993)
10. McEliece, R., Sarwate, D.: On sharing secrets and Reed-Solomon codes. Commun. ACM **24**(9), 583–584 (1981)

11. Mehta, S., Saraswat, V., Sen, S.: Secret sharing using near-MDS codes. In: Carlet, C., Guilley, S., Nitaj, A., Souidi, E.M. (eds.) C2SI 2019. LNCS, vol. 11445, pp. 195–214. Springer, Cham (2019). https://doi.org/10.1007/978-3-030-16458-4_12
12. Pieprzyk, J., Zhang, X.-M.: Ideal threshold schemes from MDS codes. In: Lee, P.J., Lim, C.H. (eds.) ICISC 2002. LNCS, vol. 2587, pp. 253–263. Springer, Heidelberg (2003). https://doi.org/10.1007/3-540-36552-4_18
13. Shamir, A.: How to share a secret. Commun. ACM **22**(11), 612–613 (1979)
14. Viswanath, G., Rajan, B.S.: Matrix characterization of generalized Hamming weights. In: IEEE International Symposium on Information Theory, p. 61. IEEE (2001)

Exploiting Linearity of Modular Multiplication

Hamdi Murat Yıldırım[(✉)]

Department of Computer Technology and Information Systems,
Bilkent University, 06800 Ankara, Turkey
hmurat@bilkent.edu.tr

Abstract. The XOR $\oplus$ and the addition $\boxplus$ operations have been widely used as building blocks for many cryptographic primitives. These operations and the multiplication $\odot$ operation are successively used in the design of IDEA and the MESH block ciphers. This work presents several interesting algebraic properties of the multiplication operation. By fixing one operand, we obtain vector valued function g_Z on $\mathbb{Z}_2^n$, associated with $\odot$. In this paper we show that the nonlinearity of g_Z remains the same under some transformations of Z and moreover we give an upper bound for the nonlinearity of g_Z when Z is a power of 2. Under weak-key assumptions, we furthermore present a list of new linear relations for 1-round IDEA cipher, some of directly derived and others algorithmically generated using these relations and known ones. We extend the largest linear weak key class for IDEA cipher with size 2^{23} to derive such a class with sizes 2^{24}. Under the independent key subblocks (subkeys) and weak-key assumptions we derive many linear relations for IDEA cipher using linear relations for 1-round IDEA cipher.

Keywords: IDEA cipher · Nonlinearity · Modular multiplication · Boolean functions · Cryptanalysis

1 Introduction

Block ciphers can be used to build other cryptographic primitives such as stream ciphers, hash functions, message authentication codes and cryptographically secure pseudorandom number generators. Both block ciphers and stream ciphers provide confidentiality, which ensures that information is accessible only to those authorized for access, one of the goals of information security. The addition modulo 2^n ($\boxplus$) and exclusive-OR (XOR) ($\oplus$, bitwise addition on modulo 2) are operations and have been widely used as building blocks in many cryptosystems: in RC6, Twofish, MARS, FEAL, SAFER as block ciphers and in ChaCha, Phelix, Snow as stream ciphers. The design of both the International Data Encryption Algorithm (IDEA) [4], the MESH block ciphers [9], WIDEA [3] cipher and RIDEA cipher [12] are based on the successive use of these operations and the multiplication modulo $2^{16} + 1$ ($\odot$) operation. Extensive survey of such

© Springer Nature Switzerland AG 2020
D. Slamanig et al. (Eds.): MACIS 2019, LNCS 11989, pp. 249–269, 2020.
https://doi.org/10.1007/978-3-030-43120-4_19

block ciphers whose design following the Lai-Massey design paradigm and their analyses are provided by Nakahara [8]. IDEA was used in Pretty Good Privacy (PGP), which is a widely used computer program that provides confidentiality, authentication and data integrity. There are other applications of multiplication modulo $2^{16} + 1$ ($\odot$), which are encountered in residue number systems and Fermat number transform and studies about improving its efficiency [1,6] Some algebraic properties of the operations $\boxplus$, $\odot$ and $\oplus$ have already been exploited to cryptanalyze the first 2-round of IDEA in [5]. 15 linear relations for 1-round IDEA cipher, which are derived by considering the linearity of both XOR $\oplus$ and the addition $\boxplus$ operation and also linearity of the multiplication $\odot$ for values 0 and 1, are used to derive the linear weak key class for IDEA cipher with size 2^{23} [2]. In this respect, nonlinearity is one of the well-known criterion for evaluating cryptographic Boolean functions. Note that the nonlinearity of both addition and multiplication is considered as high because of their polynomial expressions according to Theorem 3 and 4 in [4]. This is one of the reasons they are used in IDEA cipher. On the other hand, we consider the widely known and accepted measurement for nonlinearity based on the Hamming distance presented in [10] to study the nonlinearity of the multiplication operation. It is proved that this type of nonlinearity of $\odot$ is zero for six cases for $n \geq 2$ [12].

1.1 Contribution

In this paper we view each operation of IDEA cipher as a vector valued boolean function from $\mathbb{Z}_2^n \times \mathbb{Z}_2^n$ to $\mathbb{Z}_2^n$. Note that the designer of IDEA cipher just considers the case $n = 16$. We fix one operand of each operation to have a vector valued function from $\mathbb{Z}_2^n$ to $\mathbb{Z}_2^n$ and we use the nonlinearity measurement in [10]. We give an upper bound for its nonlinearity when $Z = 2^k$, $2 \leq k \leq \lceil (n-1)/2 \rceil$. This means that the nonlinearity of the operation $\odot$ is low for small values of k. In fact, it is expected that the nonlinearity of such building blocks of block ciphers should be high. In Sect. 3 for the operation $\odot$, we construct a family of transformations that leaves nonlinearity invariant. In Sect. 4, in addition to 15 linear relations holding with probability one for 1-round IDEA cipher given in [2], we use all cases making nonlinearity of IDEA cipher's operations zero in order to derive such extra 39 linear relations. Moreover, we devise an algorithm to derive 201 more such linear relation considering these 54 relations. Section 5 presents one linear weak key class for IDEA cipher with size 2^{24}, which is extended from a largest linear weak key class for IDEA cipher with size 2^{23} presented in [2] and a method for 438 linear relations for IDEA cipher considering subkeys chosen independently and 255 linear relations for 1-round IDEA cipher.

2   Preliminaries

We shall use the following notations throughout the rest of the paper:

- $x \oplus y = x + y \pmod 2$ for $x, y \in \mathbb{Z}_2$;
- $\mathbb{Z}_2^n = \mathbb{Z}_2 \times \ldots \times \mathbb{Z}_2$ (n-times) denotes the n-dimensional vector space over $\mathbb{Z}_2$;

- When $\boldsymbol{A} = (a_n, a_{n-1}, \ldots, a_1)$ and $\boldsymbol{X} = (x_n, x_{n-1}, \ldots, x_1) \in \mathbb{Z}_2^n$,
  - $\boldsymbol{A} \oplus \boldsymbol{X} = (a_n \oplus x_n, a_{n-1} \oplus + x_{n-1}, \ldots, a_1 \oplus x_1)$.
  - the dot product $\boldsymbol{A} \cdot \boldsymbol{X} = (\sum_{i=1}^n a_i x_i) \pmod 2 = a_n x_n \oplus a_{n-1} x_{n-1} \oplus \ldots \oplus a_1 x_1$.
  - for $\lambda \in \mathbb{Z}_2$, $l_{A,\lambda} : \mathbb{Z}_2^n \to \mathbb{Z}_2$ be the function defined by $l_{A,\lambda}(\boldsymbol{X}) = \boldsymbol{A} \cdot \boldsymbol{X} \oplus \lambda$ is called an affine function (respectively linear) if $\lambda \neq 0$ (respectively $\lambda = 0$).

- $\mathcal{A} = \{l_{A,\lambda} \mid \boldsymbol{A} \in \mathbb{Z}_2^n, \lambda \in \mathbb{Z}_2\}$ denotes the set of all affine functions on $\mathbb{Z}_2^n$.
- $|S|$ denotes the cardinality of the set S.

It is easy to introduce the addition $\boxplus$, the multiplication $\odot$ and XOR $\oplus$ operations for any positive integer n as functions from $\mathbb{Z}_2^n \times \mathbb{Z}_2^n \to \mathbb{Z}_2^n = \mathbb{Z}_2 \times \ldots \times \mathbb{Z}_2$ (n-times) as follows:
Let $\mathbb{Z}_{2^n} = \{0, 1, \ldots, 2^n - 1\}$, $\mathbb{Z}_{2^n+1}^* = \{1, 2, \ldots, 2^n\}$, and let
$v : \mathbb{Z}_{2^n} \to \mathbb{Z}_2^n$ be a function defined by $v(X) = \boldsymbol{X}$,
where $\boldsymbol{X} = (x_n, \ldots, x_2, x_1)$ is a bit representation of $X = \sum_{i=1}^n x_i 2^{i-1} \in \mathbb{Z}_{2^n}$
and
$d : \mathbb{Z}_{2^n+1}^* \to \mathbb{Z}_{2^n}$ be a function defined by $d(X) = X$ if $X \neq 2^n$ and $d(2^n) = 0$.
With this convention, the addition $\pmod{2^n}$, $\boxplus$, the multiplication, $\odot$, $\pmod{2^n + 1}$ and the XOR $\oplus$ operations produce the three functions $\boldsymbol{f}$, $\boldsymbol{g}$ and $\boldsymbol{h} : \mathbb{Z}_2^n \times \mathbb{Z}_2^n \to \mathbb{Z}_2^n$:

The addition operation $\boxplus$; $\boldsymbol{f}(\boldsymbol{X}, \boldsymbol{Z}) = \boldsymbol{X} \boxplus \boldsymbol{Z} = v(X + Z \pmod{2^n}))$.
The multiplication operation $\odot$; $\boldsymbol{g}(\boldsymbol{X}, \boldsymbol{Z}) = \boldsymbol{X} \odot \boldsymbol{Z} = v(d(d^{-1}(X) d^{-1}(Z) \pmod{2^n + 1})))$, where d^{-1} is the inverse d.
The XOR operation $\oplus$; $\boldsymbol{h}(\boldsymbol{X}, \boldsymbol{Z}) = \boldsymbol{X} \oplus \boldsymbol{Z} = (x_n \oplus z_n, x_{n-1} \oplus z_{n-1}, \ldots, x_1 \oplus z_1)$.

Notation: for any $Z \in \mathbb{Z}_{2^n}$, $v(Z) = \boldsymbol{Z} \in \mathbb{Z}_2^n$, we denote by $\boldsymbol{f}_Z$, $\boldsymbol{g}_Z$ and $\boldsymbol{h}_Z$ the following vector valued functions $\mathbb{Z}_2^n \to \mathbb{Z}_2^n$: $\boldsymbol{f}_Z(\boldsymbol{X}) = f(\boldsymbol{X}, \boldsymbol{Z})$, $\boldsymbol{g}_Z(\boldsymbol{X}) = g(\boldsymbol{X}, \boldsymbol{Z})$ and $\boldsymbol{h}_Z(\boldsymbol{X}) = h(\boldsymbol{X}, \boldsymbol{Z})$.
Let $f : \mathbb{Z}_2^n \to \mathbb{Z}_2$ be any function and let $H(f)$ denote the Hamming distance from f to the set of all affine functions $\mathcal{A}$ on $\mathbb{Z}_2^n$. Namely,

$$H(f) = \min\{E_{A,\lambda}(f) \mid \boldsymbol{A} \in \mathbb{Z}_2^n, \lambda \in \mathbb{Z}_2\}$$

where $E_{A,\lambda}(f) = |\{\boldsymbol{X} \in \mathbb{Z}_2^n \mid f(\boldsymbol{X}) \neq l_{A,\lambda}(\boldsymbol{X}) = \boldsymbol{A} \cdot \boldsymbol{X} \oplus \lambda\}|$.
This non-negative integer $H(f)$ attached to $f : \mathbb{Z}_2^n \to \mathbb{Z}_2$ is called the nonlinearity of f.
It is clear that $H(f) = 0$ iff f is an affine function. The concept of nonlinearity of arbitrarily vector function $\boldsymbol{F} : \mathbb{Z}_2^n \to \mathbb{Z}_2^k$ was introduced in [10] as follows:
Let $\boldsymbol{F} = (f_k, \ldots, f_1)$, $f_i : \mathbb{Z}_2^n \to \mathbb{Z}_2$, where $1 \leq i \leq k$.

Definition 1.

$$N(\boldsymbol{F}) = \min_{C = (c_1, \ldots, c_k) \in \mathbb{Z}_2^k \setminus \{0\}} \{H(\boldsymbol{C} \cdot \boldsymbol{F} = c_k f_k \oplus c_{k-1} f_{k-1} \oplus \ldots \oplus c_1 f_1)\}$$

Definition 2. Let f be a function from $\mathbb{Z}_2^n$ to $\mathbb{Z}_2$. The truth table of f is an ordered 2^n-tuple $(f(0), f(1), \ldots, f(2^n - 1)) \in \mathbb{Z}_2^{2^n}$, which is denoted by T_f.

3 Nonlinearity of Multiplication Operation

It is a well-known fact that for every $Z \in \mathbb{Z}_{2^n}$, the nonlinearity $N(\boldsymbol{f}_Z)$ and $N(\boldsymbol{h}_Z)$ of $\boldsymbol{f}_Z$ and $\boldsymbol{h}_Z$ are equal to 0. However, the nonlinearity $N(\boldsymbol{g}_Z)$ of the vector function $\boldsymbol{g}_Z$ is not zero for every $Z \in \mathbb{Z}_{2^n}$. The following theorem, which is proved in [12], gives a list of Z values such that $N(\boldsymbol{g}_Z)$ is zero.

Theorem 1. *For $n \geq 2$, the nonlinearity $N(\boldsymbol{g}_Z)$ of the vector function $\boldsymbol{g}_Z(\boldsymbol{X}) = g(\boldsymbol{X}, \boldsymbol{Z})$ is zero for $Z = 0, 1, 2, 2^{n-1}, 2^{n-1} + 1, 2^n - 1$.*

Remark 1. When $n \leq 12$, we checked that the values of Z in *Theorem 1* were the only ones for which $N(\boldsymbol{g}_Z) = 0$. It is an open problem whether this is the case for $n > 12$.

Using the following proposition, it is enough to calculate $N(\boldsymbol{g}_Z)$ for given Z value to determine one, two or three related values for the vector function of the multiplication operation having the same nonlinearity.

Proposition 1

(1) For $n \in \mathbb{Z}_+$ such that $\gcd(A, 2^n + 1) = 1$, we have $N(\boldsymbol{g}_A) = N(\boldsymbol{g}_B)$ when $AB \equiv 1 \pmod{2^n + 1}$.
(2) $N(\boldsymbol{g}_A) = N(\boldsymbol{g}_B)$ when $A + B \equiv 0 \pmod{2^n + 1}$.
(3) $N(\boldsymbol{g}_{2^k}) = N(\boldsymbol{g}_{2^s})$ when $k + s = n$ for $k, s \geq 0$.

Proof. For part 1, we have $\boldsymbol{g}_B(\boldsymbol{X}) = \boldsymbol{g}_{A^{-1}}(\boldsymbol{X})$ since $AB \equiv 1 \pmod{2^n + 1}$. $N(\boldsymbol{g}_A) = N((\boldsymbol{g}_A)^{-1}) = N(\boldsymbol{g}_B)$ follows from *Theorem 1* in [10].

For part 2, the case $A = B = 0$ is trivial. For other (A, B) pairs, one can use the obvious relation $v^{-1}(\boldsymbol{g}_A(\boldsymbol{X})) + v^{-1}(\boldsymbol{g}_B(\boldsymbol{X})) \equiv 0 \pmod{2^n + 1}$ to complete the proof of this part.

For part 3, for $k + s = n$, we obtain that $2^s(2^k + 2(2^s)^{-1}) \equiv 2^n + 2 \equiv 1 \pmod{2^n + 1}$. Here $(2^s)^{-1} \equiv 2^k + 2(2^s)^{-1} \pmod{2^n + 1}$ and we have $(2^s)^{-1} + 2^k \equiv 0 \pmod{2^n + 1}$. By part 2, we get $N(\boldsymbol{g}_{(2^s)^{-1}}) = N(\boldsymbol{g}_{2^k})$. From *Theorem 1* in [10], we know that $N(\boldsymbol{g}_{(2^s)^{-1}}) = N(\boldsymbol{g}_{2^s})$. This completes the proof. $\square$

Since there is no efficient algorithm to compute $N(\boldsymbol{g}_Z)$ in general, we can look for an upper bound for some values of Z. The following theorem gives a partial solution to the problem:

Theorem 2. *For $n \geq 3$ and $2 \leq k \leq \lceil (n-1)/2 \rceil$, we have $N(\boldsymbol{g}_Z) \leq 2^{k-1}$ when*

$$\text{(i) } Z = 2^k \text{ and } Z = 2^{n-k}.$$

$$\text{(ii) } Z + 2^k \equiv 0 \pmod{2^n + 1}.$$

$$\text{(iii) } Z2^k \equiv 1 \pmod{2^n + 1}.$$

Proof. Assume that $n \geq 3$ and $2 \leq k \leq \lceil (n-1)/2 \rceil$. For every $\boldsymbol{X} \in \mathbb{Z}_2^n$, let $\boldsymbol{g}_{2^k}(\boldsymbol{X}) = (\boldsymbol{g}_{2^k}{}^{(n)}(\boldsymbol{X}), \ldots, \boldsymbol{g}_{2^k}{}^{(2)}(\boldsymbol{X}), \boldsymbol{g}_{2^k}{}^{(1)}(\boldsymbol{X}))$, and $\boldsymbol{g}_{2^k}{}^{(i)}(\boldsymbol{X})$ be i^{th} coordinate function of $\boldsymbol{g}_{2^k}(\boldsymbol{X})$.

Since $g_2(0) = 2^n - 1$, $g_2(2^{n-1}) = 0$ and $g_2(2j)$ is even and $g_2(2j+1)$ is odd for all $j \in \{1, \ldots, 2^{n-1} - 1\}$, the truth table of $g_2^{(1)}$, $T_{g_2^{(1)}} = S^{2^n}$, where $S^{2^n} = (s_{2^n}, \ldots, s_1) = (1, 0, \ldots, 0, 0, 1, \ldots, 1) \in \mathbb{Z}_2^{2^n}$, $s_{2^n} = 1$, $s_{2^{n-1}} = 0$, $s_{2^{n-1}+m} = 0$ and $s_{2^{n-1}-m} = 1$ for all $m \in \{1, \ldots, 2^{n-1} - 1\}$. Then the truth table of $T_{g_{2^k}^{(1)}}$ becomes $(\underbrace{S^{2^{n-k+1}}, \ldots, S^{2^{n-k+1}}}_{(2^k-1)-times})$. Therefore, $g_2^{(1)}(X) = \overline{x_1} \, \overline{x_2} \ldots \overline{x_{n-1}} \oplus x_n$ and

$g_{2^k}^{(1)}(X) = \overline{x_1} \, \overline{x_2} \ldots \overline{x_{n-k}} \oplus x_{n-k+1}$ according to their truth tables, where $\overline{x_i} = x_i \oplus 1$. We know that $g_{2^k}^{(1)}(X) \oplus g_{2^k}^{(2)}(X) = g_{2^{k-1}}^{(1)}(X)$ since by the proof of *Theorem 1*, $y_2 \oplus y_1 = x_1$ for $g_2(X) = Y$. The hamming distance between $g_{2^k}^{(1)}(X)$ and x_{n-k+1} is 2^k.

This implies that $N(g_{2^k}^{(1)}(X)) \leq 2^k$. By *Theorem 12* in [13], $2^k \leq N(g_{2^k}^{(1)}(X))$ since the term $x_1 \ldots x_{n-k}$ is not properly covered (see *Definition 9* in [13]) by any other terms in $g_{2^k}^{(1)}(X)$. Then, $N(g_{2^k}^{(1)}(X)) = 2^k$ and we get $N(g_{2^k}^{(1)}(X) \oplus g_{2^k}^{(2)}(X)) = N(g_{2^{k-1}}^{(1)}(X)) = 2^{k-1}$. Hence, $N(g_{2^k}(X)) \leq 2^{k-1}$ by using *Definition 1*.

The remaining parts of this theorem can be easily proven by *Proposition 1*. $\qquad\square$

Remark 2. When $n \leq 16$, we checked that the upper bound was tight, namely $N(g_Z) = 2^{k-1}$, for the choices of Z above. It is an open problem whether this is the case when $n > 16$.

4   Linear Relations for 1-Round IDEA

4.1   Linear Relations for Operations

For a fixed operation $\bowtie \in \{\boxplus, \odot, \oplus\}$ and $z \in \mathbb{Z}_{2^n}$, we consider mapping $\mathbb{Z}_2^n \to \mathbb{Z}_2^n$ defined by $X \to X \bowtie Z = Y$ ($Z = v(z)$).

We have discussed the nonlinearity of this vector valued multiplication function for some special cases. When $\bowtie$ is the XOR operation $\oplus$, it is clear that the dot product is distributive over $\oplus$, and therefore we get $A \cdot (X \oplus Z) = A \cdot X \oplus A \cdot Z = A \cdot Y$, or equivalently

$$A \cdot X \oplus A \cdot Y \oplus A \cdot Z = 0 \text{ for every } A \in \mathbb{Z}_2^n \qquad (1)$$

Similarly for $\bowtie = \boxplus$, it is easy to see that $1 \cdot (X \boxplus Z) = 1 \cdot X \oplus 1 \cdot Z = 1 \cdot Y$, or equivalently

$$1 \cdot X \oplus 1 \cdot Y \oplus 1 \cdot Z = 0 \qquad (2)$$

So for $X \bowtie Z = Y$ it makes sense to search relations in the form

$$A \cdot X \oplus B \cdot Y \oplus C \cdot Z \oplus \lambda = 0 \text{ for some } A, B, C \in \mathbb{Z}_2^n \text{ and } \lambda \in \mathbb{Z}_2. \qquad (3)$$

As it can be seen from the proof of Theorem 1 [12], we get the following linear relations for every $X = v(x) \in \mathbb{Z}_2^n$ such that $X \odot Z = Y$:

$$1 \cdot X \oplus 1 \cdot Y \oplus 1 \cdot Z \oplus 1 = 0 \text{ for } z \in \{0, 1\} \qquad (4)$$

254 H. M. Yıldırım

$$3 \cdot \mathbf{X} \oplus 1 \cdot \mathbf{Y} \oplus 1 \cdot \mathbf{Z} \oplus 1 = 0 \text{ for } z \in \{2^{n-1}, 2^{n-1} + 1\} \tag{5}$$

$$1 \cdot \mathbf{X} \oplus 3 \cdot \mathbf{Y} \oplus 1 \cdot \mathbf{Z} = 0 \text{ for } z \in \{2, 2^n - 1\}, \tag{6}$$

where $v(z) = \mathbf{Z}$.

4.2 A New List of Linear Relations

For 1-round IDEA, 15 linear relations hold with probability one are derived due to the linearity of operations of IDEA (see equations in 1, 2, 4) in paper [2]. These relations marked by (*) are given in Table 1. Note that for each round of IDEA, four of the six 16-bit key subblocks $\mathbf{Z}_i$'s ($i = \{1, 4, 5, 6\}$) are involved by the multiplication operation $\odot$. In order to derive each of these linear relation, at least one of those key subblocks were restricted to take 0 and 1 (see Example 1 and Table 1). Additional key values, $2, 2^n - 1, 2^{n-1}$ and $2^{n-1} + 1$, making the nonlinearity of the vector valued function $\mathbf{g}_z$ of $\odot$ zero were discovered in [12]. Similar to the work in paper [2], we take into account 0, 1 or these key values as round multiplicative keys to derive extra 39 linear relations, which are not marked by (*) in Table 1. All these 54 linear relations (holding with probability one) with the related key subblocks restrictions are listed in Table 1. Notice that each linear relation for 1-round IDEA should be based on linear relations for the operations used in IDEA cipher. Hence under some round key subblocks restrictions (weak key assumptions), we can express a linear relation for 1-round IDEA as:

$$\phi \star Z \oplus \psi \star X \oplus \omega \star Y \oplus \lambda = 0$$

where Z, X and Y are round key, input and output of 1-round IDEA, respectively and $\lambda \in \mathbb{Z}_2$, $\phi \star Z = \phi_1 \cdot \mathbf{Z}_1 \oplus \ldots \oplus \phi_6 \cdot \mathbf{Z}_6$, $\psi \star X = \psi_1 \cdot \mathbf{X}_1 \oplus \ldots \oplus \psi_4 \cdot \mathbf{X}_4$ and $\omega \star Y = \omega_1 \cdot \mathbf{Y}_1 \oplus \ldots \oplus \omega_4 \cdot \mathbf{Y}_4$ such that $\phi = (\phi_1, \ldots, \phi_6)$, $\psi = (\psi_1, \ldots, \psi_4)$ and $\omega = (\omega_1, \ldots, \omega_4)$ for ϕ_i, ψ_i and $\omega_i \in \mathbb{Z}_2^{16}$. Here ϕ_i, ψ_i and ω_i are masks for $\mathbf{Z}_i = v(z_i)$, $\mathbf{X}_i = v(x_i)$ and $\mathbf{Y}_i = v(y_i)$, respectively and $x_i, y_i, z_i \in \mathbb{Z}_{2^n}$.

For the sake of clarity, let us derive the 24th linear relation in Table 1, one of 15 linear relations found in [2]:

Example 1: Adding first two output of 1-round IDEA, namely $\mathbf{Y}_1$ and $\mathbf{Y}_2$ (see Fig. 2 in Appendix A), we have

$$\mathbf{Y}_1 \oplus \mathbf{Y}_2 = (\mathbf{X}_1 \oplus \mathbf{Z}_1) \oplus (\mathbf{X}_3 \boxplus \mathbf{Z}_3)$$

When $\mathbf{Z}_1 = (0, \ldots, 0)$ or $\mathbf{Z}_1 = (1, \ldots, 1)$, the least significant bit of $\mathbf{Y}_1 = \mathbf{X}_1 \odot \mathbf{Z}_1$ is $1 \cdot \mathbf{Y}_1 = 1 \cdot \mathbf{X}_1 \oplus 1 \cdot \mathbf{Z}_1 \oplus 1$ from the Eq. 4 and the least significant bit of $\mathbf{Y}_3 = \mathbf{X}_3 \boxplus \mathbf{Z}_3$ is $1 \cdot \mathbf{Y}_3 = 1 \cdot \mathbf{X}_3 \oplus 1 \cdot \mathbf{Z}_3$ from the Eq. 2. The addition of $1 \cdot \mathbf{Y}_1$ and $1 \cdot \mathbf{Y}_2$ becomes

$$1 \cdot \mathbf{Y}_1 \oplus 1 \cdot \mathbf{Y}_2 = 1 \cdot \mathbf{X}_1 \oplus 1 \cdot \mathbf{Z}_1 \oplus 1 \cdot \mathbf{X}_3 \oplus 1 \cdot \mathbf{Z}_3 \oplus 1 \tag{7}$$

When $\mathbf{Z}_1 = (0, \ldots, 0)$ or $(1, \ldots, 1)$, one can represent this equation as a linear relation for 1-round IDEA

$$(1, 0, 1, 0, 0, 0) \star Z \oplus (1, 0, 1, 0) \star X \oplus (1, 1, 0, 0) \star Y \oplus 1 = 0$$

Table 1. List of linear relations for 1-round IDEA given in [2] (indicated by *) and derived. Here k is a non-negative integer, $-1 \equiv 0 \bmod (2^{16}+1)$, $-2^{15} \equiv 2^{15}+1 \bmod (2^{16}+1)$ and $-2 \equiv 2^{16}-1 \bmod (2^{16}+1)$.

	ϕ	ψ	ω	λ	z_1	z_2	z_3	z_4	z_5	z_6	# of free bits
1	* $(0,0,0,1,0,1)$	$(0,0,0,1)$	$(0,0,1,0)$	0	–	–	–	∓ 1	–	∓ 1	66
2	$(0,0,0,1,0,1)$	$(0,0,0,3)$	$(0,0,1,0)$	0	–	–	–	$\mp 2^{15}$	–	∓ 1	66
3	* $(0,0,1,0,1,1)$	$(0,0,1,0)$	$(1,0,1,1)$	0	–	–	–	–	∓ 1	∓ 1	66
4	$(0,0,2,0,1,1)$	$(0,0,3,0)$	$(3,0,1,1)$	1	∓ 2	–	$2k$	–	$\mp 2^{15}$	∓ 2	48
5	$(0,0,2,1,1,1)$	$(0,2,3,1)$	$(3,0,3,3)$	1	∓ 2	$2k$	$2k$	∓ 2	$\mp 2^{15}$	∓ 2	31
6	* $(0,0,1,1,1,0)$	$(0,0,1,1)$	$(1,0,0,1)$	0	–	–	–	∓ 1	∓ 1	–	66
7	$(0,0,1,1,1,0)$	$(0,0,1,3)$	$(1,0,0,1)$	0	–	–	–	$\mp 2^{15}$	∓ 1	–	66
8	* $(1,0,0,0,0,1)$	$(0,1,0,0)$	$(0,0,0,1)$	1	–	–	–	–	–	∓ 1	82
9	* $(1,0,0,1,0,0)$	$(0,1,0,1)$	$(0,0,1,1)$	1	–	–	–	∓ 1	–	–	81
10	$(0,2,0,1,0,0)$	$(0,3,0,1)$	$(0,0,3,3)$	0	–	$2k$	–	∓ 2	–	–	79
11	$(0,1,0,1,0,0)$	$(0,1,0,3)$	$(0,0,3,3)$	1	–	–	–	$\mp 2^{15}$	–	–	81
12	* $(0,1,1,0,1,0)$	$(0,1,1,0)$	$(1,0,1,0)$	1	–	–	–	–	∓ 1	–	81
13	* $(0,1,1,1,1,1)$	$(0,1,1,1)$	$(1,0,0,0)$	1	–	–	–	∓ 1	∓ 1	∓ 1	51
14	$(0,1,1,1,1,1)$	$(0,1,1,3)$	$(1,0,0,0)$	1	–	–	–	$\mp 2^{15}$	∓ 1	∓ 1	51
15	$(0,1,2,1,1,1)$	$(0,1,3,1)$	$(3,0,0,0)$	0	–	∓ 2	$2k$	∓ 1	$\mp 2^{15}$	∓ 2	33
16	* $(1,0,0,0,0,1)$	$(1,0,0,0)$	$(0,1,1,1)$	1	∓ 1	–	–	–	∓ 1	∓ 1	51
17	$(1,0,0,0,0,1)$	$(1,0,0,0)$	$(0,3,1,1)$	1	∓ 2	–	$2k$	–	$\mp 2^{15}$	∓ 1	49
18	* $(1,0,0,1,1,0)$	$(1,0,0,1)$	$(0,1,0,1)$	1	∓ 1	–	–	∓ 1	∓ 1	–	51
19	$(1,0,0,1,1,0)$	$(1,0,0,3)$	$(0,1,0,1)$	1	∓ 1	–	–	$\mp 2^{15}$	∓ 1	–	51
20	$(1,0,0,1,1,0)$	$(3,0,0,1)$	$(0,1,0,1)$	1	$\mp 2^{15}$	–	–	∓ 1	∓ 1	–	51
21	$(1,0,0,1,1,0)$	$(3,0,0,3)$	$(0,1,0,1)$	1	$\mp 2^{15}$	–	–	$\mp 2^{15}$	∓ 1	–	51
22	$(1,0,2,1,1,0)$	$(1,0,2,1)$	$(0,1,0,1)$	0	∓ 2	–	$2k$	∓ 1	$\mp 2^{15}$	–	49
23	$(1,0,2,1,1,0)$	$(1,0,2,3)$	$(0,1,0,1)$	0	∓ 2	–	$2k$	$\mp 2^{15}$	$\mp 2^{15}$	–	49
24	* $(1,0,1,0,0,0)$	$(1,0,1,0)$	$(1,1,0,0)$	1	∓ 1	–	–	–	–	–	81
25	$(1,0,2,0,0,0)$	$(1,0,3,0)$	$(3,3,0,0)$	0	∓ 2	–	$2k$	–	–	–	79
26	$(1,0,1,0,0,0)$	$(3,0,1,0)$	$(1,1,0,0)$	1	$\mp 2^{15}$	–	–	–	–	–	81
27	* $(1,0,1,1,0,1)$	$(1,0,1,1)$	$(1,1,1,0)$	1	∓ 1	–	–	∓ 1	–	∓ 1	51
28	$(1,0,1,1,0,1)$	$(1,0,1,3)$	$(1,1,1,0)$	1	∓ 1	–	–	$\mp 2^{15}$	–	∓ 1	51
29	$(1,0,2,1,0,1)$	$(1,0,3,1)$	$(3,3,3,0)$	0	∓ 2	–	$2k$	∓ 1	–	∓ 1	49
30	$(1,0,2,1,0,1)$	$(1,0,3,3)$	$(3,3,3,0)$	0	∓ 2	–	$2k$	$\mp 2^{15}$	–	∓ 1	49
31	$(1,0,1,1,0,1)$	$(3,0,1,1)$	$(1,1,1,0)$	1	$\mp 2^{15}$	–	–	∓ 1	–	∓ 1	51
32	$(1,0,1,1,0,1)$	$(3,0,1,3)$	$(1,1,1,0)$	1	$\mp 2^{15}$	–	–	$\mp 2^{15}$	–	∓ 1	51
33	* $(1,1,0,0,1,0)$	$(1,1,0,0)$	$(0,1,1,0)$	0	∓ 1	–	–	–	∓ 1	–	66
34	$(1,1,0,0,1,0)$	$(3,1,0,0)$	$(0,1,1,0)$	0	$\mp 2^{15}$	–	–	–	∓ 1	–	66
35	$(1,1,2,0,1,0)$	$(1,1,2,0)$	$(0,1,1,0)$	1	∓ 2	–	$2k$	–	$\mp 2^{15}$	–	64
36	* $(1,1,0,1,1,1)$	$(1,1,0,1)$	$(0,1,0,0)$	0	∓ 1	–	–	∓ 1	∓ 1	∓ 1	36
37	$(1,1,2,1,1,1)$	$(1,1,2,1)$	$(0,1,0,0)$	1	∓ 2	–	$2k$	∓ 1	$\mp 2^{15}$	∓ 1	34
38	$(1,1,2,1,1,1)$	$(1,1,2,3)$	$(0,1,0,0)$	1	∓ 2	–	$2k$	$\mp 2^{15}$	$\mp 2^{15}$	∓ 1	34
39	$(1,1,0,1,1,1)$	$(3,1,0,1)$	$(0,1,0,0)$	0	$\mp 2^{15}$	–	–	∓ 1	∓ 1	∓ 1	36

(continued)

Table 1. (*continued*)

	ϕ	ψ	ω	λ	z_1	z_2	z_3	z_4	z_5	z_6	# of free bits
40	$(1,1,0,1,1,1)$	$(3,1,0,3)$	$(0,1,0,0)$	0	$\mp 2^{15}$	–	–	$\mp 2^{15}$	∓ 1	∓ 1	36
41	$(1,1,0,1,1,1)$	$(1,1,0,1)$	$(0,3,0,0)$	0	∓ 2	–	–	∓ 1	$\mp 2^{15}$	∓ 2	34
42	$(1,1,0,1,1,1)$	$(1,1,0,3)$	$(0,3,0,0)$	0	∓ 2	–	–	$\mp 2^{15}$	$\mp 2^{15}$	∓ 2	34
43	* $(1,1,1,0,0,1)$	$(1,1,1,0)$	$(1,1,0,1)$	0	∓ 1	–	–	–	–	∓ 1	66
44	$(1,1,1,0,0,1)$	$(3,1,1,0)$	$(1,1,0,1)$	0	$\mp 2^{15}$	–	–	–	–	∓ 1	66
45	$(1,1,2,0,0,1)$	$(1,1,3,0)$	$(3,3,0,1)$	1	∓ 2	–	$2k$	–	–	∓ 1	64
46	* $(1,1,1,1,0,0)$	$(1,1,1,1)$	$(1,1,1,1)$	0	∓ 1	–	–	∓ 1	–	–	66
47	$(1,1,1,1,0,0)$	$(1,1,1,3)$	$(1,1,1,1)$	0	∓ 1	–	–	$\mp 2^{15}$	–	–	66
48	$(1,1,1,1,0,0)$	$(3,1,1,1)$	$(1,1,1,1)$	0	$\mp 2^{15}$	–	–	∓ 1	–	–	66
49	$(1,1,1,1,0,0)$	$(3,1,1,3)$	$(1,1,1,1)$	0	$\mp 2^{15}$	–	–	$\mp 2^{15}$	–	–	66
50	$(1,1,2,1,0,0)$	$(1,1,3,1)$	$(3,3,1,1)$	1	∓ 2	–	$2k$	∓ 1	–	–	64
51	$(1,1,2,1,0,0)$	$(1,1,3,3)$	$(3,3,1,1)$	1	∓ 2	–	$2k$	$\mp 2^{15}$	–	–	64
52	$(1,2,1,1,0,0)$	$(1,3,1,1)$	$(1,1,3,3)$	1	∓ 1	$2k$	–	∓ 2	–	–	64
53	$(1,2,1,1,0,0)$	$(3,3,1,1)$	$(1,1,3,3)$	1	$\mp 2^{15}$	$2k$	–	∓ 2	–	–	64
54	$(1,2,2,1,0,0)$	$(1,3,3,1)$	$(3,3,3,3)$	1	∓ 2	$2k$	$2k$	∓ 2	–	–	62

Example 2: From the Table 1, when $\mathbf{Z}_j = v(z_j), z_1 = \mp 2, z_4 = \mp 2^{15}, z_5 = \mp 2^{15}$ and $z_6 = \mp 2$ for $\phi = (1,1,0,1,1,1)$, $\psi = (1,1,0,3)$, $\omega = (0,3,0,0)$ and $\lambda = 0$ we have

$$1 \cdot \mathbf{Z}_1 \oplus 1 \cdot \mathbf{Z}_2 \oplus 1 \cdot \mathbf{Z}_4 \oplus 1 \cdot \mathbf{Z}_5 \oplus 1 \cdot \mathbf{Z}_6 \oplus 1 \cdot \mathbf{X}_1 \oplus 1 \cdot \mathbf{X}_2 \oplus 3 \cdot \mathbf{X}_4 = 3 \cdot \mathbf{Y}_2$$

This relation, one of new 39 linear relations derived, is the 42th linear relation in Table 1.

4.3 New Linear Relations Algorithmically Generated

Let us consider the 35th and the 45th linear relations for 1-round IDEA in Table 1 to obtain a new relation which is not listed in Table 1.

For the 35th linear relation $(1,1,2,0) \rightarrow (0,1,1,0)$ with key subblocks restrictions $z_1 = \mp 2$, $z_3 = 2k$ and $z_5 = \mp 2^{15}$ and the 45th linear relation $(1,1,3,0) \rightarrow (3,3,0,1)$ with restrictions $z_1 = \mp 2$, $z_3 = 2k$ and $z_6 = \mp 1$, we have two corresponding Eqs. (8) and (9) respectively

$$1 \cdot \mathbf{Z}_1 \oplus 1 \cdot \mathbf{Z}_2 \oplus 2 \cdot \mathbf{Z}_3 \oplus 1 \cdot \mathbf{Z}_5 \oplus 1 \cdot \mathbf{X}_1 \oplus 1 \cdot \mathbf{X}_2 \oplus 2 \cdot \mathbf{X}_3 \oplus 1 \cdot \mathbf{Y}_2 \oplus 1 \cdot \mathbf{Y}_3 \oplus 1 = 0 \quad (8)$$

$$1 \cdot \mathbf{Z}_1 \oplus 1 \cdot \mathbf{Z}_2 \oplus 2 \cdot \mathbf{Z}_3 \oplus 1 \cdot \mathbf{Z}_6 \oplus 1 \cdot \mathbf{X}_1 \oplus 1 \cdot \mathbf{X}_2 \oplus 3 \cdot \mathbf{X}_3 \oplus 3 \cdot \mathbf{Y}_1 \oplus 3 \cdot \mathbf{Y}_2 \oplus 1 \cdot \mathbf{Y}_4 \oplus 1 = 0 \quad (9)$$

Equations (8) and (9) key subblocks restrictions do not give any conflicts and they can be combined (by adding them in mod 2) to obtain the following linear relation candidate:

$$1 \cdot \mathbf{Z}_5 \oplus 1 \cdot \mathbf{Z}_6 \oplus 1 \cdot \mathbf{X}_3 \oplus 3 \cdot \mathbf{Y}_1 \oplus 2 \cdot \mathbf{Y}_2 \oplus 1 \cdot \mathbf{Y}_3 \oplus 1 \cdot \mathbf{Y}_4 \oplus 1 = 0 \quad (10)$$

We have used many inputs for 1-round IDEA to check that linear relation in (10) holds with probability one under the key subblocks restrictions $z_1 = \mp 2$, $z_3 = 2k$, $z_5 = \mp 2^{15}$ and $z_6 = \mp 1$. In fact, we have observed that only key restrictions $z_5 = \mp 2^{15}$ and $z_6 = \mp 1$ are enough to make this linear relation hold with probability one according to our experiments. Hence we have devised a new algorithm to find new linear relations for 1-round IDEA based on a set of 54 linear relations for 1-round IDEA in Table 1. Considering these known linear relations, we found additional 201 new linear relations for 1-round IDEA (see Table 5, Appendix B) using the following algorithm:

Algorithm 1. *An algorithm for finding new linear relations for 1-round IDEA based on existing linear ones:*

Let S be the set of linear relations with their key subblocks restrictions.
Step 1 All pair of S whose key subblocks values coincided are chosen.
Step 2 Any chosen pairs are also combined (directly added in mod 2).
Step 3 Each linear relation candidates in Step 2 is tested using 10 million test vectors to check whether it is a linear relation or not.
Step 4 The ones (i.e. candidate linear relations) passing Step 3 added to S.
Step 5 Previous steps are repeated until there is no increase in the number of the elements of the set S.
Step 6 Key restrictions of each linear relation in S are checked to remove unnecessary restrictions using 50000 test vectors.

We note that the last step has been added as a result of comments provided by Nakahara [7]. All 54 linear relations in Table 1 can be derived by hand calculation considering all combinations of subblock outputs of 1-round IDEA, $\mathbf{Y}_i$ and subblock keys of 1-round IDEA, $\mathbf{Z}_i$ which give us linear relations for the operations used in IDEA cipher. By using Algorithm 1, it is possible to obtain linear relations that can not be derived in this way.

5   Linear Weak Key Classes for IDEA

As indicated in Table 2, three linear relations, namely the 24th, the 33th and the 12th relations in Table 1 were successively used to find a linear relation for 8,5-round IDEA holding with probability one [2]. Because of key subblocks restrictions done in each round, this linear relation is satisfied for all 64-bit plaintexts provided that ranges of zero key bits' indices of a 128-bit master key bits are between 0–25, 29–71, and 75–110. Such key is a member of a class of weak keys with size 2^{23} since each of the remaining 23 bits of the master key can take 0 or 1.

Note that this has been the largest known class of weak keys based on a linear relation for 8,5-round IDEA. Hence this linear relation can be regarded as the best linear relation for 8,5-round IDEA. Based on this linear relation, we have found a new class of weak keys with cardinality 2^{24}. For this construction, we replace the first round linear relation $(1, 0, 1, 0) \rightarrow (1, 1, 0, 0)$ with $(\{1, 3\}, 0, 1, 0) \rightarrow (1, 1, 0, 0)$ (see Table 3). For the former and latter relations, $\mathbf{Z}_1^{(1)}$ is chosen $\mathbf{0} = (0, \ldots, 0)$ or $\mathbf{1} = (1, \ldots, 1)$ and $\mathbf{Z}_1^{(1)}$ is restricted

Table 2. Each round linear relation and ranges for indices of zero key bits of IDEA master key are considered to derive the linear relation $(1, 0, 1, 0) \rightarrow (0, 1, 1, 0)$ for 8,5-round IDEA satisfied by a linear weak key class with cardinality 2^{23}.

Round i	Linear relation $\psi \rightarrow \omega$	$\mathbf{Z}_1^{(i)}$	$\mathbf{Z}_5^{(i)}$
1	$(1, 0, 1, 0) \rightarrow (1, 1, 0, 0)$	0–14	–
2	$(1, 1, 0, 0) \rightarrow (0, 1, 1, 0)$	96–110	57–71
3	$(0, 1, 1, 0) \rightarrow (1, 0, 1, 0)$	–	50–64
4	$(1, 0, 1, 0) \rightarrow (1, 1, 0, 0)$	82–96	–
5	$(1, 1, 0, 0) \rightarrow (0, 1, 1, 0)$	75–89	11–25
6	$(0, 1, 1, 0) \rightarrow (1, 0, 1, 0)$	–	4–18
7	$(1, 0, 1, 0) \rightarrow (1, 1, 0, 0)$	36–50	–
8	$(1, 1, 0, 0) \rightarrow (0, 1, 1, 0)$	29–44	93–107
8,5	$(0, 1, 1, 0) \rightarrow (0, 1, 1, 0)$	–	–

Table 3. Each round linear relation and ranges for indices of zero key bits of IDEA master key are considered to derive the linear relation $(\{1, 3\}, 0, 1, 0) \rightarrow (0, 1, 1, 0)$ for 8,5-round IDEA satisfied by a linear weak key class with cardinality 2^{24}.

Round i	Linear relation $\psi \rightarrow \omega$	$\mathbf{Z}_1^{(i)}$	$\mathbf{Z}_5^{(i)}$
1	$(\{1, 3\}, 0, 1, 0) \rightarrow (1, 1, 0, 0)$	1–15	–
2	$(1, 1, 0, 0) \rightarrow (0, 1, 1, 0)$	96–110	57–71
3	$(0, 1, 1, 0) \rightarrow (1, 0, 1, 0)$	–	50–64
4	$(1, 0, 1, 0) \rightarrow (1, 1, 0, 0)$	82–96	–
5	$(1, 1, 0, 0) \rightarrow (0, 1, 1, 0)$	75–89	11–25
6	$(0, 1, 1, 0) \rightarrow (1, 0, 1, 0)$	–	4–18
7	$(1, 0, 1, 0) \rightarrow (1, 1, 0, 0)$	36–50	–
8	$(1, 1, 0, 0) \rightarrow (0, 1, 1, 0)$	29–44	93–107
8,5	$(0, 1, 1, 0) \rightarrow (0, 1, 1, 0)$	–	–

to $\mathbf{0}$ or $\mathbf{2^{15}}$, respectively. Note that $(\{1, 3\}, 0, 1, 0) = (1, 0, 1, 0)$ (respectively $(\{1, 3\}, 0, 1, 0) = (3, 0, 1, 0)$) if $\mathbf{Z}_1^{(1)}$ is equal to $\mathbf{0}$ (respectively $\mathbf{Z}_1^{(1)} = \mathbf{2^{15}}$). Therefore, zero key bits' indices of a 128-bit key are between 1–25, 29–71, and 75–110. Then linear relation $(\{1, 3\}, 0, 1, 0) \rightarrow (0, 1, 1, 0)$ for the 8,5-round IDEA holds with probability one (Table 3) and there are 2^{24} such keys. We haven't discovered other linear relations in Tables 1 and 5 similar to the best linear relation giving a large class of weak keys because of the following reasons:

- If we compare Table 1 with Table 5 in Appendix B, then it can be seen that for most cases, linear relations in Table 1 derived in [2] have less key restrictions than others.
- In Table 1, each of linear relations numbered with 8, 9, 12, 24, 26 has one key subblock restriction and each of linear relations numbered with 1, 2, 3, 6, 7, 10,

$25, 34, 43, 44, 46, 47, 48, 49$ has two key subblocks restrictions. There aren't any linear relations with one key subblock restriction in Table 5, but there are linear relations numbered with $98, 125, 159, 185$ and 216 having two key subblocks restrictions in Table 5. In order to find a linear relation for 8,5-round IDEA providing a large class of weak keys, it is better to use those relations (with less key subblocks restrictions) listed above. However, it is not possible to derive such linear relation for 8,5-round IDEA using these relations and linear relations with key subblocks ∓ 2 or $\mp 2^{15}$ restrictions other than those derived in [2] in both Tables 1 and 5. Because

(i) we faced with key subblocks restrictions giving conflicts, that is, some bits of the master 128-bit of IDEA are both 0 and 1 due to key subblocks restrictions of two linear relations considered for two different rounds, especially when a key subblock of one linear relation is equal to 0 or 1 and a key subblock of other one is chosen as ∓ 2 or $\mp 2^{15}$;

(ii) we haven't found successive linear relations for many linear relations with key subblock restriction like ∓ 2 or $\mp 2^{15}$ while deriving multi round linear relation. For example, for the 75th linear relation in Table 5, namely $(\mathbf{3}, \mathbf{3}, \mathbf{0}, \mathbf{1}) \rightarrow (\mathbf{2}, \mathbf{3}, \mathbf{2}, \mathbf{2})$ there aren't any linear relations whose input mask is equal to $(\mathbf{2}, \mathbf{3}, \mathbf{2}, \mathbf{2})$ in both Tables 1 and 5.

Because these limitations to derive new linear relations the block cipher, we assume that key subblocks (subkeys) are independent. Then under weak-key assumptions we consider each linear relation for 1-round IDEA cipher from Tables 1 and 5 as two vertices connected by a single edge having a direction. In this manner we have a directed graph and using suitable functions of Digraph module from SageMath [11] we find many paths with length 8 and then consider last 0.5 round's relations in order to get 438 linear relations for 8.5-round IDEA cipher. In Table 6 (Appendix B), 50 of them with less number of key bits restriction for the master key, whose size is 832-bits (considering all 52 16-bit key subblocks) are listed. Note that second relation in this table $(\mathbf{1}, \mathbf{1}, \mathbf{0}, \mathbf{0}) \longrightarrow (\mathbf{3}, \mathbf{1}, \mathbf{0}, \mathbf{0})$ is a linear relation for 8.5-round IDEA cipher and associated with a class of weak keys with the cardinality 2^{586} whenever key subblocks (subkeys) are chosen independently. Note that the key space with size 2^{832} is extremely larger than this class.

6   Conclusion

In this paper we give several new properties on the nonlinearity of the multiplication operation $\odot$. Using its invariance properties, it is possible to calculate the nonlinearity just for one value of the associated vector function to learn one, two or three different values giving the same the nonlinearity. Furthermore, we give an upper bound for its nonlinearity when values are power of two. It is low for small powers. In fact, it is expected that the nonlinearity of such building blocks of block ciphers should be high. We devise an algorithm to find a new set of linear relations for 1-round IDEA using a set of linear relations directly derived and a set of known linear relations. We present one linear weak key class slightly bigger than one known in the literature. Assuming that all key subblocks are chosen independently, we generate a new set of linear relations for full IDEA cipher using linear relations for 1-round IDEA. All these findings extend the related work done by Daemen et al. and they are meaningful to understand how properties of building components of a cipher are related to its security.

A   Appendix: IDEA Block Cipher

The graph of the encryption of IDEA can be seen in Fig. 1. The key scheduling algorithm and the list of all 16-bit key subblocks (Table 4) are given in Appendix.

A.1   Key Schedule and Decryption Algorithm

For a given 128-bit key, 52 16-bit key subblocks are generated for the encryption. For the construction of these subblocks, the first step is to partition given 128-bit key into 8 pieces and assign them as the first 8 key subblocks of the 52 subblocks: $Z_1^{(1)}, Z_2^{(1)}, .., Z_6^{(1)}, Z_1^{(2)}, Z_2^{(2)}, .., Z_6^{(2)}, .., Z_1^{(8)}, Z_2^{(8)}, .., Z_6^{(8)}, Z_1^{(9)}, Z_2^{(9)}, Z_3^{(9)}, Z_4^{(9)}$.

Then the key under the consideration is cyclically shifted to the left by 25 positions. The resulting key block is again partitioned into eight subblocks that are assigned to the next eight subblock keys. This process is repeated until all 52 subblock keys are derived.

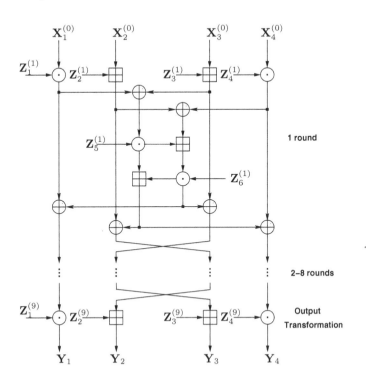

Fig. 1. Computational graph for the encryption process of the IDEA cipher

Table 4. 128-bit IDEA master key bits indices starts from 0 and ends with 127 (indexed left to right). Range of indices of this key used for each of 52 subblock keys generated by the key scheduling algorithm

r	$\mathbf{Z}_1$	$\mathbf{Z}_2$	$\mathbf{Z}_3$	$\mathbf{Z}_4$	$\mathbf{Z}_5$	$\mathbf{Z}_6$
1	0–15	16–31	32–47	48–63	64–79	80–95
2	96–111	112–127	25–40	41–56	57–72	73–88
3	89–104	105–120	121–8	9–24	50–65	66–81
4	82–97	98–113	114–1	2–17	18–33	34–49
5	75–90	91–106	107–122	123–10	11–26	27–42
6	43–58	59–74	100–115	116–3	4–19	20–35
7	36–51	52–67	68–83	84–99	125–12	13–28
8	29–44	45–60	61–76	77–92	93–108	109–124
9	22–37	38–53	54–69	70–85	–	–

A.2   The MA-Structure and 1-Round IDEA Cipher

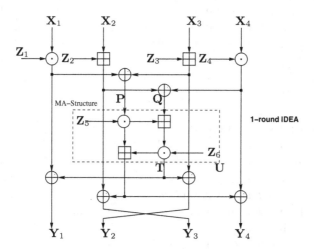

Fig. 2. Computational graph for the encryption process of 1-round IDEA cipher

Let us denote round key, input and output for the 1-round IDEA block cipher (see Fig. 2) as $Z = (\mathbf{Z}_1, \ldots, \mathbf{Z}_6)$, $X = (\mathbf{X}_1, \mathbf{X}_2, \mathbf{X}_3, \mathbf{X}_4)$ and $Y = (\mathbf{Y}_1, \mathbf{Y}_2, \mathbf{Y}_3, \mathbf{Y}_4)$, where $\mathbf{Z}_i$, $\mathbf{X}_i$, $\mathbf{Y}_i \in \mathbb{Z}_2^{16}$, respectively. Then we have:

$$\mathbf{Y}_1 = (\mathbf{X}_1 \odot \mathbf{Z}_1) \oplus \mathbf{T}. \qquad \mathbf{Y}_2 = (\mathbf{X}_3 \boxplus \mathbf{Z}_3) \oplus \mathbf{T}. \qquad (11)$$
$$\mathbf{Y}_3 = (\mathbf{X}_2 \boxplus \mathbf{Z}_2) \oplus \mathbf{U}. \qquad \mathbf{Y}_4 = (\mathbf{X}_4 \odot \mathbf{Z}_4) \oplus \mathbf{U}.$$

We have the following equations for two input subblocks of the MA-structure $\mathbf{P}$ and $\mathbf{Q}$ and two output subblocks of the MA-structure $\mathbf{U}$ and $\mathbf{T}$ (see Fig. 2):

$$\mathbf{P} = (\mathbf{X}_1 \odot \mathbf{Z}_1) \oplus (\mathbf{X}_3 \boxplus \mathbf{Z}_3) \text{ and } \mathbf{Q} = (\mathbf{X}_2 \boxplus \mathbf{Z}_2) \oplus (\mathbf{X}_4 \odot \mathbf{Z}_4). \tag{12}$$

$$\mathbf{U} = (\mathbf{P} \odot \mathbf{Z}_5) \boxplus \mathbf{T} \text{ and } \mathbf{T} = [(\mathbf{P} \odot \mathbf{Z}_5) \boxplus \mathbf{Q}] \odot \mathbf{Z}_6. \tag{13}$$

It is easy to see that $\mathbf{Y}_1 \oplus \mathbf{Y}_2 = \mathbf{P}$ and $\mathbf{Y}_3 \oplus \mathbf{Y}_4 = \mathbf{Q}$.

B Appendix: New Linear Relations for 1-Round IDEA and 8.5-Round IDEA

Table 5. List of new linear relations for 1-round IDEA, based on linear relations of Table 1, generated by Algorithm 1. Here k is a non-negative integer, $-1 \equiv 0 \bmod (2^{16} + 1)$, $-2^{15} \equiv 2^{15} + 1 \bmod (2^{16} + 1)$ and $-2 \equiv 2^{16} - 1 \bmod (2^{16} + 1)$.

	ϕ	ψ	ω	λ	z_1	z_2	z_3	z_4	z_5	z_6	# of free bits
55	$(1,2,2,1,0,0)$	$(1,2,2,1)$	$(3,3,3,3)$	0	∓2	$2k+1$	$2k+1$	∓2	–	–	62
56	$(0,1,0,1,1,1)$	$(0,1,1,1)$	$(3,2,0,0)$	1	–	–	$2k$	∓1	$\mp2^{15}$	∓1	50
57	$(1,1,2,1,1,1)$	$(1,1,3,1)$	$(0,3,0,0)$	1	∓1	–	$2k+1$	∓1	∓1	∓2	34
58	$(0,1,3,1,1,1)$	$(0,1,3,1)$	$(1,2,0,0)$	0	–	–	$2k$	∓1	∓1	∓2	49
59	$(1,1,1,1,1,1)$	$(3,1,0,3)$	$(2,3,0,0)$	0	$\mp2^{15}$	–	$2k$	$\mp2^{15}$	$\mp2^{15}$	∓1	35
60	$(1,3,0,1,0,1)$	$(1,3,1,1)$	$(3,1,3,2)$	0	∓2	$2k+1$	$2k+1$	∓2	–	∓2	46
61	$(0,0,0,0,1,1)$	$(0,0,1,0)$	$(3,2,1,1)$	1	–	–	$2k+1$	–	$\mp2^{15}$	∓1	65
62	$(1,1,1,1,1,1)$	$(1,1,0,1)$	$(2,1,0,0)$	1	∓2	–	$2k+1$	∓1	∓1	∓2	33
63	$(1,0,3,0,1,1)$	$(3,0,2,0)$	$(2,1,1,1)$	0	$\mp2^{15}$	–	$2k$	–	$\mp2^{15}$	∓2	49
64	$(0,1,2,1,1,1)$	$(0,1,2,1)$	$(3,0,0,0)$	1	–	–	$2k+1$	∓1	$\mp2^{15}$	∓2	49
65	$(1,2,3,1,1,1)$	$(3,2,3,1)$	$(2,1,2,2)$	0	$\mp2^{15}$	$2k+1$	$2k+1$	∓2	$\mp2^{15}$	∓2	32
66	$(1,2,2,1,0,0)$	$(1,3,2,1)$	$(3,3,3,3)$	1	∓2	$2k$	$2k+1$	∓2	–	–	62
67	$(1,2,3,1,1,1)$	$(1,2,2,1)$	$(2,3,2,2)$	1	∓2	$2k+1$	$2k$	∓2	∓1	∓1	32
68	$(1,2,3,1,1,1)$	$(1,3,2,1)$	$(2,3,2,2)$	0	∓2	$2k$	$2k$	∓2	∓1	∓1	32
69	$(0,0,2,1,0,1)$	$(0,0,3,1)$	$(0,2,1,0)$	1	–	–	$2k+1$	∓1	–	∓2	64
70	$(0,0,0,1,1,0)$	$(0,0,1,1)$	$(3,2,0,1)$	1	–	–	$2k+1$	∓1	$\mp2^{15}$	–	65
71	$(1,0,3,1,0,1)$	$(3,0,3,3)$	$(1,3,1,0)$	0	$\mp2^{15}$	–	$2k$	$\mp2^{15}$	–	∓2	49
72	$(1,0,3,1,1,0)$	$(1,0,2,3)$	$(2,3,0,1)$	0	∓2	–	$2k$	$\mp2^{15}$	∓1	–	49
73	$(1,1,3,1,1,1)$	$(1,1,3,1)$	$(2,1,0,0)$	0	∓1	–	$2k+1$	∓1	$\mp2^{15}$	∓2	34
74	$(1,2,1,1,1,1)$	$(3,2,0,1)$	$(2,3,2,2)$	1	$\mp2^{15}$	$2k+1$	$2k+1$	∓2	$\mp2^{15}$	∓1	33
75	$(1,2,1,1,1,1)$	$(3,3,0,1)$	$(2,3,2,2)$	0	$\mp2^{15}$	$2k$	$2k+1$	∓2	$\mp2^{15}$	∓1	33
76	$(1,2,3,1,1,1)$	$(3,3,3,1)$	$(2,1,2,2)$	1	$\mp2^{15}$	$2k$	$2k+1$	∓2	$\mp2^{15}$	∓2	32
77	$(1,0,1,1,1,0)$	$(3,0,0,3)$	$(2,3,0,1)$	0	$\mp2^{15}$	–	$2k+1$	$\mp2^{15}$	$\mp2^{15}$	–	50
78	$(0,1,2,1,1,1)$	$(0,1,3,3)$	$(3,0,0,0)$	0	–	–	$2k$	$\mp2^{15}$	$\mp2^{15}$	∓2	49

(*continued*)

Table 5. (*continued*)

	ϕ	ψ	ω	λ	z_1	z_2	z_3	z_4	z_5	z_6	# of free bits
79	$(1,1,2,1,1,1)$	$(1,1,2,3)$	$(0,3,0,0)$	1	∓1	$-$	$2k$	$\mp2^{15}$	∓1	∓2	34
80	$(1,1,1,1,1,1)$	$(1,1,0,1)$	$(2,3,0,0)$	1	∓1	$-$	$2k+1$	∓1	$\mp2^{15}$	∓1	35
81	$(1,3,1,1,1,0)$	$(1,3,0,1)$	$(2,3,2,3)$	1	∓1	$2k+1$	$2k$	∓2	$\mp2^{15}$	$-$	48
82	$(1,3,3,1,1,0)$	$(1,2,3,1)$	$(2,3,2,3)$	0	∓2	$2k$	$2k+1$	∓2	∓1	$-$	47
83	$(1,0,1,0,1,1)$	$(1,0,0,0)$	$(2,1,1,1)$	1	∓2	$-$	$2k$	$-$	∓1	∓2	48
84	$(1,1,3,0,1,0)$	$(1,1,2,0)$	$(2,3,1,0)$	1	∓2	$-$	$2k$	$-$	∓1	$-$	64
85	$(1,2,2,1,1,1)$	$(1,3,3,1)$	$(0,1,2,2)$	0	∓2	$2k$	$2k+1$	∓2	$\mp2^{15}$	∓1	32
86	$(1,0,0,1,0,1)$	$(1,0,1,1)$	$(3,1,1,0)$	0	∓2	$-$	$2k+1$	∓1	$-$	∓2	48
87	$(0,0,2,1,0,1)$	$(0,0,2,3)$	$(0,2,1,0)$	1	$-$	$-$	$2k$	$\mp2^{15}$	$-$	∓2	64
88	$(1,1,2,1,1,1)$	$(3,1,3,1)$	$(0,3,0,0)$	1	$\mp2^{15}$	$-$	$2k+1$	∓1	∓1	∓2	34
89	$(1,3,0,1,1,0)$	$(1,3,0,1)$	$(0,1,2,3)$	1	∓1	$2k+1$	$-$	∓2	∓1	$-$	49
90	$(1,3,3,1,0,1)$	$(1,2,2,1)$	$(1,3,3,2)$	1	∓1	$2k$	$2k+1$	∓2	$-$	∓2	47
91	$(1,1,1,0,1,0)$	$(3,1,0,0)$	$(2,3,1,0)$	1	$\mp2^{15}$	$-$	$2k+1$	$-$	$\mp2^{15}$	$-$	65
92	$(1,1,2,1,0,0)$	$(1,1,2,1)$	$(3,3,1,1)$	0	∓2	$-$	$2k+1$	∓1	$-$	$-$	64
93	$(1,2,3,1,1,1)$	$(1,2,2,1)$	$(2,1,2,2)$	1	∓1	$2k+1$	$2k$	∓2	$\mp2^{15}$	∓2	32
94	$(0,1,3,1,1,1)$	$(0,1,2,3)$	$(1,2,0,0)$	0	$-$	$-$	$2k+1$	$\mp2^{15}$	∓1	∓2	49
95	$(1,2,3,1,1,1)$	$(1,3,2,1)$	$(2,1,2,2)$	0	∓1	$2k$	$2k$	∓2	$\mp2^{15}$	∓2	32
96	$(1,1,3,1,1,1)$	$(1,1,3,3)$	$(2,3,0,0)$	0	∓2	$-$	$2k+1$	$\mp2^{15}$	∓1	∓1	34
97	$(1,0,1,0,1,1)$	$(1,0,0,0)$	$(2,3,1,1)$	1	∓1	$-$	$2k$	$-$	$\mp2^{15}$	∓1	50
98	$(0,2,0,1,0,0)$	$(0,2,0,1)$	$(0,0,3,3)$	1	$-$	$2k+1$	$-$	∓2	$-$	$-$	79
99	$(0,3,1,1,1,0)$	$(0,2,1,1)$	$(1,0,2,3)$	1	$-$	$2k$	$-$	∓2	∓1	$-$	64
100	$(1,1,3,1,1,1)$	$(3,1,3,1)$	$(2,1,0,0)$	0	$\mp2^{15}$	$-$	$2k+1$	∓1	$\mp2^{15}$	∓2	34
101	$(1,3,2,1,0,1)$	$(1,3,3,1)$	$(3,3,3,2)$	0	∓2	$2k+1$	$2k$	∓2	$-$	∓1	47
102	$(1,1,3,1,1,1)$	$(1,1,2,1)$	$(2,3,0,0)$	1	∓2	$-$	$2k$	∓1	∓1	∓1	34
103	$(1,3,1,1,0,1)$	$(1,2,1,1)$	$(1,1,3,2)$	0	∓1	$2k$	$-$	∓2	$-$	∓1	49
104	$(1,1,2,0,1,0)$	$(1,1,3,0)$	$(0,1,1,0)$	1	∓2	$-$	$2k+1$	$-$	$\mp2^{15}$	$-$	64
105	$(1,1,3,0,0,1)$	$(1,1,2,0)$	$(1,3,0,1)$	1	∓1	$-$	$2k+1$	$-$	$-$	∓2	64
106	$(1,0,1,1,1,0)$	$(1,0,0,1)$	$(2,3,0,1)$	1	∓1	$-$	$2k$	∓1	$\mp2^{15}$	$-$	50
107	$(1,1,2,1,1,1)$	$(3,1,2,3)$	$(0,3,0,0)$	1	$\mp2^{15}$	$-$	$2k$	$\mp2^{15}$	∓1	∓2	34
108	$(1,3,2,1,0,1)$	$(1,2,2,1)$	$(3,3,3,2)$	0	∓2	$2k$	$2k+1$	∓2	$-$	∓1	47
109	$(1,1,1,1,1,1)$	$(3,1,0,1)$	$(2,3,0,0)$	1	$\mp2^{15}$	$-$	$2k+1$	∓1	$\mp2^{15}$	∓1	35
110	$(1,2,1,1,0,0)$	$(3,2,1,1)$	$(1,1,3,3)$	0	$\mp2^{15}$	$2k+1$	$-$	∓2	$-$	$-$	64
111	$(1,3,1,1,1,0)$	$(3,3,0,1)$	$(2,3,2,3)$	1	$\mp2^{15}$	$2k+1$	$2k$	∓2	$\mp2^{15}$	$-$	48
112	$(1,3,0,1,0,1)$	$(1,2,1,1)$	$(3,1,3,2)$	0	∓2	$2k$	$2k$	∓2	$-$	∓2	46
113	$(1,3,1,1,1,0)$	$(1,2,0,1)$	$(2,3,2,3)$	1	∓1	$2k$	$2k+1$	∓2	$\mp2^{15}$	$-$	48
114	$(1,2,2,1,1,1)$	$(1,2,3,1)$	$(0,1,2,2)$	1	∓2	$2k+1$	$2k+1$	∓2	$\mp2^{15}$	∓1	32
115	$(1,0,2,1,1,0)$	$(1,0,3,3)$	$(0,1,0,1)$	0	∓2	$-$	$2k+1$	$\mp2^{15}$	$\mp2^{15}$	$-$	49
116	$(1,3,3,1,0,1)$	$(3,2,2,1)$	$(1,3,3,2)$	1	$\mp2^{15}$	$2k$	$2k+1$	∓2	$-$	∓2	47
117	$(0,2,0,1,1,1)$	$(0,3,1,1)$	$(3,2,2,2)$	1	$-$	$2k$	$2k+1$	∓2	$\mp2^{15}$	∓1	48

(*continued*)

Table 5. (*continued*)

	ϕ	ψ	ω	λ	z_1	z_2	z_3	z_4	z_5	z_6	# of free bits
118	$(0,3,0,1,0,1)$	$(0,2,0,1)$	$(0,0,3,2)$	1	–	$2k$	–	∓ 2	–	∓ 1	64
119	$(1,1,2,0,0,1)$	$(1,1,2,0)$	$(3,3,0,1)$	0	∓ 2	–	$2k+1$	–	–	∓ 1	64
120	$(0,3,0,1,1,0)$	$(0,2,1,1)$	$(3,2,2,3)$	1	–	$2k$	$2k$	∓ 2	$\mp 2^{15}$	–	63
121	$(1,2,3,1,1,1)$	$(3,3,2,1)$	$(2,1,2,2)$	0	$\mp 2^{15}$	$2k$	$2k$	∓ 2	$\mp 2^{15}$	∓ 2	32
122	$(0,0,2,1,0,1)$	$(0,0,3,3)$	$(0,2,1,0)$	1	–	–	$2k+1$	$\mp 2^{15}$	–	∓ 2	64
123	$(1,1,3,1,1,1)$	$(1,1,3,3)$	$(2,1,0,0)$	0	∓ 1	–	$2k+1$	$\mp 2^{15}$	$\mp 2^{15}$	∓ 2	34
124	$(1,1,0,0,0,1)$	$(1,1,1,0)$	$(3,1,0,1)$	0	∓ 2	–	$2k$	–	–	∓ 2	63
125	$(0,1,0,0,1,0)$	$(0,1,1,0)$	$(3,2,1,0)$	0	–	–	$2k+1$	–	$\mp 2^{15}$	–	80
126	$(1,0,1,0,1,1)$	$(3,0,0,0)$	$(2,3,1,1)$	1	$\mp 2^{15}$	–	$2k$	–	$\mp 2^{15}$	∓ 1	50
127	$(0,0,3,0,1,1)$	$(0,0,2,0)$	$(1,2,1,1)$	1	–	–	$2k+1$	–	∓ 1	∓ 2	64
128	$(1,3,1,1,0,1)$	$(3,2,1,1)$	$(1,1,3,2)$	0	$\mp 2^{15}$	$2k$	–	∓ 2	–	∓ 1	49
129	$(1,2,1,1,1,1)$	$(1,2,0,1)$	$(2,1,2,2)$	0	∓ 2	$2k+1$	$2k$	∓ 2	∓ 1	∓ 2	31
130	$(1,1,3,0,0,1)$	$(3,1,2,0)$	$(1,3,0,1)$	1	$\mp 2^{15}$	–	$2k+1$	–	–	∓ 2	64
131	$(1,0,1,1,1,0)$	$(3,0,0,1)$	$(2,3,0,1)$	1	$\mp 2^{15}$	–	$2k$	∓ 1	$\mp 2^{15}$	–	50
132	$(1,0,2,0,1,1)$	$(1,0,2,0)$	$(0,3,1,1)$	0	∓ 1	–	$2k$	–	∓ 1	∓ 2	49
133	$(1,0,0,0,1,1)$	$(3,0,0,0)$	$(0,1,1,1)$	1	$\mp 2^{15}$	–	–	–	∓ 1	∓ 1	51
134	$(1,3,3,1,0,1)$	$(1,2,3,1)$	$(1,3,3,2)$	1	∓ 1	$2k$	$2k$	∓ 2	–	∓ 2	47
135	$(0,1,0,1,1,1)$	$(0,1,1,3)$	$(3,2,0,0)$	1	–	–	$2k$	$\mp 2^{15}$	$\mp 2^{15}$	∓ 1	50
136	$(1,0,0,1,0,1)$	$(1,0,1,3)$	$(3,1,1,0)$	0	∓ 2	–	$2k+1$	$\mp 2^{15}$	–	∓ 2	48
137	$(0,2,0,1,1,1)$	$(0,2,1,1)$	$(3,2,2,2)$	0	–	$2k+1$	$2k+1$	∓ 2	$\mp 2^{15}$	∓ 1	48
138	$(1,1,2,1,1,1)$	$(1,1,3,3)$	$(0,3,0,0)$	1	∓ 1	–	$2k+1$	$\mp 2^{15}$	∓ 1	∓ 2	34
139	$(1,2,1,1,1,1)$	$(1,3,0,1)$	$(2,1,2,2)$	1	∓ 2	$2k$	$2k$	∓ 2	∓ 1	∓ 2	31
140	$(0,1,3,1,1,1)$	$(0,1,3,3)$	$(1,2,0,0)$	0	–	–	$2k$	$\mp 2^{15}$	∓ 1	∓ 2	49
141	$(1,3,3,1,0,1)$	$(1,3,2,1)$	$(1,3,3,2)$	0	∓ 1	$2k+1$	$2k+1$	∓ 2	–	∓ 2	47
142	$(1,1,1,1,1,1)$	$(1,1,0,3)$	$(2,1,0,0)$	1	∓ 2	–	$2k+1$	$\mp 2^{15}$	∓ 1	∓ 2	33
143	$(0,1,2,1,1,1)$	$(0,1,2,3)$	$(3,0,0,0)$	1	–	–	$2k+1$	$\mp 2^{15}$	$\mp 2^{15}$	∓ 2	49
144	$(1,3,3,1,1,0)$	$(1,3,3,1)$	$(2,3,2,3)$	1	∓ 2	$2k+1$	$2k+1$	∓ 2	∓ 1	–	47
145	$(1,2,1,1,1,1)$	$(1,2,0,1)$	$(2,3,2,2)$	0	∓ 1	$2k+1$	$2k$	∓ 2	$\mp 2^{15}$	∓ 1	33
146	$(0,3,2,1,0,1)$	$(0,2,2,1)$	$(0,2,3,2)$	0	–	$2k$	$2k$	∓ 2	–	∓ 2	62
147	$(1,3,0,1,1,0)$	$(3,3,0,1)$	$(0,1,2,3)$	1	$\mp 2^{15}$	$2k+1$	–	∓ 2	∓ 1	–	49
148	$(1,1,2,1,1,1)$	$(1,1,3,1)$	$(0,1,0,0)$	1	∓ 2	–	$2k+1$	∓ 1	$\mp 2^{15}$	∓ 1	34
149	$(1,2,3,1,1,1)$	$(3,2,2,1)$	$(2,1,2,2)$	1	$\mp 2^{15}$	$2k+1$	$2k$	∓ 2	$\mp 2^{15}$	∓ 2	32
150	$(0,0,0,1,1,0)$	$(0,0,1,3)$	$(3,2,0,1)$	1	–	–	$2k+1$	$\mp 2^{15}$	$\mp 2^{15}$	–	65
151	$(1,3,2,1,1,0)$	$(1,2,2,1)$	$(0,1,2,3)$	1	∓ 2	$2k$	$2k$	∓ 2	$\mp 2^{15}$	–	47
152	$(1,1,3,1,1,1)$	$(3,1,3,3)$	$(2,1,0,0)$	0	$\mp 2^{15}$	–	$2k+1$	$\mp 2^{15}$	$\mp 2^{15}$	∓ 2	34
153	$(1,1,3,0,0,1)$	$(1,1,3,0)$	$(1,3,0,1)$	1	∓ 1	–	$2k$	–	–	∓ 2	64
154	$(0,3,1,1,1,0)$	$(0,3,1,1)$	$(1,0,2,3)$	0	–	$2k+1$	–	∓ 2	∓ 1	–	64
155	$(1,1,3,1,1,1)$	$(1,1,2,1)$	$(2,1,0,0)$	1	∓ 1	–	$2k$	∓ 1	$\mp 2^{15}$	∓ 2	34
156	$(1,2,1,1,1,1)$	$(1,3,0,1)$	$(2,3,2,2)$	1	∓ 1	$2k$	$2k$	∓ 2	$\mp 2^{15}$	∓ 1	33

(*continued*)

Table 5. (*continued*)

	ϕ	ψ	ω	λ	z_1	z_2	z_3	z_4	z_5	z_6	# of free bits
157	$(1,0,1,1,1,0)$	$(1,0,0,3)$	$(2,3,0,1)$	1	∓1	–	$2k$	$\mp2^{15}$	$\mp2^{15}$	–	50
158	$(1,1,1,1,1,1)$	$(1,1,0,3)$	$(2,3,0,0)$	1	∓1	–	$2k+1$	$\mp2^{15}$	$\mp2^{15}$	∓1	35
159	$(0,1,2,0,0,1)$	$(0,1,2,0)$	$(0,2,0,1)$	0	–	–	$2k$	–	–	∓2	79
160	$(1,0,2,0,1,1)$	$(3,0,2,0)$	$(0,3,1,1)$	0	$\mp2^{15}$	–	$2k$	–	∓1	∓2	49
161	$(1,0,3,0,1,1)$	$(1,0,3,0)$	$(2,3,1,1)$	1	∓2	–	$2k+1$	–	∓1	∓1	49
162	$(1,1,1,0,1,0)$	$(1,1,0,0)$	$(2,3,1,0)$	0	∓1	–	$2k$	–	$\mp2^{15}$	–	65
163	$(1,2,0,1,1,1)$	$(1,3,0,1)$	$(0,1,2,2)$	1	∓1	$2k$	–	∓2	∓1	∓1	34
164	$(1,3,1,1,0,1)$	$(1,3,1,1)$	$(1,1,3,2)$	1	∓1	$2k+1$	–	∓2	–	∓1	49
165	$(1,3,3,1,0,1)$	$(3,2,3,1)$	$(1,3,3,2)$	1	$\mp2^{15}$	$2k$	$2k$	∓2	–	∓2	47
166	$(1,3,3,1,1,0)$	$(1,2,2,1)$	$(2,3,2,3)$	1	∓2	$2k$	$2k$	∓2	∓1	–	47
167	$(1,0,3,1,0,1)$	$(1,0,2,1)$	$(1,3,1,0)$	0	∓1	–	$2k+1$	∓1	–	∓2	49
168	$(1,0,2,0,1,1)$	$(1,0,3,0)$	$(0,3,1,1)$	0	∓1	–	$2k+1$	–	∓1	∓2	49
169	$(1,0,3,1,1,0)$	$(1,0,3,1)$	$(2,3,0,1)$	1	∓2	–	$2k+1$	∓1	∓1	–	49
170	$(1,1,2,1,1,1)$	$(3,1,3,3)$	$(0,3,0,0)$	1	$\mp2^{15}$	–	$2k+1$	$\mp2^{15}$	∓1	∓2	34
171	$(1,3,2,1,0,1)$	$(1,3,2,1)$	$(3,3,3,2)$	1	∓2	$2k+1$	$2k+1$	∓2	–	∓1	47
172	$(1,3,3,1,0,1)$	$(3,3,2,1)$	$(1,3,3,2)$	0	$\mp2^{15}$	$2k+1$	$2k+1$	∓2	–	∓2	47
173	$(1,3,1,1,1,0)$	$(3,2,0,1)$	$(2,3,2,3)$	1	$\mp2^{15}$	$2k$	$2k+1$	∓2	$\mp2^{15}$	–	48
174	$(1,1,2,1,0,0)$	$(1,1,2,3)$	$(3,3,1,1)$	0	∓2	–	$2k+1$	$\mp2^{15}$	–	–	64
175	$(0,0,2,0,1,1)$	$(0,0,2,0)$	$(3,0,1,1)$	0	–	–	$2k+1$	–	$\mp2^{15}$	∓2	64
176	$(0,3,0,1,0,1)$	$(0,3,0,1)$	$(0,0,3,2)$	0	–	$2k+1$	–	∓2	–	∓1	64
177	$(0,3,0,1,1,0)$	$(0,3,1,1)$	$(3,2,2,3)$	0	–	$2k+1$	$2k$	∓2	$\mp2^{15}$	–	63
178	$(0,0,3,0,1,1)$	$(0,0,3,0)$	$(1,2,1,1)$	1	–	–	$2k$	–	∓1	∓2	64
179	$(1,1,3,0,0,1)$	$(3,1,3,0)$	$(1,3,0,1)$	1	$\mp2^{15}$	–	$2k$	–	–	∓2	64
180	$(0,3,2,1,0,1)$	$(0,2,3,1)$	$(0,2,3,2)$	0	–	$2k$	$2k+1$	∓2	–	∓2	62
181	$(1,1,3,1,1,1)$	$(1,1,2,3)$	$(2,3,0,0)$	1	∓2	–	$2k$	$\mp2^{15}$	∓1	∓1	34
182	$(1,2,0,1,1,1)$	$(1,2,0,1)$	$(0,1,2,2)$	0	∓1	$2k+1$	–	∓2	∓1	∓1	34
183	$(1,1,3,1,1,1)$	$(3,1,2,1)$	$(2,1,0,0)$	1	$\mp2^{15}$	–	$2k$	∓1	$\mp2^{15}$	∓2	34
184	$(1,3,2,1,1,0)$	$(1,2,3,1)$	$(0,1,2,3)$	1	∓2	$2k$	$2k+1$	∓2	$\mp2^{15}$	–	47
185	$(1,0,2,0,0,0)$	$(1,0,2,0)$	$(3,3,0,0)$	1	∓2	–	$2k+1$	–	–	–	79
186	$(1,0,3,0,1,1)$	$(1,0,3,0)$	$(2,1,1,1)$	1	∓1	–	$2k+1$	–	$\mp2^{15}$	∓2	49
187	$(0,2,2,1,1,1)$	$(0,2,3,1)$	$(3,0,2,2)$	0	–	$2k+1$	$2k$	∓2	$\mp2^{15}$	∓2	47
188	$(1,2,2,1,1,1)$	$(1,3,2,1)$	$(0,3,2,2)$	0	∓1	$2k$	$2k$	∓2	∓1	∓2	32
189	$(1,3,3,1,0,1)$	$(1,3,3,1)$	$(1,3,3,2)$	0	∓1	$2k+1$	$2k$	∓2	–	∓2	47
190	$(1,0,2,0,1,1)$	$(1,0,2,0)$	$(0,1,1,1)$	0	∓2	–	$2k$	–	$\mp2^{15}$	∓1	49
191	$(1,3,1,1,0,1)$	$(3,3,1,1)$	$(1,1,3,2)$	1	$\mp2^{15}$	$2k+1$	–	∓2	–	∓1	49
192	$(1,0,2,1,0,1)$	$(1,0,2,1)$	$(3,3,1,0)$	1	∓2	–	$2k+1$	∓1	–	∓1	49
193	$(1,0,3,1,0,1)$	$(3,0,2,1)$	$(1,3,1,0)$	0	$\mp2^{15}$	–	$2k+1$	∓1	–	∓2	49
194	$(1,0,2,0,1,1)$	$(3,0,3,0)$	$(0,3,1,1)$	0	$\mp2^{15}$	–	$2k+1$	–	∓1	∓2	49

(*continued*)

Table 5. (*continued*)

	ϕ	ψ	ω	λ	z_1	z_2	z_3	z_4	z_5	z_6	# of free bits
195	$(0,2,2,1,1,1)$	$(0,3,3,1)$	$(3,0,2,2)$	1	–	$2k$	$2k$	∓ 2	$\mp 2^{15}$	∓ 2	47
196	$(1,1,2,1,1,1)$	$(1,1,3,3)$	$(0,1,0,0)$	1	∓ 2	–	$2k+1$	$\mp 2^{15}$	$\mp 2^{15}$	∓ 1	34
197	$(1,3,2,1,1,0)$	$(1,3,2,1)$	$(0,1,2,3)$	0	∓ 2	$2k+1$	$2k$	∓ 2	$\mp 2^{15}$	–	47
198	$(1,2,0,1,1,1)$	$(1,2,0,1)$	$(0,3,2,2)$	0	∓ 2	$2k+1$	–	∓ 2	$\mp 2^{15}$	∓ 2	32
199	$(0,2,3,1,1,1)$	$(0,2,2,1)$	$(1,2,2,2)$	0	–	$2k+1$	$2k+1$	∓ 2	∓ 1	∓ 2	47
200	$(1,1,3,1,1,1)$	$(1,1,2,3)$	$(2,1,0,0)$	1	∓ 1	–	$2k$	$\mp 2^{15}$	$\mp 2^{15}$	∓ 2	34
201	$(1,0,3,0,1,1)$	$(1,0,2,0)$	$(2,3,1,1)$	0	∓ 2	–	$2k$	–	∓ 1	∓ 1	49
202	$(0,3,2,1,0,1)$	$(0,3,2,1)$	$(0,2,3,2)$	1	–	$2k+1$	$2k$	∓ 2	–	∓ 2	62
203	$(1,2,0,1,1,1)$	$(3,2,0,1)$	$(0,1,2,2)$	0	$\mp 2^{15}$	$2k+1$	–	∓ 2	∓ 1	∓ 1	34
204	$(1,2,0,1,1,1)$	$(1,3,0,1)$	$(0,3,2,2)$	1	∓ 2	$2k$	–	∓ 2	$\mp 2^{15}$	∓ 2	32
205	$(0,2,3,1,1,1)$	$(0,3,2,1)$	$(1,2,2,2)$	1	–	$2k$	$2k+1$	∓ 2	∓ 1	∓ 2	47
206	$(1,0,3,1,0,1)$	$(1,0,3,1)$	$(1,3,1,0)$	0	∓ 1	–	$2k$	∓ 1	–	∓ 2	49
207	$(1,0,3,0,1,1)$	$(3,0,3,0)$	$(2,1,1,1)$	1	$\mp 2^{15}$	–	$2k+1$	–	$\mp 2^{15}$	∓ 2	49
208	$(1,2,2,1,1,1)$	$(1,2,2,1)$	$(0,3,2,2)$	1	∓ 1	$2k+1$	$2k$	∓ 2	∓ 1	∓ 2	32
209	$(1,2,3,1,1,1)$	$(1,3,3,1)$	$(2,3,2,2)$	1	∓ 2	$2k$	$2k+1$	∓ 2	∓ 1	∓ 1	32
210	$(1,2,2,1,1,1)$	$(3,3,2,1)$	$(0,3,2,2)$	0	$\mp 2^{15}$	$2k$	$2k$	∓ 2	∓ 1	∓ 2	32
211	$(1,3,3,1,1,0)$	$(1,3,2,1)$	$(2,3,2,3)$	0	∓ 2	$2k+1$	$2k$	∓ 2	∓ 1	–	47
212	$(1,0,3,1,0,1)$	$(1,0,2,3)$	$(1,3,1,0)$	0	∓ 1	–	$2k+1$	$\mp 2^{15}$	–	∓ 2	49
213	$(1,3,3,1,0,1)$	$(3,3,3,1)$	$(1,3,3,2)$	0	$\mp 2^{15}$	$2k+1$	$2k$	∓ 2	–	∓ 2	47
214	$(1,2,0,1,1,1)$	$(3,3,0,1)$	$(0,1,2,2)$	1	$\mp 2^{15}$	$2k$	–	∓ 2	∓ 1	∓ 1	34
215	$(1,2,2,1,0,0)$	$(1,2,3,1)$	$(3,3,3,3)$	1	∓ 2	$2k+1$	$2k$	∓ 2	–	–	62
216	$(0,1,2,0,0,1)$	$(0,1,3,0)$	$(0,2,0,1)$	0	–	–	$2k+1$	–	–	∓ 2	79
217	$(1,2,2,1,1,1)$	$(1,3,3,1)$	$(0,3,2,2)$	0	∓ 1	$2k$	$2k+1$	∓ 2	∓ 1	∓ 2	32
218	$(0,2,1,1,1,1)$	$(0,3,1,1)$	$(1,0,2,2)$	0	–	$2k$	–	∓ 2	∓ 1	∓ 1	49
219	$(1,1,3,0,1,0)$	$(1,1,3,0)$	$(2,3,1,0)$	0	∓ 2	–	$2k+1$	–	∓ 1	–	64
220	$(0,2,1,1,1,1)$	$(0,2,1,1)$	$(1,0,2,2)$	1	–	$2k+1$	–	∓ 2	∓ 1	∓ 1	49
221	$(0,0,2,1,0,1)$	$(0,0,2,1)$	$(0,2,1,0)$	1	–	–	$2k$	∓ 1	–	∓ 2	64
222	$(1,0,3,0,1,1)$	$(1,0,2,0)$	$(2,1,1,1)$	0	∓ 1	–	$2k$	–	$\mp 2^{15}$	∓ 2	49
223	$(1,1,3,1,1,1)$	$(3,1,2,3)$	$(2,1,0,0)$	1	$\mp 2^{15}$	–	$2k$	$\mp 2^{15}$	$\mp 2^{15}$	∓ 2	34
224	$(1,3,2,1,1,0)$	$(1,3,3,1)$	$(0,1,2,3)$	0	∓ 2	$2k+1$	$2k+1$	∓ 2	$\mp 2^{15}$	–	47
225	$(1,2,3,1,1,1)$	$(1,2,3,1)$	$(2,3,2,2)$	0	∓ 2	$2k+1$	$2k+1$	∓ 2	∓ 1	∓ 1	32
226	$(0,1,3,1,1,1)$	$(0,1,2,1)$	$(1,2,0,0)$	0	–	–	$2k+1$	∓ 1	∓ 1	∓ 2	49
227	$(1,2,3,1,1,1)$	$(1,3,3,1)$	$(2,1,2,2)$	1	∓ 1	$2k$	$2k+1$	∓ 2	$\mp 2^{15}$	∓ 2	32
228	$(1,0,3,1,0,1)$	$(3,0,3,1)$	$(1,3,1,0)$	0	$\mp 2^{15}$	–	$2k$	∓ 1	–	∓ 2	49
229	$(1,0,3,1,1,0)$	$(1,0,2,1)$	$(2,3,0,1)$	0	∓ 2	–	$2k$	∓ 1	∓ 1	–	49
230	$(1,0,2,1,0,1)$	$(1,0,2,3)$	$(3,3,1,0)$	1	∓ 2	–	$2k+1$	$\mp 2^{15}$	–	∓ 1	49
231	$(1,0,3,1,0,1)$	$(3,0,2,3)$	$(1,3,1,0)$	0	$\mp 2^{15}$	–	$2k+1$	$\mp 2^{15}$	–	∓ 2	49
232	$(1,2,2,1,1,1)$	$(3,2,2,1)$	$(0,3,2,2)$	1	$\mp 2^{15}$	$2k+1$	$2k$	∓ 2	∓ 1	∓ 2	32
233	$(1,0,3,1,1,0)$	$(1,0,3,3)$	$(2,3,0,1)$	1	∓ 2	–	$2k+1$	$\mp 2^{15}$	∓ 1	–	49

(*continued*)

Table 5. (*continued*)

	ϕ	ψ	ω	λ	z_1	z_2	z_3	z_4	z_5	z_6	# of free bits
234	$(1,2,2,1,1,1)$	$(1,2,3,1)$	$(0,3,2,2)$	1	∓ 1	$2k+1$	$2k+1$	∓ 2	∓ 1	∓ 2	32
235	$(1,1,2,1,1,1)$	$(1,1,2,1)$	$(0,3,0,0)$	1	∓ 1	$-$	$2k$	∓ 1	∓ 1	∓ 2	34
236	$(1,3,0,1,1,0)$	$(1,2,0,1)$	$(0,1,2,3)$	0	∓ 1	$2k$	$-$	∓ 2	∓ 1	$-$	49
237	$(0,2,3,1,1,1)$	$(0,2,3,1)$	$(1,2,2,2)$	0	$-$	$2k+1$	$2k$	∓ 2	∓ 1	∓ 2	47
238	$(1,0,2,0,1,1)$	$(1,0,3,0)$	$(0,1,1,1)$	0	∓ 2	$-$	$2k+1$	$-$	$\mp 2^{15}$	∓ 1	49
239	$(0,2,2,1,1,1)$	$(0,2,2,1)$	$(3,0,2,2)$	1	$-$	$2k+1$	$2k+1$	∓ 2	$\mp 2^{15}$	∓ 2	47
240	$(0,2,2,1,1,1)$	$(0,3,2,1)$	$(3,0,2,2)$	0	$-$	$2k$	$2k+1$	∓ 2	$\mp 2^{15}$	∓ 2	47
241	$(0,2,3,1,1,1)$	$(0,3,3,1)$	$(1,2,2,2)$	1	$-$	$2k$	$2k$	∓ 2	∓ 1	∓ 2	47
242	$(1,0,2,1,1,0)$	$(1,0,3,1)$	$(0,1,0,1)$	0	∓ 2	$-$	$2k+1$	∓ 1	$\mp 2^{15}$	$-$	49
243	$(1,2,3,1,1,1)$	$(1,2,3,1)$	$(2,1,2,2)$	0	∓ 1	$2k+1$	$2k+1$	∓ 2	$\mp 2^{15}$	∓ 2	32
244	$(1,0,3,1,0,1)$	$(1,0,3,3)$	$(1,3,1,0)$	0	∓ 1	$-$	$2k$	$\mp 2^{15}$	$-$	∓ 2	49
245	$(1,1,0,1,1,1)$	$(1,1,0,3)$	$(0,1,0,0)$	0	∓ 1	$-$	$-$	$\mp 2^{15}$	∓ 1	∓ 1	36
246	$(0,3,2,1,0,1)$	$(0,3,3,1)$	$(0,2,3,2)$	1	$-$	$2k+1$	$2k+1$	∓ 2	$-$	∓ 2	62
247	$(1,1,3,1,1,1)$	$(1,1,3,1)$	$(2,3,0,0)$	0	∓ 2	$-$	$2k+1$	∓ 1	∓ 1	∓ 1	34
248	$(1,3,2,1,0,1)$	$(1,2,3,1)$	$(3,3,3,2)$	1	∓ 2	$2k$	$2k$	∓ 2	$-$	∓ 1	47
249	$(1,2,2,1,1,1)$	$(1,3,2,1)$	$(0,1,2,2)$	0	∓ 2	$2k$	$2k$	∓ 2	$\mp 2^{15}$	∓ 1	32
250	$(1,2,2,1,1,1)$	$(1,2,2,1)$	$(0,1,2,2)$	1	∓ 2	$2k+1$	$2k$	∓ 2	$\mp 2^{15}$	∓ 1	32
251	$(1,2,1,1,0,0)$	$(1,2,1,1)$	$(1,1,3,3)$	0	∓ 1	$2k+1$	$-$	∓ 2	$-$	$-$	64
252	$(1,2,2,1,1,1)$	$(3,3,3,1)$	$(0,3,2,2)$	0	$\mp 2^{15}$	$2k$	$2k+1$	∓ 2	∓ 1	∓ 2	32
253	$(1,1,2,1,1,1)$	$(3,1,2,1)$	$(0,3,0,0)$	1	$\mp 2^{15}$	$-$	$2k$	∓ 1	∓ 1	∓ 2	34
254	$(1,2,2,1,1,1)$	$(3,2,3,1)$	$(0,3,2,2)$	1	$\mp 2^{15}$	$2k+1$	$2k+1$	∓ 2	∓ 1	∓ 2	32
255	$(1,3,0,1,1,0)$	$(3,2,0,1)$	$(0,1,2,3)$	0	$\mp 2^{15}$	$2k$	$-$	∓ 2	∓ 1	$-$	49

Table 6. 50 linear relations with less number of key bits restriction for 8.5-round IDEA cipher. Here each row is associated with one such relation, a linear mask for each round input and one for the last round output, namely ciphertext are provided. Last column shows the number of key bits from the master key that are not restricted, that is, each such bit can be either 0 or 1. Note that mask $(\mathbf{a}, \mathbf{b}, \mathbf{c}, \mathbf{d})$ is denoted by abcd. When $832 - 556 = 276$ key bits are restricted according to Tables 1 and 2, twenty second row of this table gives a linear relation for 8.5-round IDEA cipher involving plaintext bit $(0,1,0,0) \star (\mathbf{X}_1^0, \mathbf{X}_2^0, \mathbf{X}_3^0, \mathbf{X}_4^0) = 1 \cdot \mathbf{X}_2^0$ and ciphertext bits added $(1,2,1,3) \star (\mathbf{Y}_1, \mathbf{Y}_2, \mathbf{Y}_3, \mathbf{Y}_4) = 1 \cdot \mathbf{Y}_1 \oplus 2 \cdot \mathbf{Y}_2 \oplus 1 \cdot \mathbf{Y}_3 \oplus 3 \cdot \mathbf{Y}_4$ (see Sect. 4.2 and Fig. 1 in Appendix A).

#	1st round's input mask	2nd round's input mask	3rd round's input mask	4th round's input mask	5th round's input mask	6th round's input mask	7th round's input mask	8th round's input mask	Last 0.5 round's input mask	Cipher text mask	# of free key bits
1	1100	0110	0110	1010	1100	0110	1010	1100	0110	0110	586
2	1010	1100	0110	0110	1010	1100	0110	1010	1100	3100	586
3	1010	1100	0110	0110	1010	1100	0110	1010	1100	1100	586
4	0110	1010	1100	0110	0110	1010	1100	0110	1010	1010	586
5	0110	1010	1100	0110	0110	1010	1100	0110	1010	3010	585

(*continued*)

Table 6. (*continued*)

#	1st round's input mask	2nd round's input mask	3rd round's input mask	4th round's input mask	5th round's input mask	6th round's input mask	7th round's input mask	8th round's input mask	Last 0.5 round's input mask	Cipher text mask	# of free key bits
6	0100	0001	0010	1011	1110	1101	0100	0001	0010	0010	579
7	1001	0101	0011	1001	0101	0011	1001	0101	0011	0011	577
8	0101	0011	1001	0101	0011	1001	0101	0011	1001	1001	577
9	1001	0101	0011	1001	0101	0011	1001	0101	0011	0013	576
10	0101	0011	1001	0101	0011	1001	0101	0011	1001	3001	576
11	0101	0011	1001	0101	0011	1001	0101	0011	1001	1003	576
12	0100	0001	0010	1011	1110	3101	0100	0001	0010	0010	576
13	0101	0011	1001	0101	0011	1001	0101	0011	1001	3003	575
14	1111	1111	1111	1111	1111	1111	1111	1111	1111	1111	562
15	0011	1001	0101	0011	1001	0101	0011	1001	0101	0101	562
16	1111	1111	1111	1111	1111	1111	1111	1111	1111	3111	561
17	1111	1111	1111	1111	1111	1111	1111	1111	1111	1113	561
18	0011	1001	0101	0011	1001	0101	0011	1001	0101	0103	561
19	1111	1111	1111	1111	1111	1111	1111	1111	1111	3113	560
20	1133	0100	0001	0010	3211	1133	0100	0001	0010	0010	557
21	0100	0001	0010	3211	1133	0100	0001	0010	3211	1211	557
22	0100	0001	0010	3211	1133	0100	0001	0010	3211	1213	556
23	3311	1133	3311	1133	3311	1133	3311	1133	3311	1311	545
24	1133	3311	1133	3311	1133	3311	1133	3311	1133	1131	545
25	3311	1133	3311	1133	3311	1133	3311	1133	3311	1313	544
26	1133	3311	1133	3311	1133	3311	1133	3311	1133	3133	544
27	3211	1133	0100	0001	0010	3211	1133	0100	0001	0001	540
28	3211	1133	0100	0001	0010	3211	1133	0100	0001	0003	539
29	0001	0010	3211	1133	0100	0001	0010	3211	1133	1131	539
30	0010	3211	1133	0100	0001	0010	3211	1133	0100	0100	538
31	0001	0010	3211	1133	0100	0001	0010	3211	1133	3133	538
32	1101	0100	0001	0010	1011	1110	1101	0100	0001	0001	534
33	1110	1101	0100	0001	0010	1011	1110	1101	0100	0100	533
34	1101	0100	0001	0010	1011	1110	1101	0100	0001	0003	533
35	0010	1011	1110	1101	0100	0001	0010	1011	1110	1110	533
36	0001	0010	1011	1110	1101	0100	0001	0010	1011	1011	533
37	0010	1011	1110	1101	0100	0001	0010	1011	1110	3110	532
38	0001	0010	1011	1110	1101	0100	0001	0010	1011	3011	532
39	0001	0010	1011	1110	1101	0100	0001	0010	1011	1013	532
40	3101	0100	0001	0010	1011	1110	3101	0100	0001	0001	531
41	0001	0010	1011	1110	1101	0100	0001	0010	1011	3013	531
42	3101	0100	0001	0010	1011	1110	3101	0100	0001	0003	530
43	0010	1011	1110	3101	0100	0001	0010	1011	1110	1110	530
44	0001	0010	1011	3110	1101	0100	0001	0010	1011	1011	530
45	0001	0010	1011	1110	3101	0100	0001	0010	1011	1011	530
46	0010	1011	1110	3101	0100	0001	0010	1011	1110	3110	529
47	0001	0010	1011	3110	1101	0100	0001	0010	1011	3011	529
48	0001	0010	1011	3110	1101	0100	0001	0010	1011	1013	529
49	0001	0010	1011	1110	3101	0100	0001	0010	1011	3011	529
50	0001	0010	1011	1110	3101	0100	0001	0010	1011	1013	529

References

1. Chaves, R., Sousa, L.: Improving residue number system multiplication with more balanced moduli sets and enhanced modular arithmetic structures. IET Comput. Digital Tech. **1**(5), 472–480 (2007)
2. Daemen, J., Govaerts, R., Vandewalle, J.: Weak keys for IDEA. In: Stinson, D.R. (ed.) CRYPTO 1993. LNCS, vol. 773, pp. 224–231. Springer, Heidelberg (1994). https://doi.org/10.1007/3-540-48329-2_20
3. Junod, P., Macchetti, M.: Revisiting the IDEA philosophy. In: Dunkelman, O. (ed.) FSE 2009. LNCS, vol. 5665, pp. 277–295. Springer, Heidelberg (2009). https://doi.org/10.1007/978-3-642-03317-9_17
4. Lai, X.: On the Design and Security of Block Cipher. ETH Series in Information Processing, vol. 1. Hartung-Gorre Verlag, Konstanz (1992)
5. Meier, W.: On the security of the IDEA block cipher. In: Helleseth, T. (ed.) EUROCRYPT 1993. LNCS, vol. 765, pp. 371–385. Springer, Heidelberg (1994). https://doi.org/10.1007/3-540-48285-7_32
6. Modugu, R., Choi, M., Park, N.: A fast low-power modulo $2^n + 1$ multiplier design. In: IEEE Instrumentation and Measurement Technology Conference, I2MTC 2009, pp. 951–956. IEEE (2009)
7. Nakahara Jr., J.: Personal communication, November 2004
8. Nakahara Jr., J.: Lai-Massey Cipher Designs: History. Design Criteria and Cryptanalysis. Springer, Cham (2018). https://doi.org/10.1007/978-3-319-68273-0
9. Nakahara Jr., J., Rijmen, V., Preneel, B., Vandewalle, J.: The MESH block ciphers. In: Chae, K.-J., Yung, M. (eds.) WISA 2003. LNCS, vol. 2908, pp. 458–473. Springer, Heidelberg (2004). https://doi.org/10.1007/978-3-540-24591-9_34
10. Nyberg, K.: On the construction of highly nonlinear permutations. In: Rueppel, R.A. (ed.) EUROCRYPT 1992. LNCS, vol. 658, pp. 92–98. Springer, Heidelberg (1993). https://doi.org/10.1007/3-540-47555-9_8
11. SageMath, the Sage Mathematics Software System (Version 6.7). The Sage Developers (2015). http://www.sagemath.org
12. Yıldırım, H.M.: Nonlinearity properties of the mixing operations of the block cipher IDEA. In: Johansson, T., Maitra, S. (eds.) INDOCRYPT 2003. LNCS, vol. 2904, pp. 68–81. Springer, Heidelberg (2003). https://doi.org/10.1007/978-3-540-24582-7_5
13. Zhang, X.-M., Zheng, Y., Imai, H.: Duality of Boolean functions and its cryptographic significance. In: Han, Y., Okamoto, T., Qing, S. (eds.) ICICS 1997. LNCS, vol. 1334, pp. 159–169. Springer, Heidelberg (1997). https://doi.org/10.1007/BFb0028472

Combinatorics, Codes, Designs and Graphs

On a Weighted Spin of the Lebesgue Identity

Ali Kemal Uncu$^{(\boxtimes)}$

Austrian Academy of Sciences, Johann Radon Institute for Computational
and Applied Mathematics, Altenbergerstrasse 69, 4040 Linz, Austria
akuncu@risc.jku.at

Abstract. Alladi studied partition theoretic implications of a two variable generalization of the Lebesgue identity. In this short note, we focus on a slight variation of the basic hypergeometric sum that Alladi studied. We present two new partition identities involving weights.

Keywords: Lebesgue identity · Generalized Lebesgue identities · Heine transformation · Weighted partition identities

1 Introduction

One of the fundamental identities in the theory of partitions and q-series is the Lebesgue identity:

$$\sum_{n\geq 0} \frac{(-aq)_n}{(q)_n} q^{\frac{n(n+1)}{2}} = \frac{(-aq^2; q^2)_\infty}{(q; q^2)_\infty}, \tag{1}$$

where a and q are variables and the q-Pochhammer symbol is defined as follows

$$(a)_n := (a; q)_n := \prod_{i=0}^{n-1} (1 - aq^i),$$

for any $n \in \mathbb{Z} \cup \{\infty\}$. Some combinatorial implications of this result were studied by Alladi [3]. In the same paper, he also did a partition theoretic study of a summation formula due to Ramanujan [5, (1.3.13), p. 13]

$$\sum_{n\geq 0} \frac{(-b/a)_n a^n q^{n(n+1)/2}}{(q)_n (bq)_n} = \frac{(-aq)_\infty}{(bq)_\infty}. \tag{2}$$

Alladi called this identity and it's dilated forms *Generalized Lebesgue identities*.

Research of the author is supported by the Austrian Science Fund FWF, SFB50-07, SFB50-09 and SFB50-11 Projects.

D. Slamanig et al. (Eds.): MACIS 2019, LNCS 11989, pp. 273–279, 2020.
https://doi.org/10.1007/978-3-030-43120-4_20

We would like to study a similar function that is not directly related to (2) or that satisfies a summation formula, but that still manifest beautiful relations. Let a, z and q be variables and define

$$\mathcal{F}(a, z, q) := \sum_{n \geq 0} \frac{(za)_n}{(q)_n (zq)_n} z^n q^{\frac{n(n+1)}{2}}. \tag{3}$$

Looking at $\mathcal{F}(b, a, q)$ it is clear that this sum is—so to speak—a sibling of the Generalized Lebesgue identity (2), and $\mathcal{F}(-aq, 1, q)$ is a cousin of the original Lebesgue identity (1) with an extra q-factorial, $1/(q)_n$, in the summand. This extra factor will be the source of the weights in the combinatorial/partition theoretic study of the identities related to the (3). For other references related to weighted partition identities of this spirit one can refer to [2,6,12], and in a wider perspective some other recent weighted partition identities can be found in [1,7,9].

Before any combinatorial study, we would like to note the following theorem.

Theorem 1. *For variables a, z and q, we have*

$$\sum_{n \geq 0} \frac{(za)_n (zq^{n+1})_\infty}{(q)_n} z^n q^{\frac{n(n+1)}{2}} = \sum_{n \geq 0} \frac{(-za)_n (-zq^{n+1})_\infty}{(q)_n} (-z)^n q^{\frac{n(n+1)}{2}}. \tag{4}$$

Please note that the only difference between the left- and right-hand sides of (4) is $z \mapsto -z$. In other words, the object is even in the variable z. In author's view, the observed symmetry makes this identity visually highly pleasing.

The following sections are arranged as follows. In Sect. 2, we give a proof of Theorem 1 and note some Corollaries of this result. In Sect. 3, we study the partition theoretic interpretations of the results in Sect. 2.

2   Proof of Theorem 1

We require two main ingredients for the proof of (4). First, it is a known fact that

$$\lim_{\rho \to \infty} \frac{(\rho)_n}{\rho^n} = (-1)^n q^{\frac{n(n-1)}{2}}, \tag{5}$$

and, second, Heine Transformation [10, p. 241, III.2]

$$\sum_{n \geq 0} \frac{(a)_n (b)_n}{(q)_n (c)_n} z^n = \frac{(c/b)_\infty (bz)_\infty}{(c)_\infty (z)_\infty} \sum_{n \geq 0} \frac{(abz/c)_n (b)_n}{(q)_n (bz)_n} \left(\frac{c}{b}\right)^n \tag{6}$$

Proof (Proof of Theorem 1). The function $\mathcal{F}(a, z, q)$ can be written as the following due to (5):

$$\mathcal{F}(a, z, q) = \lim_{\rho \to \infty} \sum_{n \geq 0} \frac{(za)_n (\rho)_n}{(q)_n (zq)_n} \left(-\frac{zq}{\rho}\right)^n.$$

Then we can directly apply the Heine transformation (6), and after tending $\rho \to \infty$, one gets

$$\mathcal{F}(a, z, q) = \frac{(-zq)_\infty}{(zq)_\infty} \mathcal{F}(a, -z, q). \tag{7}$$

Multiplying both sides of (7) with $(zq)_\infty$, carrying the infinite q-Pochhammers inside the sums, and doing elementary simplifications in the summand level finishes the proof.

It is evident that some special cases of (4) (such as $(a, z, q) = (q, 1, q)$) can be summed by utilizing simple summation formulas (such as [10, II.2, p. 354] and shown to be equal to $(q^2; q^2)_\infty$). This is not our motivation. We would like to look at special cases of (4) to extract some combinatorial information. The $(a, z, q) = (q, 1, q)$ and $(-q, 1, q)$ cases are presented in Corollary 1.

Corollary 1. *Let q be a variable, we have*

$$\sum_{n \geq 0} (q^{n+1})_\infty q^{\frac{n(n+1)}{2}} = \sum_{n \geq 0} (-q^{n+1})_\infty \frac{(-q)_n}{(q)_n} (-1)^n q^{\frac{n(n+1)}{2}}, \tag{8}$$

$$\sum_{n \geq 0} (-q^{n+1})_\infty (-1)^n q^{\frac{n(n+1)}{2}} = \sum_{n \geq 0} (q^{n+1})_\infty \frac{(-q)_n}{(q)_n} q^{\frac{n(n+1)}{2}}. \tag{9}$$

Another interesting corollary can be seen by picking $a = z = 1$ in (7) and using Jacobi Triple Product identity [10, p. 239, II.2],

$$\sum_{n=-\infty}^{\infty} z^n q^{n^2} = (-zq; q^2)_\infty (-q/z; q^2)_\infty (q^2; q^2)_\infty. \tag{10}$$

Corollary 2. *We have*

$$\sum_{n \geq 1} (-1)^n q^{n^2} = \sum_{n \geq 1} \frac{(-1)^n q^{\frac{n(n+1)}{2}}}{(q)_n (1 + q^n)}.$$

Proof. It is clear that only the $n = 0$ term of the sum on the left-hand side of (7) is non-zero when $a = z = 1$, and the total sum on the left hand side is 1:

$$1 = \frac{(-q)_\infty}{(q)_\infty} \sum_{n \geq 0} \frac{(-1)_n}{(q)_n (-q)_n} (-1)^n q^{\frac{n(n+1)}{2}}.$$

We multiply both sides of this equation by $(q)_\infty / (-q)_\infty$ and observe that

$$\frac{(q)_\infty}{(-q)_\infty} = \frac{(q; q^2)_\infty (q^2; q^2)_\infty}{(-q)_\infty} = \frac{(q; q^2)_\infty (q)_\infty (-q)_\infty}{(-q)_\infty} = (q; q^2)_\infty^2 (q^2; q^2)_\infty.$$

The right-hand side of the last line is the same as the right-hand side of (10) with $z = -1$. This yields

$$\sum_{n=-\infty}^{\infty} (-1)^n q^{n^2} = \sum_{n \geq 0} \frac{(-1)_n}{(q)_n (-q)_n} (-1)^n q^{\frac{n(n+1)}{2}}, \tag{11}$$

where the left-hand side is coming from (10) and the right-hand side is $\mathcal{F}(1, -1, q)$. Splitting the bilateral sum on the left-hand side and using simple cancellations on $(-1)_n/(-q)_n$, using the definition of the q-factorials, on the right-hand side, we get

$$1 + 2\sum_{n\geq 1}(-1)^n q^{n^2} = 1 + 2\sum_{n\geq 1}\frac{(-1)^n q^{\frac{n(n+1)}{2}}}{(q)_n(1+q^n)}. \tag{12}$$

This shows claim.

Another proof of this result appears in the author's joint paper with Berkovich as Lemma 4.1 [7]. Combinatorial interpretation of this identity was done by Bessenrodt–Pak [8] and later by Alladi [3].

3   Partition Theoretic Interpretations of Corollary 1

We would like to interpret the identities (8) and (9) as weighted partition identities. To that end, we need to define what a partition is and some related statistics. A *partition* (in *frequency notation* [4]) is a list of the form

$$(1^{f_1}, 2^{f_2}, 3^{f_3}, \dots)$$

where $f_i \in \mathbb{N} \cup \{0\}$ and all but finitely many f_i are non-zero. When writing example partitions down, one tends to drop the zero frequency parts to keep the notation clean.

If none of the *frequencies* f_i are greater than 1, we call these partitions *distinct*. One can define the *size* of a partition π as

$$|\pi| = \sum_{i\geq 1} i \cdot f_i,$$

and the sum of all f_i is the number of parts in a partition, we denote this by $\#(\pi)$. The partition with $f_i \equiv 0$ for all $i \in \mathbb{N}$ is the only partition of 0 with 0 parts.

Let $t(\pi)$ be the number of non-zero frequencies of a partition π starting from f_1. In other words, one can think of $t(\pi)$ as the length of the initial frequency chain. The length of the initial frequency chain seems to be an underutilized statistic in interpretations of q-series identities, the only other closely related statistic that the author knows of is used in [7, Thm 3.1]. Let $p_j(\pi)$ be the maximum index i such that $f_i \geq j$ in π and for all $k \geq i$ has the property $f_k < j$, if no positive value satisfies this we define $p_j(\pi) = 0$. Let $r_j(\pi)$ be the number of different parts with frequencies $\geq j$.

To exemplify the statistics defined, let $\pi = (1^4, 2^2, 3^4, 5^1, 6^1)$ then $|\pi| = 31$, $\#(\pi) = 12$, $t(\pi) = 3$, $p_1(\pi) = 6$, $p_2(\pi) = 3$, $p_3(\pi) = 3$, $p_4(\pi) = 3$, $p_5(\pi) = 0, \dots$, $r_1(\pi) = 5$, $r_2(\pi) = 3$, $r_3(\pi) = 2$, $r_4(\pi) = 2$, $r_5(\pi) = 0, \dots$.

With the statistics defined above, one can interpret Corollary 1 as a weighted partition theorem, where $i = 1$ corresponds to (8) and $i = 2$ refers to (9), as follows.

Theorem 2. *Let* $\mathcal{D}$ *be the set of distinct partitions and let* $\mathcal{A}$ *be the set of partitions where all the partitions* $\pi \in \mathcal{A}$ *satisfy* $p_2(\pi) \leq t(\pi)$. *Then for* $i = 1$ *and 2, we have*

$$\sum_{\pi \in \mathcal{D}} w_i(\pi) q^{|\pi|} = \sum_{\pi \in \mathcal{A}} \hat{w}_i(\pi) q^{|\pi|}, \tag{13}$$

where

$$w_i(\pi) = \left[1 - f_1 \left(\frac{1 - (-1)^{t(\pi)}}{2} \right) \right] (-1)^{i \# (\pi)}, \tag{14}$$

$$\hat{w}_i(\pi) = 2^{r_2(\pi)} \left(\frac{(-1)^{t(\pi)} + (-1)^{p_2(\pi)}}{2} \right) (-1)^{(i-1)(r_1(\pi) + t(\pi) + p_2(\pi))}. \tag{15}$$

We would like to exemplify Theorem 2 with relevant partitions of 6 in Table 1.

Table 1. Partitions of 6 from $\mathcal{D}$ and $\mathcal{A}$ and the related weights w_i and $\hat{w}_i$ to exemplify 13.

$\pi \in \mathcal{D}$	$t(\pi)$	$w_1(\pi)$	$w_2(\pi)$	$\pi \in \mathcal{A}$	$t(\pi)$	$p_2(\pi)$	$r_2(\pi)$	$\hat{w}_1$	$r_1(\pi)$	$\hat{w}_2$
(6^1)	0	-1	1	(6^1)	0	0	0	1	1	-1
$(1^1, 5^1)$	1	0	0	$(1^1, 5^1)$	1	0	0	0	2	0
$(2^1, 4^1)$	0	1	1	$(2^1, 4^1)$	0	0	0	1	2	1
$(1^1, 2^1, 3^1)$	3	0	0	$(1^2, 4^1)$	1	1	1	-2	2	-2
				$(1^1, 2^1, 3^1)$	3	0	0	0	3	0
				$(1^3, 3^1)$	1	1	1	-2	2	-2
				$(1^4, 2^1)$	2	1	1	0	2	0
				$(1^2, 2^2)$	2	2	2	4	2	4
				(1^6)	1	1	1	-2	1	2
Total:		0	2					0		2

One key observation is that $w_2(\pi) = |w_1(\pi)| \geq 0$ for all distinct partitions. This proves that the series in (9), which is the analytic version of (13) with $i = 2$, have non-negative coefficients. We write this as a theorem using an equivalent form of the left-hand side series of (9).

Theorem 3. *We have*

$$(-q; q)_\infty \sum_{n \geq 0} \frac{(-1)^n q^{\frac{n(n+1)}{2}}}{(-q, q)_n} \succcurlyeq 0,$$

where $\succcurlyeq 0$ *is used to indicate that the series coefficients are all greater or equal than 0.*

The sum in Theorem 3 is a false theta function that Rogers studied [11]. Although this series has alternating signs, its product with the manifestly positive factor $(-q; q)_\infty$ has non-negative coefficients and the above key observation is a combinatorial explanation of this fact.

Broadly speaking, connections of false/partial theta functions and their implications in the theory of partitions have been studied in various places. Interested readers can refer to [1,6].

Proof (Proof of Theorem 2). This theorem is a consequence of Corollary 1, the $i = 1$ and 2 cases correspond to the combinatorial interpretations of (8) and (9), respectively.

First we focus on the left-hand side summands. For a fixed n and $\varepsilon_1 = \pm 1$, $(\varepsilon_1 q^{n+1})_\infty$ is the generating function for the distinct partitions π_d where every part is $\geq n + 1$ counted with the weight $(-\varepsilon_1)^{\#(\pi_d)}$. We also interpret the q-factor, $\varepsilon_2^n q^{n(n+1)/2}$ as the partition $\pi_i = (1^1, 2^1, \ldots, n^1)$ counted with the weight ε_2^n, where $\varepsilon_2 = \pm 1$. We can combine (add the frequencies of both partitions) π_d and π_i into a distinct partition π.

In the sum,

$$\sum_{n \geq 0} (\varepsilon_1 q^{n+1})_\infty \varepsilon_2^n q^{\frac{n(n+1)}{2}},$$

there are $t(\pi) + 1$ possible pairs (π_d, π_i) that can yield π, and one needs to count the weights of these accordingly. Note that if $t(\pi) \geq 1$ since π is a distinct partition $f_1 = 1$. For the total weight of π, one needs to sum from $k = 0$ to $t(\pi)$ of the alternating weights $(-\varepsilon_1)^{\#(\pi)-k} \varepsilon_2^k$:

$$\sum_{k=0}^{t(\pi)} (-\varepsilon_1)^{\#(\pi)-k} \varepsilon_2^k.$$

By reducing the summations of alternating weights, one finds that $w_i(\pi)$ can be represented as in (14) for $i = 1$ and 2, where $\varepsilon_1 = \varepsilon_2 = 1$ and $\varepsilon_1 = \varepsilon_2 = -1$, respectively.

We interpret the right-hand side summand similar to the left-hand side's interpretation. For a fixed n, once again the parts $(\varepsilon_1 q^{n+1})_\infty \varepsilon_2^n q^{n(n+1)/2}$ can be interpreted as the generating function for the partition pairs (π_d, π_i) counted by some weights dependent of ε_1 and ε_2. The new factor $(-q)_n/(q)_n$ is the generating function for the number of overpartitions π_o, into parts $\leq n$. Overpartitions are the same as partitions counted with the weight $2^{r_1(\pi)}$. When we combine π_d, π_i and π_o, we end up with a partition π where some parts may repeat.

Any repetition of the parts in π comes from the overpartition π_o and these repetitions can only appear for parts $\leq t(\pi)$. Note that π_i, has a single copy of every part size up to $t(\pi)$ and π_o may add more occurrences of these parts. This modifies the overpartition related weight a little and we need to take the first occurrence of a part for granted. On the other hand, if a part appears more than once the repetition should be counted with the weight $2^{r_2(\pi)}$.

Here the summation bounds are slightly different than the previous case. One needs to sum all the possible ε_1 and ε_2 related weights from $k = p_2(\pi)$ to $t(\pi)$. Different than the previous one, $\#(\pi)$ is replaced by the number of non-repeating parts above the initial chain $t(\pi)$, which is $r_1(\pi) - t(\pi)$. Moreover, one needs to replace k by $k - p_2(\pi)$ to eliminate the effect of the parity of $p_2(\pi)$ on the alternating sum. Hence, the sum to reduce here is

$$\sum_{k=p_2(\pi)}^{t(\pi)} (-\varepsilon_1)^{r_1(\pi)-t(\pi)-p_2(\pi)-k} \varepsilon_2^k.$$

These sums, once reduced, can be seen to yield $\hat{w}_i(\pi)$ for $i = 1$ and 2, where $\varepsilon_1 = \varepsilon_2 = -1$ and $\varepsilon_1 = \varepsilon_2 = 1$, respectively.

Acknowledgement. The author would like to thank the SFB50-07, SFB50-09 and SFB50-11 Projects of the Austrian Science Fund FWF for supporting his research.

References

1. Alladi, K.: A partial theta identity of Ramanujan and its number-theoretic interpretation. Ramanujan J. **20**, 329–339 (2009)
2. Alladi, K.: Partition identities involving gaps and weights. Trans. Am. Math. Soc. **349**(12), 5001–5019 (1997)
3. Alladi, K.: Analysis of a generalized Lebesgue identity in Ramanujan's Lost Notebook. Ramanujan J. **29**, 339–358 (2012)
4. Andrews, G.E.: The Theory of Partitions. Cambridge Mathematical Library. Cambridge University Press, Cambridge (1998). Reprint of the 1976 Original. MR1634067 (99c:11126)
5. Andrews, G.E., Berndt, B.C.: Ramanujan's Lost Notebook: Part II. Springer, New York (2009). https://doi.org/10.1007/b13290
6. Berkovich, A., Uncu, A.K.: Variation on a theme of Nathan Fine. New weighted partition identities. J. Number Theory **176**, 226–248 (2017)
7. Berkovich, A., Uncu, A.K.: New weighted partition theorems with the emphasis on the smallest part of partitions. In: Andrews, G.E., Garvan, F. (eds.) ALLADI60 2016. SPMS, vol. 221, pp. 69–94. Springer, Cham (2017). https://doi.org/10.1007/978-3-319-68376-8_6
8. Bessenrodt, C., Pak, I.: Partition congruences by involutions. Eur. J. Comb. **25**, 1139–1149 (2004)
9. Dixit, A., Maji, B.: Partition implications of a three-parameter q-series identity. Ramanujan J. (2019). https://doi.org/10.1007/s11139-019-00177-6
10. Gasper, G., Rahman, M.: Basic Hypergeometric Series, vol. 96. Cambridge University Press, Cambridge (2004)
11. Rogers, L.J.: On two theorems of combinatory analysis and some allied identities. Proc. Lond. Math. Soc. **s2–16**(1), 315–336 (1917)
12. Uncu, A.K.: Weighted Rogers-Ramanujan partitions and Dyson Crank. Ramanujan J. **46**(2), 579–591 (2018)

Edge-Critical Equimatchable Bipartite Graphs

Yasemin Büyükçolak[1]($\boxtimes$)(iD), Didem Gözüpek[2](iD), and Sibel Özkan[1](iD)

[1] Department of Mathematics, Gebze Technical University,
Gebze, Kocaeli, Turkey
{y.buyukcolak,s.ozkan}@gtu.edu.tr
[2] Department of Computer Engineering, Gebze Technical University,
Gebze, Kocaeli, Turkey
didem.gozupek@gtu.edu.tr

Abstract. A graph is called equimatchable if all of its maximal matchings have the same size. Lesk et al. [6] provided a characterization of equimatchable bipartite graphs. Since this characterization is not structural, Frendrup et al. [4] also provided a structural characterization for equimatchable graphs with girth at least five; in particular, a characterization for equimatchable bipartite graphs with girth at least six. In this work, we extend the partial characterization of Frendrup et al. [4] to equimatchable bipartite graphs without any restriction on girth. For an equimatchable graph, an edge is said to be critical-edge if the graph obtained by removal of this edge is not equimatchable. An equimatchable graph is called edge-critical if every edge is critical. Reducing the characterization of equimatchable bipartite graphs to the characterization of edge-critical equimatchable bipartite graphs, we give two characterizations of edge-critical equimatchable bipartite graphs.

Keywords: Equimatchable · Bipartite graphs · Edge-critical

1 Introduction

All graphs in this paper are finite, simple, and undirected. For a graph $G = (V(G), E(G))$, $V(G)$ and $E(G)$ denote the set of vertices and edges in G, respectively. An edge joining the vertices u and v in G will be denoted by uv. A *bipartite graph* G is a graph whose point set $V(G)$ can be partitioned into two subsets V_1 and V_2 such that every edge of G joins V_1 with V_2. If $|V_1| = |V_2|$, then we say that G is *balanced*. For a vertex v in G and a subset $X \subseteq V(G)$, $N_G(v)$ denotes the set of neighbors of v in G, while $N_G(X)$ denotes the set of all vertices adjacent to at least one vertex of X in G. We omit the subscript G when it is clear from the context. The *order* of G is denoted by $|V(G)|$ and the *degree* of a vertex v

This work is supported by the Scientific and Technological Research Council of Turkey (TUBITAK) under grant no. 118E799. The work of Didem Gözüpek was supported by the BAGEP Award of the Science Academy of Turkey.

© Springer Nature Switzerland AG 2020
D. Slamanig et al. (Eds.): MACIS 2019, LNCS 11989, pp. 280–287, 2020.
https://doi.org/10.1007/978-3-030-43120-4_21

of G is denoted by $d(v)$. A vertex of degree one is called a *leaf* and a vertex adjacent to a leaf is called a *stem*. For a graph G and $U \subseteq V(G)$, the subgraph induced by U is denoted by $G[U]$. The *difference* $G \backslash H$ of two graphs G and H is defined as the subgraph induced by the difference of their vertex sets, that is, $G \backslash H = G[V(G) \backslash V(H)]$. For a graph G and a vertex v of G, the subgraph induced by $V(G) - v$ is denoted by $G - v$ for the sake of brevity. We also denote by $G \backslash e$ the graph $G(V, E \backslash \{e\})$. The cycle and complete graph on n vertices are denoted by C_n and K_n, respectively, while the complete bipartite graph with partite sets of sizes n and m is denoted by $K_{n,m}$. The length of a shortest cycle in G is called the *girth* of G. For a graph G, $c(G)$ denotes the number of components in G. A set of vertices S of a graph G such that $c(G \backslash S) > c(G)$ is called a *cut-set*. A vertex v is called a *cut-vertex* if $\{v\}$ is a cut-set. A graph is called *2-connected* if its cut-sets have at least 2 vertices.

A *matching* in a graph G is a set $M \subseteq E(G)$ of pairwise nonadjacent edges of G. A vertex v of G is *saturated by* M if $v \in V(M)$ and *exposed by* M otherwise. A matching M is called *maximal* in G if there is no other matching of G that contains M. A matching is called a *maximum* matching of G if it is a matching of maximum size. The size of a maximum matching of G is denoted by $\nu(G)$. A matching M in G is a *perfect matching* if M saturates all vertices in G, that is, $V(M) = V(G)$. For a vertex v, a matching M is called a *matching isolating v* if $\{v\}$ is a component of $G \backslash V(M)$. A graph G is *equimatchable* if every maximal matching of G is a maximum matching, that is, every maximal matching has the same cardinality. A graph G is *randomly matchable* if it is an equimatchable graph admitting a perfect matching. A graph G is *factor-critical* if $G - v$ has a perfect matching for every vertex v of G. A factor-critical graph cannot be bipartite, since if you choose a vertex from the small partite set (or from any partite set if their cardinalities are equal) there cannot be a perfect matching in the rest of the graph.

In the literature, the structure of equimatchable graphs are extensively studied by several authors, see [5,7,9]. In 1984, Lesk et al. [6] formally introduced equimatchable graphs and provided a characterization of equimatchable graphs via Gallai-Edmonds decomposition, yielding a polynomial-time recognition algorithm. In [10], Sumner characterized the equimatchable graphs with a perfect matching, i.e., randomly matchable graphs, whereas the work in [6] provided a characterization for general equimatchable graphs. Particularly, [6] provided a characterization of equimatchable bipartite graphs in terms of subsets of neighborhoods of vertices in smaller partite set. Although this characterization is valid for all equimatchable bipartite graphs, the structure of these graphs is not completely understood. In 2010, Frendrup et al. [4] gave a structural characterization of equimatchable graphs with girth at least five. However, the work in [4] provides a partial characterization for equimatchable bipartite graphs; namely, a characterization for equimatchable bipartite graphs with girth at least six.

Motivated by lack of a structural characterization for all equimatchable bipartite graphs, we investigate the structure of equimatchable bipartite graphs in this work. For an equimatchable graph, an edge is said to be *critical-edge* if the graph obtained by removal of this edge is not equimatchable. An equimatchable graph is called *edge-critical* if every edge is critical. Notice that each edge-critical equimatchable bipartite graph can be obtained from some equimatchable bipartite graphs having the same vertex partition by recursively removing non-critical edges. Conversely, each equimatchable bipartite graph can also be constructed from some edge-critical equimatchable bipartite graphs by joining some non-adjacent vertices from different partite sets. Therefore, we focus on the structure of bipartite edge-critical equimatchable bipartite graphs instead of the structure of equimatchable bipartite graphs.

In Sect. 2, we provide some structural results for equimatchable bipartite graphs by using Gallai-Edmonds decomposition. Particularly, we extend the partial characterization of Frendrup et al. [4] to all equimatchable bipartite graphs without any girth condition. In Sect. 3, we discuss the structure of edge-critical equimatchable bipartite graphs. We first show that every connected edge-critical equimatchable bipartite graph is 2-connected. Afterwards, we provide two characterizations for edge-critical equimatchable bipartite graphs.

2   Equimatchable Bipartite Graphs

In this section, we would like to investigate connected equimatchable bipartite graphs, more simply EB-graphs. Since a graph is equimatchable if and only if all of its components are equimatchable, it suffices to focus on connected EB-graphs. In the literature, there exist some characterizations for equimatchable bipartite graphs. For instance, the characterization of randomly matchable graphs, not necessarily bipartite, was provided in [10] as follows:

Theorem 1 [10]. *A connected graph is randomly matchable if and only if it is isomorphic to K_{2n} or $K_{n,n}$, $n \geq 1$.*

The following characterization of equimatchable graphs with girth at least five, not necessarily bipartite, was provided in [4]:

Theorem 2 [4]. *Let G be a connected equimatchable graph with girth at least 5. Then $G \in \mathcal{F} \cup \{C_5, C_7\}$, where $\mathcal{F}$ is the family of graphs containing K_2 and all connected bipartite graphs with bipartite sets V_1 and V_2 such that all vertices in V_1 are stems and no vertex from V_2 is a stem.*

Note here that this characterization is only a partial characterization for EB-graphs although it completely reveals the structure of EB-graphs with girth at least six. On the other hand, the work in [6] provides a general characterization for EB-graphs as follows:

Theorem 3 [6]. *A connected bipartite graph $G = (U \cup V, E)$ with $|U| \leq |V|$ is equimatchable if and only if for all $u \in U$, there exists a non-empty $X \subseteq N(u)$ such that $|N(X)| \leq |X|$.*

The following result is a reformulation of the characterization of EB-graphs in Theorem 3 by using well-known Hall's Theorem saying that in a bipartite graph $G = (A \cup B, E)$ with $|A| \leq |B|$, there exists a matching saturating all vertices in A if and only if for all subset $S \subseteq A$, we have $|N(S)| \geq |S|$.

Theorem 4 [3]. *Let $G = (U \cup V, E)$ be a connected bipartite graph with $|U| \leq |V|$. Then G is equimatchable if and only if every maximal matching of G saturates all vertices in U.*

Although Theorem 3 provides a complete characterization for EB-graphs, it does not explicitly reveal the structure of EB-graphs. The lack of structural characterization for all EB-graphs motivated us to study the structure of EB-graphs.

The following well-known structural result, which is called *Gallai-Edmonds decomposition*, provides an important characterization for general graphs, not necessarily bipartite, based on maximum matchings as follows:

Theorem 5 [8]. *For any graph G, let $D(G)$ denote the set of vertices which are exposed by at least one maximum matching of G and $A(G)$ be the vertices of $V(G) \backslash D(G)$ which are neighbors of at least one vertex of $D(G)$. Let $C(G) = V(G) \backslash (D(G) \cup A(G))$. Then:*

1. *Every component of the graph $G[D(G)]$ is factor-critical,*
2. *$G[C(G)]$ has a perfect matching,*
3. *every maximum matching of G matches every vertex of $A(G)$ to a vertex of distinct component of $G[D(G)]$.*

It is easy to observe that if a graph G admits a perfect matching then $C(G) = V(G)$, and if G is a connected equimatchable graph with no perfect matching then $C(G) = \emptyset$ and $A(G)$ is an independent set in G.

We focus on the case where G is an equimatchable graph with no perfect matching and $A(G)$ is nonempty. It can be easily seen that all EB-graphs with no perfect matching are non-factor-critical; however, all equimatchable non-factor-critical graphs with no perfect matching are not bipartite. The following result is not explicitly given in [6], but it is an immediate consequence of Theorems 3 and 4 in [6]:

Lemma 1 [6]. *Let G be a connected equimatchable non-factor-critical graph with no perfect matching. G is bipartite if and only if each component of $G[D]$ is K_1.*

Corollary 1. *Let G be a connected EB-graph with no perfect matching. Then, $C(G) = \emptyset$, and each of $D(G)$ and $A(G)$ are a nonempty independent set.*

For the rest of the paper, $G = (U \cup V, E)$ denotes a connected EB-graph with $|U| < |V|$. By Theorem 1, the only connected EB-graph with equal partite sets is $K_{n,n}$ where $n \geq 1$. The next result shows that the parts U and V of G correspond to the sets $A(G)$ and $D(G)$, respectively, where $C(G)$ is empty:

Lemma 2. *Let $G = (U \cup V, E)$ with $|U| < |V|$ be a connected EB-graph and let (D, A, C) be its Gallai-Edmonds decomposition. Then we have $C = \emptyset$, $A = U$ and $D = V$.*

Corollary 2. *Let* $G = (U \cup V, E)$ *with* $|U| < |V|$ *be a connected EB-graph. Then there exists an isolating matching for each* $v \in V$ *and there is no isolating matching for any* $u \in U$.

Corollary 3. *Let* $G = (U \cup V, E)$ *with* $|U| \leq |V|$ *be a connected EB-graph with Gallai-Edmonds decomposition* (D, A, C).

- *If* G *admits a perfect matching, that is,* $|U| = |V|$, *then* $C = V(G)$, $D = \emptyset$ *and* $A = \emptyset$. *In particular,* G *is* $K_{n,n}$ *where* $n \geq 1$.
- *If* G *admits no perfect matching, that is,* $|U| < |V|$, *then* $C = \emptyset$, $A = U$ *and* $D = V$.

The next result extends the characterization of Frendrup et al. [4] to all EB-graphs by eliminating the girth condition:

Theorem 6. *Let* $G = (U \cup V, E)$ *with* $|U| < |V|$ *be a connected EB-graph. Then each vertex* $u \in U$ *satisfies at least one of the followings:*

(i) u *is a stem in* G,
(ii) u *is included in a subgraph* $K_{2,2}$ *in* G.

Proof. Let $G = (U \cup V, E)$ with $|U| < |V|$ be a connected EB-graph. Let $u \in U$ and $N(u) = \{v_1, v_2, ..., v_n\}$ be the set of neighbors of u in V. If one vertex in $N(u)$ is a leaf in G, then we are done. Assume to the contrary that none of the vertices in $N(u)$ is a leaf in G. If u is not included in a subgraph $K_{2,2}$ in G, then there is no pair of vertices in $N(u)$ having a common neighbor except u. It implies that there exists a matching isolating u in G. Since it contradicts with Corollary 2, we deduce that there exists at least one pair, say $\{v_1, v_2\}$, of vertices in $N(u)$ having a common neighbor except u, say u^*. It follows that the vertices $\{u, v_1, u^*, v_2\}$ induce a $K_{2,2}$ in G, as desired. $\qquad\square$

Therefore, an extension of Theorem 2 can be derived as a corollary of Theorem 6 in the following way:

Corollary 4. *Let* G *be a connected EB-graph with girth at least* 6. *Then* $G \in \mathcal{F}$, *where* $\mathcal{F}$ *is the family of graphs containing* K_2 *and all connected bipartite graphs with bipartite sets* V_1 *and* V_2 *with* $|V_1| \leq |V_2|$ *such that all vertices in* V_1 *are stems and no vertex from* V_2 *is a stem.*

Finally, in Lemma 4, we extend the following known result about the cut vertices in equimatchable graphs to EB-graphs as in the following way:

Lemma 3 [1]. *Let* G *be a connected equimatchable graph with a cut vertex* c, *then each component of* $G - c$ *is also equimatchable.*

Lemma 4. *Let* $G = (U \cup V, E)$ *with* $|U| < |V|$ *be a connected EB-graph with a cut vertex* c, *then each component of* $G - c$ *is also an EB-graph. Furthermore, if* $H = (U_H \cup V_H, E_H)$ *with* $|U_H| \leq |V_H|$ *is a component of* $G - c$, *then* $U_H \subseteq U$ *and* $V_H \subseteq V$.

The theorem says that for each cut vertex c of G, the components of $G - c$ are induced EB-subgraphs preserving (U, V)-partitions of G.

3   Edge-Critical Equimatchable Bipartite Graphs

The aim of this section is to characterize a generating subclass of EB-graphs, namely *edge-critical EB-graphs*. Recall that for an equimatchable graph, an edge is a *critical-edge* if the graph obtained by removal of this edge is not equimatchable, and an equimatchable graph is *edge-critical*, if every edge is critical.

For a connected EB-graph $G = (U \cup V, E)$ with $|U| < |V|$, Theorem 4 implies that any bipartite supergraph of G obtained by joining some pair of non-adjacent vertices $u \in U$ and $v \in V$ of G is also a connected EB-graph with the same vertex sets $U \cup V$. Intuitively, we consider EB-subgraphs of G with the same vertex set $U \cup V$. In fact, the smallest such EB-subgraph of G is an edge-critical EB-graph with the same vertex set $U \cup V$. It follows that each edge-critical EB-graph can be obtained from some EB-graphs having the same vertex set by recursively removing non-critical edges. Therefore, in order to characterize all EB-graphs we only need to characterize all edge-critical EB-graphs. It means that the class of edge-critical EB-graphs form a generating subclass of EB-graphs. Since a graph is equimatchable if and only if each of its components is equimatchable, it suffices to focus on connected edge-critical EB-graphs. Notice that the complete graph K_2 (or equivalently the complete bipartite graph $K_{1,1}$) is equimatchable but not edge-critical.

The following results about edge-critical equimatchable graphs, not necessarily bipartite, are frequently used in our arguments:

Lemma 5 [2]. *Let $G \neq K_2$ be a connected equimatchable graph. Then $\nu(G) = \nu(G \backslash e)$ for every non-critical edge $e \in E(G)$.*

Lemma 6 [2]. *Let $G \neq K_2$ be a connected equimatchable graph. Then $uv \in E(G)$ is critical if and only if there is a matching of G containing uv and saturating $N_{G \backslash uv}(\{u, v\})$.*

Corollary 5. *A connected equimatchable graph $G \neq K_2$ is edge-critical if and only if there is a matching containing uv and saturating $N(\{u, v\})$ for every $uv \in E(G)$.*

Corollary 6. *All randomly matchable graphs except K_2 are edge-critical.*

The next result shows that edge-critical EB-graphs cannot have a cut vertex.

Lemma 7. *Let $G = (U \cup V, E)$ with $|U| \leq |V|$ be a edge-critical EB-graph. Then G is 2-connected; i.e. G has no cut vertex.*

Proof. Let $G = (U \cup V, E)$ with $|U| \leq |V|$ be a edge-critical EB-graph. In the case where $|U| = |V|$, G is $K_{n,n}$ for some $n \geq 2$ by Corollary 6. It is clear that $K_{n,n}$, $n \geq 2$, is 2-connected and we are done. We then suppose that $|U| < |V|$. Assume to the contrary that G has a cut vertex c. Let $H_1, H_2, ..., H_k$ ($k \geq 2$) be connected components of $G - c$ such that $d_i \in H_i$ where $d_1, d_2, ..., d_k \in N(c)$ and $i \in [k]$. By Lemma 4, each H_i is an EB-subgraph preserving (U, V)-partitions of G where $i \in [k]$. By Lemma 6, for each edge $e_i = cd_i$, there exists a matching M_j

in H_j saturating all vertices in $N_{H_j}(c)$ for $j \in [k]$ with $j \neq i$. Then, it follows that there exists a matching $M = \bigcup_{l=1}^k M_l$ isolating c in G. By Corollary 2, we have $c \in V$ and $d_1, d_2, ..., d_k \in U$. On the other hand, by Lemma 6, for each edge $e_i = cd_i$, there also exists a matching M_i' in H_i saturating all vertices in $N_{H_i}(d_i)$ for $i \in [k]$. It is easy to see that each M_i' is indeed a matching isolating d_i in H_i for $i \in [k]$. By Corollary 2, we conclude that $d_i \notin U_{H_i}$, which contradicts with Lemma 4.   □

The next result provides a characterization for edge-critical EB-graphs as follows:

Theorem 7. *A connected bipartite graph $G = (U \cup V, E)$ with $|U| \leq |V|$ is an edge-critical EB-graph if and only if for every $u \in U$, $|N(S)| \geq |S|$ holds for any subset $S \subseteq N(u)$. In particular, the equality holds only for $S = N(u)$.*

Proof. Let $G = (U \cup V, E)$ with $|U| \leq |V|$ be a connected bipartite graph. In the case where $|U| = |V|$, by Corollary 6, G is $K_{n,n}$ for some $n \geq 2$. The theorem holds and we are done. Hence, we then suppose that $|U| < |V|$.

($\Rightarrow$) Suppose that G is a edge-critical EB-graph. Assume to the contrary that there exists $u \in U$ such that $|N(S)| < |S|$ holds for some $S \subseteq N(u)$. Since $|N(S)| < |S|$, there is no matching saturating all vertices in S. It implies that for any $w \in S$, there is no matching saturating $N(\{u, w\})$ and containing uw. By Lemma 6, the edge uw is not critical, contradicting with G being an ECE-graph.

($\Leftarrow$) Suppose that for every $u \in V(G)$, $|N(S)| \geq |S|$ holds for any subset $S \subseteq N(u)$ and the equality holds only for $S = N(u)$. Since $S = N(u)$ satisfies $|N(S)| = |S|$, G is an equimatchable graph by Theorem 3. Assume to the contrary that $uv \in E(G)$ is not critical, where $u \in U$ and $v \in N(u)$; that is, $G \backslash uv$ is equimatchable. Then, by Theorem 3, there exists $X \subseteq N_{G \backslash uv}(u)$ such that $|N_{G \backslash uv}(X)| \leq |X|$. Note also that $X \subseteq N_{G \backslash uv}(u) \subset N(u)$. It follows that $N(X) = N_{G \backslash uv}(X) \leq |X|$. However, since $X \neq N(u)$, it contradicts with the assumption that $|N(S)| \geq |S|$ holds for any subset $S \subseteq N(u)$ and the equality holds only for $S = N(u)$.   □

The next theorem provides another characterization for edge-critical EB-graphs in terms of induced subgraphs as follows:

Theorem 8. *A connected bipartite graph $G = (U \cup V, E)$ with $|U| \leq |V|$ is an edge-critical EB-graph if and only if for any $u \in U$, the subgraph $H = (U_H \cup V_H, E_H)$ of G induced by the vertices $N(u)$ and $N(N(u))$ is a 2-connected balanced bipartite subgraph of G with a perfect matching.*

Proof. Let $G = (U \cup V, E)$ with $|U| \leq |V|$ be a connected bipartite graph.

($\Rightarrow$) Suppose that G is a edge-critical EB-graph. Let $u \in U$ and H be the subgraph of G induced by the vertices $N(u)$ and $N(N(u))$. By Theorem 7, we have $|N(u)| = |N(N(u))|$, implying that $|U_H| = |V_H|$. That is, H is a balanced bipartite subgraph of G. By Corollary 5, for any neighbor v of u, there exists a matching containing uv and saturating all other neighbors of u in G. It implies that there exists a perfect matching containing uv in H since H is a balanced bipartite graph.

Notice that u cannot be nonstem cut vertex because otherwise it contradicts with the fact that for any neighbor v of u, there exists a perfect matching containing uv in H. By definition of H, u is a dominating vertex in H; that is, u is adjacent to all vertices of V_H. Hence, it is easy to see that H cannot have any other nonstem cut vertex. We now show that H has no stem cut vertex; that is, there is no leaf in H. By definition of H, all neighbors of vertices of V_H in G are included in U_H. Then there is no leaf in V_H since G is 2-connected by Lemma 7. If there exists a leaf in U_H, say u^*, then there exists a stem v^* in V_H such that $u^*v^* \in E(G)$. By Corollary 5, there exists a matching containing the edge uv^* and saturating all neighbors of u in G. Note that u^* has no neighbor in V_H other than v^* and u is a dominating vertex in H. Hence, there is no such matching by the cardinalities of $V_H \backslash v^*$ and $U_H \backslash u$ since H is a balanced bipartite graph. Therefore, H is a 2-connected balanced bipartite subgraph of G with a perfect matching.

($\Leftarrow$) Suppose that for any $u \in U$, the subgraph $H = (U_H \cup V_H, E_H)$ of G induced by the vertices $N(u)$ and $N(N(u))$ is a 2-connected balanced bipartite subgraph of G with a perfect matching. Since H is a 2-connected balanced bipartite graph with a perfect matching, $|N(S)| \geq |S|$ holds for any subset $S \subseteq N(u)$ while the equality holds only for $S = N(u)$. Hence, by Theorem 7, we deduce that G is an edge-critical EB-graph. $\square$

Corollary 7. *Let $G = (U \cup V, E)$ with $|U| \leq |V|$ be a connected edge-critical EB-graph. If $H = (U_H \cup V_H, E_H)$ is a subgraph of G induced by the vertices $N(u)$ and $N(N(u))$ for any $u \in U$, then all vertices in $U_H \backslash \{u\}$ form a C_4 with u.*

References

1. Akbari, S., Ghodrati, A.H., Hosseinzadeh, M.A., Iranmanesh, A.: Equimatchable regular graphs. J. Graph Theory **87**, 35–45 (2018)
2. Deniz, Z., Ekim, T.: Critical equimatchable graphs. Preprint
3. Deniz, Z., Ekim, T.: Edge-stable equimatchable graphs. Discrete Appl. Math. **261**, 136–147 (2019)
4. Frendrup, A., Hartnell, B., Preben, D.: A note on equimatchable graphs. Australas. J. Comb. **46**, 185–190 (2010)
5. Grünbaum, B.: Matchings in polytopal graphs. Networks **4**, 175–190 (1974)
6. Lesk, M., Plummer, M.D., Pulleyblank, W.R.: Equi-matchable graphs. In: Graph Theory and Combinatorics (Cambridge, 1983), pp. 239–254. Academic Press, London (1984)
7. Lewin, M.: M-perfect and cover-perfect graphs. Israel J. Math. **18**, 345–347 (1974)
8. Lovász, L., Plummer, M.D.: Matching Theory, vol. 29, Annals of Discrete Mathematics edn. North-Holland, Amsterdam (1986)
9. Meng, D.H.-C.: Matchings and coverings for graphs. Ph.D. thesis. Michigan State University, East Lansing, MI (1974)
10. Sumner, D.P.: Randomly matchable graphs. J. Graph Theory **3**, 183–186 (1979)

Determining the Rank of Tensors in $\mathbb{F}_q^2 \otimes \mathbb{F}_q^3 \otimes \mathbb{F}_q^3$

Nour Alnajjarine[✉] and Michel Lavrauw[✉]

Sabanci University, Istanbul, Turkey
{nour,mlavrauw}@sabanciuniv.edu

Abstract. Let $\mathbb{F}_q$ be a finite field of order q. This paper uses the classification in [7] of orbits of tensors in $\mathbb{F}_q^2 \otimes \mathbb{F}_q^3 \otimes \mathbb{F}_q^3$ to define two algorithms that take an arbitrary tensor in $\mathbb{F}_q^2 \otimes \mathbb{F}_q^3 \otimes \mathbb{F}_q^3$ and return its orbit, a representative of its orbit, and its rank.

Keywords: Tensor rank · Rank distribution · Tensor decomposition

1 Introduction and Preliminaries

The study of tensors of order at least three has been an active area in recent years, with numerous applications in representation theory, algebraic statistics and complexity theory [5,6]. For example, the problem of determining the complexity of matrix multiplication can be rephrased as the problem of determining the rank of a particular tensor (the matrix multiplication operator). This problem has only been solved for 2×2-matrices (see Strassen and Winograd), and we refer to [6, Chap. 2, Sect. 4] for more on this topic.

Determining the decomposition of a tensor A is a notoriously hard problem that arises in many other applications such as psychometrics, chemometrics, numerical linear algebra and numerical analysis [5]. In many tensor decomposition problems, the first issue to resolve is to determine the rank of the tensor, which is not always an easy task.

Let $Sym(2)$ denote the symmetric group of order 2 and $V := \mathbb{F}_q^2 \otimes \mathbb{F}_q^3 \otimes \mathbb{F}_q^3$, where $\mathbb{F}_q$ is the finite field of order q. Consider then the two natural actions on V of the group G and its subgroup H, where $G \cong \mathrm{GL}(\mathbb{F}_q^2) \times (\mathrm{GL}(\mathbb{F}_q^3) \wr Sym(2))$, as a subgroup of $\mathrm{GL}(V)$ stabilising the set of fundamental tensors in V, and $H \cong \mathrm{GL}(\mathbb{F}_q^2) \times \mathrm{GL}(\mathbb{F}_q^3) \times \mathrm{GL}(\mathbb{F}_q^3)$. In this paper, we study tensors in V under the action of G to present the algorithms *"RankOfTensor"* and *"OrbitOfTensor"*, which take an arbitrary tensor in V and return its orbit, a representative of its orbit, and its rank.

We follow the notation and terminology from [8]. Let A be a tensor in V. The *rank* of A, $Rank(A)$, is defined to be the smallest integer r such that

$$A = \sum_{i=1}^{r} A_i \tag{1}$$

D. Slamanig et al. (Eds.): MACIS 2019, LNCS 11989, pp. 288–294, 2020.
https://doi.org/10.1007/978-3-030-43120-4_22

with each A_i, a rank one tensor in V. Recall that the set of rank one tensors (fundamental tensors) in V is the set $\{v_1 \otimes v_2 \otimes v_3 : v_1 \in \mathbb{F}_q^2 \setminus \{0\}, \ v_2, v_3 \in \mathbb{F}_q^3 \setminus \{0\}\}$. It is clear from this definition that the rank of a tensor is a *projective property* in the vector space V. In other words, the rank of A does not change when A is multiplied by a nonzero scalar. For this reason, to dispose of the unneeded information, it makes sense to consider the problem of rank and decomposition in the projective space $\mathrm{PG}(V)$.

The Segre Variety. Projectively, the set of nonzero tensors of rank one corresponds to the set of points on the Segre variety $S_{1,2,2}(\mathbb{F}_q)$, which is the image of the Segre embedding $\sigma_{1,2,2}$ defined as:

$$\sigma_{1,2,2} : \mathrm{PG}(\mathbb{F}_q^2) \times \mathrm{PG}(\mathbb{F}_q^3) \times \mathrm{PG}(\mathbb{F}_q^3) \longrightarrow \mathrm{PG}(V)$$
$$(\langle v_1 \rangle, \langle v_2 \rangle, \langle v_3 \rangle) \mapsto \langle v_1 \otimes v_2 \otimes v_3 \rangle.$$

For any projective point in $\mathrm{PG}(V)$, we define its rank to be the rank of any corresponding tensor.

Contraction Spaces. For $A \in V$, we define the *first contraction space of A* to be the following subspace of $\mathbb{F}_q^3 \otimes \mathbb{F}_q^3$:

$$A_1 := \langle u_1^\vee(A) : u_1^\vee \in \mathbb{F}_q^{2^\vee} \rangle \tag{2}$$

where $\mathbb{F}_q^{2^\vee}$ denotes the dual space of $\mathbb{F}_q^2$, and where the *contraction* $u_1^\vee(A)$ is defined by its action on the fundamental tensors as follows:

$$u_1^\vee(v_1 \otimes v_2 \otimes v_3) = u_1^\vee(v_1)v_2 \otimes v_3. \tag{3}$$

Similarly, the *second* and *third* contraction spaces, A_2 and A_3, can be defined. Note that we are considering in this study the projective subspaces $\mathrm{PG}(A_1)$, $\mathrm{PG}(A_2)$ and $\mathrm{PG}(A_3)$ of $\mathrm{PG}(\mathbb{F}_q^3 \otimes \mathbb{F}_q^3)$, $\mathrm{PG}(\mathbb{F}_q^2 \otimes \mathbb{F}_q^3)$ and $\mathrm{PG}(\mathbb{F}_q^2 \otimes \mathbb{F}_q^3)$, respectively, where we have $\mathrm{PG}(\mathbb{F}_q^3 \otimes \mathbb{F}_q^3) \cong \mathrm{PG}(8, q)$ and $\mathrm{PG}(\mathbb{F}_q^2 \otimes \mathbb{F}_q^3) \cong \mathrm{PG}(5, q)$. Also, remark that the rank of any *contraction* coincides with the usual matrix rank.

Rank Distributions. For $1 \le i \le 3$, define the *i-th rank distribution of A*, R_i, to be the 3-tuple whose j-th entry is the number of rank j points in the i-th contraction space $\mathrm{PG}(A_i)$. Consider now the canonical basis of F_q^ℓ, $\{e_1, \ldots, e_\ell\}$, for $\ell = 2, 3$. We define the canonical basis of V as $\{e_i \otimes e_j \otimes e_k : 1 \le i \le 2 \text{ and } 1 \le j, k \le 3\}$. By writing $A \in V$ as $A = \sum A_{i,j,k} e_i \otimes e_j \otimes e_k$, one can view A as a $2 \times 3 \times 3$ rectangular solid whose entries are the $A_{i,j,k}$'s. This solid can be decomposed into slices that completely determine A. For example, we may view A as a collection of 2 size 3×3 matrices: $(A_{1,j,k}), (A_{2,j,k})$, which are called the *horizontal slices* of A, or a collection of 3 matrices $(A_{i,1,k}), (A_{i,2,k}), (A_{i,3,k})$ called the *lateral slices* of A, or a collection of 3 matrices $(A_{i,j,1}), (A_{i,j,2}), (A_{i,j,3})$ called the *frontal slices* of A.

Proposition 1 *(Corollary 2.2 in [8]).* Let $G_1 = \mathrm{GL}(\mathbb{F}_q^3) \wr Sym(2)$ *and* $H_1 := \mathrm{GL}(\mathbb{F}_q^3) \times \mathrm{GL}(\mathbb{F}_q^3)$. *Then, two tensors A and C in V are G-equivalent if and only* if A_1 is G_1-equivalent to C_1, if and only if A is H-equivalent to one of $\{C, C^T\}$, where T is the map on V defined by sending $c_1 \otimes c_2 \otimes c_3$ to $c_1 \otimes c_3 \otimes c_2$ and expanding linearly.

Theorem 1 *(Theorem 3.10 in [8]). There are* 21 *H-orbits and* 18 *G-orbits of tensors in V.*

Note that since we are working projectively, the trivial orbit containing the zero tensor will be ignored.

For the convenience of the reader, we have collected some information from [8] about each G-orbit in V and their contraction spaces including representatives of orbits, the *tensor rank* of each orbit and rank distributions on the webpage [2], to which we will refer as Table 1.

2   The Algorithms

In this section, we present a GAP function that takes an arbitrary tensor in V and returns its orbit number (see Table 1) and a representative of its orbit. The construction of this function is mainly based on the rank distributions of the projective contraction spaces associated with tensors in V, and the fact that tensors of the same orbit have the same rank distributions (see Proposition 1). We follow for this purpose the classification of G-orbits of tensors in V [7] as summarized in [2].

We start with a series of auxiliary functions that will be needed to construct our main function. The calling of most of these functions in GAP requires the usage of the GAP-package *FinInG* [3,4].

2.1   Auxiliary Functions

1. *MatrixOfPoint*: turns a point of a projective space into an $(m \times n)-$matrix containing the coordinates.
2. *RankOfPoint*: returns the rank of *MatrixOfPoint(x,m,n)*.
3. *RankDistribution*: returns the rank distribution of a subspace by considering its points as $m \times n$ matrices using the *RankOfPoint* function.
4. *CubicalArrayFromPointInTensorProductSpace*: returns the horizontal slices of a tensor in PG(V) where in our case we have $n_1 = 2$, $n_2 = 3$ and $n_3 = 3$. Notice that this function depends on how we choose the coordinates.
5. *ContractionOfPointInTensorProductSpace*: returns the projective contraction $vec^\vee(point)$; recall that in our case a point represents a tensor in PG(V).
6. *SubspaceOfContractions*: returns the projective contraction spaces associated with a projective point in PG(V).
7. *Rank1PtsOftheContractionSubspace*: returns the set of rank 1 points of PG(A_i) using the *RankOfPoint* function.
8. *RepO10odd*: returns a representative of o_{10} if q is odd.
9. *AlternativeRepresentationOfFiniteFieldElements*: gives an alternative way of representing finite fields' elements.

10. *RepO10even*: returns a representative of o_{10} if q is even.
11. *RepO15odd*: returns a representative of o_{15} if q is odd.
12. *RepO15even*: returns a representative of o_{15} if q is even.

2.2   OrbitOfTensor

The *OrbitOfTensor* function takes an arbitrary tensor A in PG(V) and by using the above auxiliary functions, it calculates the rank distribution of the first contraction space of A, R_1, and compares it with the results in Table 1 to specify the orbit number containing A. In some cases, R_1 is not enough to distinguish between orbits. For example, orbits o_{10}, o_{11} and o_{12} (resp. o_6 and o_7) have the same R_1. In this case, we calculate R_2 and R_3 to differentiate among them. But since the orbits o_4, o_7 and o_{11} are the only G-orbits of tensors which split under the action of H to o_i and o_i^T [8], we can see that a direct comparison between R_2 and R_3 from Table 1 will not be enough to distinguish between o_{10}, o_{11} and o_{12} (resp. o_6 and o_7). For this reason, we consider (algorithmically) some extra possible cases of R_2 and R_3 to insure that if $A \in o_j$ then $A^T \in o_j$, where $j \in \{7, 11\}$ [2]. Notice that, we do not have to do a similar work for o_4 since it is completely determined by R_1.

Although rank distributions are sufficient to specify the tensor's orbit in most cases, they are not helpful in distinguishing o_{15} and o_{16} as they have the same rank distributions. For this purpose, we use *Lemma 1* to distinguish between them.

Lemma 1. *Consider the two G-orbits of tensors in V, o_{15} and o_{16}. In both cases* PG(A_1) *is a line with rank distribution* $[0, 1, q]$. *Let x_2 be the unique rank 2 point on* PG(A_1) *and x_1 be a point among the q points of rank 3 on* PG(A_1). *Then, there exists a unique solid V containing x_2 which intersects $S_{3,3}(\mathbb{F}_q)$ in a subvariety $Q(x_2)$ equivalent to a Segre variety $S_{2,2}(\mathbb{F}_q)$. Furthermore, there is no rank one point in $U \setminus Q(x_2)$ for o_{16} where $U := \langle V, x_1 \rangle$, and there is one for o_{15}.*

Proof. The first result is a direct application of [8, Lemma 2.4]. The second one uses the two possible cases of having 2 points y_i, $i = 1, 2$ of rank i such that x_1 is on the line $\langle y_1, y_2 \rangle$ and $Q(x_2) = Q(y_2)$ or no such points exist, which were used in [8, Theorem 3.1 case(4)] to define o_{15} and o_{16}, respectively.   □

For the same reason, we consider the case $q = 2$ separately. In this case, as R_1 is the same for the orbits o_{10}, o_{12} and o_{14}, we distinguish between o_{10} and o_{14} using R_2. However, as o_{12} and o_{14} have the same rank distributions, we differentiate between them using the geometric description of the second contraction space. In particular, the difference between o_{12} and o_{14} is that for o_{14} the 3 points of rank one in the second contraction space (which is a plane) span the space, while for o_{12} they do not (see Table 1).

In most of the cases, except for o_{10}, o_{15} and o_{17}, the orbit representative is directly deduced from Table 1 and it is defined by its two horizontal slices.

For example, a representative of o_{11} is $e_1 \otimes (e_1 \otimes e_1 + e_2 \otimes e_2) + e_2 \otimes (e_1 \otimes e_2 + e_2 \otimes e_3)$ (see Table 1), and this can be represented by its horizontal slices as

$$\left\{ \begin{bmatrix} 1 & 0 & 0 \\ 0 & 1 & 0 \\ 0 & 0 & 0 \end{bmatrix}, \begin{bmatrix} 0 & 1 & 0 \\ 0 & 0 & 1 \\ 0 & 0 & 0 \end{bmatrix} \right\}.$$

Representative for o_{17}. We know that the orbit o_{17} has representatives of the form $e_1 \otimes (e) + e_2 \otimes (e_1 \otimes e_2 + e_2 \otimes e_3 + e_3 \otimes (\alpha e_1 + \beta e_2 + \gamma e_3))$ where $\lambda^3 + \gamma \lambda^2 - \beta \lambda + \alpha \neq 0$ for all $\lambda \in \mathbb{F}_q$ and $e = e_1 \otimes e_1 + e_2 \otimes e_2 + e_3 \otimes e_3$ (see Table 1). Instead of computing these parameters for every q (which would become computationally infeasible for very large q), we will give an explicit construction which does not require any computation at all. First, observe that o_{17} is the only orbit of lines in $\mathrm{PG}(\mathbb{F}_q^3 \otimes \mathbb{F}_q^3) \cong \mathrm{PG}(8, q)$ consisting entirely of points of rank 3 (see [7]). Thus, to obtain a representative for the orbit o_{17} it suffices to construct such a line of *constant rank* 3. In order to do so, consider the cubic extension $\mathbb{F}_{q^3}$ of $\mathbb{F}_q$ as an $\mathbb{F}_q$-vector space W and the set $U = \{M_\alpha : \alpha \in \mathbb{F}_{q^3}\}$ where M_α is the matrix representative of the linear operator on W defined by: $x \to \alpha x$. Clearly, U is a 3-dimensional $\mathbb{F}_q$-vector space consisting of the zero matrix and $q^3 - 1$ matrices of rank 3. Any 2-dimensional $\mathbb{F}_q$-subspace of U will give us a representative of o_{17}. Furthermore, a basis of this subspace gives us the two horizontal slices of the representative. In particular, we consider the 2-dimensional $\mathbb{F}_q$-subspace generated by the identity matrix and the companion matrix of the minimal polynomial of a primitive element w of the cubic extension.

Representatives for o_{10} and o_{15}. By Table 1, we can see that $e_1 \otimes (e_1 \otimes e_1 + e_2 \otimes e_2 + u e_1 \otimes e_2) + e_2 \otimes (e_1 \otimes e_2 + v e_2 \otimes e_1)$ and $e_1 \otimes (e_1 \otimes e_1 + e_2 \otimes e_2 + e_3 \otimes e_3 + u e_1 \otimes e_2) + e_2 \otimes (e_1 \otimes e_2 + v e_2 \otimes e_1)$ are representatives of o_{10} and o_{15} respectively, where $v\lambda^2 + uv\lambda - 1 \neq 0$ for all $\lambda \in \mathbb{F}_q$ and $u, v \in \mathbb{F}_q^*$. Similar to the previous case, we give an explicit construction of o_{10}, which does not require any computations. It follows from [9] that o_{10} has a representative line of constant 2-rank 2×2-matrices, which is an external line to a conic in $\mathcal{V}_3(\mathbb{F}_q)$, where $\mathcal{V}_3(\mathbb{F}_q)$ is the image of the map $\nu_3 : \mathrm{PG}(2, q) \to \mathrm{PG}(5, q)$ induced by the mapping sending $v \in \mathbb{F}_q^3$ to $v \otimes v$. Thus, by constructing such a line and taking any 2 points on it, we obtain the required representative. First, recall that interior points of the conic $(C) : X_0 X_2 - X_1^2 = 0$ in $\mathrm{PG}(2, q)$ are (x, y, z) where $xz - y^2$ are non-squares. Hence, if q is odd, one can start with a primitive root in $\mathbb{F}_q$ (which is a non-square in $\mathbb{F}_q$). Then, by considering its image under the polarity α associated to (C), we obtain an external line to (C) in $\mathrm{PG}(2, q)$. This line can be seen in $\mathrm{PG}(8, q)$ by embedding $\mathrm{PG}(2, q)$ as the set of points with last column and last row equal to zero. If q is even, a similar argument works. In this case, we can start with the minimal polynomial of a generator of the multiplicative group of $\mathbb{F}_{q^2}$ to obtain an irreducible quadratic polynomial over $\mathbb{F}_q$. The coefficients can then be used as the dual coordinates of a line in $\mathrm{PG}(2, q)$ disjoint from the conic consisting of the points (a^2, ab, b^2) with $(a, b) \in \mathrm{PG}(1, q)$. Once we have that line, we can map it to a line in $\mathrm{PG}(8, q)$ by embedding $\mathrm{PG}(2, q)$ as the set of points with last column and last row equal to zero. Now, by using a representative of o_{10}, we can find the above u and v, which gives us directly a representative of o_{15}.

2.3 RankOfTensor

The *RankOfTensor* function takes an arbitrary tensor A in PG(V) and uses the *OrbitOfTensor* function to specify the G-orbit of the tensor and returns the tensor's rank. The code of all of these functions can be found in [1].

3 Computations and Summary

Example 1.
```
gap> q:=397; sv:=SegreVariety([PG(1,q),PG(2,q),PG(2,q)]);
397
Segre Variety in ProjectiveSpace(17, 397)
gap> m:=Size(Points(sv));
9936552395502
gap> pg:=AmbientSpace(sv);
ProjectiveSpace(17, 397)
gap> n:=Size(Points(pg));
151542321438098147995655901146938756967526078
gap> A:=VectorSpaceToElement(pg,[Z(397)^0,Z(397)^336,Z(397)^339,
Z(397)^37,Z(397)^233,Z(397)^56,Z(397)^268,Z(397)^363,Z(397)^342,
Z(397)^297,Z(397)^146,Z(397)^71,Z(397)^57,Z(397)^84,Z(397)^33,
Z(397)^203,Z(397)^229,Z(397)^191]);
gap> OrbitOfTensor(A)[1]; time;
14
94
gap> RankOfTensor(A);
3
gap> time;
141
gap> NrCombinations([1..m], 3);
163514371865202881474954561407873423500
```

Summary. The *RankOfTensor* is an efficient tool to compute tensor ranks of points in PG(V). Without this algorithm, it is computationally infeasible to do this. For example, consider q, sv, pg and A from *Example 1*. The space pg has n points. Among these we have m points of rank 1, which gives a 38-order of magnitude number of possible 3-combinations of points of rank 1, which might generate a plane containing A. This reflects how hard it would be to compute the rank without this algorithm.

Acknowledgement. The second author acknowledges the support of *The Scientific and Technological Research Council of Turkey*, TÜBİTAK (project no. 118F159).

References

1. Alnajjarine, N., Lavrauw, M.: Determining the rank of tensors in $\mathbb{F}_q^2 \otimes \mathbb{F}_q^3 \otimes \mathbb{F}_q^3$. http://people.sabanciuniv.edu/mlavrauw/T233/T233_paper.html
2. Alnajjarine, N., Lavrauw, M.: Projective description and properties of the G-orbits of $(2 \times 3 \times 3)$-tensors. http://people.sabanciuniv.edu/mlavrauw/T233/table1.html
3. Bamberg, J., Betten, A., Cara, Ph., Beule, J. De., Lavrauw, M., Neunhöffer, M.: FinInG: Finite Incidence Geometry: FinInG - a GAP package. http://www.fining.org. Accessed 31 Mar 2018
4. The GAP Group: GAP Groups, Algorithms, and Programming. https://www.gap-system.org. Accessed 19 June 2019
5. Kolda, T.G., Bader, B.W.: Tensor decompositions and applications. SIAM Rev. **51**(3), 455–500 (2009)
6. Landsberg, J.M.: Tensors: Geometry and Applications: Geometry and Applications, 2nd edn. American Mathematical Society, Providence (2011)
7. Lavrauw, M., Sheekey, J.: Classification of subspaces in $\mathbb{F}^2 \otimes \mathbb{F}^3$ and orbits in $\mathbb{F}^2 \otimes \mathbb{F}^3 \otimes \mathbb{F}^r$. J. Geom. **108**(1), 5–23 (2017)
8. Lavrauw, M., Sheekey, J.: Canonical forms of $2 \times 3 \times 3$ tensors over the real field, algebraically closed fields, and finite fields. Linear Algebra Appl. **476**, 133–47 (2015)
9. Lavrauw, M., Popiel, T.: The symmetric representation of lines in PG $(\mathbb{F}_q^2 \otimes \mathbb{F}_q^3)$. Discrete Math. (to appear)

Second Order Balance Property on Christoffel Words

Lama Tarsissi[1,2]([✉]) and Laurent Vuillon[3]

[1] LAMA, Université Gustave Eiffel, CNRS, 77454 Marne-la-Vallée, France
[2] LIGM, Université Gustave Eiffel, CNRS, ESIEE Paris,
77454 Marne-la-Vallée, France
lama.tarsissi@esiee.fr
[3] LAMA, Univ. Grenoble Alpes, Univ. Savoie Mont Blanc, CNRS,
73000 Chambéry, France
laurent.vuillon@univ-smb.fr

Abstract. In this paper we study the *balance matrix* that gives the order of balance of any binary word. In addition, we define for Christoffel words a new matrix called *second order balance matrix*. This matrix gives more information on the balance property of a word that codes the number of occurrences of the letter 1 in successive blocks of the same length for the studied Christoffel word. By taking the maximum of the Second order balance matrix we define the second order of balance and we are able to order the Christoffel words according to these values. Our construction uses extensively the continued fraction associated with the slope of each Christoffel word, and we prove a recursive formula based on fine properties of the Stern-Brocot tree to construct second order matrices.

Keywords: Balance property · Second order balance property · Christoffel words · Stern-Brocot tree · Continued fractions

1 Introduction

Balanced words appear in many developments of combinatorics on words and the balance property is considered as a fine tool to investigate the structure of words [16,20]. As a typical example of infinite balanced words, Sturmian words could be constructed equivalently by discretizations of irrational slope lines in a square grid [10,29], by billiard words in a square [2,23] or by coding of irrational rotations on a unit circle with a partition in two intervals [10]. The finite balanced words are given by discretizations of rational slope lines in a square grid and have been studied in particular by Christoffel [9]. Interestingly, finite and infinite balanced words show up in specific optimization problems [1,25,29] and for example optimal schedules for job-shop problems with two

This work was partly funded by the French Programme d'Investissements d'Avenir (LabEx Bézout, ANR-10-LABX-58) and ANR-15-CE40-0006.

D. Slamanig et al. (Eds.): MACIS 2019, LNCS 11989, pp. 295–312, 2020.
https://doi.org/10.1007/978-3-030-43120-4_23

tasks are exactly given by balanced words [13,18,29]. Furthermore, particular solutions of job-shop problems with k tasks sharing the same ressource [1] are given by finite or infinite balanced words on a k−letters alphabet where the balanced property is checked on each letter of the alphabet [1,22,28]. More precisely, for $k = 2$ the solutions of the job-shop problem is coded by an infinite word which is either a periodic balanced word or an aperiodic balanced word [13,18]. The situation gets more complicated for $k > 2$, which leads to the famous Fraenkel's conjecture [11,12]. It is restated in combinatorics on words terms: An infinite word on a k−letters alphabet balanced on each letter of the alphabet and with all letters frequencies pairwise distinct is given by an infinite periodic word constructed on an unique period word FR_k (up to a permutation of letter and circular permutation) by the recursive formula $FR_k = FR_{k-1}kFR_{k-1}$ with $FR_3 = 1213121$. Many researchers have worked on the general problem of infinite balanced words on an alphabet with one letter [15] or equivalently to cover integers by Beatty sequences [12,26]. The conjecture is proved for $k = 3$ by Morikawa [19], for $k = 4$ by Altman, Gaujal and Hordijk [26], for $k = 5$ and $k = 6$ by Tijdeman [28] and for $k = 7$ by Barat et Varju [14] and the conjecture is still open for $k > 7$. Indeed, in order to investigate new discrete tools that allow us to deeply understand the structure of balanced words, we propose a second order balance property for Christoffel words that gives a refinement for the balance property. In fact, we define for Christoffel words a new matrix called *second order balance matrix* which gives information on the balance property of a word that codes the number of 1's of successive blocks of same length in the studied Christoffel word. Thus we investigate balance property on successive blocks instead of balance property on letters for second order balance property. The main idea, to go further in the resolution of the Fraenkel's conjecture, is to consider synchronization of the blocks instead of synchronization of letters and this is why we introduce the notion of second order of balance.

    In Sect. 2, we recall some properties of the Christoffel words. In Sect. 3, we define the balance matrix which gives information on the number of occurrences of a given letter in all factors of a given binary circular word. This balance matrix gives us the order of balance, for binary words. Afterwards, in Sect. 4, we introduce the *second order balance matrix* of a given Christoffel word by computing the balance matrix for the rows of the associated balance matrix. We show in Sect. 5 that this matrix has many symmetries and is constructed by using properties of continued fractions and the Stern-Brocot tree. We present in Sect. 6 a recursive construction for the second order balance matrix by considering the properties of the continued fraction expansion for the slope of each Christoffel word. Section 7 is left for the perspectives of this work. Remark that all the proofs can be found on the long version on hal-02433984.

2   Notation and Christoffel Words

Let A be an alphabet of cardinality m, the word w is the concatenation of letters of this alphabet and we write $w \in A^*$, where A^* represents the set of all the words formed by the alphabet A. We denote by $n = |w|$ the length of the word and by $|w|_a$ the number of occurrences of the letter a in the word w. The notation $w[i \ldots j]$ refers to the factor of the word w from position i to position j. The notation w^ω represents: $w^\omega = ww \cdots w \cdots$ and named "circular word associated with w". By convention, $w^0 = \epsilon$ and a word w is said primitive if it is not the power of a nonempty word. Two words w and w' are *conjugate* of order k if and only if there exist u, v such that $|u| = k$ *with* $w = uv$ *and* $w' = vu$ and we denote: $w \equiv_k w'$. When the exact value of k is not relevant, we simply write $w \equiv w'$ and we say that the two words are conjugate. A positive integer p is a *period* of w if $w[i] = w[i + p]$; for all $1 \le i \le |w| - p$. Given a word $w = aw'$ where a is a letter, we note $a^{-1}w = w'$ that is the removal of the letter a at the beginning of w. If w ends with letter a, then the notation wa^{-1} is defined accordingly. Let $A = \{a_0, a_1, \ldots, a_{m-1}\}$ be an alphabet, we let $\tilde{\ }$ be the anti-morphism such that: $\overline{a_0} = a_{m-1}, \overline{a_1} = a_{m-2} \ldots \overline{a_i} = a_{m-1-i}$. A word $w \in \{0, 1\}^*$ is $k-$balanced if and only if for all factors u, v of w, we get: $|u| = |v| \implies ||u|_1 - |v|_1|| \le k$. The word w is called balanced if $k = 1$.

Christoffel words [9] have many equivalent definitions and characterizations. The following geometrical definition is taken from [7] (see [5] for a self-contained survey). The lower Christoffel path of slope $\frac{a}{b}$, where a and b are relatively prime, is the path from $(0, 0)$ to (b, a) in the integer lattice $\mathbb{Z} \times \mathbb{Z}$ that satisfies the following conditions:

1. The path lies below the line segment that begins at the origin and ends at (b, a).
2. The region enclosed by the path and the line segment contains no other points of $\mathbb{Z} \times \mathbb{Z}$ besides those of the path.

We encode the lower Christoffel path (or simply Christoffel word) by means of a word in the alphabet A using 0 (resp. 1) for any unit horizontal (vertical) step. We get the *Christoffel word* of slope $\frac{a}{b}$ denoted: $C(\frac{a}{b})$, see Fig. 1. Equivalently, the Christoffel word $w = C\left(\frac{a}{b}\right)$ is obtained by calculating the elements of the sequence $(r_i)_{0 \le i \le n}$, where $n = a + b$ as follows: $r_i = ia \mod n$. Each letter, $w[i]$, $\forall 1 \le i \le n$, of the word w and length n is obtained by computing:
$$w[i] = \begin{cases} 0 & \text{if } r_{i-1} < r_i, \\ 1 & \text{otherwise.} \end{cases}$$

Example 1. Let $(a, b) = (3, 5)$, the sequence $(r_i)_{0 \le i \le 8} = (0, 3, 6, 1, 4, 7, 2, 5, 0)$ defines the Christoffel word $C\left(\frac{3}{5}\right) = 00100101$.

3   Balance Matrix

In this section, we introduce a new matrix used to obtain the order of balance for any binary word in an explicit way. The ith row of the matrix M, $M[i]$, is seen as a word where each entry of the matrix is a letter. Given a word $w \in A^*$ of length n, we let S_w be the $n \times n$ matrix defined by $S_w[i, j] = w[j] + \ldots w[i+j]$ over the circular word w. By definition, we have that w is δ-balanced if

$$\delta = \max_i \left(\max(S_w[i]) - \min(S_w[i]) \right).$$

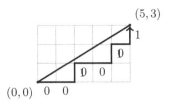

Fig. 1. Illustration of the geometrical definition of Christoffel words. The Christoffel path goes from (0, 0) to (5, 3) and $C\left(\frac{3}{5}\right) = 00100101$.

The *balance matrix* B_w, is defined from S_w by subtracting the minimum value on each row,

$$B_w[i,j] = S_w[i,j] - \min(S_w[i]).$$

Obviously, we have that w is $(\max B_w)$-balanced.

It is clear that by construction, if w is not a sequence of $1's$, then the first row of B_w is equal to w. If $k < |w|$ is a period of w^ω, again by construction we have that the row $B_w[k]$ contains only zeros. The converse is also true, an integer k such that the row $B_w[k]$ contains only zeros is a period of w^ω. Consequently, a row of zeros is called a *period row*. Note that $n = |w|$ is always a period of w^ω and therefore we define the matrix B_w as a $(n-1) \times n$ matrix since the n-th row would not add any information.

For the sake of simplicity, when working with the Christoffel word $C\left(\frac{a}{b}\right)$, the balance matrix $B_{C\left(\frac{a}{b}\right)}$ is simply denoted $B_{\frac{a}{b}}$.

Example 2. Let us consider the rational number $\frac{5}{2}$ and $n = 7$. By writing the Christoffel word $w = C(\frac{5}{2}) = 0110111$, the elements of the balance matrix B_w and S_w are the following:

$$S_w = \begin{pmatrix} 0\,1\,1\,0\,1\,1\,1 \\ 1\,2\,1\,1\,2\,2\,1 \\ 2\,2\,2\,2\,3\,2\,2 \\ 2\,3\,3\,3\,3\,3\,3 \\ 3\,4\,4\,3\,4\,4\,3 \\ 4\,5\,4\,4\,5\,4\,4 \\ 5\,5\,5\,5\,5\,5\,5 \end{pmatrix}, \quad B_{\frac{5}{2}} = \begin{pmatrix} 0\,1\,1\,0\,1\,1\,1 \\ 0\,1\,0\,0\,1\,1\,0 \\ 0\,0\,0\,0\,1\,0\,0 \\ 0\,1\,1\,1\,1\,1\,1 \\ 0\,1\,1\,0\,1\,1\,0 \\ 0\,1\,0\,0\,1\,0\,0 \end{pmatrix}.$$

3.1   Properties of the Balance Matrix

In this section, we present some properties of the matrix B_w, where w is a binary word of length $|w| = n$, allowing us to restrict the work to the upper half of the balance matrix since the lower part will be obtained by symmetry.

Definition 1. *Two words w and w' are* complementary *if $w \equiv_k \overline{w'}$ for some k. Moreover, a word is said to be* autocomplementary *if it is complementary to itself.*

Example 3. The word $w = 0110110010$ is an autocomplementary word since we have: $w \equiv_5 \overline{w}$. While $w = 0000100$ and $w' = 0111111$ are such that $w \equiv_4 \overline{w'}$.

We show some basic combinatorial properties of the matrix S_w that will be used in a further proof.

Property 1. For any binary word w of length n, and $1 \le i \le n$, the matrix S_w satisfies:
$$\max S_w[i] + \min S_w[n - i] = |w|_1, \ \min S_w[i] + \max S_w[n - i] = |w|_1.$$
In particular, if w is a 1-balanced Christoffel word, we have:
$$\min S_w[i] + \min S_w[n - i] = |w|_1 - 1, \ \max S_w[i] + \max S_w[n - i] = |w|_1 + 1.$$

By applying the results of Property 1 and the definition of S_w, we can notice that the lower half of B_w is deduced from its upper half part, as we can see in Property 2.

Property 2. The balance matrix is such that for all $1 \le i < n$, $B_w[i] \equiv_{n-i} \overline{B_w[n - i]}$.

If $B_{\frac{a}{b}}$ has a middle row, then this row is autocomplementary as we can see in Example 4.

Corollary 1. *If n is an even number, $B_{\frac{a}{b}}[\frac{n}{2}]$ is an autocomplementary row.*

Example 4. Let us consider the rational number $3/7$ with $n = 10$. The balance matrix $B_{\frac{3}{7}}$ shows that $B_{\frac{3}{7}}[5]$ is an autocomplementary row.

$$B_{\frac{3}{7}} = \begin{pmatrix} 0\,0\,0\,1\,0\,0\,1\,0\,0\,1 \\ 0\,0\,1\,1\,0\,1\,1\,0\,1\,1 \\ 0\,1\,1\,1\,1\,1\,1\,1\,1\,1 \\ 0\,0\,0\,1\,0\,0\,1\,0\,0\,0 \\ 0\,0\,1\,1\,0\,1\,1\,0\,0\,1 \\ 0\,1\,1\,1\,1\,1\,1\,0\,1\,1 \\ 0\,0\,0\,1\,0\,0\,0\,0\,0\,0 \\ 0\,0\,1\,1\,0\,0\,1\,0\,0\,1 \\ 0\,1\,1\,1\,0\,1\,1\,0\,1\,1 \end{pmatrix}.$$

Remark 1. In the case where $a = 1$, the balance matrix is such that: $B_{\frac{1}{b}}[i] = 0^{(n-i)}1^i$.

3.2   Construction of the Balance Matrix for Christoffel Words

In this section, we are interested in giving a direct construction for the balance matrix of a Christoffel word, by determining for each row of B_w, the positions of the letter 1. For that, we start by defining the set of positions for the letter 1 in $C(\frac{a}{b})$.

Definition 2. *Let w be a Christoffel word of slope a/b. The set of* decreasing positions *of w, denoted $D(a, b)$, is the set of the positions of the occurrences of the letter 1 in w. More formally, $D(a, b) = \{1 \le i \le n \mid w[i] = 1\}$.*

The following theorem is a reformulation of Paquin and Reutenauer's result ([21], Corollary 3.2), that describes the set of decreasing positions of a Christoffel word.

Theorem 1. *Let α be such that $\alpha a \equiv -1 \mod n$, then the set $D(a, b)$ is:*

$$D(a, b) = \{(i\alpha \mod n) + 1 \mid i = 1 \ldots a\}.$$

Example 5. Let us consider the rational number $a/b = 3/5$ with $n = 8, \alpha = 5$ and $w = 00100101$. The set of occurrences of the letter 1 is $D(3, 5) = \{(5i \mod 8) + 1 \mid i = 1 \ldots 3\} = \{3, 6, 8\}$.

Using Theorem 1, we can conclude that $B_{\frac{a}{b}}$ and $B_{\frac{b}{a}}$ are not equal but complementary as Lemma 1 shows.

Lemma 1. *The balance matrices $B_{\frac{a}{b}}$ and $B_{\frac{b}{a}}$ are conjugate in the sense that:*

$$B_{\frac{a}{b}}[i] \equiv_\alpha \overline{B_{\frac{b}{a}}[i]} \ \forall \ 1 \le i < n; \ where \ \alpha a \equiv -1 \bmod n.$$

For this part, w is a Christoffel word and we give a recursive construction of B_w by identifying for each row i, the set of positions of the 1's. This set is denoted by D_i, where for each row i in B_w, $D_i = \{j \mid B_w[i, j] = 1\}$, with $n = |w|$ and $1 \le j \le n$. For any set of integers S and any integer k, we denote $S + k = \{a + k \mid a \in S\}$.

Theorem 2. *If $a < b$ then the sets $(D_i)_{1 \le i \le n-1}$ are recursively obtained as follows: $D_1 = D(a, b)$ and for each i from 2 to $n - 1$:*

$$D_i = \begin{cases} D_{i-1} \cup (D_1 - (i-1) \bmod n) & if \ i \notin D_1 \\ D_{i-1} \cap (D_1 - (i-1) \bmod n) & if \ i \in D_1 \end{cases} \quad (1)$$

Example 6. Let us consider the rational number $2/7 \mod 9$. By calculating the set of decreasing values for each row of $B_{\frac{2}{7}}$, we get the following matrix:

$$
\begin{aligned}
D_1 &= \{5, 9\} \\
D_2 &= D_1 \cup \{4, 8\} = \{4, 5, 8, 9\} \\
D_3 &= D_2 \cup \{3, 7\} = \{3, 4, 5, 7, 8, 9\} \\
D_4 &= D_3 \cup \{2, 6\} = \{2, 3, 4, 5, 6, 7, 8, 9\} \\
D_5 &= D_4 \cap \{1, 5\} = \{5\} \\
\\
D_6 &= D_5 \cup \{4, 9\} = \{4, 5, 9\} \\
D_7 &= D_6 \cup \{3, 8\} = \{3, 4, 5, 8, 9\} \\
D_8 &= D_7 \cup \{2, 7\} = \{2, 3, 4, 5, 7, 8, 9\} \\
D_9 &= D_8 \cap \{1, 6\} = \{\}.
\end{aligned}
\qquad
B_{\frac{2}{7}} =
\begin{pmatrix}
0 & 0 & 0 & 0 & 1 & 0 & 0 & 0 & 1 \\
0 & 0 & 0 & 1 & 1 & 0 & 0 & 1 & 1 \\
0 & 0 & 1 & 1 & 1 & 0 & 1 & 1 & 1 \\
0 & 1 & 1 & 1 & 1 & 1 & 1 & 1 & 1 \\
0 & 0 & 0 & 0 & 1 & 0 & 0 & 0 & 0 \\
0 & 0 & 0 & 1 & 1 & 0 & 0 & 0 & 1 \\
0 & 0 & 1 & 1 & 1 & 0 & 0 & 1 & 1 \\
0 & 1 & 1 & 1 & 1 & 0 & 1 & 1 & 1
\end{pmatrix}.
$$

From this construction, we can get a relation between the number of occurrences of 1 in each row of the balance matrix of Christoffel words and the numerator of the slope related to this word.

Lemma 2. *For the balance matrix* $B_{\frac{a}{b}}$, *we have:* $|D_i| = i.a \bmod n$, *where* $|D_i|$ *is the cardinal of the set* D_i.

Note: This Lemma confirms that the period row is made only of zeros since $|D_n| = n.a \bmod n = 0$, hence we have no occurrences for the letter 1 in this row.

4   Second Order Balance Matrix

Let w^ω be a 1-balanced circular word associated with w; by computing the balanced property on each row of B_w, we get a refinement of the balanced property for w. For any factor v of length j, we can find k_j or $k_j + 1$ occurrences of the letter 1. The second order balance is the repartition of these blocks in a binary balanced word. This second order balance is computed via a matrix called the *second order balance matrix*. In other words, we are studying the balance of each row of B_w.

For a pair of integers i, j, where $1 \le i, j \le |w| - 1$, we consider the word $B_w[i]$ and we list all its factors of length j. Among these factors, we choose p, a factor that maximizes the number of occurrences of the letter 1 and q, a factor that minimizes it. The entry $U_w[i, j]$ is given by $|p|_1 - |q|_1$. Equivalently, if $L_\ell(w)$ is the restriction of the language of w^ω to words of length ℓ, then:

$$U_w[i, j] = \max_{v \in L_j(B_w[i])} |v|_1 - \min_{v \in L_j(B_w[i])} |v|_1.$$

In other words, we can define the second order balance matrix U_w by:

Definition 3. *Let* w *be a word such that* w^ω *is 1-balanced or, equivalently, that* B_w *is a binary matrix. The second order balance matrix* $U_w = (u_{ij})_{1 \le i, j \le n-1}$ *where* $U_w[i, j] = \max(B_{(B_w[i])}[j])$.

Definition 4. *The second order of balance of a circular 1-balanced word* w *is* $\delta^2(w) = \max(U_w)$.

Once again, in order to lighten the notation, when working with the Christoffel word $C\left(\frac{a}{b}\right)$, the second order balance matrix $U_{C\left(\frac{a}{b}\right)}$ is simply denoted $U_{\frac{a}{b}}$.

For the rest of the paper, we let w be a Christoffel word of slope a/b and length $n = a + b$, such that: $w = C(\frac{a}{b})$. The second order balance matrix of a Christoffel word of slope $\frac{a}{b}$, $U_{\frac{a}{b}}$, is of dimension $(n-1) \times (n-1)$.

Example 7. Let us consider the rational number $\frac{a}{b} = \frac{3}{7}$ with $n = 10$. The balance matrix $B_{\frac{3}{7}}$ was calculated previously and $B_{\frac{3}{7}}[5] = [0, 0, 1, 1, 0, 1, 1, 0, 0, 1]$. By computing the balance matrix for this word and taking the blocks of length 5 we get the 5th row of $S_{B_{\frac{3}{7}}}[5]$ where the difference between the maximum and the minimum values of each row of $S_{B_{\frac{3}{7}}}[5]$ determines the entries of $U_{\frac{3}{7}}[5]$. Hence, with these two blocks $\overline{00|11011|001}$, we obtain the element $U_{\frac{3}{7}}[5,5] = 4 - 1 = 3$ as we can see in the following second order balance matrix of $\frac{3}{7}$, where we also get $\delta^2(C(\frac{3}{7})) = 3$.

$$U_{\frac{3}{7}} = \begin{pmatrix} 1\,1\,1\,1\,1\,1\,1\,1\,1 \\ 1\,2\,1\,1\,2\,1\,1\,2\,1 \\ 1\,1\,1\,1\,1\,1\,1\,1\,1 \\ 1\,1\,1\,2\,2\,2\,1\,1\,1 \\ 1\,2\,1\,2\,3\,2\,1\,2\,1 \\ 1\,1\,1\,2\,2\,2\,1\,1\,1 \\ 1\,1\,1\,1\,1\,1\,1\,1\,1 \\ 1\,2\,1\,1\,2\,1\,1\,2\,1 \\ 1\,1\,1\,1\,1\,1\,1\,1\,1 \end{pmatrix}.$$

Properties of the Matrix $U_{\frac{a}{b}}$

   Now we give some properties of the second order balance matrix in order to show that $U_{\frac{a}{b}} = U_{\frac{b}{a}}$. Hence, we can restrict our study to the irreducible fractions a/b with $a < b$. But before that, we prove the three symmetries that appear in this matrix. From Sect. 3.1, we have that the rows of the upper half of $B_{\frac{a}{b}}$ are complementary to the rows of its lower half, which induces the symmetries in the matrix $U_{\frac{a}{b}}$. More precisely, the second order balance matrix $U_{\frac{a}{b}}$ of dimension $(n-1) \times (n-1)$, has horizontal, vertical and diagonal symmetries. The axis of symmetry are at position $\frac{n}{2}$ or between $\frac{n-1}{2}$ and $\frac{n+1}{2}$ depending on the parity of n.

Property 3. For any position (i,j), $U_{\frac{a}{b}}[i,j] = U_{\frac{a}{b}}[n-i,j] = U_{\frac{a}{b}}[i,n-j] = U_{\frac{a}{b}}[n-i,n-j]$.

   Moreover, $U_{\frac{a}{b}}$ has an extra diagonal symmetry;

Property 4. For the Christoffel word of slope a/b, we have $(U_{\frac{a}{b}})^T = U_{\frac{a}{b}}$.

   After those two properties, we are able to prove that $U_{\frac{a}{b}} = U_{\frac{b}{a}}$.

5   More About Christoffel Words

Let w be a Christoffel word of length at least 2, the standard factorization is obtained by writing $w = (w_1, w_2)$ in a unique way, where w_1, w_2 are two Christoffel words by [7]. The *Christoffel tree* is an infinite tree whose vertices are all the standard factorizations of Christoffel words (see [5], Section 3.2). It uses the fact that given a standard factorization (w_1, w_2), the pairs (w_1, w_1w_2) and (w_1w_2, w_2) are also standard factorizations. Let ϕ_0, ϕ_1 be the two functions from $A^* \times A^*$ into itself defined by: $\phi_0(w_1, w_2) = (w_1, w_1w_2)$; $\phi_1(w_1, w_2) = (w_1w_2, w_2)$. We have that any Christoffel word can be obtained in a unique way by iteration of these two functions on $(0,1)$. Consequently, the Christoffel tree is defined as follows: the root is $(0,1)$, the Christoffel word of slope 1. Then each node (w_1, w_2)

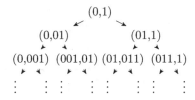

Fig. 2. The first levels of the Christoffel tree.

has two sons: $\phi_0(w_1, w_2)$ on the left and $\phi_1(w_1, w_2)$ on the right. See Fig. 2 for an illustration.

Definition 5. *Let* $w = C(\frac{a}{b})$ *be a non-trivial Christoffel word, the directive sequence of* w*, denoted* $\Delta(\frac{a}{b})$*, is the word* $\Delta(\frac{a}{b}) = i_1 i_2 \cdots i_n \in A^*$ *such that* $w = (\phi_{i_n} \circ \cdots \circ \phi_{i_2} \circ \phi_{i_1})(0, 1)$.

Note that a directive sequence $\Delta(\frac{a}{b}) = i_1 \cdots i_n$ describes the path from the root of the Christoffel tree to the Christoffel word $C(\frac{a}{b})$ as follows: at step k, if $i_k = 0$ then go left, otherwise, if $i_k = 1$ then go right.

5.1 Stern-Brocot Tree and Continued Fractions

In this section, we introduce the *Stern-Brocot tree* that contains all the reduced fractions $\frac{a}{b}$. It was first introduced by a German mathematician Moritz Abraham Stern and a French clockmaker Achille Brocot in the 19th century [8]. In order to construct recursively the Stern-Brocot tree, we need to introduce the *mediant* of two fractions $\frac{a}{b}$ and $\frac{c}{d}$, that is $\frac{a}{b} \oplus \frac{c}{d} = \frac{a+c}{b+d}$. In addition to that, we have to define the recursive sequence s_i, that is obtained from s_{i-1}, by completing with the mediant of each two consecutive fractions in s_{i-1}, where s_0 is given by: $s_0 = \left(\frac{0}{1}, \frac{1}{0}\right)$. Note that, $\frac{1}{0}$, is considered as a normal fraction: $s_0 = \left(\frac{0}{1}, \frac{1}{0}\right)$, $s_1 = \left(\frac{0}{1}, \frac{1}{1}, \frac{1}{0}\right)$, $s_2 = \left(\frac{0}{1}, \frac{1}{2}, \frac{1}{1}, \frac{2}{1}, \frac{1}{0}\right)$, $s_3 = \left(\frac{0}{1}, \frac{1}{3}, \frac{1}{2}, \frac{2}{3}, \frac{1}{1}, \frac{3}{2}, \frac{2}{1}, \frac{3}{1}, \frac{1}{0}\right)$.

The mediants added in each new step to the sequence s_i are the fractions that appear on the ith level of the Stern-Brocot tree. For example, on the third level, we have the fractions: $\frac{1}{3}, \frac{2}{3}, \frac{3}{2}, \frac{3}{1}$ that are extracted from the sequence s_3. We call consecutive fractions, two fractions that belong to the same set s_i and are next to each other, like $\frac{1}{1}$, and $\frac{2}{1}$ in s_2. Some properties about these consecutive fractions will be given in the next section. In order to simplify the notation, we denote $\frac{a}{b} \oplus \frac{c}{d}$ by $\frac{a}{b}\frac{c}{d}$ and $\frac{a}{b} \oplus \ldots \oplus \frac{a}{b}$ repeated p times by $(\frac{a}{b})^p$. The Christoffel tree is isomorphic to the Stern-Brocot tree where each vertex of the Christoffel tree of the form (u,v) is associated to the fraction $\frac{|uv|_1}{|uv|_0}$, see Fig. 3.

The continued fraction of a rational number $\frac{a}{b} \geq 0$ is the sequence of integers $\frac{a}{b} = [a_0, \ldots, a_z]$, with $a_0 \geq 0$; $a_i \geq 1$ for $1 \leq i \leq z$ and if $z \geq 2$ then $a_z \geq 2$.

$$\frac{a}{b} = a_0 + \cfrac{1}{a_1 + \cfrac{1}{\cdots + a_z}}.$$

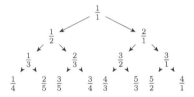

Fig. 3. The first levels of the Stern-Brocot tree.

In the following part, we will explain how to pass from the continued fraction of a rational number a/b to the Christoffel word of slope a/b. For that, we introduce the following theorem by Henry J.S in 1876. This theorem gives an additional characterization for the Christoffel words. In fact, he showed that the Christoffel word can be obtained by a recursive expression using the partial denominators of the rational number.

Theorem 3. *[27] Let $\frac{a}{b} = [a_0, a_1 \ldots, a_z]$, a word $w = 0u1$ is a Christoffel word of slope $\frac{a}{b}$ if and only if $u01$ or $u10$ is equal to s_{n+1}, where s_{n+1} is defined recursively by: $s_{-1} = 0$, $s_0 = 1$ and $s_{n+1} = s_n^{a_n} s_{n-1}$ for all $n \geq 0$.*

Moreover, in 1987, Berstel introduced, in a report for LITP, the following relation between the partial denominators of a rational number and two particular matrices based on the Theorem 2.1 in George Raney's paper [24] in 1973, that was published in 1990 [3]. Using the two matrices $A = \begin{pmatrix} 1 & 1 \\ 0 & 1 \end{pmatrix}$ and $B = \begin{pmatrix} 1 & 0 \\ 1 & 1 \end{pmatrix}$, and for the particular case of the rational number a/b where $a < b$, we have:

Proposition 1. *[3] For $a < b$, we associate for the rational number $\frac{a}{b} = [0, a_1, a_2, \ldots, a_n]$, the following matrix: $M(\frac{a}{b}) = A^{a_1-1} B^{a_2} A^{a_3} \ldots C^{a_n-1} D^{a_n-1}$, where C and D are respectively A, B or B, A depending on the parity of n. We have: $M(\frac{a}{b}).\begin{pmatrix} 1 \\ 1 \end{pmatrix} = \begin{pmatrix} b-a \\ a \end{pmatrix}$.*

Furthermore, in 1993, Borel and Laubie, followed by de Luca in 1997 then Berthé, de Luca and Reutenauer in 2008, gave the following theorem, where they linked these two particular matrices to the Christoffel word of slope a/b. Let the function Pal be the iterative palindromic closure function defined recursively in [4] by $Pal(ua) = (Pal(u)a)^{(+)}$ and $Pal(\epsilon) = \epsilon$, where a is a letter and u a word and $(w)^+$ is the palindromic closure of w i.e. $(w)^+ = ww'$ with ww' is the shortest palindrome having w as a prefix.

Theorem 4. *[4, 6, 7] Let w be a Christoffel word of slope a/b, there exists a unique word v such that $w = 0Pal(v)1$, if (w_1, w_2) is the standard factorization of w then we define the multiplicative monoid morphism $\mu : \{0,1\}^* \longrightarrow SL_2(\mathbb{Z})$ such that: $\mu(0) = \begin{pmatrix} 1 & 1 \\ 0 & 1 \end{pmatrix} = A$ and $\mu(1) = \begin{pmatrix} 1 & 0 \\ 1 & 1 \end{pmatrix} = B$, where: $\mu(v) = \begin{pmatrix} |w_1|_0 & |w_2|_0 \\ |w_1|_1 & |w_2|_1 \end{pmatrix}$.*

In [3,5,17,24] we can find results allowing us to write, in an explicit way, the relation between the continued fraction of a rational number and its directive sequence as we can see in the following theorem (see [4]).

Theorem 5. Let $\frac{a}{b} = [a_0, \ldots, a_z]$, we have: $\Delta(\frac{a}{b}) = 1^{a_0} 0^{a_1} 1^{a_2} \ldots p^{a_z-1}$ where $p \in \{0,1\}$. The Christoffel word of slope a/b is written: $C\left(\frac{a}{b}\right) = 0w'1$, where w' is a palindrome and $w' = Pal(\Delta(\frac{a}{b}))$.

6 Recursive Construction of the Second Order Balance Matrix

Due to the isomorphism and the recursive construction of the Stern-Brocot tree and the Christoffel tree, we can conclude that there must exist a recursive construction for the second order balance matrix. In order to show and to prove this recursivity, we let $U_{\frac{a}{b}}$ where $a/b = [a_0, \ldots, a_z] = [a_0, a_1, \ldots, a_z - 1, 1]$ be the second order balance matrix of the Christoffel word $C\left(\frac{a}{b}\right)$. We introduce the following terminology for some specific rational numbers on the Stern-Brocot tree and that will be used for the rest of the paper. See Fig. 4 for an illustration.

Definition 6. Given $\frac{a}{b} = [a_0, \ldots, a_z]$,
The top branch fraction of $\frac{a}{b}$, denoted $\mathrm{TBF}(\frac{a}{b})$, is the fraction $[a_0, \ldots, a_{z-1}+1]$,
The first reduced fraction of $\frac{a}{b}$, denoted $\mathrm{FRF}(\frac{a}{b})$, is the fraction $[a_0, \ldots, a_{z-1}]$,
The first extended fraction of $\frac{a}{b}$, denoted $\mathrm{FEF}(\frac{a}{b})$, is the fraction $[a_0, a_1, \ldots, a_z + 1]$,
The first deviation fraction of $\frac{a}{b}$, denoted $\mathrm{FDF}(\frac{a}{b})$, is the fraction $[a_0, a_1, \ldots, a_z - 1, 2]$,
The first parallel fraction of $\frac{a}{b}$, denoted $\mathrm{FPF}(\frac{a}{b})$, is either $[a_0, a_1, \ldots, a_{z-1}-1, 2]$ if $a_{z-1} \neq 1$ or $[a_0, a_1, \ldots, a_{z-2} + 2]$ if $a_{z-1} = 1$.
The second unidirectional father of $\frac{a}{b}$, denoted $\mathrm{SUF}(\frac{a}{b})$, is either $[a_0, a_1, \ldots, a_z - 2]$ if $a_z > 2$ or $[a_0, a_1, \ldots, a_{z-2}]$ if $a_z = 2$.

Note that $\mathrm{SUF}(\frac{a}{b})$ is not defined for fractions $\frac{1}{1}$, $\frac{1}{2}$ and $\frac{2}{1}$. Using Theorem 5 we can get the directive sequence of each of these fractions.

Example 8. Let $\frac{a}{b} = \frac{3}{5} = [0,1,1,2]$, from Definition 6, we get: $\mathrm{TBF}(3/5) = 2/3$, $\mathrm{FEF}(3/5) = 4/7$, $\mathrm{FDF}(3/5) = 5/8$, $\mathrm{FPF}(3/5) = 1/3$ and $\mathrm{SUF}(3/5) = 1/1$. See Fig. 4 for the positions of theses fractions in the Stern-Brocot tree.

General Form of the Second Order Balance Matrix
To construct $U_{\frac{a}{b}}$, we start by placing 4 rows of separation that divide the matrix into 9 blocks. Due to the symmetries proved in Properties 3 and 4, it is sufficient to know three of these blocks to deduce the others. These blocks are denoted α, β and γ and are represented in the matrix as follows:

$$U_{\frac{a}{b}} = \begin{pmatrix} \alpha & \cdot & \cdot \\ \hline \gamma & \beta & \cdot \\ \hline \cdot & \cdot & \cdot \end{pmatrix},$$

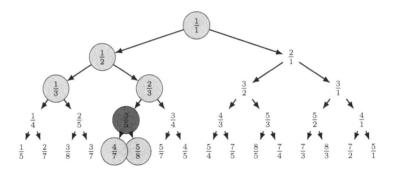

Fig. 4. Illustration of the specific rational numbers related to $3/5 = [0, 1, 1, 2]$ on the Stern-Brocot tree. We have: $\mathrm{TBF}(3/5) = 2/3$, $\mathrm{FRF}(3/5) = 1/2$, $\mathrm{FEF}(3/5) = 4/7$, $\mathrm{FDF}(3/5) = 5/8$, $\mathrm{FPF}(3/5) = 1/3$ and $\mathrm{SUF}(3/5) = 1/1$.

In the following section, we show that the blocks α, β, γ are described by the second order balance matrices of simpler fractions. More precisely, α is deduced from $U_{\mathrm{TBF}(\frac{a}{b})}$ while β is given by adding one to each entry of $U_{\mathrm{SUF}(\frac{a}{b})}$. Finally, the construction of γ depends on the position of $\frac{a}{b}$ in the Stern-Brocot tree, where the fraction is, relatively to its father, either a deviation (first deviation fraction) or an extension (first extended fraction).

6.1   The Construction of $U_{\frac{a}{b}}$

In this section, we start to explain how we can obtain the recursive construction of the second order balance matrix of the Christoffel word of slope $a/b = [a_0, \ldots, a_z]$. Figure 6 displays the rational numbers needed for the construction. Since $U_{\frac{a}{b}} = U_{\frac{b}{a}}$, we reduce the work to the first half of the Stern-Brocot tree, that contains all the irreducible fractions with $a \leq b$. Given $\frac{a}{b} = [a_0, \ldots, a_z]$, we consider separately the cases $z < 2$ and $z \geq 2$.

The Trivial Cases: $z \in \{0, 1\}$
For $z = 0$, we have $U_{\frac{0}{1}} = U_{\frac{1}{0}} = [\ ]$ and $U_{\frac{1}{1}} = [1]$. For $z = 1$, we have $\frac{a}{b} = [0, a_1]$ which implies that $a = 1$ and $b = a_1$. In this case, Remark 1, states that $B_w[i] = 0^{b+1-i}1^i$ and from Definition 3, we have: $U_w[i, j] = \max(B_{B_w[i]}[j])$. Hence, we consider the first quarter of the matrix U_w which is sufficient from Properties 3 and 4. Due to the diagonal symmetry, we will consider only $j \leq i$ where $i, j \leq \frac{n}{2}$ if n is even or $i, j \leq \frac{n-1}{2}$ if n is odd.

Proposition 2. *If $a = 1$, then for all $j \leq i$ and $i, j \leq \frac{n}{2}$ we have: $U_{\frac{a}{b}}[i, j] = j$.*

We can observe that this matrix can also be constructed using a recursive form where the first (resp. last) row and the first (resp. last) column are all 1's and in the middle we have the matrix of $\mathrm{SUF}(\frac{1}{b})$ where its elements are increased by 1 (see Fig. 5).

$$U_{\frac{1}{b}} = \begin{pmatrix} 1|1 & & .. & 1|1 \\ 1 & & & 1 \\ . & & & . \\ . & U_{[0,a_1-2]}+1 & & . \\ . & & & . \\ 1 & & & 1 \\ 1|1 & & .. & 1|1 \end{pmatrix}$$

Fig. 5. The general form of the matrix $U_{\frac{1}{b}}$.

The General Case: $z \geq 2$

Now we assume that $z \geq 2$, in order to lighten the presentation we define the following fractions, let: $\frac{u}{v} = \mathrm{TBF}(\frac{a}{b})$, $\frac{x}{y} = \mathrm{FRF}(\frac{a}{b})$, $\frac{c}{d} = \mathrm{FDF}(\frac{u}{v})$, $\frac{e}{f} = \mathrm{FPF}(\frac{c}{d})$, $\frac{g}{h} = \mathrm{SUF}(\frac{u}{v})$, $\frac{p}{q} = \mathrm{TBF}(\frac{x}{y})$, $\frac{s}{t} = \mathrm{FRF}(\frac{u}{v})$. See Fig. 6 for an illustration of their relative positions on the Stern-Brocot tree.

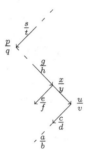

Fig. 6. Position of the fractions $\frac{u}{v}, \frac{x}{y}, \frac{c}{d}, \frac{e}{f}, \frac{g}{h}, \frac{p}{q}$ and $\frac{s}{t}$ relatively to $\frac{a}{b}$, for the case where $a_{z-1} \geq 2$ and z odd. By definition, we have: $\frac{a}{b} = [a_0 \ldots, a_z]$, $\frac{u}{v} = [a_0 \ldots, a_{z-1}+1]$, $\frac{x}{y} = [a_0, \ldots, a_{z-1}]$, $\frac{p}{q} = [a_0, \ldots, a_{z-2}+1]$, $\frac{s}{t} = [a_0, \ldots, a_{z-2}]$, $\frac{e}{f} = [a_0, \ldots, a_{z-1}-1,2]$, $\frac{c}{d} = [a_0, \ldots, a_{z-1},2]$, $\frac{g}{h} = [a_0, \ldots, a_{z-1}-1]$. With respect to the Farey addition we have: $\frac{u}{v} = \frac{x}{y} \frac{s}{t}$ and $\frac{c}{d} = \frac{x}{y} \frac{u}{v}$.

Lemma 3. *The Christoffel words of slope $\frac{a}{b}, \frac{g}{h}$ and $\frac{e}{f}$ can be written as follows:*
$$C(\tfrac{a}{b}) = \left(C(\tfrac{x}{y})\right)^{a_z-1} C(\tfrac{u}{v}) = \left(C(\tfrac{x}{y})\right)^{a_z} C(\tfrac{s}{t}), \quad C(\tfrac{g}{h}) = C(\tfrac{g}{h})C(\tfrac{s}{t}) \text{ and } C(\tfrac{e}{f}) = C(\tfrac{g}{h})C(\tfrac{x}{y}).$$

Separation Rows. For the rest of this section, the fractions mentioned in Fig. 6 are used to prove the construction of $U_{\frac{a}{b}}$. In the following part we prove that separation rows allow the decomposition of $U_{\frac{a}{b}}$ into 9 blocks. In Definition 7, we give a characteristic for the consecutive fractions of each sequence s_i, used to construct the Stern-Brocot tree.

Definition 7. *Let $\frac{a}{b}$ and $\frac{a'}{b'}$ be two consecutive fractions of a certain sequence s_i, $\forall i > 0$. They respect the following relation: $a'b - ab' = 1$.*

Property 5. Let $\frac{a}{b}$ and $\frac{a'}{b'}$ be two consecutive fractions in the Stern-Brocot tree, we have:

$$(a' + b').a = a'.(a + b) - 1.$$

The proof of this property is obtained by an arithmetic calculation based on Definition 7.

In the following lemma, we prove that in each $U_{\frac{a}{b}}$, we have at least 2 rows and columns full of 1's.

Lemma 4. *Let $k = x + y$, for the rational number $\frac{a}{b}$ with $n = a + b$, we have:*

$$U[i, k] = U[i, n - k] = U[k, j] = U[n - k, j] = 1, \ \forall \ i, j \in \{1, \ldots, n - 1\}.$$

Remark 2. From Lemma 4, we get that the row (respectively column) k and $n - k$ contain only values of 1's. Therefore, we define the separation rows to be between the rows (resp. columns) $(k; k+1)$ and between the rows (resp. columns) $(n - k - 1; n - k)$ which divide $U_{\frac{a}{b}}$ into 9 blocks. See Fig. 7.

Now we prove that each block α, β and γ of $U_{\frac{a}{b}}$ can be decomposed in some smaller second order balanced matrices. In fact, to construct the matrix $U_{\frac{a}{b}}$ we first place the rows of separation. This shows that $U_{\frac{a}{b}}$ is composed of 9 blocks where the three blocks $\alpha_{k \times k}$, $\beta_{(n-2k-1) \times (n-2k-1)}$ and $\gamma_{(n-2k-1) \times k}$ are constructed while the others are obtained by symmetry.
In the following part, we construct respectively each of the blocks α, γ and β to get $U_{\frac{a}{b}}$.

α-*block:* Recall that $z \geq 2$ and we let for the rest of this section z to be an odd number where the even case is obtained in a similar way. Let $\frac{s}{t}$, $\frac{g}{h}$ and $\frac{x}{y}$ be the fractions defined at the beginning of Sect. 6.1. The α-block is formed of k rows and columns where $k = x + y = |\,\mathrm{FRF}(\frac{a}{b})\,|$.

Lemma 5. *The α-block of $U_{\frac{a}{b}}$ is exactly the first $k - 1$ rows and columns of the matrix $U_{\frac{u}{v}}$ where $\frac{u}{v} = \mathrm{TBF}(\frac{a}{b})$.*

γ-*block:* The second block of $U_{\frac{a}{b}}$ is the γ-block of dimension $(n - 2k - 1) \times k$. This block is located between the $k + 1$ and the $(n - k - 1)^{th}$ rows and bounded by the row of separation at the kth column. This part of the second order balance matrix admits a recursive form. If we consider the concatenation of matrices as a vertical stacking, then $\gamma(\frac{a}{b})$ is given by

$$\gamma\left(\frac{a}{b}\right) = \begin{cases} \gamma\left(\frac{c}{d}\right)\left[\gamma\left(\frac{u}{v}\right)\gamma\left(\frac{c}{d}\right)\right]^{a_z - 2} & \text{if } a_z \geq 3 \\[2mm] \gamma\left(\frac{e}{f}\right) & \text{if } a_z = 2 \end{cases} \qquad (2)$$

Let $\frac{a}{b} = \frac{5}{8}$, the 9 blocks are represented as follows:

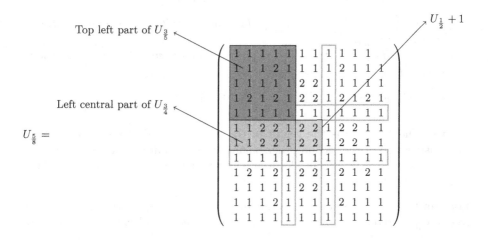

Fig. 7. The decomposition of the matrix $U_{\frac{5}{8}}$ into 9 blocks. We have $\mathrm{TBF}(\frac{5}{8}) = \frac{3}{5}$, $\mathrm{SUF}(\frac{5}{8}) = \frac{1}{2}$, $\mathrm{FPF}(\frac{5}{8}) = \frac{3}{4}$, $\frac{x}{y} = \frac{2}{3}$ and $k = 5$.

We mentioned that we are considering the case where $z \geq 2$ and with z an odd value. In fact, the parity of z determines if a/b is a right or left child. Furthermore, it helps us to know the value of α, where α is the value needed to obtain $D(a, b)$.

Lemma 6. *Let $\frac{a}{b} = [a_0, \ldots, a_z]$, and $\mathrm{FRF}(\frac{a}{b}) = \frac{x}{y}$ with $x + y = k$. The set $D(a, b) = \{i\alpha + 1 \mod n | 1 \leq i \leq a\}$ where $n = a + b$ has:*

$$\begin{cases} \alpha = k, & \text{if } z \text{ is even} \\ \alpha = n - k, & \text{if } z \text{ is odd.} \end{cases}$$

In order to prove the construction of the γ–block, we need to introduce some additional properties of the balance matrix $B_{\frac{a}{b}}$ used in the technical proof of the γ–block.

Property 6. Let $w = 0^x 1^y$ such that $x, y > 0$. For $i \leq x$ and $i \leq y$, we have: $B_w[i] = 0^{x-i+1}12\cdots(i-2)(i-1) \cdot i^{y-i+1} \cdot (i-1)(i-2)\cdots 21$.

Lemma 7. *Let $\frac{a}{b} = [a_0, \ldots, a_z]$ and $\frac{x}{y} = \mathrm{FRF}(\frac{a}{b})$ such that $|C(\frac{x}{y})| = x + y = k$.*

$$\begin{cases} B_{\frac{a}{b}}[k] = 0^\alpha 10^{n-\alpha-1}; & B_{\frac{a}{b}}[2k] = 0^{n-2k}10^{k-1}10^{k-1} \text{if } z \text{ is odd,} \\ B_{\frac{a}{b}}[n - k] = 0^\alpha 10^{n-\alpha-1}; & B_{\frac{a}{b}}[n - 2k] = 0^{n-2k}10^{k-1}10^{k-1} \text{if } z \text{ is even.} \end{cases}$$

$$(3)$$

The following lemma states the recursive construction of the γ–block in a formal way.

Lemma 8. *The* $\gamma-block$ *is obtained depending on the position of the fraction in the Stern-Brocot tree.*

- *If* $a_z \geq 3$ *then* $\gamma\left(\frac{a}{b}\right)$ *is obtained by stacking vertically* $\gamma(\frac{c}{d})$ *over* $a_z - 2$ *copies of* $\gamma(\frac{u}{v})$ *and* $\gamma(\frac{c}{d})$.
- *Otherwise, if* $a_z = 2$, *then* $\gamma\left(\frac{a}{b}\right)$ *is given by the extension of* $\gamma(\frac{e}{f})$ *to the* k-*th column of* $U_{\frac{e}{f}}$.

where $\frac{u}{v} = \mathrm{TBF}(\frac{a}{b})$; $\frac{c}{d} = \mathrm{FDF}(\frac{u}{v})$; $\frac{e}{f} = \mathrm{FPF}(\frac{a}{b})$.

β-*block*: The $\beta-$block is the center of $U_{\frac{a}{b}}$, the last block needed to complete the construction of the second order balance matrix. This block is of dimension $(n - 2k - 1) \times (n - 2k - 1)$ and located between the rows (respectively columns) k and $n - k$.

Lemma 9. *The* $\beta-block$ *of* $U_{\frac{a}{b}}$ *is exactly the second order balance matrix of the fraction* $\frac{\rho}{\theta} = \mathrm{SUF}(\frac{a}{b})$ *in the Stern-Brocot tree where its elements are increased by* 1.

7   Perspectives

In the introduction, we stated that there were connections between the second order of balance and the Fraenkel conjecture, and we mentioned the importance of the second order of balance to construct balanced words on a $k-$letters alphabet. Hence, we have to show some hints on how to use the second order balance matrix U_w to have new information on the synchronisation of words. Of course the Fraenkel conjecture is far from being solved because of its own complexity nevertheless we have built a new tool to synchronize balanced words by considering synchronization of blocs instead of letters. The following example is based on the word $FR_3 = 1213121$ on a three letter alphabet, which is obtained from the synchronization of the 3 Christoffel words of slope $1/6$ associated with the word $C_3 = ***3***$, of slope $2/5$ associated with the word $C_2 = *2***2*$ and of slope $3/4$ associated with the word $C_1 = 1*1*1*1$. Remark that in each position we have only one value and the other symbols are stars. We are able to read this synchronization on rows of matrices in Fig. 8:

If we name $M_{\frac{a}{b}}$ the set of indices of the rows with a maximal values in $U_{\frac{a}{b}}$ and $m_{\frac{a}{b}}$ the set of indices of the rows with minimal values in $U_{\frac{a}{b}}$ (which are non previously chosen rows), we could extract the following set of indices: $M_{\frac{1}{6}} = \{3,4\}$ and $m_{\frac{1}{6}} = \{1,6\}$; $M_{\frac{2}{5}} = \{2,5\}$ and $m_{\frac{2}{5}} = \{3,4\}$; $M_{\frac{3}{4}} = \{3,4\}$ and $m_{\frac{3}{4}} = \{2,5\}$. We can observe that starting with the elements of the set $m_{\frac{3}{4}}$, if we double the values and take modulo 7, we get the elements of $m_{\frac{2}{5}}$. The same note can be also given for the sets $m_{\frac{2}{5}}$ and $m_{\frac{1}{6}}$. $2.2 = 4$; $5.2 = 10 \equiv_7 3$, where $\{3,4\} = m_{\frac{2}{5}}$. $4.2 = 8 \equiv_7 1$; $3.2 = 6$, where $\{1,6\} = m_{\frac{1}{6}}$. Except for the rows 1 and 6, we can see the synchronization between the sets $m_{\frac{a}{b}}$ and $M_{\frac{a}{b}}$, more precisely, if a row appears as a minimum for the set m it appears as a maximum for the set M, which allows us to read all information for the synchronization in

$$U_{\frac{1}{6}} = \begin{pmatrix} 1\,1\,1\,1\,1\,1 \\ 1\,2\,2\,2\,2\,1 \\ 1\,2\,3\,3\,2\,1 \\ 1\,2\,3\,3\,2\,1 \\ 1\,2\,2\,2\,2\,1 \\ 1\,1\,1\,1\,1\,1 \end{pmatrix}, U_{\frac{2}{5}} = \begin{pmatrix} 1\,1\,1\,1\,1\,1 \\ 1\,2\,1\,1\,2\,1 \\ 1\,1\,1\,1\,1\,1 \\ 1\,1\,1\,1\,1\,1 \\ 1\,2\,1\,1\,2\,1 \\ 1\,1\,1\,1\,1\,1 \end{pmatrix} \text{ and } U_{\frac{4}{3}} = U_{\frac{3}{4}} = \begin{pmatrix} 1\,1\,1\,1\,1\,1 \\ 1\,1\,1\,1\,1\,1 \\ 1\,1\,2\,2\,1\,1 \\ 1\,1\,2\,2\,1\,1 \\ 1\,1\,1\,1\,1\,1 \\ 1\,1\,1\,1\,1\,1 \end{pmatrix}.$$

Fig. 8. The three matrices $U_{\frac{1}{6}}$, $U_{\frac{2}{5}}$ and $U_{\frac{3}{4}}$ used for the synchronization of the triplet $(1, 2, 4)$ mod 7.

the second order balance matrix. This will be a keypoint for our next research, in order to study the synchronization of Christoffel words over a k−letters alphabet and to try to tackle the Fraenkel's conjecture.

In addition, many research problems are still open in this study. For example, how to extend the second order balance matrix construction for a binary word that are not Christoffel words?

We could also investigate the Stern-Brocot tree in order to find the structure of infinite paths that minimize the second order balance and find a combinatorial construction of these paths.

To end the perspectives, we could notice that considering the balance matrix is equivalent to compute balance property of a given word. This computation appears in another context in order to compute Parikh vectors that are k−dimensional vectors defined for a finite word w on a k−letters alphabet. Nevertheless, the computation of the second order balance matrix using Parikh vectors is not straightforward and will be considered in a future article.

References

1. Altman, E., Gaujal, B., Hordijk, A.: Balanced sequences and optimal routing. J. ACM **47**(4), 752–775 (2000)
2. Baryshnikov, Y.: Complexity of trajectories in rectangular billiards. Commun. Math. Phys. **174**(1), 43–56 (1995)
3. Berstel, J.: Tracé de droites, fractions continues et morphismes itérés. In: Mots, pp. 298–309. Lang. Raison. Calc., Hermès (1990, incollection)
4. Berstel, J., De Luca, A.: Sturmian words, Lyndon words and trees. Theor. Comput. Sci. **178**(1), 171–203 (1997)
5. Berstel, J., Lauve, A., Reutenauer, C., Saliola, F.: Combinatorics on Words: Christoffel Words and Repetition in Words, vol. 27, p. 147. American Mathematical Society, Providence (2008)
6. Berthé, V., De Luca, A., Reutenauer, C.: On an involution of Christoffel words and sturmian morphisms. Eur. J. Comb. **29**(2), 535–553 (2008)
7. Borel, J.P., Laubie, F.: Quelques mots sur la droite projective réelle. J. de théorie des nombres de Bordeaux **5**(1), 23–51 (1993)
8. Brocot, A.: Calcul des rouages par approximation. Revue chronométrique J. des horlogers, scientifique et pratique **3**, 186–194 (1861)

9. Christoffel, E.B.: Observatio arithmetica. Annali di Matematica Pura ed Applicata (1867-1897) **6**(1), 148–152 (1873)
10. Fogg, N.P.: Substitutions in Dynamics, Arithmetics and Combinatorics. Springer, Heidelberg (2002). https://doi.org/10.1007/b13861
11. Fraenkel, A.S.: The bracket function and complementary sets of integers. Can. J. Math. **21**, 6–27 (1969)
12. Fraenkel, A.S., Levitt, J., Shimshoni, M.: Characterization of the set of values $f(n) = [na]$, $n = 1, 2, \ldots$. Discrete Math. **2**(4), 335–345 (1972)
13. Gaujal, B.: Optimal allocation sequences of two processes sharing a resource. Discrete Event Dyn. Syst. **7**(4), 327–354 (1997)
14. Gaujal, B., Varjú, P.P.: Partitioning the positive integers into seven Beatty sequences. Indag. Math. N.S. **12**, 149–161 (2003)
15. Hubert, P.: Suites équilibrées. Theor. Comput. Sci. **242**(1), 91–108 (2000)
16. Lothaire, M.: Algebraic Combinatorics on Words, Encyclopedia of Mathematics and its Applications, vol. 90. Cambridge University Press, Cambridge (2002)
17. de Luca, A., Mignosi, F.: Some combinatorial properties of sturmian words. Theor. Comput. Sci. **136**(2), 361–385 (1994)
18. Mairesse, J., Vuillon, L.: Asymptotic behavior in a heap model with two pieces. Theor. Comput. Sci. **270**(1), 525–560 (2002)
19. Morikawa, R.: On eventually covering families generated by the bracket function IV. Nat. Sci. **25**, 1–8 (1985)
20. Morse, M., Hedlund, G.A.: Symbolic dynamics II. Sturmian trajectories. Am. J. Math. **62**(1), 1–42 (1940)
21. Paquin, G., Reutenauer, C.: On the superimposition of Christoffel words. Theor. Comput. Sci. **412**(4), 402–418 (2011)
22. Paquin, G., Vuillon, L.: A characterization of balanced episturmian sequences. Electron. J. Comb. **14**(1), R33, 12 (2007)
23. Provençal, X., Vuillon, L.: Discrete segments of Z^3 constructed by synchronization of words. Discrete Appl. Math. **183**, 102–117 (2014)
24. Raney, G.N.: On continued fractions and finite automata. Mathematische Annalen **206**(4), 265–283 (1973)
25. Sidorov, N.: Optimizing properties of balanced words, pp. 240–246 (2011)
26. Skolem, T.: On certain distributions of integers in pairs with given differences. Math. Scand. **5**, 57–68 (1957)
27. Smith, H.J.S.: Note on continued fractions. Messenger Math. **6**, 1–14 (1876)
28. Tijdeman, R.: Fraenkel's conjecture for six sequences. Discrete Math. **222**(1), 223–234 (2000)
29. Vuillon, L.: Balanced words. Bull. Belg. Math. Soc. Simon Stevin **10**(5), 787–805 (2003)

IPO-Q: A Quantum-Inspired Approach to the IPO Strategy Used in CA Generation

Michael Wagner, Ludwig Kampel, and Dimitris E. Simos[(✉)]

SBA Research, 1040 Vienna, Austria
{mwagner,lkampel,dsimos}@sba-research.org

Abstract. Covering arrays are combinatorial structures, that can be considered generalizations of orthogonal arrays and find application in the field of automated software testing amongst others. The construction of covering arrays is a highly researched topic, with existing works focusing on heuristic, metaheuristic and combinatorial algorithms to successfully construct covering arrays with a small number of rows. In this paper, we introduce the IPO-Q algorithm which combines a recently introduced quantum-inspired evolutionary algorithm with the widely used in-parameter order (IPO) strategy for covering array generation. We implemented different versions of this algorithm and evaluate them, by means of selected covering array instances, against each other and against an algorithm implementing the IPO strategy.

Keywords: Covering arrays · Quantum-inspired algorithms · IPO strategy · Algorithmic generation

1 Introduction

Covering arrays (CAs) are discrete combinatorial structures and can be considered a generalization of orthogonal arrays. We (informally) introduce *binary covering arrays* as binary $N \times k$ arrays $M = (\mathbf{m}_1, \ldots, \mathbf{m}_k)$, denoted as $\mathsf{CA}(N; t, k)$, with the property that any array $(\mathbf{m}_{i_1}, \ldots, \mathbf{m}_{i_t})$, with $\{i_1, \ldots, i_t\} \subseteq \{1, \ldots, k\}$, comprised of any t different columns of M has the property that each binary t-tuple in $\{0,1\}^t$ appears at least once as a row. See also [2]. The parameter t is also called the *strength* of the CA, N, t and k are called the *parameters* of the CA. In the literature CAs are also defined for arbitrary alphabets, however we restrict our attention to binary CAs in this work. An example of a CA is given in Example 1.

Alternatively, binary CAs can be defined as binary $N \times k$ arrays which rows *cover* all binary *t-way interactions*, where for a given strength t and a number of columns k, a *t-way interaction* is a set of t pairs $\{(p_1, v_1), \ldots, (p_t, v_t)\}$ with $1 \le p_1 < p_2 < \ldots < p_t \le k$, and $v_i \in \{0,1\}$ for all $i = 1, \ldots, t$. As the value k is usually clear from the context, it is mostly omitted. We say the t-way interaction

© Springer Nature Switzerland AG 2020
D. Slamanig et al. (Eds.): MACIS 2019, LNCS 11989, pp. 313–323, 2020.
https://doi.org/10.1007/978-3-030-43120-4_24

$\{(p_1, v_1), \ldots, (p_t, v_t)\}$ is *covered* by an array A, if there exists a row in A that has the value v_i in position p_i for all $i = 1, \ldots, t$. Then a $\mathsf{CA}(N; t, k)$ is characterized by covering all t-way interactions.

Their properties make CAs attractive for application in, amongst others, the field of automated software testing, serving as key ingredient for the construction of test sets. The interested reader may have a look at [6]. For practical applications, it is desired to construct CAs with a small number of rows. This leads to the formal problem of generating a CA with the smallest possible number of rows, for a given number of columns. For given values t and k, the smallest integer N for which a $\mathsf{CA}(N; t, k)$ exists is called *covering array number* and is denoted as $\mathsf{CAN}(t, k)$. The problem of determining $\mathsf{CAN}(t, k)$ remains unsolved in general (for arbitrary strengths and alphabets), is algorithmically challenging to solve and is tightly coupled with NP-hard problems, see for example [4] and references therein.

Example 1. A $\mathsf{CA}(4; 2, 3)$ is given by the following array

$$A = \begin{pmatrix} 0 \ 0 \ 0 \\ 0 \ 1 \ 1 \\ 1 \ 0 \ 1 \\ 1 \ 1 \ 0 \end{pmatrix}. \tag{1}$$

As can be verified by the reader, when selecting any two columns of A, in the resulting 4×2 array, each binary pair appears once as a row. Put differently, each of the 12 binary 2-way interactions $\{(1, 0), (2, 0)\}, \{(1, 0), (2, 1)\}, \{(1, 1), (2, 0)\}, \{(1, 1), (2, 1)\}, \{(1, 0), (3, 0)\}, \{(1, 0), (3, 1)\}, \{(1, 1), (3, 0)\}, \{(1, 1), (3, 1)\}, \{(2, 0), (3, 0)\}, \{(2, 0), (3, 1)\}, \{(2, 1), (3, 0)\}$ and $\{(2, 1), (3, 1)\}$, is covered by the rows of A.

There exist many algorithmic approaches dedicated to the construction of CAs with a *small* number of rows, including, amongst others, greedy heuristics [5,9] and metaheuristics such as [14] and [15]. For a survey of CA generation methods, the interested reader may also have a look in [13].

This paper proposes an algorithm for CA generation, that combines the IPO strategy [9] with the metaheuristic approach introduced in [15], and is structured as follows. In Sect. 2 we provide the necessary preliminaries for this work. In Sect. 3 we introduce the IPO-Q algorithm and the different versions of it considered in this paper. An experimental evaluation is given in Sects. 4 and 5. Finally, Sect. 6 concludes this paper.

2   Preliminaries

In this section we briefly summarize the main concepts, the IPO strategy and quantum inspired evolutionary algorithms for CA generation, that will serve as starting point for our newly introduced IPO-Q algorithm.

2.1   Review of the IPO Strategy

The In-Parameter-Order (IPO) strategy was first introduced in [9] for the generation of strength $t = 2$ covering arrays, and was later generalized for higher strengths [7,8] and arbitrary alphabets. The IPO strategy is the base for CA generation tools like ACTS [16] or CAgen [10], that are also used in industrial applications like [3] and [12]. Algorithms implementing the IPO strategy [7,8] are greedy algorithms that grow CAs in two dimensions, horizontally and vertically. The input to these algorithms is the desired strength t as well as the number of columns k and the alphabet of the desired CA.

As we are concerned with binary CAs in this work, the IPO strategy can be summarized as follows. Starting with the $2^t \times t$ array $\{0, 1\}^t$ of all row vectors over $\{0, 1\}$ of length t, algorithms implementing the IPO strategy proceed iteratively in two phases that alternate each other, the horizontal extension (adding a column) and the vertical extension (adding rows), until the desired CA with k columns has been constructed. A schematic overview of this strategy is given in Fig. 2. In the horizontal extension step the entries of a newly added column are specified in a greedy manner, maximizing the number of newly covered t-way interactions in each step. In the vertical extension step additional rows are added to the array until all t-way interactions are covered, i.e. the array is a covering array for strength t and the current number of columns.

2.2   Review of the QIEAFORCA Algorithm for CA Generation

Recently, a quantum-inspired evolutionary algorithm for covering array construction (QIEAFORCA) was proposed in [15], which we review briefly and gather the notions necessary for the algorithm introduced in Sect. 3. QIEAFORCA takes as input the parameters N, t and k of a desired CA. The underlying idea is to consider an $N \times k$ array of qubits, which states $|0\rangle$ and $|1\rangle$ are identified with the numerical values 0 and 1. Thus, when observing the individual qubits, they collapse to either state and an $N \times k$ array over $\{0, 1\}$ is attained. To obtain an actual covering array, the states of the qubits are iteratively evolved.

For *classical* computational realization, in [15], a reduced qubit representation is used,

$$|\Psi\rangle = \cos\Theta \,|0\rangle + \sin\Theta \,|1\rangle, \text{ with } \Theta \in [0, \frac{\pi}{2}],$$

where the state of a qubit is completely specified by its angle Θ, see Fig. 1a. An angle of $0°$ corresponds to the state $|0\rangle$ and $90°$ to $|1\rangle$ respectively. Measurement of a qubit yields state $|0\rangle$ with probability $(\cos\Theta)^2$ and state $|1\rangle$ otherwise. *In the following we use the term qubit synonymous for the reduced qubit representation.*

In [15], an initial array of qubits in the state corresponding to an angle of $45°$ is generated, representing a uniform distribution of the possible states, hence representing a *neutral state*. The modification of qubits is realized in rounds, until either a CA is found, or a set number of iterations passed. In each round, a new candidate solution is generated by measuring every qubit. The resulting array is evaluated in terms of the number of covered t-way interactions. The best array over all rounds is stored as base for modification. In contrast to actual quantum computation, we can preserve the state $|\Psi\rangle = \cos\Theta\,|0\rangle + \sin\Theta\,|1\rangle$ of a qubit after measurement. To update the state of a qubit, it gets rotated by a specific angle towards state $|0\rangle$ or $|1\rangle$, see also Fig. 1b. The direction of the rotation is based on the corresponding entry in the current best array. The qubit rotations serve to guide the search towards a promising subset of the search space, while the probabilistic nature of the qubit measurement serves as an exploration mechanism.

Further, the concept of *mutation* is used as a constraint on the rotation of qubits, by only allowing them to approach $|0\rangle$ or $|1\rangle$ up to a certain ϵ, see again Fig. 1a. This prevents qubits from completely converging to one of the states $|0\rangle$ or $|1\rangle$ and thus offers means for individual qubits to escape local minima. For more details the interested reader is referred to [15].

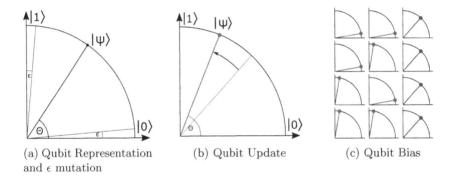

(a) Qubit Representation
and ϵ mutation

(b) Qubit Update

(c) Qubit Bias

Fig. 1. Reduced qubit representation (a) and used update mechanics (b) and (c).

3   IPO-Q: A Quantum-Inspired IPO Algorithm

In this section we introduce the quantum-inspired extension strategy IPO-Q for CA generation. It combines the IPO strategy with ideas of quantum-inspired evolutionary algorithms. Like other algorithms implementing the IPO strategy, IPO-Q consists of horizontal and vertical extension steps. The algorithm acts on an array of qubits that is iteratively extended. Based on this qubit array, in each extension step QIEAFORCA is used to generate an array optimizing the number of covered t-way interactions. IPO-Q is given by means of a pseudocode in Algorithm 1 and can be described as follows.

The algorithm starts with a $2^t \times t$ array Q_0 of qubits in neutral state, from which an initial binary array is generated, using QIEAFORCA. Afterwards, when in the i-th step a CA was found by QIEAFORCA, based on the array of qubits Q_i, IPO-Q enters the horizontal extension step and adds a new column q_{i+1} of qubits, yielding $Q_{i+1} = (Q_i, q_{i+1})$. The newly added qubits in q_{i+1} are initialized in neutral state, while the qubits of Q_i are biased based on the previously computed

CA. This bias is realized by rotating qubits from the neutral state towards the state corresponding to the values in the previously found CA. Figure 1c depicts the resulting qubit array, when biasing the qubits in the first two columns, colored in red, towards the values given in the first two columns of the CA given in Eq. (1) and the newly added qubits in neutral state on the right, colored in blue. Hence, the *new* qubits are left open for exploration, while the *old* qubits are more likely to measure the values of the previous CA with i columns. Then QIEAFORCA is used to maximize the number of covered t-way interactions based on the array of qubits Q_{i+1}.

If QIEAFORCA does not find a CA based upon Q_{i+1}, IPO-Q performs a vertical extension step. A new row of qubits is added to Q_{i+1}. The new qubits are initialized in the neutral state, while all other qubits get biased towards the corresponding value in the previous best solution. Again, QIEAFORCA is used attempting to find a CA based on Q_{i+1} with the increased number of rows. If it fails, additional rows are added to Q_{i+1} one by one, as the process is repeated, until a CA is found.

Once a CA with $(i+1)$ columns is generated, IPO-Q enters the next horizontal extension phase. These steps are repeated until a CA with the desired number of columns is found.

Remark 1. The main advantage of the proposed IPO-Q algorithm in comparison to the QIEAFORCA algorithm, proposed in [15] is that all returned arrays are CAs. For example, when initializing QIEAFORCA with CA parameters N, t, k where $N < \mathsf{CAN}(t, k)$ it is impossible to find a CA, whereas the IPO-Q algorithm, on input t, k, can add additional rows of qubits until a CA can be found.

Algorithm 1. IPO-Q

1: INPUT: k, t, QIEAFORCA settings
2: Generate initial $2^t \times t$ array using QIEAFORCA
3: **for** $i \leftarrow t, ..., k$ **do**
4:     Add new column of qubits in neutral state   ▷ Horizontal Extension
5:     Bias old qubits towards previous solution
6:     Apply QIEAFORCA to maximize the number of covered t-way interactions
7:     **while** any t-way interactions are uncovered **do**
8:         Add new row of qubits in neutral state   ▷ Vertical Extension
9:         Bias old qubits towards previous solution
10:         Apply QIEAFORCA to maximize the number of covered t-way interactions
11:     **end while**
12: **end for**
13: **return** generated CA

New Concepts for IPO-Q. In the following we describe two new realizations of the concepts of mutation and bias, that adapt to the extension strategy used in IPO-Q.

First, during our experiments (see Sect. 4), we noticed that constant angles ϵ used to implement the concept of mutation negatively affect the solution quality. Further, once the number of columns k and the number of rows N of the desired arrays and thus the number of qubits, gets too large, mutation rates that work well for less columns make the system too unstable. Therefore, in addition to the mutation types presented in [15], we introduce `variable mutation`, which sets the angle for mutation to

$$\epsilon = \begin{cases} g, & N \times k \leq 100 \\ \frac{100 \times g}{N \times k}, & N \times k > 100 \end{cases} \tag{2}$$

where g is an initial parameter called `base mutation`.

Second, we introduce a concept further referred to as `onion` extension. In this version, the bias of all qubits is increased by a set amount at the beginning of every horizontal extension step, until the bias reaches $45°$, i.e. the qubits are fixed to state $|0\rangle$ or $|1\rangle$. This results in an onion-like structure of the bias values of qubits in Q_i, where "newer" qubits are left more open for exploration, while "older" qubits allow less exploration.

Figure 2 illustrates the different bias layers caused by the `onion` concept for $CA(12; 3, 7)$. After the initial $CA(8; 3, 3)$ is generated, IPO-Q performs a horizontal extension. At the beginning of the extension step, the bias of the old qubits is increased by a given angle Δb. Afterwards, QiEAforCA is used to find a CA with 4 columns, resulting in a successful horizontal extension. Hence, the next horizontal extension step starts and the bias of all qubits is increased. Note that the bias of the qubits in Q_0 is increased for the second time, therefore its qubits are biased more towards the previous solution than the qubits in Q_1. If a horizontal extension fails to construct a CA, the rows added by means of vertical extension have the same bias as the added column, see e.g. Q_2 and Q_3.

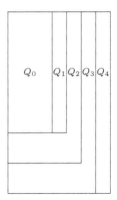

Fig. 2. Onion concept.

4   Evaluation of Different Configurations of the IPO-Q Algorithm

To evaluate the effect of different settings of IPO-Q, we compared the following selected configurations:

1. `Global Mutation eps=0` uses no mutation, i.e. $\epsilon = 0°$;
2. `Global Mutation eps=5` uses a mutation angle of $\epsilon = 5°$, that is applied to every qubit;
3. `Variable Mutation eps=5` uses `variable mutation` with a `base mutation` of $g = 5°$, see Eq. (2);
4. `Onion eps=10` implements the `onion` concept with $\Delta b = 2°$, using `variable mutation` and a `base mutation` of $g = 10°$.

We compared these configurations by means of computing a $CA(N; 3, k)$ for $k = 2, \ldots, 100$ and comparing the resulting values for N. For each configuration, we conducted five runs for every $CA(N; 3, k)$ instance and recorded the minimal number of rows N of the five generated CAs. The results are depicted in Fig. 3, where the horizontal axis represents the number of columns k and the vertical axis represents the number of rows N. As one aims to minimize the number of rows in a CA, smaller values indicate better results.

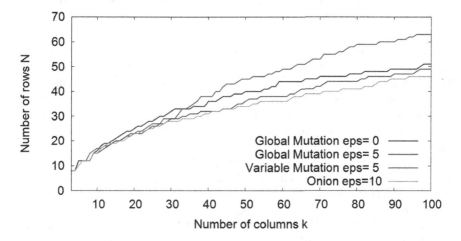

Fig. 3. Parameter evaluation for IPO-Q on the instance $CA(N; 3, k)$ (smaller values are better).

We can see that configuration Global Mutation eps=0 performs worse than the other configurations for up to $k = 30$ columns, i.e. it generates CAs with a higher number of rows, but manages to find acceptable solutions for a higher number of columns. Interpreting these results, we believe this is due to Global Mutation eps=0 using no mutation and hence lacking means of exploration to optimize smaller instances.

Configuration Global Mutation eps=5, on the other hand, finds CAs with a smaller number of rows compared to Global Mutation eps=0, for $k \le 30$, but becomes unstable for higher values of k.

Configuration Variable Mutation eps=5 improves on both, Global Mutation eps=0 and Global Mutation eps=5, for all values of k. This shows nicely that the concept of variable mutation, decreasing the mutation angles proportional to the number of qubits $N \times k$, combines the advantages of exploration of small instances (Global Mutation eps=5) and exploitation of previous solutions for larger instances (Global Mutation eps=0), respectively.

Finally, Onion eps=10 further improves the results of Variable Mutation eps=5, yielding the best results of the considered IPO-Q configurations. In the general IPO-Q algorithm approach without the onion concept, due to iterative horizontal and vertical extension steps, inner parts of the array are optimized multiple times. While, the concept of `variable mutation` can only decrease exploration on a global scale, `Onion eps=10` reflects the core idea of the original IPO strategy, as depicted in [9]. The inner parts of the qubit array, that already experienced multiple optimization rounds, are fixed by means of increasing the bias up to $45°$, while the high initial mutation angle of $10°$ allows for exploration for newly added qubits.

5   Comparison with IPOG-F and Best Known Upper Bounds for CAN

In a second phase of experimental evaluation, we compared the results of the IPO-Q algorithm against the best known upper bounds for covering array numbers [1] and the numbers provided by NIST [11] using the IPOG-F algorithm. For that purpose we considered the configuration of IPO-Q, that had the best performance in the experiments reported in the previous section, i.e. in the following evaluation IPO-Q refers to the `Onion eps=10` configuration described above.

The graphs in Fig. 4 compare the number of rows N of generated CAs, on the vertical axes, for a given number of columns k, on the horizontal axes. Furthermore, selected numerical values are highlighted in Table 1. For strength two, IPO-Q consistently finds CAs with less rows and approximates the covering array number CAN. For higher strengths t, IPO-Q finds smaller CAs for up to $k = 81$ columns for strength $t = 3$ and for up to $k = 21$ columns for strength $t = 4$, see Fig. 4. For higher number of columns, IPOG-F produces better results. We believe these results reflect the probabilistic nature of the Quantum-inspired algorithm very well. For small strengths t and number of columns k, IPO-Q can fully utilize the probabilistic search.

Table 1. Results for selected CAs $CA(N; t, k)$. Values for upper bounds for covering array numbers in the column headed by "CAN$\leq$" are taken from [1].

CA	IPO-Q	IPOG-F	CAN$\leq$
$CA(N; 2, 100)$	10	13	10
$CA(N; 2, 500)$	14	17	10
$CA(N; 2, 1500)$	16	20	14
$CA(N; 3, 20)$	23	25	18
$CA(N; 3, 50)$	34	36	28
$CA(N; 3, 100)$	46	45	33
$CA(N; 4, 10)$	34	41	24
$CA(N; 4, 20)$	66	65	39
$CA(N; 4, 35)$	91	85	64

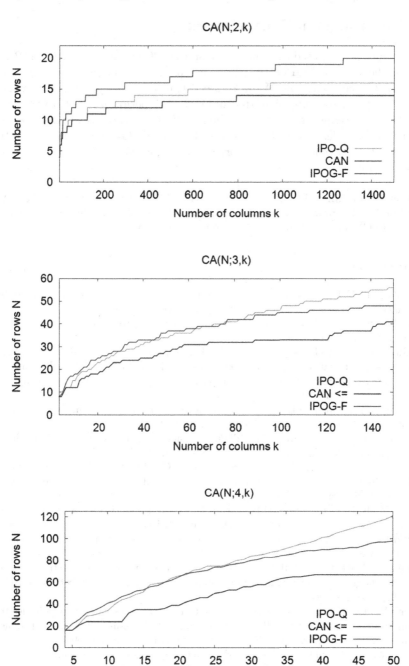

Fig. 4. Comparison of the number of rows of generated CAs by IPO-Q and IPOG-F with the best known upper bounds for CAN, maintained at [1], denoted as "CAN <=", for strengths $t = 2, 3, 4$ from top to bottom (smaller values are better).

6   Conclusion and Future Work

In this paper we proposed the quantum-inspired IPO extension algorithm IPO-Q, merging the ideas of the IPO strategy with a quantum inspired evolutionary algorithm for CA computation. We introduced two new concepts, called `variable mutation` and `onion` extension, that improved the performance of IPO-Q in our evaluation. The experiments further showed, that in some cases IPO-Q can construct CAs with less rows than the well known IPOG-F algorithm. As future work, we want to generalize the algorithm for higher alphabets as well as mixed covering arrays.

Acknowledgements. This research was carried out partly in the context of the Austrian COMET K1 program and publicly funded by the Austrian Research Promotion Agency (FFG) and the Vienna Business Agency (WAW).

References

1. Colbourn, C.J.: Covering Array Tables for t = 2, 3, 4, 5, 6. http://www.public. asu.edu/~ccolbou/src/tabby/catable.html. Accessed 18 Sept 2019
2. Colbourn, C.J., Dinitz, J.H.: Handbook of Combinatorial Designs. CRC Press, Boca Raton (2006)
3. Ghandehari, L.S.G., Bourazjany, M.N., Lei, Y., Kacker, R.N., Kuhn, D.R.: Applying combinatorial testing to the siemens suite. In: 2013 IEEE Sixth International Conference on Software Testing, Verification and Validation Workshops, pp. 362–371, March 2013
4. Kampel, L., Simos, D.E.: A survey on the state of the art of complexity problems for covering arrays. Theor. Comput. Sci. **800**, 107–124 (2019)
5. Kleine, K., Simos, D.E.: An efficient design and implementation of the in-parameter-order algorithm. Math. Comput. Sci. **12**(1), 51–67 (2018)
6. Kuhn, D., Kacker, R., Lei, Y.: Introduction to Combinatorial Testing. Chapman & Hall/CRC Innovations in Software Engineering and Software Development Series. Taylor & Francis, London (2013)
7. Lei, Y., Kacker, R., Kuhn, D.R., Okun, V., Lawrence, J.: IPOG: a general strategy for T-way software testing. In: 14th Annual IEEE International Conference and Workshops on the Engineering of Computer-Based Systems (ECBS 2007), pp. 549–556, March 2007
8. Lei, Y., Kacker, R., Kuhn, D.R., Okun, V., Lawrence, J.: IPOG/IPOG-D: efficient test generation for multi-way combinatorial testing. Softw. Test. Verif. Reliab. **18**(3), 125–148 (2008)
9. Lei, Y., Tai, K.C.: In-parameter-order: a test generation strategy for pairwise testing. In: Proceedings Third IEEE International High-Assurance Systems Engineering Symposium (Cat. No. 98EX231), pp. 254–261, November 1998
10. Wagner, M., Kleine, K., Simos, D.E., Kuhn, R., Kacker, R.: CAgen: a fast combinatorial test generation tool with support for constraints and higher-index arrays. In: 2020 IEEE International Conference on Software Testing, Verification and Validation Workshops (ICSTW) (to appear)
11. NIST: Covering arrays generated by IPOG-F. National Institute of Standards and Technology. https://math.nist.gov/coveringarrays/ipof/ipof-results. html. Accessed 18 Sept 2019

12. Smith, R., et al.: Applying combinatorial testing to large-scale data processing at adobe. In: 2019 IEEE International Conference on Software Testing, Verification and Validation Workshops (ICSTW), pp. 190–193, April 2019
13. Torres-Jimenez, J., Izquierdo-Marquez, I., Avila-George, H.: Methods to construct uniform covering arrays. IEEE Access **7**, 42774–42797 (2019)
14. Torres-Jimenez, J., Rodriguez-Tello, E.: New bounds for binary covering arrays using simulated annealing. Inf. Sci. **185**(1), 137–152 (2012)
15. Wagner, M., Kampel, L., Simos, D.E.: Quantum-inspired evolutionary algorithms for covering arrays of arbitrary strength. In: Kotsireas, I., Pardalos, P., Parsopoulos, K.E., Souravlias, D., Tsokas, A. (eds.) SEA 2019. LNCS, vol. 11544, pp. 300–316. Springer, Cham (2019). https://doi.org/10.1007/978-3-030-34029-2_20
16. Yu, L., Lei, Y., Kacker, R.N., Kuhn, D.R.: ACTS: a combinatorial test generation tool. In: 2013 IEEE Sixth International Conference on Software Testing, Verification and Validation, pp. 370–375, March 2013

A Fast Counting Method for 6-Motifs with Low Connectivity

Taha Sevim📖, Muhammet Selçuk Güvel📖, and Lale Özkahya(✉)📖

Department of Computer Engineering, Hacettepe University, Ankara, Turkey
tahasevim231@gmail.com, selcukguvel@gmail.com, laleozkahya@gmail.com

Abstract. A k-motif (or graphlet) is a subgraph on k nodes in a graph or network. Counting of motifs in complex networks has been a well-studied problem in network analysis of various real-word graphs arising from the study of social networks and bioinformatics. In particular, the triangle counting problem has received much attention due to its significance in understanding the behavior of social networks. Similarly, subgraphs with more than 3 nodes have received much attention recently. While there have been successful methods developed on this problem, most of the existing algorithms are not scalable to large networks with millions of nodes and edges.

The main contribution of this paper is a preliminary study that genaralizes the exact counting algorithm provided by Pinar, Seshadhri and Vishal to a collection of 6-motifs. This method uses the counts of motifs with smaller size to obtain the counts of 6-motifs with low connectivity, that is, containing a cut-vertex or a cut-edge. Therefore, it circumvents the combinatorial explosion that naturally arises when counting subgraphs in large networks.

Keywords: Social networks · Motif analysis · Subgraph counting

1 Introduction

In social network analysis, any fixed subgraph with k nodes is called a *k-motif (or graphlet)* and their analysis has been a useful method to characterize the structure of real-world graphs. It has observed particularly in social networks, that some motifs are more common than others, and the structure of the network is different than the structure of the random graphs [12,18,29]. While knowing that only the analysis of these subgraphs is not sufficient to understand the structure of the networks, it has been validated that the motif frequencies provide substantial information about the local network structure in various domains [8,9,12]. By counting the number of embeddings of each motif in a network, it is possible to create a profile of sufficient statistics that characterizes the network structure [24].

Although there has been significant amount of success and impact on areas varying from social science to biology, the search for faster and more efficient

© Springer Nature Switzerland AG 2020
D. Slamanig et al. (Eds.): MACIS 2019, LNCS 11989, pp. 324–332, 2020.
https://doi.org/10.1007/978-3-030-43120-4_25

algorithms to compute the frequencies of graph patterns continues. The main reason to study algorithms to count motifs faster is combinatorial explosion. The running time of algorithms to exactly count k-motifs on the vertex set V is of the order $O(|V|^k)$. The counts of 6-motifs are in the orders of billions to trillions for graphs with more than a few million edges. Thus, an enumeration algorithm cannot terminate in a reasonable time. The idea presented in [19] and extended to 6-motifs here uses a framework of counting with minimal enumeration. The main contribution of this paper is a preliminary study that genaralizes the exact counting algorithm provided in [19] to a collection of 6-motifs. To the best of our knowledge, this is the only study that counts 6-motifs using exact computation and performs all counts in graphs with millions of edges in minutes. As a preliminary work, we are able to exactly count the motifs shown in Fig. 2. The particular reason that this subset of motifs are chosen is that each of them contains a cut-vertex or a cut-edge, that is, removing that vertex or edge makes the motif disconnected. The main idea is to build a framework to cut each pattern of 6 nodes into smaller patterns, where each of the patterns contain that particular cutting subset, also called *cut-set*. Then, the enumeration is only needed for these smaller patterns rather than the big pattern. For our purposes, we do not carry out the enumeration and use the counts for these smaller patterns obtained in [19].

There are various approximation algorithms [4,14,20,23,31], however the results they provide are not exact and scalable for counting larger motifs with more than 4 nodes, whereas the method presented here is also scalable to very large networks. As presented in Sect. 3, our method is able to count 6-motifs in Fig. 2 for a network with 3 millions of edges under 5 min. Most of the studies on counting motifs have been focusing on smaller motifs with size at most 4. In particular, the count of triangles has been widely studied due to its importance in the analysis of social networks [28]. These results have been helpful for graph classification and often used as graph attributes. Another group of recent studies on subgraph counts are used for detecting communities and dense subgraphs, such as [2,22,25,27]. More recent algorithmic improvements on counting triangles can be found in [23,26]. Exact and approximate algorithms for computing the number of non-induced 4-motifs are proposed in [10].

It has been observed that sampling algorithms [4,20,30,31] and randomized algorithms, such as the color coding method [3,13,32], are not feasible for counting motifs of size larger than 4. One of the most recently developed algorithms in [5] estimates the number of 7-motifs on a graph with 65M nodes and 1.8B edges in around 40 min. Exact counting algorithms as in [15,16,31] exist, but they are very slow and not scalable to large graphs. The recent study in [19] showed an exact counting technique that counts all patterns with at most 5 vertices on graphs with tens of millions of edges in several minutes.

The main contribution in this paper is to cut each pattern in the chosen collection into smaller patterns and use the enumeration on these smaller patterns to count the big pattern by using the framework in [19]. Some other algorithms that used ideas to avoid enumeration can be seen in [1,6,7,11].

2   Methodology

The input graph $G = G(V, E)$ is undirected, where V and E denote the vertex set and the edge set of G, respectively. A subgraph of G is called *induced* if all edges present in the host graph exist as edges in that subgraph. Otherwise, it is called *non-induced.* In our counting method, a subgraph means a non-induced subgraph. We call a triangle with a missing edge a *wedge*, a K_4 with a missing edge a *diamond* and a triangle with an edge attached to one of its vertices, a *tailed triangle.* In our notation, for each vertex i, we use $d(i)$ and $T(i)$, (resp. $T(e)$) to denote the degree of i and the number of triangles that contain the vertex i (resp. edge e), respectively. Similarly, $C_4(i)$ (resp. $C_4(e)$) and $K_4(i)$ (resp. $K_4(e)$) indicate the number of C_4's and K_4's that contain the vertex i (resp. edge e), respectively. The number of diamonds, tailed triangles and K_4's in the graph G are denoted by $D(G)$, $TT(G)$, and $K_4(G)$, respectively. The number of wedges between two vertices i and j and ending at a vertex i are written as $W(i, j)$ and $W(i)$. The numbers of the 5-motifs given in Fig. 1 are described with N_i^5, and of the ones in Fig. 2 are described with N_i.

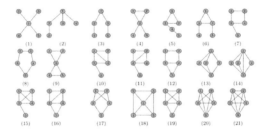

Fig. 1. The collection of connected 5-motifs [19].

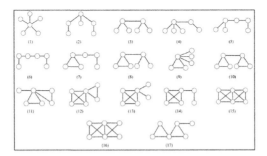

Fig. 2. The 6-motifs with low connectivity.

A standard method for counting triangles is to enumerate the wedges and find the triangles by checking whether the missing edge is there or not. By a similar idea, the formulation here uses a cut-set, say S, for each motif H, whose removal disconnects H. Let the components be C_i and let $S \cup C_i$ be H_i. There is some care needed in choosing this cut-set, however in our algorithm it is typically a vertex or an edge. The count for each possible H_i that contains S is obtained by the counts of 4-motifs and 5-motifs given in [19]. The collection of 5-motifs that are used in our counting method can be seen in Fig. 1.

2.1   Main Theorems

The exact computation for the motifs presented in Fig. 2 is obtained in the following theorems. We refer the reader to [19] for the technical details of the method used. Here, we briefly discuss two examples to present the general idea and how we apply it to obtain Theorems 1 and 2.

Theorem 1 (Cut is a vertex).

$$N_1 = \sum_{i \in V} \binom{d(i)}{5}$$
$$N_2 = \sum_{i \in V} \binom{W(i)}{2}(d(i) - 2) - N_7^5 - N_2^5 - 2N_6^5 - 2TT(G) - 6D(G)$$
$$N_9 = \sum_{i \in V} T(i)\binom{d(i)-2}{3}$$
$$N_{11} = \sum_{i \in V} \binom{T(i)}{2}(d(i) - 4) - N_{11}^5$$
$$N_{12} = \sum_{i \in V} K_4(i)(T(i) - 3) - 2N_{19}^5$$
$$N_{13} = \sum_{i \in V} K_4(i)\binom{d(i)-3}{2}$$
$$N_{14} = \sum_{i \in V} K_4(i)(W(i) - 6) - 2N_{19}^5 - 2N_{15}^5$$

For example, the expression calculating N_9 in Theorem 1 has no overcounting and it considers every vertex as a cut-vertex i and counts by pairing triangles and the three neighbors attached to i. However, in the calculation of N_{17}, Theorem 2, we subtract the number of other motifs, counted unnecessarily. Here, the cut-set is an ordered pair $e = <i, j>$. One example of overcounting occurs when the vertices labeled 1 and 3 in Fig. 3 are chosen to be the same, meaning also 5-motifs with index 10 are counted. Thus, we subtract it twice considering that i is mapped to the vertices labeled either 1 or 4 (in Fig. 1). Similarly, all subtractions remove the contributions of overcounting.

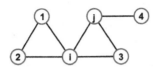

Fig. 3. The chosen cut-edge for motif-17.

Theorem 2 (Cut is an edge).
Here, $<i, j>$ indicates an ordered and (i, j) an unordered pair.

$$N_3 = \sum_{(i,j)\in E} \binom{d(i)-1}{2}\binom{d(j)-1}{2} - N_6^5 - D(G)$$
$$N_4 = \sum_{<i,j>\in E} \binom{d(i)-1}{3}(d(j)-1) - 2N_4^5$$
$$N_5 = \sum_{(i,j)\in E} \binom{d(j)-1}{2}(W(i)-2) - -2N_7^{(5)} - 2N_6^{(5)} - \sum_{x\in V}\binom{d(x)}{4}$$
$$N_6 = \sum_{(i,j)\in E} [W(i)-(d(j)-1)][W(j)-(d(i)-1)] - 2N_4^5 - 5N_8^5 - 2N_7^5 - 2N_5^5 - TT(G) - 3T(G)$$
$$N_7 = \sum_{e=(i,j)\in E} (T(i)-T(e))(W(j)-(d(i)-1)) - 2N_{12}^5 - 4N_9^5 - 8D(G)$$
$$N_8 = \sum_{e=<i,j>\in E}(T(i)-T(e))\binom{d(j)-1}{2} - 2N_{11}^5 - 6K_4(G)$$
$$N_{10} = \sum_{e=(i,j)\in E}(T(i)-T(e))(T(j)-T(e)) - N_{16}^5 - 6K_4(G)$$
$$N_{15} = \sum_{e\in E}\binom{K_4(e)}{2} - 3N_{20}^5$$
$$N_{16} = \sum_{e\in E} K_4(e)\binom{T(e)-2}{2}$$
$$N_{17} = \sum_{e=<i,j>\in E}(t(i)-t(e))t(e)(d(j)-2) - 2N_{10}^5 - 12K_4(G) - 2N_{16}^5$$

3 Experimental Results and Conclusions

Experiments are performed on a computer that has 2.7 GHz dual-core Intel Core i5 processor, 3 MB L3 Cache and 8 GB 1867 MHz LPDDR3 memory. Our counting formulas are implemented with C++ using ESCAPE [19] framework. The datasets are taken from [17,21].

The input graph $G = G(V,E)$ is undirected and has n vertices and m edges, where multiple edges and loops are ignored. The input graph is stored as an adjacency list, where each list is a hash table. Thus, edge queries can be made in constant time.

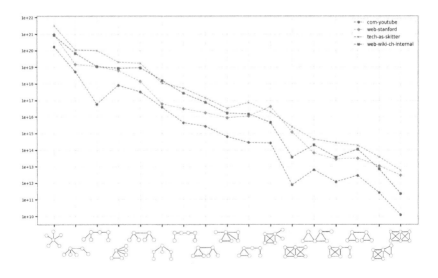

Fig. 4. The counts of 6-motifs in the given networks.

The motifs studied here are not induced, however it is still possible to observe the behavior of the relationships in the corresponding network by the motif analysis obtained in Fig. 4. As expected, the most common motif is the 5-star and the tree motifs occur more frequently. One exception to that is the 5-star with an edge added. This is not surprising, since this and the 5-star are two graphs, abundant at the hub vertices with very high degrees in most social networks. Also, Fig. 4 indicates that when the clique number of a motif is higher, the count of that motif is less.

In Table 1, the runtimes of the algorithm together with the size of each network are provided. The fourth column shows the runtimes obtained in [19] to evaluate the counts of motifs with 4 and 5 nodes. The runtime spent only for the counts of 6-motifs by our algorithm is listed in the last column. The runtimes to count smaller motifs were predicted for any network in [19]. Similarly, we obtain predictions using the runtimes in the last column of Table 1, as shown in Fig. 5. All counts in Theorems 1 and 2 can be computed in time $O(n+m)$, where $n = |V|$ and $m = |E|$. As social networks are sparse graphs and $|E| = O(n)$, our prediction is $-0.1476 + 1.6204|E|$ seconds for any network with $|E|$ edges. As observed in Table 1, our algorithm is able to execute the counts of all 6-motifs in Fig. 2 under 20 s, excluding the runtime spent to obtain the counts of smaller motifs.

Table 1. The runtimes in seconds for the motif counts of various networks

Network	$|V|$	$|E|$	4–5 motifs [19]	6-motif
com-youtube	1.1M	2.9M	168.880	4.896
web-wiki-ch-internal	1.9M	8.9M	2017.165	17.047
web-stanford	281.9K	1.9M	222.296	3.233
tech-as-skitter	1.7M	11.1M	1401.271	15.991
soc-brightkite	56.7K	212.9K	6.629	0.242
tech-RL-caida	190.9K	607.6K	4.719	0.729
flickr	757.2K	1.4M	13.008	1.886
com-amazon	334.8K	925.8K	2.908	1.272
web-google-dir	875.5K	4.3M	63.511	5.589
ia-email-EU-dir	265.0K	364.4K	6.537	0.479

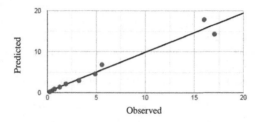

Fig. 5. The prediction of runtimes in seconds.

4 Conclusions

In this study, we presented a preliminary work that genaralizes the exact counting method for motifs of networks in [19] to a collection of 6-motifs with lower connectivity. We performed experiments to analyze the motif structure in real-world graphs and analyzed the runtime efficiency for the computations. The idea of counting 6-motifs by using algorithms based on the enumeration of smaller motifs results in much shorter runtime compared to other state-of-the-art algorithms. In a future study, we plan to extend this counting method to the remaining connected 6-motifs and use this data to obtain the counts of induced 6-motifs.

Acknowledgements. The research of the third author was supported by the BAGEP Award of the Science Academy of Turkey.

References

1. Ahmed, N.K., Neville, J., Rossi, R.A., Duffield, N.: Efficient graphlet counting for large networks. In: 2015 IEEE International Conference on Data Mining, pp. 1–10. IEEE (2015)
2. Benson, A.R., Gleich, D.F., Leskovec, J.: Higher-order organization of complex networks. Science **353**(6295), 163–166 (2016)
3. Betzler, N., Van Bevern, R., Fellows, M.R., Komusiewicz, C., Niedermeier, R.: Parameterized algorithmics for finding connected motifs in biological networks. IEEE/ACM Trans. Comput. Biol. Bioinform. (TCBB) **8**(5), 1296–1308 (2011)
4. Bhuiyan, M.A., Rahman, M., Rahman, M., Al Hasan, M.: GUISE: uniform sampling of graphlets for large graph analysis. In: 2012 IEEE 12th International Conference on Data Mining, pp. 91–100. IEEE (2012)
5. Bressan, M., Leucci, S., Panconesi, A.: Motivo: fast motif counting via succinct color coding and adaptive sampling (2019). https://arxiv.org/pdf/1906.01599.pdf
6. Elenberg, E.R., Shanmugam, K., Borokhovich, M., Dimakis, A.G.: Beyond triangles: a distributed framework for estimating 3-profiles of large graphs. In: Proceedings of the 21th ACM SIGKDD International Conference on Knowledge Discovery and Data Mining, pp. 229–238. ACM (2015)
7. Elenberg, E.R., Shanmugam, K., Borokhovich, M., Dimakis, A.G.: Distributed estimation of graph 4-profiles. In: Proceedings of the 25th International Conference on World Wide Web, pp. 483–493. International World Wide Web Conferences Steering Committee (2016)
8. Faust, K.: A puzzle concerning triads in social networks: graph constraints and the triad census. Soc. Netw. **32**(3), 221–233 (2010)
9. Frank, O.: Triad count statistics. Ann. Discrete Math. **38**, 141–149 (1988)
10. Gonen, M., Shavitt, Y.: Approximating the number of network motifs. Internet Math. **6**(3), 349–372 (2009)
11. Hočevar, T., Demšar, J.: A combinatorial approach to graphlet counting. Bioinformatics **30**(4), 559–565 (2014)
12. Holland, P.W., Leinhardt, S.: Local structure in social networks. Sociol. Methodol. **7**, 1–45 (1976)
13. Hormozdiari, F., Berenbrink, P., Pržulj, N., Sahinalp, S.C.: Not all scale-free networks are born equal: the role of the seed graph in PPI network evolution. PLoS Comput. Biol. **3**(7), e118 (2007)

14. Jha, M., Seshadhri, C., Pinar, A.: Path sampling: a fast and provable method for estimating 4-vertex subgraph counts. In: Proceedings of the 24th International Conference on World Wide Web, pp. 495–505. International World Wide Web Conferences Steering Committee (2015)

15. Kashani, Z.R.M., et al.: Kavosh: a new algorithm for finding network motifs. BMC Bioinform. **10**(1), 318 (2009)

16. Kashtan, N., Itzkovitz, S., Milo, R., Alon, U.: Efficient sampling algorithm for estimating subgraph concentrations and detecting network motifs. Bioinformatics **20**(11), 1746–1758 (2004)

17. Leskovec, J., Krevl, A.: Stanford large network dataset collection (2014). https://snap.stanford.edu/data

18. Milo, R., Shen-Orr, S., Itzkovitz, S., Kashtan, N., Chklovskii, D., Alon, U.: Network motifs: simple building blocks of complex networks. Science **298**(5594), 824–827 (2002)

19. Pinar, A., Seshadhri, C., Vishal, V.: ESCAPE: efficiently counting all 5-vertex subgraphs. In: Proceedings of the 26th International Conference on World Wide Web, pp. 1431–1440. International World Wide Web Conferences Steering Committee (2017)

20. Rahman, M., Bhuiyan, M.A., Al Hasan, M.: GRAFT: an efficient graphlet counting method for large graph analysis. IEEE Trans. Knowl. Data Eng. **26**(10), 2466–2478 (2014)

21. Rossi, R., Ahmed, N.: Network data repository (2012). https://networkrepository.com

22. Sariyuce, A.E., Seshadhri, C., Pinar, A., Catalyurek, U.V.: Finding the hierarchy of dense subgraphs using nucleus decompositions. In: Proceedings of the 24th International Conference on World Wide Web, pp. 927–937. International World Wide Web Conferences Steering Committee (2015)

23. Seshadhri, C., Pinar, A., Kolda, T.G.: Fast triangle counting through wedge sampling. In: Proceedings of the SIAM Conference on Data Mining, vol. 4, p. 5 (2013)

24. Shervashidze, N., Vishwanathan, S., Petri, T., Mehlhorn, K., Borgwardt, K.: Efficient graphlet kernels for large graph comparison. In: Artificial Intelligence and Statistics, pp. 488–495 (2009)

25. Tsourakakis, C.: The K-clique densest subgraph problem. In: Proceedings of the 24th International Conference on World Wide Web, pp. 1122–1132. International World Wide Web Conferences Steering Committee (2015)

26. Tsourakakis, C.E., Kolountzakis, M.N., Miller, G.L.: Triangle sparsifiers. J. Graph Algorithms Appl. **15**(6), 703–726 (2011)

27. Tsourakakis, C.E., Pachocki, J., Mitzenmacher, M.: Scalable motif-aware graph clustering. In: Proceedings of the 26th International Conference on World Wide Web, pp. 1451–1460. International World Wide Web Conferences Steering Committee (2017)

28. Ugander, J., Backstrom, L., Kleinberg, J.: Subgraph frequencies: mapping the empirical and extremal geography of large graph collections. In: Proceedings of the 22nd International Conference on World Wide Web, pp. 1307–1318. ACM (2013)

29. Watts, D.J., Strogatz, S.H.: Collective dynamics of 'small-world' networks. Nature **393**(6684), 440 (1998)

30. Wernicke, S.: Efficient detection of network motifs. IEEE/ACM Trans. Comput. Biol. Bioinform. (TCBB) **3**(4), 347–359 (2006)
31. Wernicke, S., Rasche, F.: FANMOD: a tool for fast network motif detection. Bioinformatics **22**(9), 1152–1153 (2006)
32. Zhao, Z., Wang, G., Butt, A.R., Khan, M., Kumar, V.A., Marathe, M.V.: SAHAD: subgraph analysis in massive networks using hadoop. In: 2012 IEEE 26th International Parallel and Distributed Processing Symposium, pp. 390–401. IEEE (2012)

LaserTank is NP-Complete

Per Alexandersson and Petter Restadh[✉]

KTH The Royal Institute of Technology, 100 44 Stockholm, Sweden
per.w.alexandersson@gmail.com, petterre@kth.se

Abstract. We show that the classical game *LaserTank* is NP-complete, even when the tank movement is restricted to a single column and the only blocks appearing on the board are mirrors and solid blocks. We show this by reducing 3-SAT instances to LaserTank puzzles.

Keywords: NP-completeness · LaserTank · 3-SAT

1 Introduction

From Wikipedia: "*LaserTank (also known as Laser Tank) is a computer puzzle game requiring logical thinking to solve a variety of levels*".[1] It was first released on the Windows platform in 1995, and a similar game was released in 1998 for the graphing calculator Texas Instruments Ti-83, under the name *Laser Mayhem*.[2] To our knowledge, the complexity of LaserTank has not been studied before, while several other classical games have been shown to be (co)NP-complete, NP-hard or PSPACE-complete. For example, *Sokoban* [3], *Tetris* [2], *Rush Hour* [4], and *Minesweeper* [5,6] to list a few.

In this short note, we prove the following.

Theorem 1. *LaserTank is* NP-*complete.*

It should be noted that one can perhaps apply more general meta-theoretical approaches for puzzle games and planning games in particular, to prove NP-completeness. It is likely that the framework by Viglietta [7]—which can be applied to games such as *Boulder Dash, Pipe Mania* and *Starcraft*—can successfully be applied to LaserTank as well. We opted for a self-contained hands-on approach where 3-SAT is reduced to LaserTank. Furthermore, we only use a small subset of the available pieces in the original game, as well as restrict the movement of the tank in two directions. These restrictions have the benefit that they imply that the *Laser Mayhem* variant is also NP-complete.

1.1 Short Background on 3-SAT

A 3-*SAT expression E* is a conjunction of clauses, where each clause involves exactly three distinct literals. A literal is either a boolean variable, or its

---

[1] https://en.wikipedia.org/wiki/LaserTank.

[2] https://www.ticalc.org/archives/files/fileinfo/95/9532.html.

© Springer Nature Switzerland AG 2020
D. Slamanig et al. (Eds.): MACIS 2019, LNCS 11989, pp. 333–338, 2020.
https://doi.org/10.1007/978-3-030-43120-4_26

negation. The 3-*SAT problem* states: Determine if E is *satisfiable*—that is, there is an assignment of truth values to the variables that makes E true. For example, $E = (x \vee y \vee \neg z) \wedge (\neg x \vee \neg y \vee \neg z) \wedge (x \vee \neg y \vee z)$ is such a conjunction, and the assignment $x, z = \texttt{true}$, $y = \texttt{false}$ shows that E is satisfiable. Determining satisfiability of a 3-SAT expression is an NP-complete problem [1].

2 LaserTank

LaserTank is a turn-based single-player puzzle game played on a 2-dimensional grid (the *board*), where in each turn, the player either moves the tank, or fires a laser from the tank. The laser interacts with different *pieces* on the board, and the goal is to hit a certain piece with the laser. The pieces[3] we use are *mirrors* $\{\triangle, \triangledown, \triangle, \triangledown\}$, *solid blocks* ■, *movable blocks* ▢, the *tank* ▶, and the *goal* ⊕. The tank is the only piece directly controlled by the player, and the laser exits the tank from the front, which is the pointy end of ▶. In our version, the tank is restricted to sideways movement only, see Example 1. In our final notes we explain how we take care of the case where the movement of the tank is not restricted. The tank can fire a laser from the front. If the laser hits a mirror on a slanted edge it is reflected. When a mirror is hit on one of the two (non-reflective) short edges by the laser, the mirror is pushed in the direction of the laser. A movable block is pushed one step if it is hit by the laser. A movable block or a mirror is only pushed if the tile directly behind it is empty. The aim of the puzzle is to hit the goal piece with the laser. The solid blocks do not allow lasers or the tank to pass through and they do not move when hit by the laser. The following example shows all game mechanics in action.

Example 1. Here is a small instance of the problem, with a step-by-step solution. The tank fires a laser which moves a mirror (1), then takes one step sideways, (2). It then shoots a laser at the movable block (3), and finally moves in position to have a clear shot of the goal (4).

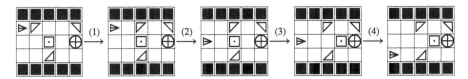

Our goal is now to construct puzzles which imitates an instance of 3-SAT. We employ so called *gadgets* that emulate boolean operations. Below, we let $\dashrightarrow$, $\downarrow$ indicate the inputs to the gadgets (considered as boolean variables), and $\{\overset{\dashrightarrow}{*}, \overset{\downarrow}{*}\}$ indicate inputs that are always available as clear shots from the tank. The latter are used for producing the output of the gadget.

---

[3] For a complete list of pieces available in the official game, see https://lasertan ksolutions.blogspot.com/p/in-my-opinionlaser-tank-is-best-logic.html.

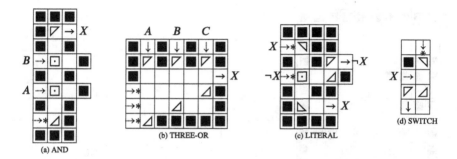

Fig. 1. The and-gadget, three-or-gadget, literal-gadget, and switch-gadget.

The and Gadget. The configuration in Fig. 1a serves as our **and**-gadget. We need to shoot through both A and B in order to allow for $A \wedge B = X$ as output. Notice that the two movable blocks can only be moved up, right and down. If we want the gadget to produce an output through X, all movable blocks must be moved out of the way. This can only be accomplished if the movable block must have been moved to the right via activation from both A and B, which shows that the gadget is indeed an **and**-gadget. The **and**-gadget can easily be generalized to more than two inputs.

The three-or gadget is depicted in Fig. 1b. If either of the inputs A, B or C are available, then X allows for output. The only way to produce output from X is to move a ▽ to the same row as X. The ▽ can only be moved into that row from above and thus we must have some input from A, B or C in order for a laser to pass out through X. Thus the **three-or** gadget works the way intended.

The literal gadget is depicted in Fig. 1c. This gadget emulates a literal, with two different mutually exclusive outputs depending on the choice of value of the literal. To unlock X as output, fire once through $\neg X$ first. This moves the movable block out of the way but prevents $\neg X$ from being available as output. Similarly for $\neg X$.

The switch gadget is depicted in Fig. 1d. The **switch**-gadget is our main building block for encoding an instance of a 3-SAT problem. It allows for the input X to be available first as output to the right, then redirected down. This allows X to be used in multiple **or**-clauses.

Example 2. In the puzzle in Fig. 2, only a single "input", X, is available. However, with the **switches** we can redirect input X to activate the **and** gadget. Notice the two ◺ pieces that are required to activate the **switches** and that the rightmost switch gadget must be used first in order to solve the puzzle. This is also true in the general setup, where switches should be used from right to left.

2.1   The Reduction

A 3-SAT expression may now be encoded as a LaserTank puzzle as follows. There is one **literal**-gadget for each variable appearing in the expression, a

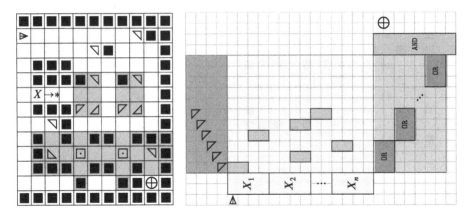

Fig. 2. Left: A small puzzle showing the use of two `switch`-gadgets and one `and`-gadget. Right: Layout of a general 3-SAT puzzle. Above each `or`-gadget are three switches, corresponding to the three literals involved on the or-clause.

`three-or-gadget` for each or-clause, and a single `and`-gadget with multiple inputs is used to bind all together. The puzzle is designed such that the output of the `and`-gadget is the only way to hit the goal. The general layout of such a puzzle is shown in Fig. 2. For each three-or-clause in the 3-SAT expression, three switches are placed on the board corresponding to the three literals involved. In other words, the clauses of the 3-SAT expression are encoded via `switch`-gadgets. The switches can always be activated via the ◸ pieces at the top of the board as in Fig. 2. As a concrete example, the expression $(A \lor B \lor \neg C) \land (A \lor \neg B \lor C)$ is encoded as the puzzle shown in Fig. 3.

The following lemma shows that solving LaserTank puzzles can be done in polynomial time with a non-deterministic Turing machine. Hence LaserTank is in NP.

Lemma 1. *A solution consisting of k steps to a LaserTank puzzle on a board of size n can be verified in time $O(kn)$.*

Proof. It is straightforward to show that the laser movement is time-reversible. This implies that it is impossible for a laser shot by the tank to end up in an "infinite loop" while being reflected by mirrors. Remember also that the laser stops as soon as it hits a solid block, a movable block, or moves a mirror. It follows that after firing the laser, it takes less than $4n$ steps before the laser finds its final destination, where n is the number of tiles on the board. Simulating a sequence of k moves thus requires $O(kn)$ time.

From our construction, it is a straightforward calculation to see that given a 3-SAT expression with V variables and C clauses gives a puzzle contained on a board with size $(7V + 9C + 4)(7C + 10)$. This is evidently polynomial in the size of the expression.

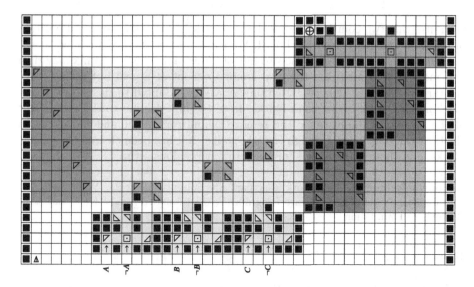

Fig. 3. The puzzle corresponding to the expression $(A \lor B \lor \neg C) \land (A \lor \neg B \lor C)$. Note that if $A = \text{true}$, the expression is satisfied. Thus this particular puzzle can be solved without deciding truth values for the variables B and C, and the movable blocks in the B and C variable gadgets do not need to be moved.

Proof (of Theorem 1). A 3-SAT problem can be converted to a LaserTank puzzle in polynomial time since the board size is a polynomial in the number of variables and clauses. Furthermore, a solution to such a LaserTank puzzle can easily be translated back to a solution of the original 3-SAT problem in polynomial time, by simply performing all the steps and reading how the `variable`-gadgets have been used. Note that a LaserTank puzzle solution might not decide the truth value of some variables (see caption of Fig. 3), in which case, one may simply let these values be `true`. Given a solution to the 3-SAT problem we can easily construct a solution to the corresponding LaserTank instance. Thus the 3-SAT problem is satisfiable if and only if the corresponding LaserTank puzzle has a solution. According to Lemma 1, the translation of a puzzle solution to a 3-SAT solution only requires a polynomial time in the input size (number of steps). This shows that LaserTank is at least as hard as 3-SAT. Finally, Lemma 1 shows that a solution can be verified in polynomial time and hence LaserTank is NP-complete.

Notice that in both the **and**– and `literal`-gadget, each movable block can be replaced with a $\triangledown$-mirror without changing the behavior of the gadget. Thus Theorem 1 is valid even in the case when restricting to puzzles without movable blocks. Furthermore, we can extend Theorem 1 to the case where the tank can turn and move in all four directions. To do so, we need to make sure the tank only has access to the same inputs as in the previous setup. This can be done by inserting additional rows in the puzzle such that every other row is empty, and then inserting two columns in with the pattern between the initial position of the tank and the rest of the board. We leave the details to the reader.

Finally, we note that in some variants of LaserTank where the tank may turn and move in all directions, the goal is not to shoot at a goal but rather move the tank to the goal. Our construction can also be adapted to this situation, by blocking off a path (using solid blocks) from the first column to the **and**-gadget at the bottom of the instance and remove the mirrors in the **and**-gadget. Then the tank can only move through the **and**-gadget if it has been cleared of movable blocks, and the setup is now equivalent to our previous setup.

Acknowledgements. This work was partially supported by the Wallenberg AI, Autonomous System and Software Program (WASP) founded by the Knut and Alice Wallenberg Foundation.

References

1. Cook, S.A.: The complexity of theorem-proving procedures. In: Proceedings of the Third Annual ACM Symposium on Theory of Computing, STOC 1971, pp. 151–158. ACM, New York (1971). https://doi.org/10.1145/800157.805047
2. Demaine, E.D., Hohenberger, S., Liben-Nowell, D.: Tetris is hard, even to approximate. Technical report, MIT-LCS-TR-865. MIT, Cambridge (2002). https://arxiv.org/abs/cs/0210020
3. Dor, D., Zwick, U.: SOKOBAN and other motion planning problems. Comput. Geom. **13**(4), 215–228 (1999). https://doi.org/10.1016/s0925-7721(99)00017-6
4. Flake, G.W., Baum, E.B.: Rush Hour is PSPACE-complete, or "Why you should generously tip parking lot attendants". Theor. Comput. Sci. **270**(1–2), 895–911 (2002). https://doi.org/10.1016/s0304-3975(01)00173-6
5. Kaye, R.: Minesweeper is NP-complete. Math. Intell. **22**(2), 9–15 (2000). https://doi.org/10.1007/bf03025367
6. Scott, A., Stege, U., van Rooij, I.: Minesweeper may not be NP-complete but is hard nonetheless. Math. Intell. **33**(4), 5–17 (2011). https://doi.org/10.1007/s00283-011-9256-x
7. Viglietta, G.: Gaming is a hard job, but someone has to do it!. Theory Comput. Syst. **54**(4), 595–621 (2013). https://doi.org/10.1007/s00224-013-9497-5

Data Modeling and Machine Learning

Improved Cross-Validation for Classifiers that Make Algorithmic Choices to Minimise Runtime Without Compromising Output Correctness

Dorian Florescu and Matthew England$^{(\boxtimes)}$

Faculty of Engineering, Environment and Computing, Coventry University, Coventry CV1 5FB, UK
{Dorian.Florescu,Matthew.England}@coventry.ac.uk

Abstract. Our topic is the use of machine learning to improve software by making choices which do not compromise the correctness of the output, but do affect the time taken to produce such output. We are particularly concerned with computer algebra systems (CASs), and in particular, our experiments are for selecting the variable ordering to use when performing a cylindrical algebraic decomposition of n-dimensional real space with respect to the signs of a set of polynomials.

In our prior work we explored the different ML models that could be used, and how to identify suitable features of the input polynomials. In the present paper we both repeat our prior experiments on problems which have more variables (and thus exponentially more possible orderings), and examine the metric which our ML classifiers targets. The natural metric is computational runtime, with classifiers trained to pick the ordering which minimises this. However, this leads to the situation where models do not distinguish between any of the non-optimal orderings, whose runtimes may still vary dramatically. In this paper we investigate a modification to the cross-validation algorithms of the classifiers so that they do distinguish these cases, leading to improved results.

Keywords: Machine Learning · Cross-validation · Computer algebra · Symbolic computation · Cylindrical algebraic decomposition

1 Introduction

1.1 Background and Main Thesis

Machine Learning (ML), that is statistical techniques to give computer systems the ability to *learn* rules from data, is a topic that has found great success in a diverse range of fields over recent years. ML is most attractive when the underlying functional relationship to be modelled is complex or not well understood. With regards to the creation of software itself, while ML has a history of use for testing and security analysis [26] it is less often used in the actual algorithms.

© Springer Nature Switzerland AG 2020
D. Slamanig et al. (Eds.): MACIS 2019, LNCS 11989, pp. 341–356, 2020.
https://doi.org/10.1007/978-3-030-43120-4_27

On the surface, this would be especially true for software that prizes mathematical correctness, such as computer algebra systems (CASs). Here, a thorough understanding of the underlying relationships would seem to be a pre-requisite.

However, CAS developers would acknowledge that their software actually comes with a range of options that, while having no effect on the correctness of the end result, can have a great effect on how long it takes to produce the result and exactly what form that result takes. These choices range from the low level (in what order to perform a search that may terminate early) to the high (which of a set of competing exact algorithms to use for this problem instance).

A well-known example is the choice of monomial ordering for a Gröbner Basis. This choice is actually quite abnormal in that there has been much study devoted to it and there are some clear pieces of advice to follow (e.g. that `degrevlex` ordering is the easiest to compute, and that if a `lex` ordering is needed it would be best to first compute a `degrevlex` basis and then convert). A better example of the choices we consider would be the underlying variable order that is required to define any monomial ordering, for which there exists no such clear advice.

In practice these less understood choices are usually either left entirely to the user, taken by human-made heuristics based on some experimentation (e.g. [19]), or made according to *magic constants* where crossing a single threshold changes system behaviour [11]. Our main thesis is that many of these decisions could be improved by allowing ML algorithms to analyse the data.

1.2   Outline of the Paper and Contribution

Our experiments concern variable orderings for another prominent symbolic computation algorithm: Cylindrical Algebraic Decomposition (CAD). CAD is an expensive procedure, with the choice of ordering affecting not only computation time but often the tractability of even considering a problem. We introduce the necessary background on CAD and its orderings in Sect. 2. We describe our prior work using ML to make this choice [23,25,28,29] in Sect. 3 which includes experimenting with a range of ML models, and developing techniques to generate suitable features from the input data. This prior work was all conducted on a large dataset of 3-variable problems (a choice from 6 orderings).

The new contributions of the present paper are two-fold. First, we have applied our prior methodology to a dataset of 4-variable problems (choice from 24 orderings) and we report on how it handled this increased complexity. Secondly, we examine and improve the training goal of our ML classifiers. The natural metric for this problem is runtime, and our old classifiers are trained to pick the ordering which minimises this for a given CAD input. However, this meant our training did not distinguish between any of the non-optimal orderings even though the difference between these could be huge. In Sect. 4 we report on a new cross-validation approach for our classifiers which aims to make them aware of these *different shades of wrong* and thus make choices which reduce the overall runtime even if the number of problems where the classifiers pick the absolute best runtime is unchanged.

In Sects. 5 and 6 we describe the methodology and results respectively for our new experiments on choosing the variable ordering for 4-variable CAD problems, both with and without the new cross-validation approach. We also compare against the best known human-made heuristics.

2   Background on Variable Ordering for CAD

2.1   Cylindrical Algebraic Decomposition

A *Cylindrical Algebraic Decomposition* (CAD) is a *decomposition* of ordered $\mathbb{R}^n$ space into cells arranged *cylindrically*: the projections of any pair of cells with respect to the variable ordering are either equal or disjoint. I.e. the projections all lie within cylinders over the cells of an induced CAD of the lower dimensional space. All these cells are (semi)-algebraic meaning each can be described with a finite sequence of polynomial constraints.

A CAD is usually produced to be *truth-invariant* for a logical formula, meaning the formula is either true or false on each cell. Such a decomposition can then be used to analyse the formula, and for example, perform Quantifier Elimination (QE) over the reals. I.e. given a quantified Tarski formula in prenex normal form we can find an equivalent quantifier free formula over the reals by building a CAD for the quantifier-free part of the formula, querying a finite number of sample points (one from each cell), and then using the corresponding cell descriptions. For example, QE could transform $\exists x, ax^2 + bx + c = 0 \land a \neq 0$ to the equivalent unquantified statement $b^2 - 4ac \geq 0$ by building a CAD of (x, a, b, c). In practice, the quantifier free equivalent would come as the conjunction of several parts (one from each cell) which logically simplify to the stated result.

CAD was introduced by Collins in 1975 [16] and works relative to a set of polynomials. Collins' CAD produces a decomposition so that each polynomial has constant sign on each cell (thus truth-invariant for any formula built with those polynomials). The algorithm first projects the polynomials into smaller and smaller dimensions; and then uses these to lift − to incrementally build decompositions of larger and larger spaces according to the polynomials at that level. There have been a great many developments in the theory and implementation of CAD since Collins' original work which we do not describe here. The collection [12] summarises the work up to the mid-90s while the second author's journal articles [5,21] attempt summaries of CAD progress since in their introduction and background sections. CAD is the backbone of all QE implementations as it is the only implemented complete procedure for the problem. QE has numerous applications throughout science and engineering[1] [38] which would in turn benefit from faster CAD. Our work also speeds up independent applications of CAD, such as reasoning with multi-valued functions [18], motion planning [40], and identifying multistationarity in biological networks [3,4].

---

[1] Recently even economics too [35,36].

2.2 Variable Ordering

The definition of cylindricity and both stages of the CAD algorithm are relative to an ordering of the variables. For example, given polynomials in variables ordered as $x_n \succ x_{n-1} \succ \dots, \succ x_2 \succ x_1$ we first project away x_n and so on until we are left with polynomials univariate in x_1. We then start lifting by decomposing the x_1-axis, and then the (x_1, x_2)-plane and so on. The cylindricity condition refers to projections of cells in $\mathbb{R}^n$ onto a space $(x_1, \dots, x_m)$ where $m < n$. As noted above there have been numerous advances to CAD since its inception but the need for a fixed variable ordering remains.

Depending on the application, the variable ordering may be determined, constrained, or free. QE requires that quantified variables are eliminated first and that variables are eliminated in the order in which they are quantified. However, variables in blocks of the same quantifier (and the free variables) can be swapped, so there is partial freedom. In the example discussed in Sect. 2.1 we may use any variable ordering that projects the quantified variable x first to perform the QE and discover the discriminant. A CAD for the quadratic polynomial under ordering $a \prec b \prec c$ has only 27 cells, but we need 115 for the reverse ordering.

This choice of variable ordering can have a great effect on the time and memory use of CAD, and the number of cells in the output (how course or fine the decomposition is). In fact, Brown and Davenport presented a class of problems in which one variable ordering gave output of double exponential complexity in the number of variables and another output of a constant size [10].

Heuristics have been developed to choose a variable ordering, with Dolzmann et al. [19] giving the best known study. After analysing a variety of metrics they proposed a heuristic, sotd, which constructs the full set of projection polynomials for each permitted ordering and selects the ordering whose corresponding set has the lowest sum of total degrees for each of the monomials in each of the polynomials. The second author demonstrated examples for which that heuristic could be misled in [6]; and then later showed that tailoring to an implementation could improve performance [22]. These heuristics all involved potentially costly projection operations on the input polynomials.

Another human-made heuristic was proposed by Brown in his ISSAC 2004 tutorial notes [9]. This chooses a variable ordering according to the following criteria, starting with the first and breaking ties with successive ones.

(1) Eliminate a variable first if it appears with the lowest overall individual degree in the input.
(2) For each variable calculate the maximum total degree (i.e. sum of the individual degrees) for the set of terms in the input in which it occurs. Eliminate first the variable for which this is lowest.
(3) Eliminate a variable first if there is a smaller number of terms in the input which contain the variable.

The Brown heuristic is far cheaper than the sotd heuristic (because the latter performs projections before measuring degrees). Surprisingly, our experiments on CAD problems in 3-variables all suggest that the Brown heuristic makes better choices than `sotd` (even before one considers the time taken to run the heuristic itself). This counter-intuitive finding does not generalise into our 4-variable problem set, as discussed later.

3 Prior ML Work on This Problem

3.1 Results from CICM 2014

The first application of ML for choosing a CAD variable ordering was [29] which used a support vector machine to select which of three human-made heuristics to follow. The SVM considered 11 simple algebraic features of the input polynomials (mostly different measures of degree and variable occurrence). The experiments were on 3-variable CAD problems and although the Brown heuristic was found to make the best choices on average, the experiments identified substantial subsets of examples for which each of the three heuristics outperformed the others. The key conclusion was that the machine learned choice did significantly better than any one heuristic overall.

3.2 Results from CICM 2019

The present authors revisited these experiments earlier this year in [23]. We used the same dataset but this time ML was used to predict directly the variable ordering for CAD, rather than choosing a heuristic. The motivation for picking a heuristic in [29] was that if the methodology were applied to problems with more variables it would still mean making a choice from 3 possibilities rather than an exponentially growing number. However, upon investigation there were many problems where none of the human-made heuristics made good choices and so savings could be made by considering all possible orderings[2].

In [23] we also considered a more diverse selection of ML methods than [29]. We experimented with four common ML classifiers: K-Nearest Neighbours (KNN); Multi-Layer Perceptron (MLP); Decision Tree (DT); and Support Vector Machine (SVM) with RBF kernel, all using the same set of 11 features from [29].

The results showed that all three of the new models performed substantially better than the SVM (the only classifier to be tried before); and that all four classifiers outperformed the human-made heuristics.

---

[2] Of course, this methodology will have to be changed to deal with higher numbers of variables but since CAD is rarely tractable with more than 5 variables this is not a particularly pressing concern. We note that there are several meta-algorithms that may be applicable to sample the possible ordering without evaluating them all. For example, a Monte Carlo tree search was used in [33] to sample the possible multivariate Horner schemes and pick an optimal one in the CAS FORM.

3.3   Results from SC-Square 2019

We next considered how to extract further information from the input data. The 11 features used in [23, 29] were inspired by Brown's heuristic [9] (e.g. measures of variable degree and frequency of occurrence). In particular, they can all be cheaply extracted from polynomials.

In [25] a new feature generation procedure was presented, based on the observation that the original features can be formalised mathematically using a small number of basic functions (average, sign, maximum) evaluated on the degrees of the variables in either one polynomial or the whole system. Considering all possible combinations of these functions led to 78 useful and independent features for our 3-variable dataset. The experiments were repeated with these, with the results showing that all four ML classifiers improved their predictions.

Using these new features the choices of the best performing classifier allowed CAD to solve all problems in the testing set with a runtime only 6% more than the best possible (i.e. the time taken if the optimal ordering were used for every problem). Using only the original features, the choices of the best ML classifier led to 14% more than the minimum runtime. Following the choices of Brown's heuristic led to runtimes 27% more than the minimum.

3.4   Related Work on ML for Mathematical Software

The work described above is the only published work on ML for choosing a CAD variable ordering. There are only a handful of other examples of ML within CASs: [27, 28] on the question of whether to precondition CAD with Gröbner Bases; [31] on deciding the order of sub-formulae solving for a QE procedure; and [33] on choosing a multivariate Horner scheme. Other areas of mathematical software have made more use of ML. For example, in the mathematical logic community the ML-selected portfolio SAT solver SATZILLA [41] is well-known, while more recently MAPLESAT views solver branching as an optimisation problem to be tackled with ML [34]. There are also several examples of ML within the automated reasoning community (see e.g. [8, 32, 39]). A survey on ML for mathematical software was presented at ICMS 2018 [20].

4   New Cross-Validation Based on Computing Times

4.1   Motivation

In all of the author's previous ML experiments for CAD [23, 25, 29], the models were optimised simply to predict which of the possible variable orderings leads to the smallest computing time for CAD. This is not an ideal approach:

– First, runtimes for CAD, like all software, will contain a degree of noise from various hardware and software factors. While it is common for a given CAD problem to have a wide range of possible runtimes depending on the ordering, that does not mean that all orderings give runtimes distinct from the others.

The runtimes commonly appear in clusters. Thus it is often the case that the smallest runtime be only slightly lower than the second smallest, and that difference could well be down to noise. Thus when training to target only the very quickest runtime we risk exaggerating the effects of such noise.

- Second, during training, when a model makes an incorrect prediction this could mean selecting an ordering that produces a runtime very close to the optimal or another that is significantly larger. The training would not distinguish between these cases – there is no distinction between picking an "almost good" ordering and a "very bad" ordering. However, from the point of view of a user judging these selections there is a big difference!

One of the traditional metrics used to evaluate an ML classifier is *accuracy*, defined as the number of test examples for which the classifier makes the correct choice. In our context, correct means picking the optimal variable ordering from the $n!$ possibilities. We recognised that for our application this definition of accuracy is not sufficient to judge the classifiers and so in our prior work we also presented the total CAD runtime for the testing set when using the variable orderings of a classifier (which we referenced in the summary above).

The anonymous referees of our earlier papers commented that perhaps accuracy could be redefined into something more appropriate for our application. For example, judge a classifier as being correct for a problem instance if it picks an ordering which produces a runtime within $x\%$ of the minimum runtime that can be achieved for that instance[3]. This led us to consider whether the training algorithms could be adapted to take account of this more nuanced definition of accuracy. We decided to introduce this in the stage of the methodology where cross-validation is used for hyperparameter selection: a single technique that is used for all of the different ML classifiers we work with.

4.2 Traditional ML Cross-Validation

We describe first the typical procedure of cross-validation used when preparing a ML classifier which sets the parameters and hyperparameters of a model.

The parameters are variables that can be fine-tuned so that the prediction error reaches a local or global minimum in the parameter space. For example, the weights in an artificial neural network or the support vectors in an SVM. The hyperparameters are model configurations selected before training. They are often specified by the practitioner based on experience or a heuristic, e.g. the number of layers in a neural network or the value of k in a k-nearest neighbour model. The connection between the hyperparameters and the model prediction is more complex, and thus, typically, these are tuned using grid search in the hyperparameter space to minimise the prediction error.

To prevent the situation where the model returns poor results on new datasets not used in training, also known as overfitting, the hyperparameters and parameters are tuned on different datasets. The typical approach is cross-validation.

---

[3] In Sect. 5 we use $x = 20$ but we are still debating the most appropriate value.

In G-fold cross-validation (see for example the introduction of [1]), the data is split into G groups of equal size M:

$$\mathcal{D}_1 = \left\{ \boldsymbol{f}^{(k_1^1)}, \ldots, \boldsymbol{f}^{(k_1^M)} \right\}$$

$$\vdots \qquad\qquad (1)$$

$$\mathcal{D}_G = \left\{ \boldsymbol{f}^{(k_G^1)}, \ldots, \boldsymbol{f}^{(k_G^M)} \right\},$$

where each group entry is a vector of features for a problem instance:

$$\boldsymbol{f}^{(k_g^m)} = \left[f_1^{(k_g^m)}, \ldots, f_{n_f}^{(k_g^m)} \right], \qquad g = 1, \ldots, G, \quad m = 1, \ldots, M.$$

Each entry in such a vector is a scalar number and n_f denotes the number of features we derive for each instance. See [25] for details of the features we use and how they are generated from the polynomials.

Let $c^{(k_g^m)}$ denote the target class corresponding to data point $\boldsymbol{f}^{(k_g^m)}$. An ML classifier with parameters $\boldsymbol{\theta}$ is modelled as a function $\mathcal{M}^{\boldsymbol{\theta}} : \mathbb{R}^{n_f} \to \{1, 2, \ldots, n_c\}$, where n_c denotes the number of classes. In our context the number of classes is the number of CAD variable orderings acceptable for the underlying application.

Typically, the classifier also depends on a number of hyperparameters that can each take a finite number of values. Here, we will denote by H the number of all possible hyperparameter combinations, such that $\mathcal{M}_h^{\boldsymbol{\theta}}$, $h = 1, \ldots, H$, denotes the classifier with parameters $\boldsymbol{\theta}$ and hyperparameters defined by index h. The typical cross-validation procedure trains the parameters $\boldsymbol{\theta}$ of classifiers $\{\mathcal{M}_h^{\boldsymbol{\theta}}\}_{h=1}^{H}$ on each combination of $G - 1$ data groups in (1), adding up to $G \cdot H$ models.

Let $\hat{c}_h^{(k_g^m)}$ denote $\mathcal{M}_h^{\boldsymbol{\theta}_g} \left(\boldsymbol{f}^{(k_g^m)} \right)$, the class prediction of a classifier whose parameters were trained on the dataset $\mathcal{D}_1 \cup \cdots \cup \mathcal{D}_{g-1} \cup \mathcal{D}_{g+1} \cup \cdots \cup \mathcal{D}_G$. Then the optimal h_{opt} is computed by maximising the following quantity:

$$h_{opt} = \underset{h}{\mathrm{argmax}} \left(\frac{1}{G} \sum_{g=1}^{G} \mathrm{score}_h^g \right), \qquad\qquad (2)$$

where $\mathrm{score}_h^g = \mathrm{score}\left(\hat{c}_h^{(k_g^m)}, c^{(k_g^m)} \right)$, and $\mathrm{score}(\cdot, \cdot)$ denotes the F1-score of group G for the model prediction [15]. In other words, the typical cross-validation procedure identifies the hyperparameters that maximise the performance of the model at predicting the very best ordering. I.e. it does not take into account the actual computing time of the prediction — just whether it was the quickest.

4.3   Adapted ML Cross-Validation

Our change to the cross-validation procedure is to instead calculate h_{opt} as

$$h_{opt} = \underset{h}{\mathrm{argmax}} \left(\frac{1}{G} \sum_{g=1}^{G} -\mathrm{ctime}_h^g \right), \qquad\qquad (3)$$

where $\text{ctime}_h^g = \frac{1}{M} \sum_m \text{ctime}\left(k_g^m, \hat{c}^{(k_g^m)}\right)$, and $\text{ctime}\left(k_g^m, \hat{c}^{(k_g^m)}\right)$ denotes the recorded time for computing CAD on data point $f^{(k_g^m)}$ using the variable ordering given by class prediction $\hat{c}^{(k_g^m)}$. By evaluating the computing time for all data points, this cross-validation method penalises the variable orderings leading to very large computing times, but does not penalise the ones close to the optimum. Thus we do not expect the change to affect how often a classifier chooses the optimal ordering, but it should improve the choices made in cases where the optimum is missed.

5 ML Experiment Methodology

We describe a ML experiment to choose the variable ordering for CAD. The methodology used is similar to that of our recent paper [25] except that (a) we use a dataset of 4-variable problems instead of 3-variable ones; and (b) we ran the classifiers with both the original and the adapted cross-validation procedure.

5.1 Problem Set

We are working with the `nlsat` dataset[4] produced to evaluate the work in [30], thus the problems are all fully existentially quantified. Although there are CAD algorithms that reduce what is being computed based on the quantification structure (e.g. Partial CAD [17]), the conclusions we draw are likely to generalise.

We selected the 2080 problems with 4 variables, meaning each has a choice of 24 different variable orderings. We extracted only the polynomials involved, and randomly divided into two datasets for training (1546) and testing (534). Only the former is used to tune the ML model parameters and hyperparameters.

5.2 Software

We work with the CAD routine `CylindricalAlgebraicDecompose`: part of the `RegularChains` Library for MAPLE. It builds decompositions first of $\mathbb{C}^n$ before refining to a CAD of $\mathbb{R}^n$ [2,13,14]. We ran the code in Maple 2018 but used an updated version of the `RegularChains` Library (http://www.regularchains. org). Brown's heuristic and the features for ML were coded in the `sympy` package v1.3 for Python 2.7. The `sotd` heuristic was implemented in MAPLE as part of the `ProjectionCAD` package [24]. Training and evaluation of the ML models was done using the `scikit-learn` package [37] v0.20.2 for Python 2.7. In order to implement our adapted cross-validation procedure we had to rewrite a number of the standard commands within the package to both use the redefined h_{opt} in (3), and to access the data it requires during the cross-validation.

---

[4] Freely available from http://cs.nyu.edu/~dejan/nonlinear/.

5.3 Timings

Each individual CAD was constructed by a Maple script called separately from Python (to avoid any Maple caching of results). The target variable ordering for ML was defined as the one that minimises the computing time for a given problem. All CAD function calls included a time limit. For the training dataset an initial time limit of 16 s was used, which was doubled if all orderings timed out (a target variable ordering could be assigned for all problems using time limits no bigger than 32 s). The problems in the testing dataset were all processed with a single larger time limit of 64 s for all orderings, with any problems that timed out having their runtime recorded as 64 s.

5.4 Computing the Features

We computed algorithmically the set of features for 4 variables $\{f^{(i)}\}_{i=1}^{n_f}$ where $n_f = 1440$, using the procedure introduced in [25].

Given the set of problems $\{Pr_1, \ldots, Pr_N\}$, $N = 1546$, some of the features $f^{(i)}$ turn out to be constant, i.e. $f^{(i)}(Pr_1) = f^{(i)}(Pr_2) = \cdots = f^{(i)}(Pr_N)$. Such features will have no benefit for ML and are removed. Further, other features may be repetitive, i.e. $f^{(i)}(Pr_n) = f^{(j)}(Pr_n), \forall n = 1, \ldots, N$, and are merged into one single feature. After this step, we are left with 105 features.

5.5 ML Models

We choose commonly used deterministic ML models for this experiment (for details on the methods see e.g. the textbook [1]).

- The K-Nearest Neighbours (KNN) classifier [1, §2.5].
- The Decision Tree (DT) classifier [1, §14.4].
- The Multi-Layer Perceptron (MLP) classifier [1, §2.5].
- The Support Vector Machine (SVM) classifier with Radial Basis Function (RBF) kernel [1, §6.3].

We fixed the RBF kernel for SVM as it was found to produce better results than other basis functions for a similar problem of learning from algebraic features in [7], and including basis choice in cross-validation creates a much larger search space.

Each model was trained using grid search 3-fold cross-validation, i.e. the set was randomly divided into 3 and each possible combination of 2 parts was used to tune the model parameters, leaving the last part for fitting the hyperparameters with cross-validation, by optimising the average F-score. Grid searches were performed for an initially large range for each hyperparameter; then gradually decreased to home into optimal values. The optimal hyperparameters selected during cross-validation are in Table 1.

Table 1. The ML hyperparameters optimised on the training dataset using the standard cross-validation (CV) routine and the new CV routine.

Model	Hyperparameter	Value (standard CV)	Value (new CV)
Decision Tree	Criterion Maximum tree depth	Entropy 6	Gini impurity 14
K-Nearest Neighbours	Train instances weighting Algorithm Number of neighbours	Inversely proportional to distance Ball Tree 13	Inversely proportional to distance Ball Tree 14
SVM	Regularization para. C Basis para. γ	2.41 0.0097	1.66 0.0097
Multi-Layer Perceptron	Hidden layer size Activation function Regularization para. α	18 Hyperbolic Tangent $1 \cdot 10^{-4}$	17 Identity $1 \cdot 10^{-4}$

5.6 Evaluating the ML Models and Human-Made Heuristics

The ML models will be compared on two metrics: *Accuracy*, defined as the percentage of problems where a model's predicted variable ordering led to a computing time closer than 20% of the time it took the optimal ordering; and *Time* defined as the total time taken to evaluate all problems in the test set using that model's predictions for variable ordering. We note that Accuracy is defined differently in our prior work [23,25] where we measured only how often a heuristic picked the very best ordering.

We will also test the two best-known human constructed heuristics [9,19] described in Sect. 2.2. Unlike the ML models, these can end up predicting several variable orderings (when they cannot discriminate). In practice if this were to happen the heuristic would select one randomly (or perhaps lexicographically), however that final pick is not meaningful. To accommodate this we evaluate these heuristics as follows:

– For each problem, the prediction accuracy of such a heuristic is judged to be the percentage of its predicted variable orderings that are also target orderings (i.e. within 20% of the minimum). The average of this percentage over all problems in the testing dataset represents the prediction accuracy.
– Similarly, the computing time for such methods is assessed as the average computing time over all predicted orderings, and it is this that is summed up for all problems in the testing dataset.

6 Experimental Results

The results are presented in Table 2. Each ML model appears twice in the top table via its acronym with each of the following appended:

–O: for one trained with the original (and typical) ML cross-validation method based on (2) as was used in our prior work [23,25].

−N: for one trained by the new cross-validation approach described in Sect. 4.3 which is based on computing time as in (3).

The bottom table details the two human-constructed heuristics along with the outcome of a random choice between the 24 orderings. We might expect a random choice to be correct once in 24 times of the time but it is higher as for some problems there were multiple variable orderings with equally fast timings.

The minimum total computing time, achieved if we select an optimal ordering for every problem, is 2, 177 s. This is what would be achieved by the *Virtual Best Heuristic*. Choosing at random would take 8, 291 s, almost 4 times as much. The maximum time, if we selected the worst ordering for every problem (the *Virtual Worst Heuristic*), is 22, 735 s. The Decision Tree model trained with the new cross validation achieved the shortest time of all with 3, 627 s, 67% more than the minimal possible.

The recorded time taken by each model to make a prediction, which is included in the timings reported in Table 2, varied greatly between ML and the heuristics. The prediction time for the heuristics was 286 s for sotd and 23 s for Brown. In contrast, the total time taken by the ML to make predictions was less than one second for all models.

6.1 Results of New Cross-Validation Method

For each ML model the performance when trained with the new cross-validation was better (measured using either of our metrics) than when trained with the original procedure. The scale of the improvement varied: the timings of the decision tree reduced by 9.8% but those of the KNN classifier only by 1.6%.

Thus we can conclude the new methodology to be beneficial. However, we note that it is still the case that our two metrics do not agree on the best model: DT-N achieved the lowest times but KNN-N the highest accuracy. The latter is better at picking a good (within 20% of the minimal) ordering but when it fails to do so it makes mistakes of greater magnitude. So there is scope for further work to make our ML models take into account the full range of possibilities. It may be that this requires a tailored approach to the training of parameters in each different classifier.

Table 2. Performance on the testing dataset of the ML classifiers (using both the standard and new cross-validation routines), the human-made heuristics, and a random choice. The virtual best and worst solvers show the range of possibilities.

	DT-O	DT-N	KNN-O	KNN-N	MLP-O	MLP-N	SVM-O	SVM-N
Accuracy	51.7%	54.3%	53.9%	54.5%	53.6%	56.9%	53.9%	54.9%
Time (s)	4, 022	3, 627	3, 808	3, 748	3, 972	3, 784	3, 795	3, 672

	Virtual Best	Virtual Worst	random	Brown	sotd
Accuracy	100%	0%	17.0%	20.1%	47.8%
Time (s)	2, 177	22, 735	8, 291	8, 292	4, 348

6.2   Comparison of Brown and Sotd on the 4-Variable Dataset

Of the two human-made heuristics, `Brown` performed far worse than `sotd`. This is the opposite of the findings in [23,25,29] for 3-variable problems. This is not necessarily in conflict: the added information taken by `sotd` will grow in size exponentially with the variables, and thus we would expect the predictive information it carries to be more valuable. However, the cost of `sotd` will also be increasing rapidly denting this value. The time taken by `sotd` to make all the predictions is 286 s, while the time for `Brown` is less than 10% of that at 23 s. For this dataset at least, it is well worth paying the price of `sotd` as the savings over Brown's heuristic are far more substantial.

6.3   Value of ML on the 4-Variable Dataset

All heuristics (ML and human-made) are further away from the optimum on this 4-variable dataset than they were on the three variable one, to be expected given we are choosing from 24 rather than 6 orderings. Our best performing model achieves timings 67% greater than the minimum (it was 6% for 3-variable problems). However, the best human-made heuristic had timings 98% greater.

In fact, every ML model outperformed both the human constructed heuristics in regards to both metrics, and when using either the original or the new cross-validation approach. So we can easily conclude that our ML methodology generalises to 4-variable problems. However, it is also clear that there is much scope for future improvement.

7   Summary

We have demonstrated that our methodology of ML for choosing a CAD variable ordering may be applied to 4-variable problems where it continues its dominance over human-made heuristics. We have also presented an addition to the ML training methodology to better reflect our application domain and demonstrated the benefit of this experimentally. This new methodology could be applied to any ML application which seeks to make a choice to minimise computational runtime.

Acknowledgements. This work is funded by EPSRC Project EP/R019622/1: *Embedding Machine Learning within Quantifier Elimination Procedures.*

References

1. Bishop, C.: Pattern Recognition and Machine Learning. Springer, New York (2006)
2. Bradford, R., Chen, C., Davenport, J.H., England, M., Moreno Maza, M., Wilson, D.: Truth table invariant cylindrical algebraic decomposition by regular chains. In: Gerdt, V.P., Koepf, W., Seiler, W.M., Vorozhtsov, E.V. (eds.) CASC 2014. LNCS, vol. 8660, pp. 44–58. Springer, Cham (2014). https://doi.org/10.1007/978-3-319-10515-4_4

3. Bradford, R., et al.: A case study on the parametric occurrence of multiple steady states. In: Proceedings of the 2017 ACM International Symposium on Symbolic and Algebraic Computation, ISSAC 2017, pp. 45–52. ACM (2017). https://doi.org/10.1145/3087604.3087622

4. Bradford, R., et al.: Identifying the parametric occurrence of multiple steady states for some biological networks. J. Symb. Comput. **98**, 84–119 (2020). https://doi.org/10.1016/j.jsc.2019.07.008

5. Bradford, R., Davenport, J., England, M., McCallum, S., Wilson, D.: Truth table invariant cylindrical algebraic decomposition. J. Symb. Comput. **76**, 1–35 (2016). https://doi.org/10.1016/j.jsc.2015.11.002

6. Bradford, R., Davenport, J.H., England, M., Wilson, D.: Optimising problem formulation for cylindrical algebraic decomposition. In: Carette, J., Aspinall, D., Lange, C., Sojka, P., Windsteiger, W. (eds.) CICM 2013. LNCS (LNAI), vol. 7961, pp. 19–34. Springer, Heidelberg (2013). https://doi.org/10.1007/978-3-642-39320-4_2

7. Bridge, J.: Machine learning and automated theorem proving. Technical report. UCAM-CL-TR-792, University of Cambridge, Computer Laboratory (2010)

8. Bridge, J., Holden, S., Paulson, L.: Machine learning for first-order theorem proving. J. Autom. Reason. **53**, 141–172 (2014). https://doi.org/10.1007/s10817-014-9301-5

9. Brown, C.: Companion to the tutorial: cylindrical algebraic decomposition. Presented at ISSAC 2004 (2004). http://www.usna.edu/Users/cs/wcbrown/research/ISSAC04/handout.pdf

10. Brown, C., Davenport, J.: The complexity of quantifier elimination and cylindrical algebraic decomposition. In: Proceedings of the 2007 International Symposium on Symbolic and Algebraic Computation, ISSAC 2007, pp. 54–60. ACM (2007). https://doi.org/10.1145/1277548.1277557

11. Carette, J.: Understanding expression simplification. In: Proceedings of the 2004 International Symposium on Symbolic and Algebraic Computation, ISSAC 2004, pp. 72–79. ACM (2004). https://doi.org/10.1145/1005285.1005298

12. Caviness, B., Johnson, J.: Quantifier Elimination and Cylindrical Algebraic Decomposition. Texts & Monographs in Symbolic Computation. Springer, New York (1998). https://doi.org/10.1007/978-3-7091-9459-1

13. Chen, C., Moreno Maza, M.: An incremental algorithm for computing cylindrical algebraic decompositions. In: Feng, R., Lee, W., Sato, Y. (eds.) Computer Mathematics, pp. 199–221. Springer, Heidelberg (2014). https://doi.org/10.1007/978-3-662-43799-5_17

14. Chen, C., Moreno Maza, M., Xia, B., Yang, L.: Computing cylindrical algebraic decomposition via triangular decomposition. In: Proceedings of the 2009 International Symposium on Symbolic and Algebraic Computation, ISSAC 2009, pp. 95–102. ACM (2009). https://doi.org/10.1145/1576702.1576718

15. Chinchor, N.: MUC-4 evaluation metrics. In: Proceedings of the 4th Conference on Message Understanding (MUC4 1992), pp. 22–29. Association for Computational Linguistics (1992). https://doi.org/10.3115/1072064.1072067

16. Collins, G.E.: Quantifier elimination for real closed fields by cylindrical algebraic decompostion. In: Brakhage, H. (ed.) GI-Fachtagung 1975. LNCS, vol. 33, pp. 134–183. Springer, Heidelberg (1975). https://doi.org/10.1007/3-540-07407-4_17. Reprinted in the collection [12]

17. Collins, G., Hong, H.: Partial cylindrical algebraic decomposition for quantifier elimination. J. Symb. Comput. **12**, 299–328 (1991). https://doi.org/10.1016/S0747-7171(08)80152-6

18. Davenport, J., Bradford, R., England, M., Wilson, D.: Program verification in the presence of complex numbers, functions with branch cuts etc. In: 14th International Symposium on Symbolic and Numeric Algorithms for Scientific Computing, SYNASC 2012, pp. 83–88. IEEE (2012). http://dx.doi.org/10.1109/SYNASC.2012.68
19. Dolzmann, A., Seidl, A., Sturm, T.: Efficient projection orders for CAD. In: Proceedings of the 2004 International Symposium on Symbolic and Algebraic Computation, ISSAC 2004, pp. 111–118. ACM (2004). https://doi.org/10.1145/1005285.1005303
20. England, M.: Machine learning for mathematical software. In: Davenport, J.H., Kauers, M., Labahn, G., Urban, J. (eds.) ICMS 2018. LNCS, vol. 10931, pp. 165–174. Springer, Cham (2018). https://doi.org/10.1007/978-3-319-96418-8_20
21. England, M., Bradford, R., Davenport, J.: Cylindrical algebraic decomposition with equational constraints. J. Symb. Comput. (2019). https://doi.org/10.1016/j.jsc.2019.07.019
22. England, M., Bradford, R., Davenport, J.H., Wilson, D.: Choosing a variable ordering for truth-table invariant cylindrical algebraic decomposition by incremental triangular decomposition. In: Hong, H., Yap, C. (eds.) ICMS 2014. LNCS, vol. 8592, pp. 450–457. Springer, Heidelberg (2014). https://doi.org/10.1007/978-3-662-44199-2_68
23. England, M., Florescu, D.: Comparing machine learning models to choose the variable ordering for cylindrical algebraic decomposition. In: Kaliszyk, C., Brady, E., Kohlhase, A., Sacerdoti Coen, C. (eds.) CICM 2019. LNCS (LNAI), vol. 11617, pp. 93–108. Springer, Cham (2019). https://doi.org/10.1007/978-3-030-23250-4_7
24. England, M., Wilson, D., Bradford, R., Davenport, J.H.: Using the regular chains library to build cylindrical algebraic decompositions by projecting and lifting. In: Hong, H., Yap, C. (eds.) ICMS 2014. LNCS, vol. 8592, pp. 458–465. Springer, Heidelberg (2014). https://doi.org/10.1007/978-3-662-44199-2_69
25. Florescu, D., England, M.: Algorithmically generating new algebraic features of polynomial systems for machine learning. In: Abbott, J., Griggio, A. (eds.) Proceedings of the 4th Workshop on Satisfiability Checking and Symbolic Computation (SC² 2019). No. 2460 in CEUR Workshop Proceedings (2019). http://ceur-ws.org/Vol-2460/
26. Ghaffarian, S., Shahriari, H.: Software vulnerability analysis and discovery using machine-learning and data-mining techniques: a survey. ACM Comput. Surv. **50**(4) (2017). https://doi.org/10.1145/3092566
27. Huang, Z., England, M., Davenport, J., Paulson, L.: Using machine learning to decide when to precondition cylindrical algebraic decomposition with Groebner bases. In: 18th International Symposium on Symbolic and Numeric Algorithms for Scientific Computing (SYNASC 2016), pp. 45–52. IEEE (2016). https://doi.org/10.1109/SYNASC.2016.020
28. Huang, Z., England, M., Wilson, D., Bridge, J., Davenport, J., Paulson, L.: Using machine learning to improve cylindrical algebraic decomposition. Math. Comput. Sci. **13**(4), 461–488 (2019). https://doi.org/10.1007/s11786-019-00394-8
29. Huang, Z., England, M., Wilson, D., Davenport, J.H., Paulson, L.C., Bridge, J.: Applying machine learning to the problem of choosing a heuristic to select the variable ordering for cylindrical algebraic decomposition. In: Watt, S.M., Davenport, J.H., Sexton, A.P., Sojka, P., Urban, J. (eds.) CICM 2014. LNCS (LNAI), vol. 8543, pp. 92–107. Springer, Cham (2014). https://doi.org/10.1007/978-3-319-08434-3_8

30. Jovanović, D., de Moura, L.: Solving non-linear arithmetic. In: Gramlich, B., Miller, D., Sattler, U. (eds.) IJCAR 2012. LNCS (LNAI), vol. 7364, pp. 339–354. Springer, Heidelberg (2012). https://doi.org/10.1007/978-3-642-31365-3_27

31. Kobayashi, M., Iwane, H., Matsuzaki, T., Anai, H.: Efficient subformula orders for real quantifier elimination of non-prenex formulas. In: Kotsireas, I.S., Rump, S.M., Yap, C.K. (eds.) MACIS 2015. LNCS, vol. 9582, pp. 236–251. Springer, Cham (2016). https://doi.org/10.1007/978-3-319-32859-1_21

32. Kühlwein, D., Blanchette, J.C., Kaliszyk, C., Urban, J.: MaSh: machine learning for sledgehammer. In: Blazy, S., Paulin-Mohring, C., Pichardie, D. (eds.) ITP 2013. LNCS, vol. 7998, pp. 35–50. Springer, Heidelberg (2013). https://doi.org/10.1007/978-3-642-39634-2_6

33. Kuipers, J., Ueda, T., Vermaseren, J.: Code optimization in FORM. Comput. Phys. Commun. **189**, 1–19 (2015). https://doi.org/10.1016/j.cpc.2014.08.008

34. Liang, J.H., Hari Govind, V.K., Poupart, P., Czarnecki, K., Ganesh, V.: An empirical study of branching heuristics through the lens of global learning rate. In: Gaspers, S., Walsh, T. (eds.) SAT 2017. LNCS, vol. 10491, pp. 119–135. Springer, Cham (2017). https://doi.org/10.1007/978-3-319-66263-3_8

35. Mulligan, C., Bradford, R., Davenport, J., England, M., Tonks, Z.: Non-linear real arithmetic benchmarks derived from automated reasoning in economics. In: Bigatti, A., Brain, M. (eds.) Proceedings of the 3rd Workshop on Satisfiability Checking and Symbolic Computation (SC^2 2018). No. 2189 in CEUR Workshop Proceedings, pp. 48–60 (2018). http://ceur-ws.org/Vol-2189/

36. Mulligan, C.B., Davenport, J.H., England, M.: TheoryGuru: a mathematica package to apply quantifier elimination technology to economics. In: Davenport, J.H., Kauers, M., Labahn, G., Urban, J. (eds.) ICMS 2018. LNCS, vol. 10931, pp. 369–378. Springer, Cham (2018). https://doi.org/10.1007/978-3-319-96418-8_44

37. Pedregosa, F., et al.: Scikit-learn: machine learning in Python. J. Mach. Learn. Res. **12**, 2825–2830 (2011). http://www.jmlr.org/papers/v12/pedregosa11a.html

38. Sturm, T.: New domains for applied quantifier elimination. In: Ganzha, V.G., Mayr, E.W., Vorozhtsov, E.V. (eds.) CASC 2006. LNCS, vol. 4194, pp. 295–301. Springer, Heidelberg (2006). https://doi.org/10.1007/11870814_25

39. Urban, J.: MaLARea: a metasystem for automated reasoning in large theories. In: Empirically Successful Automated Reasoning in Large Theories (ESARLT 2007), CEUR Workshop Proceedings, vol. 257, p. 14. CEUR-WS (2007). http://ceur-ws.org/Vol-257/

40. Wilson, D., Davenport, J., England, M., Bradford, R.: A "piano movers" problem reformulated. In: 15th International Symposium on Symbolic and Numeric Algorithms for Scientific Computing, SYNASC 2013, pp. 53–60. IEEE (2013). http://dx.doi.org/10.1109/SYNASC.2013.14

41. Xu, L., Hutter, F., Hoos, H., Leyton-Brown, K.: SATzilla: portfolio-based algorithm selection for SAT. J. Artif. Intell. Res. **32**, 565–606 (2008). https://doi.org/10.1613/jair.2490

A Numerical Efficiency Analysis
of a Common Ancestor Condition

Luca Carlini[1,4(✉)], Nihat Ay[1,2,3], and Christiane Görgen[1]

[1] Max Planck Institute for Mathematics in the Sciences, Leipzig, Germany
{carlini,ay,goergen}@mis.mpg.de
[2] Santa Fe Institute, Santa Fe, USA
[3] Faculty of Mathematics and Computer Science, University of Leipzig,
Leipzig, Germany
[4] Università degli Studi di Genova, Genoa, Italy

Abstract. The aim of this paper is to understand if the sufficient condition for the existence of a common ancestor for some variables in a larger graph discovered by Steudel and Ay is worth checking. The goodness of this criterion will be tested with a numerical method.

Keywords: Causality · Entropy · Graphical models · Information theory · Statistical model

1 Introduction

In this paper we provide an in-depth analysis of an example of Steudel and Ay's result on common ancestors [1]. We want to estimate the volume of a region of interest, this is the set of probability distributions which satisfy a sufficient condition for the existence of a common cause of some variables. This region has an interpretation in terms of causality [2]. In fact, if the volume of this portion of our statistical model is large, then it will be possible to detect the existence of a common ancestor between some variables in a graph using this criterion. This method could be very useful because it links observed variables to the underlying causal structure. In practice it is most often impossible to observe the whole network, so it is essential to take as much information as possible from the observed variables exclusively.

At first we restrict ourselves to the most simple case considering three variables, as we can see in Fig. 1(b), and, in a second step, we will repeat the analysis for four and five variables. We can see the general case with k observed variables in Fig. 1(a). Moreover, the method could be used to distinguish between the situation in Fig. 1(b), with a common cause of all the three variables, and the situation in Fig. 1(c), with common causes of two of the variables.

Supported by ERASMUS+.

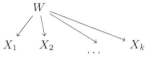
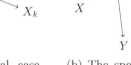
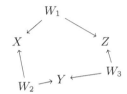

(a) The most general case, k observed random variables with one common ancestor.

(b) The special case $k = 3$, key example of this paper.

(c) Pairwise interaction between three random variables.

Fig. 1. Causal graphical models with interactions of different orders.

2   Background

In this section we present the main result we are interested in and we describe the geometrical and algebraic aspects of the configuration seen in Fig. 1(b), the key example of this paper.

Theorem 1 *(Lower bound on entropy of common ancestors [1]).*
  Let $X_1, \ldots, X_n$ be all the variables associated to a DAG G, denote the observed variables $X_{i_1}, \ldots, X_{i_k}$, and let $c \in \{1, \ldots, k-1\}$ be a constant. If

$$I^c(X_{i_1}, \ldots, X_{i_k}) = \sum_{j=1}^{k} H(X_{i_j}) - c \cdot H(X_{i_1}, \ldots, X_{i_k}) > 0, \tag{1}$$

then there are $c + 1$ nodes $X_{i_{j1}}, \ldots, X_{i_{jc+1}}$ among the observed variables which have a common ancestor.

Here, $H(X_{i_j})$ is the Shannon entropy of X_{i_j} and $H(X_{i_1}, \ldots, X_{i_k})$ is the entropy of the joint variables $X_{i_1}, \ldots, X_{i_k}$ [3]. This method provides a sufficient condition for the existence of a common ancestor, not a necessary one: the aim of the paper is to test the goodness of this criterion. In particular, at the beginning we restrict ourselves to the most simple configuration, described in Fig. 1(b). The first model to be analysed consists of all joint distributions over (X, Y, Z, W) for four binary random variables X, Y, Z, W that are Markov to the DAG seen in Fig. 1(b). We have $2^4 = 16$ possible outcomes and a corresponding 15-dimensional simplex of probability vectors. Also, there are 7 parameters: one is $P(W = 0)$ and the other six are the conditional probabilities of X, Y and Z being in state 0 given W is in state 0 and 1, respectively.

Henceforth we denote by θ_0 the probability of $W = 0$, by θ_1, θ_2 and θ_3 the probability of, respectively, X, Y and Z being in state 0 given $W = 0$ and by θ_4, θ_5, θ_6 the probability of, respectively, X, Y and Z being in state 0 given $W = 1$.

Yet, to simplify the problem and to exploit some particular properties that will be used later, we fix $\theta_0 = P(W = 0) = 1/2$; we now have 6 parameters. In this case, the sufficient condition for a common ancestor of all the three variables seen in Theorem 1 is:

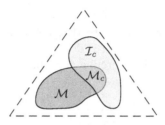

Fig. 2. A statistical model $\mathcal{M}$ inside a probability simplex, intersected with the region $\mathcal{I}_c$ where (2) is true. Our numerical approach calculates an estimate of the volume of the intersection $\mathcal{M}_c$. See Table 2 for the results.

$$H(X) + H(Y) + H(Z) - 2H(X, Y, Z) > 0. \tag{2}$$

In Fig. 2 we can see a representation of our problem. We have two sets inside a probability simplex. $\mathcal{M}$ is the statistical model, this is the set of points such that the probability distribution of the variables factorizes as follows:

$$P(X = x, Y = y, Z = z, W = w) = P(W = w)P(X = x|W = w)$$
$$\cdot P(Y = y|W = w)P(Z = z|W = w), \ \forall \ (x, y, z, w) \in \{0, 1\}^4,$$

where we have fixed $\theta_0 = P(W = 0) = 1/2$.

Table 1. For $(\theta_1, \ldots, \theta_6)$ we associate a joint probability distribution $P(X, Y, Z, W)$. Note that $\theta_0 = P(W = 0) = 1/2$.

$w\,x\,y\,z$	$P(W = w, X = x, Y = y, Z = z)$
0 0 0 0	$\theta_0 \cdot \theta_1 \cdot \theta_2 \cdot \theta_3$
0 0 1 0	$\theta_0 \cdot \theta_1 \cdot (1 - \theta_2) \cdot \theta_3$
0 0 0 1	$\theta_0 \cdot \theta_1 \cdot \theta_2 \cdot (1 - \theta_3)$
0 0 1 1	$\theta_0 \cdot \theta_1 \cdot (1 - \theta_2) \cdot (1 - \theta_3)$
0 1 0 0	$\theta_0 \cdot (1 - \theta_1) \cdot \theta_2 \cdot \theta_3$
0 1 1 0	$\theta_0 \cdot (1 - \theta_1) \cdot (1 - \theta_2) \cdot \theta_3$
0 1 0 1	$\theta_0 \cdot (1 - \theta_1) \cdot \theta_2 \cdot (1 - \theta_3)$
0 1 1 1	$\theta_0 \cdot (1 - \theta_1) \cdot (1 - \theta_2) \cdot (1 - \theta_3)$
1 0 0 0	$(1 - \theta_0) \cdot \theta_4 \cdot \theta_5 \cdot \theta_6$
1 0 1 0	$(1 - \theta_0) \cdot \theta_4 \cdot (1 - \theta_5) \cdot \theta_6$
1 0 0 1	$(1 - \theta_0) \cdot \theta_4 \cdot \theta_5 \cdot (1 - \theta_6)$
1 0 1 1	$(1 - \theta_0) \cdot \theta_4 \cdot (1 - \theta_5) \cdot (1 - \theta_6)$
1 1 0 0	$(1 - \theta_0) \cdot (1 - \theta_4) \cdot \theta_5 \cdot \theta_6$
1 1 1 0	$(1 - \theta_0) \cdot (1 - \theta_4) \cdot (1 - \theta_5) \cdot \theta_6$
1 1 0 1	$(1 - \theta_0) \cdot (1 - \theta_4) \cdot \theta_5 \cdot (1 - \theta_6)$
1 1 1 1	$(1 - \theta_0) \cdot (1 - \theta_4) \cdot (1 - \theta_5) \cdot (1 - \theta_6)$

This is the set of points which reflect the configuration seen in Fig. 1(b), meaning that X, Y, Z, W are Markov to that DAG G and X, Y and Z have a common cause W. More precisely, $\mathcal{M}$ is defined as the image of ϕ, where ϕ is a map from the six-dimensional hypercube $\Theta = [0,1]^6$ to the fifteen-dimensional simplex $\triangle_{15}$ such that:

$$\phi : \Theta \to \triangle_{15}$$
$$(\theta_1, \ldots, \theta_6) \mapsto (s_1, \ldots, s_{16}),$$

where $s_1, \ldots, s_{16}$ are the probabilities of the sixteen states computed as in the second column of Table 1, this is as the product of the marginals given the parent W.

Moreover, $\mathcal{I}_c$ is the set of probability distributions for which the ancestor condition (1) is fulfilled. In particular,

$$\mathcal{I}_c = \{p \in \triangle_{15} \mid H(X) + H(Y) + H(Z) - 2H(X,Y,Z) > 0, w.r.t. \ p\},$$

where $p = (p_1, \ldots, p_{16})$.

Lastly, $\mathcal{M}_c$ is the intersection of $\mathcal{M}$ and $\mathcal{I}_c$, which is the set of probability distributions that are in the model and for which the ancestor condition holds. We are interested in this region. We give an estimation of the volume of $\mathcal{M}_c$ in terms of sample proportion, in the sense that we will randomly generate data points over the model $\mathcal{M}$, then we will check which samples in $\mathcal{M}$ fulfil the ancestor condition, these are the data points in $\mathcal{M}_c$. The volume of this region is simply the number of samples in $\mathcal{M}_c$ divided by the number of samples in $\mathcal{M}$.

3   Examples

In this section we present some examples in which we will test the ancestor condition. In particular, we consider deterministic systems and the parametric objects related to them. Consider the variable X and the ancestor W. There are four different deterministic maps from W to X, $W \to X$:

1. The f_{id} map, this is X is the copy of W. So, when W is in state 0, also X is in state 0 and when W is in state 1, X is in state 1: $\{0 \mapsto 0, 1 \mapsto 1\}$. In terms of conditional probabilities, this can be expressed as $\theta_1 = P(X = 0|W = 0) = 1$ and $\theta_4 = P(X = 0|W = 1) = 0$;
2. The f_{rev} map, this is X is the reversal of W, in the sense that when W is in state 0, X is in state 1 and when W is in state 1, X is in state 0: $\{0 \mapsto 1, 1 \mapsto 0\}$. In terms of conditional probabilities, this can be expressed as $\theta_1 = P(X = 0|W = 0) = 0$ and $\theta_4 = P(X = 0|W = 1) = 1$;
3. The $f_{=0}$ map, this is X is deterministically 0. Whenever W is in state 0 or in state 1, X is in state 0: $\{0 \mapsto 0, 1 \mapsto 0\}$. In terms of conditional probabilities, this can be expressed as $\theta_1 = P(X = 0|W = 0) = 1$ and $\theta_4 = P(X = 0|W = 1) = 1$;
4. The $f_{=1}$ map, this is X is deterministically 1, which is analogous to the previous situation.

There is a one-to-one correspondence between the extreme points of the six-dimensional hypercube $[0,1]^6$, the parameter set, and the triplets (f_1, f_2, f_3), where $f_i \in \{f_{id}, f_{rev}, f_{=0}, f_{=1}\}$, for $i = 1, 2, 3$. For instance, the triplet (f_{id}, f_{id}, f_{id}) corresponds to the *copy vertex* $(1,1,1,0,0,0) \in [0,1]^6$.

Now, consider three f_{id} maps from W to X, Y and Z. We are going to show that in this vertex of the hypercube the ancestor condition holds. In fact,

$$P(X = 0) = P(X = 0|W = 0)P(W = 0) + P(X = 0|W = 1)P(W = 1)$$
$$= 1 \cdot 1/2 + 0 \cdot 1/2 = 1/2 = P(Y = 0) = P(Z = 0);$$
$$H(X) = P(X = 0) \ln\left(1/P(X = 0)\right) + P(X = 1) \ln\left(1/P(X = 1)\right)$$
$$= 1/2 \ln(2) + 1/2 \ln(2) = \ln(2) = H(Y) = H(Z).$$

Lastly, the joint entropy of X, Y and Z is

$$H(X, Y, Z) = \sum_{i,j,k=0}^{1} P(X = i, Y = j, Z = k) \ln\left(1/P(X=i,Y=j,Z=k)\right)$$
$$= P(X = 0, Y = 0, Z = 0) \ln\left(1/P(X=0,Y=0,Z=0)\right)$$
$$+ P(X = 1, Y = 1, Z = 1) \ln\left(1/P(X=1,Y=1,Z=1)\right)$$
$$= 1/2 \ln(2) + 1/2 \ln(2) = \ln(2).$$

Thus, the ancestor condition holds, since

$$H(X) + H(Y) + H(Z) - 2H(X, Y, Z) = \ln(2) + \ln(2) + \ln(2) - 2\ln(2) = \ln(2).$$

Moreover, we can notice that if we consider two f_{id} maps and one $f_{=0}$ or $f_{=1}$ map the condition does not hold, since the entropy of the last variable is 0.

4   Numerical Sampling

In this section we explain the technique we have used to compute the volume we are interested in by sampling uniformly from the six-dimensional hypercube. For $0 < r \leq 1$, we consider the hypercube $[1 - r, 1]^3 \times [0, r]^3$, which contains the copy vertex $(1,1,1,0,0,0)$ and whose image through the map ϕ defines a subset of the model $\mathcal{M}$, as we can see in Fig. 3.

For each value of r, we consider the area around the copy vertex of the hypercube $[1 - r, 1]^3 \times [0, r]^3$ and we sample uniformly from this area. Then, we count how many data points fulfil the ancestor condition. When r is equal to 1, we have a uniform sampling over the hypercube. We can see how the proportion of samples which satisfy the ancestor condition changes for different values of r in Fig. 4. When r becomes bigger, the proportion of data points for which the ancestor condition holds decreases (up to $r = 0.5$). In the next section we present the results we obtained by sampling uniformly 10^9 data points for $r = 1$, this is from the full hypercube, and for different values of k, this is the number of variables in the system.

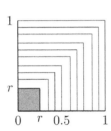

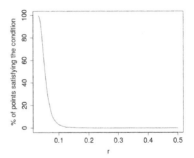

Fig. 3. 2-dimensional representation of the area around the copy vertex for different values of r.

Fig. 4. Percentage of samples fulfilling the ancestor condition for different values of r.

5   Discussion

We have repeated the same passages seen in Sect. 4 adding one and two variables in the system, in order to analyse if the ancestor condition is able to detect the existence of a common ancestor of three out of four and five variables. More precisely, we have considered two models: the first one consists of all joint distributions over (X, Y, Z, U, W) for five binary variables X, Y, Z, U, W which are Markov to the DAG in Fig. 1(a), for $k = 4$, while the second one consists of all joint distributions over (X, Y, Z, U, V, W) for six binary variables X, Y, Z, U, V, W which are Markov to the DAG in Fig. 1(a), for $k = 5$.

Looking at Table 2, we observe that the volume of $\mathcal{M}_c$ is extremely small.

Table 2. Results of the numerical analysis for $c = 2$. For each experiment, the sample size is 10^9.

Number of variables	Percentage of samples in $\mathcal{M}_c$
$k = 3$	0.0000275%
$k = 4$	0.0000234%
$k = 5$	0.0000124%

We can give three possible interpretations for this. It is possible that most of the probability distributions in the simplex can be described through the causal graphical model in Fig. 1(c). Consequently, the volume of the probability distributions which can be modeled with the DAG in Fig. 1(b) would be really small. In this case, it would be impossible to detect the existence of a common ancestor between three variables with the condition discovered by Steudel and Ay.

On the other hand, there could be a problem related to the method, which means that the common ancestor condition is too strict to detect the existence of a common ancestor and we would need better estimates of the quantities we are considering.

Lastly, another sampling approach could provide a different result. For example, it would be possible to sample uniformly from the simplex with respect to the Fisher metric employing methods from algebraic statistics [4] and information theory [5], instead of sampling uniformly from the hypercube. Moreover, an analytical approach could produce a more precise estimate of the volume of the region of interest, implementing methods from measure theory and mathematical analysis.

This paper opens the doors to other possible developments. For example, studying the extension of the volume of probability distributions which can be described with pairwise interactions in the simplex compared to the volume of probability distributions which can be modeled with the configuration we have studied in this paper would clarify if the first interpretation is correct.

Moreover, considering more complicated graphs with a larger number of variables and/or with nodes having more states could state if the second interpretation is the right one, since the models we have studied represent a very small set of all the possible systems. Thus, it is possible that in higher dimension and/or considering variables with a larger number of states the ancestor condition would be able to detect the existence of a common ancestor.

References

1. Steudel, B., Ay, N.: Information-theoretic inference of common ancestors. Entropy **17**(4), 2304–2327 (2015)
2. Pearl, J.: Causality. Models, Reasoning, and Inference, 2nd edn., xx+464 pp. Cambridge University Press, Cambridge (2009). ISBN 978-0-521-89560-6, 0-521-77362-8
3. Cover, T.M., Thomas, J.A.: Elements of Information Theory, 2nd edn., xxiv+748 pp. Wiley, Hoboken (2006). ISBN 978-0-471-24195-9, 0-471-24195-4
4. Collazo, R.A., Görgen, C., Smith, J.Q.: Chain Event Graphs. Chapman and Hall/CRC Computer Science and Data Analysis Series, xx+233 pp. CRC Press, Boca Raton (2018). ISBN 978-1-4987-2960-4
5. Ay, N., Jost, J., Lê, H.V., Schwachhöfer, L.: Information Geometry. EMGFASMSM, vol. 64. Springer, Cham (2017). https://doi.org/10.1007/978-3-319-56478-4. xi+407 pp. ISBN 978-3-319-56477-7, 978-3-319-56478-4

Optimal Transport to a Variety

Türkü Özlüm Çelik[1(✉)], Asgar Jamneshan[3], Guido Montúfar[1,3],
Bernd Sturmfels[1,2], and Lorenzo Venturello[1]

[1] Max Planck Institute for Mathematics in the Sciences, Leipzig, Germany
`Tuerkue.Celik@mis.mpg.de`
[2] University of California at Berkeley, Berkeley, USA
[3] University of California at Los Angeles, Los Angeles, USA

Abstract. We study the problem of minimizing the Wasserstein distance between a probability distribution and an algebraic variety. We consider the setting of finite state spaces and describe the solution depending on the choice of the ground metric and the given distribution. The Wasserstein distance between the distribution and the variety is the minimum of a linear functional over a union of transportation polytopes. We obtain a description in terms of the solutions of a finite number of systems of polynomial equations. The case analysis is based on the ground metric. A detailed analysis is given for the two bit independence model.

Keywords: Algebraic statistics · Linear programming · Optimal transport estimator · Polynomial optimization · Transportation Polytope · Triangulation · Wasserstein distance

1 Introduction

Density estimation in statistics is the problem of learning a hypothesis density ν based on samples $x_1, \ldots, x_N \in \Omega$ from an unknown density μ. A standard approach to solving this problem is to define a statistical model $\mathcal{M}$ of candidate hypotheses, and then select a density from $\mathcal{M}$ that minimizes some type of distance to the empirical distribution $\bar{\mu} = \frac{1}{N} \sum_i \delta_{x_i}$. An example of this is the maximum likelihood estimator [16, Chapter 7], which minimizes the Kullback-Leibler divergence between $\bar{\mu}$ and $\mathcal{M}$. This estimator selects $\nu \in \mathcal{M}$ by maximizing the log-likelihood $\sum_{i=1}^{N} \log \nu(x_i)$.

When the sample space Ω is a metric space, optimal transport defines a distance between probability distributions [17]. The corresponding estimator selects $\nu \in \mathcal{M}$ so that it assigns a high probability to points x that are close, but not necessarily equal, to samples x_i. In contrast to the maximum likelihood estimator, this incorporates the metric on Ω. One key advantage of this is that distances between distributions are well defined even when they have disjoint supports. The minimum Wasserstein distance estimator plays an important role in machine learning applications.

© Springer Nature Switzerland AG 2020
D. Slamanig et al. (Eds.): MACIS 2019, LNCS 11989, pp. 364–381, 2020.
https://doi.org/10.1007/978-3-030-43120-4_29

The key disadvantage of the optimal transport distance is that it is defined as the solution to an optimization problem. Thus, computing the minimum Wasserstein distance estimator requires solving a double minimization problem. In a few special cases, the Wasserstein distance can be given by a formula, e.g. in the case of two Gaussian distributions. However, for general ground distances and distributions, a closed formula is not available. The standard methods for numerical computation of the Wasserstein distance between two distributions have super cubic complexity in the size of the distributions [11]. Therefore, much work has been devoted to developing fast methods for optimal transport [12]. An important advance has been the introduction of entropy regularized optimal transport and iterative computations with a Sinkhorn algorithm [5], which allows for a cheaper computation and has increased the applicability of optimal transport.

In large scale problems, the exact Wasserstein distance and the minimum distance estimator remain out of reach. A very successful and popular model for obtaining implicit generative models is the Wasserstein generative adversarial network [2]. This is based on the Kantorovich dual formulation of the Wasserstein-1 distance, as a difference of expectation values of an optimal discriminative function. Training (i.e. fitting the parameters of the model) is based on estimating the expectations by sample averages, approximating the discriminator by a neural network, and following the negative gradient of the estimated distance with respect to the model parameters.

A number of works address the statistical complexity of estimating the optimal transport cost. The asymptotic behavior of the minimum Wasserstein distance estimator was studied in [3] and [4]. The convergence of the empirical distribution for increasing sample size was studied in [18].

Specifying a model beforehand allows us to focus the search for a hypothesis, reducing statistical and computational complexity. In many cases the model is given in terms of a parametrization with a small number of parameters, thus providing a compact representation of hypotheses. It can also be specified in terms of properties of interest, such as conditional independence relations. This view is taken in algebraic statistics [16]. When the model is an exponential family (a toric variety), maximum likelihood estimation is a convex optimization problem. For some exponential families, such as decomposable hierarchical models, the maximum likelihood estimator can be written explicitly (e.g. [16, Chapter 7]). Recent work characterizes such cases where the solution is rational [7]. Closed formulas are also known for some latent variable graphical models [1,14].

The present study is cast on the discrete side of algebraic statistics [16]. In our setting, the model $\mathcal{M}$ is an algebraic variety inside a probability simplex. We wish to understand fundamental properties of the minimum Wasserstein distance estimator for $\mathcal{M}$. What is the structure of the function that computes the Wasserstein distance between a given data distribution and a point in $\mathcal{M}$? How does it change depending on the ground metric that is laid on the sample space? How does it change depending on the data distribution? How does it depend on the model? Is the minimizer unique, or are there finitely many minimizers? Can we obtain a closed formula?

The optimal transport distance between two points in our simplex is the solution to a linear program over a transportation polytope. The optimal transport distance between a distribution and $\mathcal{M}$ is the minimum of a linear functional over an infinite union of transportation polytopes. Our aim is to understand the combinatorics and geometry of this parametric linear program.

This article is organized as follows. Section 2 recalls the definition of the Wasserstein distance. It also provides the relevant background in linear programming, geometric combinatorics, and commutative algebra. A key insight is that the given metric on Ω induces a regular triangulation of a product of two simplices (cf. Theorem 1), and this induces a mixed polyhedral subdivision of one simplex when μ is fixed. Section 3 presents our algorithm for computing the Wasserstein distance from a distribution μ to a model $\mathcal{M}$ in the probability simplex. The main subroutine is the optimization of linear functions over the pieces of $\mathcal{M}$ that arise from the mixed subdivision.

We illustrate Algorithm 2 by working out the geometry for the discrete ground metric on three states. This is illustrated in Fig. 1. In Sect. 4 we focus on the case of primary interest, namely when the model $\mathcal{M}$ is an algebraic variety. Here the minimum Wasserstein distance estimator is a piecewise algebraic function. We show how each piece can be represented by the hypersurface that is dual to $\mathcal{M}$ in the sense of projective geometry. In Sect. 5 we undertake a detailed case study. Namely, we determine the minimum Wasserstein estimator of a discrete independence model.

2   Geometric Combinatorics of the Wasserstein Distance

Let $\Delta_{n-1} = \{(p_1, \ldots, p_n) \in \mathbb{R}^n_{\geq 0} : \sum_{i=1}^n p_i = 1\}$ denote the simplex of probability distributions on the set $[n] = \{1, 2, \ldots, n\}$. We fix a symmetric $n \times n$ matrix $d = (d_{ij})$ with nonnegative entries. In our application, the pair $([n], d)$ will be a finite metric space, so we have $d_{ii} = 0$ and $d_{ik} \leq d_{ij} + d_{jk}$ for all i, j, k. We identify Δ_{n^2-1} with the set of nonnegative $n \times n$ matrices whose entries sum to 1.

Fix two distributions $\mu, \nu \in \Delta_{n-1}$. The associated *transportation polytope* is

$$\Pi(\mu, \nu) = \left\{ \pi \in \Delta_{n^2-1} : \sum_{i=1}^n \pi_{ij} = \mu_j \text{ for all } j \text{ and } \sum_{j=1}^n \pi_{ij} = \nu_i \text{ for all } i \right\}. \tag{1}$$

Thus, $\Pi(\mu, \nu)$ is the set of nonnegative $n \times n$-matrices with prescribed row and column sums. This polytope has dimension $(n-1)^2$, provided $\mu, \nu \in \mathrm{int}(\Delta_{n-1})$, and it is simple if μ, ν are generic.

We consider the linear programming problem on the transportation polytope $\Pi(\mu, \nu)$ with cost matrix d. This is known as the *transportation problem* for (μ, ν, d). The optimal value of this linear program is known as the Wasserstein distance between μ and ν with respect to d. Thus, the *Wasserstein distance* is

$$W(\mu, \nu) = \min_{\pi \in \Pi(\mu, \nu)} \sum_{1 \leq i, j \leq n} d_{ij} \pi_{ij}. \tag{2}$$

We are interested in the following parametric version of this linear programming problem. We fix any subset $\mathcal{M}$ of the model Δ_{n-1}. This set is our statistical model. The *Wasserstein distance* between a given distribution μ and the model $\mathcal{M}$ with respect to the metric d is defined to be

$$W(\mu, \mathcal{M}) \quad = \quad \min_{\nu \in \mathcal{M}} \min_{\pi \in \Pi(\mu,\nu)} \sum_{1 \leq i,j \leq n} d_{ij} \pi_{ij}. \tag{3}$$

Computing this quantity amounts to solving a nested optimization problem. Namely, we are minimizing the cost function d over the set $\bigcup_{\nu \in \mathcal{M}} \Pi(\mu, \nu)$. The constraint set can be thought of as a bundle of transportation polytopes over the model $\mathcal{M}$. Our goal is to understand its geometry.

The $2n$ linear constraints that define the transportation polytope $\Pi(\mu, \nu)$ can be written as $A\pi = (\mu_1, \ldots, \mu_n, \nu_1, \ldots, \nu_n)^T$ for a certain matrix $A \in \{0,1\}^{2n \times n^2}$ of rank $2n - 1$. The columns of this matrix are the vertices of the product of the standard simplices $\Delta_{n-1} \times \Delta_{n-1} \subset \mathbb{R}^n \times \mathbb{R}^n$.

Example 1. Let $n = 4$. The polytopes $\Pi(\mu, \nu)$ are 9-dimensional for $\mu, \nu \in \mathrm{int}(\Delta_3)$. They are the fibers of the linear map $\Delta_{15} \to \Delta_3 \times \Delta_3$ given by the matrix

$$A \;=\; \begin{bmatrix} 1 & 1 & 1 & 1 & 0 & 0 & 0 & 0 & 0 & 0 & 0 & 0 & 0 & 0 & 0 & 0 \\ 0 & 0 & 0 & 0 & 1 & 1 & 1 & 1 & 0 & 0 & 0 & 0 & 0 & 0 & 0 & 0 \\ 0 & 0 & 0 & 0 & 0 & 0 & 0 & 0 & 1 & 1 & 1 & 1 & 0 & 0 & 0 & 0 \\ 0 & 0 & 0 & 0 & 0 & 0 & 0 & 0 & 0 & 0 & 0 & 0 & 1 & 1 & 1 & 1 \\ 1 & 0 & 0 & 0 & 1 & 0 & 0 & 0 & 1 & 0 & 0 & 0 & 1 & 0 & 0 & 0 \\ 0 & 1 & 0 & 0 & 0 & 1 & 0 & 0 & 0 & 1 & 0 & 0 & 0 & 1 & 0 & 0 \\ 0 & 0 & 1 & 0 & 0 & 0 & 1 & 0 & 0 & 0 & 1 & 0 & 0 & 0 & 1 & 0 \\ 0 & 0 & 0 & 1 & 0 & 0 & 0 & 1 & 0 & 0 & 0 & 1 & 0 & 0 & 0 & 1 \end{bmatrix}.$$

Fix a generic matrix $d \in \mathbb{R}^{n^2}$. The optimal bases of our linear program (2), as the distributions μ, ν range over the simplex Δ_{n-1}, are the maximal simplices σ in a triangulation Σ_d of the $(2n - 2)$-dimensional polytope $\Delta_{n-1} \times \Delta_{n-1}$. Combinatorially, such a basis σ consists of the edges in a spanning tree of the complete bipartite graph on $[n] \times [n]$. Let A_σ be the submatrix of A given by the columns that are indexed by σ. For $(\mu, \nu) \in \Delta_{n-1} \times \Delta_{n-1}$, there exists a unique column vector π_σ such that $A_\sigma \cdot \pi_\sigma = (\mu, \nu)^T$. Note that the coordinates of π_σ are linear functions in (μ, ν).

Let $\tilde{\pi}_\sigma$ denote the matrix in $\mathbb{R}^{n^2}$ that agrees with π_σ in all coordinates in σ and is zero in all other coordinates. Then $\tilde{\pi}_\sigma$ is the optimal vertex of $\Pi(\mu, \nu)$ for all pairs (μ, ν) in the simplex σ. On that σ, the Wasserstein distance between our two distributions is given by the linear function

$$(\mu, \nu) \quad \mapsto \quad W(\mu, \nu) \;=\; \sum_{1 \leq i,j \leq n} d_{ij} \cdot (\tilde{\pi}_\sigma)_{ij}. \tag{4}$$

This allows us to remove the inner optimization when solving (3). For each simplex $\sigma \in \Sigma_d$, our task is to minimize the linear function (4) over the intersection

$(\mu \times \mathcal{M}) \cap \sigma$. Among these optimal solutions, one for each simplex $\sigma \in \Sigma_d$, we then select the solution with the smallest optimal value. This is the geometric idea behind the algorithm that will be presented in the next section.

We now shift gears and we discuss the study of triangulations of $\Delta_{n-1} \times \Delta_{n-1}$. This is a rich subject in geometric combinatorics, with numerous connections to optimization, tropical geometry, enumerative combinatorics, representation theory, commutative algebra, and algebraic geometry. The triangulations which appear in our context are called *regular triangulations* [6, Chapter 2]. There are various different approaches for computing these objects. The one we favor here is based on commutative algebra. Namely, we represent our objects as initial ideals of the ideal of 2×2-minors of an $n \times n$-matrix of unknowns. In the language of algebraic geometry, these are the toric degenerations of the Segre variety $\mathbb{P}^{n-1} \times \mathbb{P}^{n-1}$ in its embedding in the matrix space $\mathbb{P}^{n^2-1}$.

In the rest of this section we present relevant definitions and results. We refer to [10,15] for an extensive treatment of the subject. Fix the polynomial ring $R = K[y_{ij} : 1 \leq i, j \leq n]$ over a field K. We identify nonnegative integer vectors $\alpha \in \mathbb{N}^{n \times n}$ with monomials $y^{\alpha} = \prod_{i=1}^{n} y_{ij}^{\alpha_{i,j}}$. Let $d \in \mathbb{R}^{n \times n}$ and I an ideal in R. Consider any polynomial $f = \sum_{\alpha \in \mathbb{N}^{n \times n}} c_{\alpha} y^{\alpha} \in I$. The *initial form* of f is defined to be $\mathrm{in}_d(f) = \sum_{d \cdot \alpha = \mathbf{d}} c_{\alpha} y^{\alpha}$ with $\mathbf{d} = \max\{d \cdot \alpha : c_{\alpha} \neq 0\}$ where $\cdot$ denotes the standard dot product. The *initial ideal* of I with respect to the weight matrix d is the following ideal in R:

$$\mathrm{in}_d(I) := \langle \mathrm{in}_d(f) : f \in I \rangle.$$

For a generic choice of d, this is a monomial ideal, i.e. $\mathrm{in}_d(I)$ can be generated by monomials. In this case, we can compute (in a computer algebra system) a corresponding reduced Gröbner basis $\{g_1, g_2, \ldots, g_r\}$ of I. The initial monomials $\mathrm{in}_d(g_1), \ldots, \mathrm{in}_d(g_r)$ are minimal generators of $\mathrm{in}_d(I)$.

The connection to regular triangulations arises when I is a *toric ideal* I_A. This works for any nonnegative integer matrix A, but we here restrict ourselves to the matrix A whose columns are the vertices of $\Delta_{n-1} \times \Delta_{n-1}$, as in Example 1. The toric ideal associated to A is the *determinantal ideal*

$$I_A := \langle y^{u^+} - y^{u^-} : u \in \ker(A) \rangle = \langle \, 2 \times 2\text{-minors of the } n \times n \text{ matrix } (y_{ij}) \rangle.$$

The regular polyhedral subdivisions of the product of simplices are encoded by the initial ideals of the ideal I_A.

Theorem 1 (Sturmfels' Correspondence). [6, Theorem 9.4.5] *There is a bijection between regular subdivisions of $\Delta_{n-1} \times \Delta_{n-1}$ induced by d and the ideals $\mathrm{in}_d(I_A)$. Moreover, $\mathrm{in}_d(I_A)$ is a monomial ideal if and only if the corresponding subdivision of $\Delta_{n-1} \times \Delta_{n-1}$ is a triangulation.*

Since the matrix A is totally unimodular [15, Exercise (9), page 72], all initial monomial ideals $\mathrm{in}_d(I_A)$ are squarefree [15, Corollary 8.9]. The desired triangulation Σ_d is the simplicial complex whose Stanley-Reisner ideal equals $\mathrm{in}_d(I_A)$.

This means that the set $\mathcal{F}(\Sigma_d)$ of its maximal simplices in the triangulation is read off from the prime decomposition of the squarefree monomial ideal:

$$\text{in}_d(I_A) \;=\; \bigcap_{\sigma \in \mathcal{F}(\Sigma_d)} \langle\, y_{ij} : ij \notin \sigma \,\rangle. \tag{5}$$

For a first illustration see [15, Example 8.12], where it is shown that the *diagonal initial ideal* of the determinantal ideal I_A corresponds to the *staircase triangulation* of the polytope $\Delta_{n-1} \times \Delta_{n-1}$.

From the perspective of optimal transport, what has been accomplished so far? We wrote the Wasserstein distance between two distributions locally as a linear function. This is the function in (4). The region σ inside $\Delta_{n-1} \times \Delta_{n-1}$ on which this formula is valid is a simplex. The set of these simplices is the triangulation Σ_d. The algebraic recipe (5) serves to compute this. Thus, the associated primes of $\text{in}_d(I_A)$ are the linear formulas for the Wasserstein distance.

Remark 1. If the matrix d is special then $\text{in}_d(I_A)$ may not be a monomial ideal. This happens for the discrete metric on $[n]$ when $n \geq 4$. In such a case, we break ties with a term order to get a triangulation. Geometrically, this corresponds to replacing d by a nearby generic matrix d_ϵ. However, since the optimal value function of a linear program is piecewise linear and continuous, the limit of the optimal values for d_ϵ as $\epsilon \to 0$ is the optimal value for d.

The discussion above is concerned with the piecewise-linear structure of the Wasserstein distance $W(\mu, \nu)$ when d is fixed and μ, ν vary. The story becomes more interesting when we allow the matrix d to vary over $\mathbb{R}^{n^2}$. This brings us to the theory of *secondary polytopes*. Two generic matrices d and d' are considered *equivalent* if their triangulations coincide: $\Sigma_d = \Sigma_{d'}$. The equivalence classes are open convex polyhedral cones that partition $\mathbb{R}^{n^2}$. This partition is the *secondary fan* of our product of simplices. This fan is the normal fan of the secondary polytope $\Sigma(\Delta_{n-1} \times \Delta_{n-1})$, which is the Newton polytope of the product of all subdeterminants (all sizes) of the matrix (y_{ij}).

For a given generic matrix d, its equivalence class (a.k.a. secondary cone) can be read off from the reduced Gröbner basis $\{g_1, g_2, \ldots, g_r\}$ of I_A with respect to d. The Gröbner basis elements are binomials $g_i = y^{u_i^+} - y^{u_i^-}$, where $u_1, u_2, \ldots, u_r \in \mathbb{Z}^{n^2}$. Then the desired secondary cone equals

$$\left\{\, d \in \mathbb{R}^{n^2} : d \cdot u_i > 0 \text{ for } i = 1, 2, \ldots, r \,\right\}. \tag{6}$$

Example 2. Let $n = 3$ and fix the *discrete metric* $d \in \{0, 1\}^{3 \times 3}$, which has $d_{ii} = 0$ and $d_{ij} = 1$ if $i \neq j$. This matrix looks special but it is actually generic. The corresponding Gröbner basis equals

$$\{\, \underline{y_{12}y_{21}} - y_{11}y_{22},\ \underline{y_{12}y_{23}} - y_{13}y_{22},\ \underline{y_{12}y_{31}} - y_{11}y_{32},\ \underline{y_{13}y_{21}} - y_{11}y_{23},\ \underline{y_{13}y_{31}}$$
$$-y_{11}y_{33},\ \underline{y_{13}y_{32}} - y_{12}y_{33},\ \underline{y_{21}y_{32}} - y_{22}y_{31},\ \underline{y_{23}y_{31}} - y_{21}y_{33},\ \underline{y_{23}y_{32}} - y_{22}y_{33} \,\}.$$

The initial monomials are underlined, so the secondary cone is defined by

$$d_{12} + d_{21} > d_{11} + d_{22},\ d_{12} + d_{23} > d_{13} + d_{22},\ \ldots,\ d_{23} + d_{32} > d_{22} + d_{33}.$$

For any matrix in that secondary cone in $\mathbb{R}^{3 \times 3}$, the initial monomial ideal equals

$$
\mathrm{in}_d(I_A) = \begin{array}{c} \langle y_{12}, y_{13}, y_{21}, y_{23} \rangle \cap \langle y_{12}, y_{13}, y_{23}, y_{32} \rangle \cap \langle y_{12}, y_{13}, y_{31}, y_{32} \rangle \\ \cap \langle y_{12}, y_{21}, y_{31}, y_{32} \rangle \cap \langle y_{13}, y_{21}, y_{23}, y_{31} \rangle \cap \langle y_{21}, y_{23}, y_{31}, y_{32} \rangle. \end{array} \tag{7}
$$

This encodes the six 4-simplices that form the triangulation Σ_d of $\Delta_2 \times \Delta_2$.

3   An Algorithm and the Geometry of Triangles

We next present our algorithm for computing the Wasserstein distance to a model, $W(\mu, \mathcal{M})$. Here the model $\mathcal{M}$ is any subset of Δ_{n-1}. Our only assumption is that we have a subroutine for minimizing a linear function over intersections of $\mu \times \mathcal{M}$ with subpolytopes σ of $\Delta_{n-1} \times \Delta_{n-1}$. The case of primary interest, when $\mathcal{M}$ is an algebraic variety, will be addressed in the next section. We begin by giving an informal summary. The precise version appears in Algorithm 2.

Algorithm 1. A friendly description of the steps in Algorithm 2

Input: An $n \times n$ matrix $d = (d_{ij})$, a model $\mathcal{M} \subset \Delta_{n-1}$, and a distribution $\mu \in \Delta_{n-1}$.
Steps 1-3: Compute the triangulation of the polytope $\Delta_{n-1} \times \Delta_{n-1}$ that is given by d.
Step 4: Incorporate μ and express matrix entries as linear functions in $\nu \in \mathcal{M}$.
Step 5: For each piece, minimize a linear function over the relevant part of the model $\mathcal{M}$.
Steps 6-7: The smallest minimum found in Step 5 is the *Wasserstein distance* $W(\mu, \mathcal{M})$.

The first step in our algorithm is the computation of the regular triangulation Σ_d. This is done using the algebraic method described in Sect. 2. As before, I_A denotes the ideal of 2×2 minors of an $n \times n$ matrix of unknowns $y = (y_{ij})$. The computation of Σ_d is a preprocessing step that depends only on d. Once the triangulation is known, we can use it to treat different models $\mathcal{M}$ and different distributions μ, by starting from Step 5 of Algorithm 2.

There are two sources of complexity in Algorithm 2. First, there is the subroutine in Step 5, where we minimize a linear function over the model $\mathcal{M}$, subject to nonnegativity constraints that specify $(\mu \times \Delta_{n-1}) \cap \sigma$. When $\mathcal{M}$ is a semialgebraic set, this is a polynomial optimization problem. For an introduction to current methods see [9]. In Sect. 4 we disregard inequality constraints and focus on the case when the model $\mathcal{M}$ is a variety. Here the complexity is governed by the algebraic degree, which refers to the number of complex critical points. The other source of complexity is combinatorial, and it is governed by the number of maximal simplices in the triangulation of $\Delta_{n-1} \times \Delta_{n-1}$. This number is independent of the triangulation. We have

$$
|\mathcal{F}(\Sigma_d)| = \binom{2n-2}{n-1} = O\left(4^n n^{-1/2}\right). \tag{8}
$$

The second equation rests on Stirling's formula. This exponential complexity can be reduced when we deal with specific finite metric spaces. Namely, if d is a symmetric matrix with very special structure, then Σ_d will not be a triangulation but a coarser subdivision with far fewer cells than $\binom{2n-2}{n-1}$. This structure can be exploited systematically, in order to gain a reduction in complexity.

Algorithm 2. Computing the Wasserstein distance to a model

Input: An $n \times n$ matrix $d = (d_{ij})$, a model $\mathcal{M} \subset \Delta_{n-1}$, and $\mu \in \Delta_{n-1}$.
Output: The Wasserstein distance $W(\mu, \mathcal{M})$ and a point in $\mathcal{M}$ that attains this distance.
Step 1: Compute the initial $\mathrm{in}_d(I_A)$ for the ideal I_A of 2×2-minors.
Step 2: If $\mathrm{in}_d(I_A)$ is not a monomial ideal, then redo Step 1 with a nearby generic matrix.
Step 3: Compute the set $\mathcal{F}(\Sigma_d)$ of maximal simplices in Σ_d using (5).
Step 4: For every $\sigma \in \mathcal{F}(\Sigma_d)$, compute the matrix $\tilde{\pi}_\sigma$ whose entries are linear in $\nu \in \mathcal{M}$.
Step 5: For every $\sigma \in \mathcal{F}(\Sigma_d)$, compute the minimum of the linear function in (4) over the intersection $(\mu \times \mathcal{M}) \cap \sigma$.
Step 6: Choose the minimum value among the optimal values in Step 5.
Step 7: Output this value and $\nu^* \in \mathcal{M}$ satisfying $W(\mu, \nu^*) = W(\mu, \mathcal{M})$.

Example 3. Consider the discrete metric $d \in \{0, 1\}^{n \times n}$, which has $d_{ii} = 0$ and $d_{ij} = 1$ if $i \neq j$. The subdivision Σ_d of $\Delta_n \times \Delta_n$ has $2^n - 2$ maximal cells. So, it is not a triangulation for $n \geq 4$. Combinatorially, Σ_d is dual to the zonotope that is obtained by taking the Minkowski sum of n line segments in $\mathbb{R}^{n-1}$. This follows from the identification of triangulations of products of simplices with tropical polytopes. The tropical polytope representing the discrete metric is the $(n-1)$-dimensional *pyrope*; see [8, Equation (4)]. For instance, consider the case $n = 4$: the 3-dimensional pyrope is the rhombic dodecahedron, which has 14 vertices, 24 edges, and 12 facets [8, Figure 4].

In the remainder of this section we offer a detailed illustration of Algorithm 2 in the case $n = 3$. We fix the discrete metric d as in Examples 2 and 3, and we take $\mathcal{M}$ to be the independence model for two identically distributed binary random variables. This is the image of the parametrization

$$\varphi : [0, 1] \to \Delta_2, \quad p \mapsto \left(p^2, 2p(1-p), (1-p)^2 \right). \tag{9}$$

Thus $\mathcal{M} = \mathrm{image}(\varphi)$ is a quadratic curve inside the triangle Δ_2. This curve is known as the *Hardy-Weinberg curve* in genetics and it is shown in Fig. 1.

Fix the distribution $\mu = (1/2, 1/7, 5/14)$. This is marked in Fig. 1. The Wasserstein distance between μ and $\mathcal{M}$ is attained at $p^* = 1/\sqrt{2}$. It equals

$$W(\mu, \mathcal{M}) \; = \; \sqrt{2} - 8/7 \; = \; 0.2713564195... \; = \; W(\mu, \nu^*).$$

The corresponding optimal distribution ν^* in the model $\mathcal{M}$ equals

$$(\nu_1^*, \nu_2^*, \nu_3^*) = \left((p^*)^2, 2p^*(1-p^*), (1-p^*)^2 \right) = \left(0.5, 0.4142135..., 0.0857864... \right).$$

An optimal transportation plan is this matrix with given row and column sums:

$$
\begin{pmatrix}
\pi_{11} & 0 & 0 \\
\pi_{21} & \pi_{22} & \pi_{23} \\
0 & 0 & \pi_{33}
\end{pmatrix}
\begin{matrix}
\nu_1^* \\
\nu_2^* \\
\nu_3^*
\end{matrix}
\tag{10}
$$
$$
\begin{matrix}
\frac{1}{2} & \frac{1}{7} & \frac{5}{14}
\end{matrix}
$$

This solution was found using Algorithm 2. Steps 1, 2 and 3 were already carried out in Example 2. In Step 4, we translate each prime component in (7) into a 3×3 matrix $\tilde{\pi}_\sigma$ whose entries are linear forms. For instance, the third component in (7) corresponds to the matrix in (10) with

$$\pi_{11} = \nu_1, \ \pi_{21} = \mu_1 - \nu_1, \ \pi_{22} = \mu_2, \ \pi_{23} = \mu_3 - \nu_3, \ \pi_{33} = \nu_3.$$

We substitute $\mu = \left(\frac{1}{2}, \frac{1}{7}, \frac{5}{14} \right)$ and $\nu = \left(p^2, 2p(1-p), (1-p)^2 \right)$ into these six 3×3 matrices $\tilde{\pi}_\sigma$. As σ runs over $\mathcal{F}(\Sigma_d)$, we obtain six feasible regions $(\mu \times \Delta_2) \cap \sigma$ in the ν-triangle Δ_2. These are the triangles and the rhombi in Fig. 1. On each of these cells, the objective function $\pi_{12} + \pi_{13} + \pi_{21} + \pi_{23} + \pi_{31} + \pi_{32}$ is a quadratic function in p. This quadric appears in the leftmost column of the table below, along with the feasible region restricted to the curve $\mathcal{M}$. The third and fourth column list the optimal solutions that are computed in Step 5.

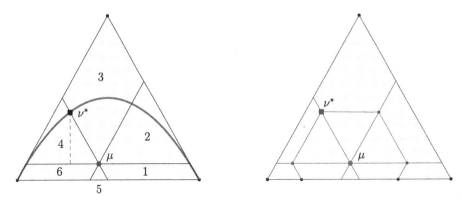

Fig. 1. The model $\mathcal{M}$ is the curve, here shown in the triangle $\mu \times \Delta_2$. It intersects five of the six cells that are obtained by restricting the 4-simplices σ in $\mathcal{F}(\Sigma_d)$ from $\Delta_2 \times \Delta_2$ to that triangle. The Wasserstein distance from μ to the curve $\mathcal{M}$ is attained by a point, labeled ν^*, that lies at the intersection of two cells. (Color figure online)

The i-th row of this table corresponds to the cell labeled i. In Step 6 of our algorithm, we identify cells 3 and 4 as those that attain the minimum value. In Step 7 we recover the optimal solution ν^*. The optimal point is marked by ν^* in Fig. 1. The geometric fact that the minimum is attained on the intersection of two cells corresponds to the algebraic fact that $\pi_{21} = 0$ in (10).

It is instructive to draw the balls in the Wasserstein metric around the point μ in Fig. 1. For small radii, these balls are regular hexagons whose sides are

Objective function	Feasible region	Solution p	Minimum value
$p^2 - 2p + 9/14$	$0 \le p \le \frac{(1-\sqrt{5/7})}{2}$	$\frac{(1-\sqrt{5/7})}{2}$	$1/14 + \sqrt{5/7}/2$
$-p^2 + 1/2$	$\frac{(1-\sqrt{5/7})}{2} \le p \le 1 - \sqrt{5/14}$	$1 - \sqrt{5/14}$	$2\sqrt{5/14} - 6/7$
$-2p^2 + 2p - \frac{1}{7}$	$1 - \sqrt{5/14} \le p \le \sqrt{1/2}$	$\sqrt{1/2}$	$\sqrt{2} - 8/7$
$-p^2 + 2p - 9/14$	$\sqrt{1/2} \le p \le \frac{(1+\sqrt{5/7})}{2}$	$\sqrt{1/2}$	$\sqrt{2} - 8/7$
$2p^2 - 2p + \frac{1}{7}$	Null set	Infeasible	
$p^2 - 1/2$	$\frac{(1+\sqrt{5/7})}{2} \le p \le 1$	$\frac{(1+\sqrt{5/7})}{2}$	$-1/14 + \sqrt{5/7}/2$

parallel to the three distinguished directions. As the radius increases, some sides of these hexagons exit the triangle. For instance, the ball around μ that contains optimal point ν^* its boundary is a non-regular hexagon containing region 5. The boundary in each of the other regions is obtained by drawing a line segment parallel to the opposite direction. For instance, in region 3, we draw a horizontal segment starting at ν^* until it hits region 2, and then we continue the boundary with a 60° turn to the right.

4   Parametric Linear Optimization over a Variety

A key step in Algorithm 2 is the repeated solution of linear optimization problems over appropriate subsets of the model $\mathcal{M}$. We now assume that $\mathcal{M}$ is an *algebraic variety* in $\Delta_{n-1} \subset \mathbb{R}^n$, i.e. $\mathcal{M}$ consists of the nonnegative real solutions of a system of polynomials $f_1, f_2, \ldots, f_k \in \mathbb{R}[x_1, \ldots, x_n]$. We tacitly assume that $f_1 = x_1 + \cdots + x_n - 1$ is the linear equation that cuts out the probability simplex. We write X for the complex algebraic variety in $\mathbb{C}^n$ defined by the same equations. Let $\bar{X}$ denote the closure of X in the complex projective space $\mathbb{P}^n$.

When computing the Wasserstein distance from μ to the model $\mathcal{M}$ with respect to d, we must minimize a linear function over $\mathcal{M}$ subject to nonnegativity constraints that specify $(\mu \times \Delta_{n-1}) \cap \sigma$. Here σ runs over all maximal simplices in the triangulation Σ_d of $\Delta_{n-1} \times \Delta_{n-1}$. Let us assume for simplicity that the minimum is attained at a smooth point of X that is in the relative interior of $(\mu \times \Delta_{n-1}) \cap \sigma$. The case when this hypothesis is violated can be modelled by adding additional linear constraints $f_i = 0$. We can phrase our problem as a parametric optimization problem:

$$\text{minimize } c_1 x_1 + \cdots + c_n x_n \text{ subject to } x \in \mathcal{M} = X \cap \Delta_{n-1}. \qquad (11)$$

Here $c_1, \ldots, c_n$ are parameters. In our applications, these c_i will be functions in the entries d_{ij} of the metric d and in the coordinates μ_k of the given point $\mu \in \Delta_{n-1}$. But, for now, let us treat the c_i as unknowns. The optimal value of the problem (11) is a function in these unknowns:

$$c_0^* = c_0^*(c_1, \ldots, c_n).$$

By [13, Section 3], the *optimal value function* $c_0^* : \mathbb{R}^n \to \mathbb{R}$ is an algebraic function in the n parameters $c_1, \ldots, c_n$. This means that there exists a polynomial $\Phi(c_0, c_1, \ldots, c_n)$ in $n+1$ variables such that $\Phi(c_0^*, c_1, \ldots, c_n)=0$. The degree of Φ in its first argument c_0 measures the algebraic complexity of our optimization problem (11). We call this number the *Wasserstein degree* of our model $\mathcal{M}$. We shall describe the Wasserstein degree geometrically and offer some bounds.

Following [13, Section 3], we consider the projective variety $\bar{X}^*$ that is dual to the variety $\bar{X}$. The dual variety $\bar{X}^*$ lives in the dual projective space $\mathbb{P}^n$, and it parametrizes hyperplanes in the ambient projective space of $\bar{X}$ that are tangent to $\bar{X}$. This dual variety $\bar{X}^*$ is typically a hypersurface, regardless of what the codimension of X is. In particular, it is a hypersurface when X is compact in $\mathbb{R}^n$. If X is irreducible then the hypersurface $\bar{X}^*$ is defined by a unique (up to scaling) irreducible homogeneous polynomial in $n+1$ unknowns $c_0, c_1, \ldots, c_n$. The degree of this hypersurface is the degree of $\bar{X}^*$. The following result is a direct consequence of [13, Theorem 3.2].

Theorem 2. *The polynomial $\Phi(-c_0, c_1, \ldots, c_n)$ is the defining equation of the hypersurface $\bar{X}^*$ that is dual to the projective variety $\bar{X}$ that represents the model $\mathcal{M}$ in Δ_{n-1}. Hence the Wasserstein degree of $\mathcal{M}$ is the degree of Φ in its first argument. This is generically equal to the degree of $\bar{X}^*$.*

For many natural classes of varieties $\bar{X}$, there are known formulas for the degree of the dual $\bar{X}^*$. This includes general complete intersections and determinantal varieties. The case of a hypersurface appears in [13, Example 2.7]. It serves as an illustration of our algebraic view on the problem (11).

Corollary 1. *Suppose that the model $\mathcal{M}$ is a hypersurface, namely, it is the zero set in the simplex Δ_{n-1} of a general polynomial of degree m. Then the Wasserstein degree of $\mathcal{M}$ equals $m(m-1)^{n-2}$.*

For instance, we have $n = m = 2$ for the Hardy-Weinberg curve (9), so this has Wasserstein degree 2. This reflects the fact that the optimal value $\sqrt{2} - 8/7$ is an algebraic number of degree 2.

Example 4. If $\mathcal{M}$ is a general curve of degree 3 in the triangle Δ_2 then its Wasserstein degree equals 6. Such an elliptic curve does not permit a rational parametrization, so we will have to consider (11) as a constrained optimization problem. For a concrete example consider the curve

$$x_1^3 + x_2^3 + x_3^3 = 4x_1 x_2 x_3.$$

Let c_0^* be the minimum of $c_1 x_1 + c_2 x_2 + c_3 x_3$ over this curve in Δ_2. This is an algebraic function of degree 6. Its minimal polynomial $\Phi(-c_0, c_1, c_2, c_3)$ is a homogeneous sextic. Namely, we have

$$\Phi = c_0^6 + (2c_1 + 2c_2 + 2c_3)c_0^5 - (65c_1^2 - 70c_1c_2 - 70c_1c_3 + 65c_2^2 - 70c_2c_3 + 65c_3^2)c_0^4$$

$$+ (208c_1^3 - 442c_1^2c_2 - 442c_1^2c_3 - 442c_1c_2^2 + 2048c_1c_2c_3 - 442c_1c_3^2 + 208c_2^3 - 442c_2^2c_3$$

$$- 442c_2c_3^2 + 208c_3^3)c_0^3$$

$$- (117c_1^4 - 546c_1^3c_2 - 546c_1^3c_3 + 1994c_1^2c_2^2 - 1024c_1^2c_2c_3 + 1994c_1^2c_3^2 - 546c_1c_2^3 - 1024c_1c_2^2c_3$$

$$- 1024c_1c_2c_3^2 - 546c_1c_3^3 + 117c_2^4 - 546c_2^3c_3 + 1994c_2^2c_3^2 - 546c_2c_3^3 + 117c_3^4)c_0^2$$

$$- (162c_1^5 - 288c_1^4c_2 - 288c_1^4c_3 + 606c_1^3c_2^2 - 1152c_1^3c_2c_3 + 606c_1^3c_3^2 + 606c_1^2c_3^3 + 352c_1^2c_2^2c_3$$

$$+ 352c_1^2c_2c_3^2 + 606c_1^2c_3^3$$

$$- 288c_1c_2^4 - 1152c_1c_2^3c_3 + 352c_1c_2^2c_3^2 - 1152c_1c_2c_3^3 - 288c_1c_3^4 + 162c_2^5 - 288c_2^4c_3 + 606c_2^3c_3^2$$

$$+ 606c_2^2c_3^3 - 288c_2c_3^4$$

$$+ 162c_3^5)c_0 - 27c_1^6 + 288c_1^4c_2c_3 - 202c_1^3c_2^3 - 202c_1^3c_3^3 - 176c_1^2c_2^2c_3^2 + 288c_1c_2^4c_3 + 288c_1c_2c_3^4$$

$$- 27c_2^6 - 202c_2^3c_3^3 - 27c_3^6.$$

For any given c_1, c_2, c_3, the optimal value is obtained by solving $\Phi = 0$ for c_0.

As we said earlier, in our application in Step 5 of Algorithm 2, the c_i depend on the matrix d and the distribution μ. We can thus consider the function that measures the Wasserstein distance:

$$\mathbb{R}^{n^2} \times \Delta_{n-1} \to \mathbb{R}, \quad (d, \mu) \mapsto c_0^*(d, \mu) = W_d(\mu, \mathcal{M}).$$

Our discussion establishes the following result about this function which depends only on $\mathcal{M}$.

Corollary 2. *The Wasserstein distance is a piecewise algebraic function of d and μ. Each piece is an algebraic function whose degree is bounded above by the degree of the hypersurface dual to $\mathcal{M}$.*

5   The Wasserstein Estimator of an Independence Model

In this section we present our solution to the problem that started this project. The task is to compute the Wasserstein estimator for the independence model on two binary random variables. Here $n = 4$ and $\mathcal{M}$ is the variety of 2×2 matrices of rank 1. This has the parametric representation

$$\begin{pmatrix} x_1 & x_2 \\ x_3 & x_4 \end{pmatrix} = \begin{pmatrix} pq & p(1-q) \\ (1-p)q & (1-p)(1-q) \end{pmatrix}, \text{ where } (p, q) \in [0, 1]^2.$$

Equivalently, $\mathcal{M}$ is the quadratic surface $\{x_1x_4 = x_2x_3\}$ in the tetrahedron Δ_3.

Our underlying metric space Ω is the square $\{0, 1\}^2$ with its Hamming distance. We identify Ω with the set $[4] = \{1, 2, 3, 4\}$ as indicated above. The ground metric is represented by the matrix

$$d = \begin{array}{c} \begin{array}{cccc} 1 & 2 & 3 & 4 \end{array} \\ \begin{bmatrix} 0 & 1 & 1 & 2 \\ 1 & 0 & 2 & 1 \\ 1 & 2 & 0 & 1 \\ 2 & 1 & 1 & 0 \end{bmatrix} \begin{array}{c} 1 \\ 2 \\ 3 \\ 4 \end{array} \end{array}$$

Given two points μ, ν in Δ_3, the transportation polytope $\Pi(\mu, \nu)$ consists of all nonnegative 4×4 matrices π with row sums ν and column sums μ. It usually is simple and has dimension 9. The Wasserstein distance between the two distributions equals $W(\mu, \nu) = \min_{\pi \in \Pi(\mu, \nu)} \sum_{1 \le i, j \le 4} d_{ij} \pi_{ij}$.

What we are interested in is the minimum Wasserstein distance from μ to any point ν in the independence model $\mathcal{M}$. This parametric linear optimization problem can be described as follows:

$$\begin{bmatrix} \pi_{11} & \pi_{12} & \pi_{13} & \pi_{14} \\ \pi_{21} & \pi_{22} & \pi_{23} & \pi_{24} \\ \pi_{31} & \pi_{32} & \pi_{33} & \pi_{34} \\ \pi_{41} & \pi_{42} & \pi_{43} & \pi_{44} \end{bmatrix} \begin{matrix} pq \\ p(1-q) \\ (1-p)q \\ (1-p)(1-q) \end{matrix}$$
$$\begin{matrix} \mu_1 & \mu_2 & \mu_3 & \mu_4 \end{matrix}$$

Here the marginal $\mu = (\mu_1, \mu_2, \mu_3, \mu_4)$ is fixed. The model $\mathcal{M}$ is parametrized by the points (p, q) in the square $[0, 1]^2$. The Wasserstein distance between μ and $\nu = \nu(p, q)$ is a continuous function on that square. The minimum value of that function is the desired Wasserstein distance $W(\mu, \mathcal{M})$.

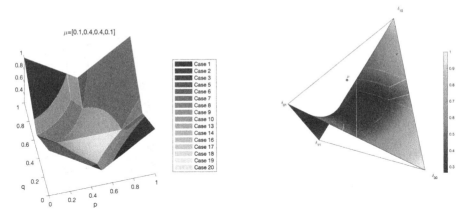

Fig. 2. Left: The graph of the distance function $[0, 1]^2 \rightarrow \mathbb{R}$, $(p, q) \mapsto W(\mu, \nu(p, q))$ for $\mu = \frac{1}{10}(1, 4, 4, 1)$. Right: The independence model $\mathcal{M}$ inside the tetrahedron Δ_3. Color corresponds to the Wasserstein distance to the target distribution μ, shown as a thick dot. The function has two global minimizers over $\mathcal{M}$. (Color figure online)

Our task is to minimize the function in Fig. 2 over the square. We see that this function is piecewise algebraic (cf. Corollary 2). Each piece is either a linear function or a quadratic function. The case distinction arises from the induced polyhedral subdivision of the 6-dimensional polytope $\Delta_3 \times \Delta_3$. This subdivision is not a triangulation, but, following Step 2 in Algorithm 2, we replace it with a nearby triangulation. That triangulation has 20 maximal simplices, as seen in (8). These are the 20 cases in Fig. 2. The graph of our function is color-coded according to these cases.

The triangulation of $\Delta_3 \times \Delta_3$ restricts to a mixed subdivision of the tetrahedron $\mu \times \Delta_3$. That subdivision consists of $20 = 4 + 12 + 4$ cells, namely 4 tetrahedra, 12 triangular prisms, and 4 parallelepipeds. After removing the 4 tetrahedra, which touch the vertices of $\mu \times \Delta_3$, we obtain a *truncated tetrahedron* which is subdivided into 16 cells. Such a subdivision is shown in Fig. 3.

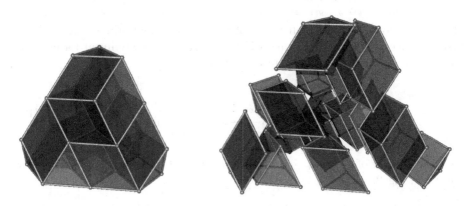

Fig. 3. A mixed subdivision of a truncated tetrahedron into $16 = 12 + 4$ cells.

The restriction of the mixed subdivision of $\mu \times \Delta_3$ divides our model $\mathcal{M}$ into regions. On each of these regions, the Wasserstein distance function $\nu \mapsto W(\mu, \nu)$ is given by a linear functional, as explained in Step 4 of Algorithm 2. The surface and this function are depicted in Fig. 2 (right).

The two images in Fig. 2 convey the same information. The piecewise linear function on the quadratic surface in Δ_3 restricts to a piecewise quadratic function on the square $[0, 1]^2$ under the parametrization of the surface. However, the color coding in the two diagrams is different. The colors in the left image in Fig. 2 show the pieces, while the right one displays a heat map. Namely, the colors here represent values of the function $\mathcal{M} \to \mathbb{R}, \nu \mapsto W(\mu, \nu)$. The two darkest points attain the minimum value $W(\mu, \mathcal{M})$. The white curve segments on the surface $\mathcal{M}$ are the boundaries between the various pieces. Each piece is the intersection of $\mathcal{M}$ with one of the polytopes in the mixed subdivision in Fig. 3.

We now discuss the computations that led to these results and pictures. The triangulation of $\Delta_3 \times \Delta_3$ and the resulting mixed subdivision of $\mu \times \Delta_3$ are computed in Steps 1–3 of Algorithm 2. These geometric objects are encoded algebraically, namely in the decomposition (5) of the ideal

$$\mathrm{in}_{d_\epsilon}(I_A) = \langle y_{11}, y_{12}, y_{14}, y_{31}, y_{32}, y_{34}, y_{41}, y_{42}, y_{44}\rangle \cap \langle y_{11}, y_{13}, y_{14}, y_{21}, y_{23}, y_{24}, y_{41}, y_{43}, y_{44}\rangle \cap$$
$$\langle y_{12}, y_{13}, y_{14}, y_{23}, y_{24}, y_{31}, y_{32}, y_{34}, y_{42}\rangle \cap \langle y_{12}, y_{13}, y_{14}, y_{23}, y_{24}, y_{32}, y_{34}, y_{42}, y_{43}\rangle \cap$$
$$\langle y_{12}, y_{13}, y_{14}, y_{21}, y_{23}, y_{24}, y_{32}, y_{34}, y_{43}\rangle \cap \langle y_{13}, y_{21}, y_{23}, y_{24}, y_{31}, y_{32}, y_{41}, y_{42}, y_{43}\rangle \cap$$
$$\langle y_{21}, y_{23}, y_{24}, y_{31}, y_{32}, y_{34}, y_{41}, y_{42}, y_{43}\rangle \cap \langle y_{12}, y_{21}, y_{23}, y_{31}, y_{32}, y_{34}, y_{41}, y_{42}, y_{43}\rangle \cap$$
$$\langle y_{13}, y_{14}, y_{21}, y_{23}, y_{24}, y_{32}, y_{34}, y_{41}, y_{43}\rangle \cap \langle y_{14}, y_{21}, y_{23}, y_{24}, y_{31}, y_{32}, y_{34}, y_{41}, y_{43}\rangle \cap$$
$$\langle y_{12}, y_{13}, y_{21}, y_{23}, y_{31}, y_{32}, y_{41}, y_{42}, y_{43}\rangle \cap \langle y_{12}, y_{13}, y_{14}, y_{21}, y_{23}, y_{32}, y_{41}, y_{42}, y_{43}\rangle \cap$$
$$\langle y_{11}, y_{13}, y_{14}, y_{21}, y_{23}, y_{24}, y_{32}, y_{41}, y_{43}\rangle \cap \langle y_{13}, y_{14}, y_{21}, y_{23}, y_{24}, y_{32}, y_{41}, y_{42}, y_{43}\rangle \cap$$
$$\langle y_{12}, y_{13}, y_{14}, y_{23}, y_{32}, y_{34}, y_{41}, y_{42}, y_{43}\rangle \cap \langle y_{12}, y_{14}, y_{23}, y_{31}, y_{32}, y_{34}, y_{41}, y_{42}, y_{43}\rangle \cap$$
$$\langle y_{12}, y_{13}, y_{14}, y_{21}, y_{23}, y_{24}, y_{31}, y_{32}, y_{34}\rangle \cap \langle y_{12}, y_{14}, y_{21}, y_{23}, y_{24}, y_{31}, y_{32}, y_{34}, y_{41}\rangle \cap$$
$$\langle y_{11}, y_{12}, y_{14}, y_{23}, y_{31}, y_{32}, y_{34}, y_{41}, y_{42}\rangle \cap \langle y_{12}, y_{14}, y_{23}, y_{24}, y_{31}, y_{32}, y_{34}, y_{41}, y_{42}\rangle.$$

Step 4 of Algorithm 2 translates each of these 20 minimal primes into a 4×4-matrix in the variety of that prime whose nonzero entries are linear forms in μ_i and ν_j. For instance, the second prime gives

$$\tilde{\pi}_\sigma = \begin{bmatrix} 0 & \nu_1 & 0 & 0 \\ 0 & \nu_2 & 0 & 0 \\ \mu_1 & \mu_2 - \nu_1 - \nu_2 - \nu_4 & \mu_3 & \mu_4 \\ 0 & \nu_4 & 0 & 0 \end{bmatrix}. \tag{12}$$

The dot product of d and $\tilde{\pi}_\sigma$ gives the Wasserstein distance on the piece labeled *Case 2* in Fig. 2:

$$d \cdot \tilde{\pi}_\sigma = \mu_1 + 2\mu_2 + \mu_4 - \nu_1 - 2\nu_2 - \nu_4 = \mu_2 - \mu_3 - \nu_2 + \nu_3 = \mu_2 - \mu_3 - p + q.$$

Hence *Case 2* is linear in p, q. The region in the square for this case is defined by the requirement that the entries of the matrix $\tilde{\pi}_\sigma$ are between 0 and 1. We only need to consider the entry in the third row and second column:

$$0 \leq \mu_2 - \nu_1 - \nu_2 - \nu_4 = (1-p)q + \mu_2 - 1 \leq 1.$$

In Fig. 2 the graph of $\mu_2 - \mu_3 - p + q$ on this region is labeled *Case 2*.

Table 1. Algebraic analysis of the Wasserstein distance function shown in Fig. 2.

Case	Objective Function	Feasible Region, $0 \leq * \leq 1$	Solution	Minimum Value	Subdivision
Quadratic pieces					
10	$2pq - p - q + \mu_2 + \mu_3$	$q - \mu_1 - \mu_3$ $\mu_3 - (1-p)q$ $(1-p)(1-q) - \mu_4$	$\left(\frac{\mu_1}{\mu_1+\mu_3}, \mu_1+\mu_3\right)$ $\left(\frac{\mu_2}{\mu_2+\mu_4}, \mu_1+\mu_3\right)$ $\left(\mu_1+\mu_2, \frac{\mu_3}{\mu_3+\mu_4}\right)$ $\left(1-\sqrt{\mu_3}, \sqrt{\mu_3}\right)$	$-\frac{\mu_1}{\mu_1+\mu_3} + \mu_1 + \mu_2$ $\frac{\mu_2}{\mu_2+\mu_4} - \mu_1 - \mu_2$ $\frac{\mu_3}{\mu_3+\mu_4} - \mu_1 - \mu_3$ $2\sqrt{\mu_3}(1-\sqrt{\mu_3}) - \mu_1 - \mu_4$	$\begin{bmatrix} * & * & * & 0 \\ 0 & * & 0 & 0 \\ 0 & 0 & * & 0 \\ 0 & * & 0 & * \end{bmatrix}$
18	$2pq - p - q + \mu_2 + \mu_3$	$pq - \mu_1$ $\mu_2 - p(1-q)$ $\mu_1 + \mu_3 - q$	$\left(\mu_1+\mu_2, \frac{\mu_1}{\mu_1+\mu_2}\right)$ $\left(\frac{\mu_2}{\mu_2+\mu_4}, \mu_1+\mu_3\right)$ $\left(\frac{\mu_1}{\mu_1+\mu_3}, \mu_1+\mu_3\right)$ $\left(\sqrt{\mu_2}, 1-\sqrt{\mu_2}\right)$	$-\frac{\mu_1}{\mu_1+\mu_2} + \mu_1 + \mu_3$ $\frac{\mu_2}{\mu_2+\mu_4} - \mu_1 - \mu_2$ $-\frac{\mu_1}{\mu_1+\mu_3} + \mu_1 + \mu_2$ $2\sqrt{\mu_2}(1-\sqrt{\mu_2}) - \mu_1 - \mu_4$	$\begin{bmatrix} * & 0 & * & 0 \\ 0 & 0 & 0 & 0 \\ 0 & 0 & 0 & 0 \\ 0 & * & * & * \end{bmatrix}$
12	$-2pq + p + q - \mu_2 - \mu_3$	$\mu_1 - pq$ $p(1-q) - \mu_2$ $q - \mu_1 - \mu_3$	$\left(\mu_1+\mu_2, \frac{\mu_1}{\mu_1+\mu_2}\right)$ $\left(\frac{\mu_1}{\mu_1+\mu_3}, \mu_1+\mu_3\right)$ $\left(\frac{\mu_2}{\mu_2+\mu_4}, \mu_1+\mu_3\right)$ $\left(\sqrt{\mu_1}, \sqrt{\mu_1}\right)$	$\frac{\mu_1}{\mu_1+\mu_2} - \mu_1 - \mu_3$ $\frac{\mu_1}{\mu_1+\mu_3} - \mu_1 - \mu_2$ $-\frac{\mu_2}{\mu_2+\mu_4} + \mu_1 + \mu_2$ $2\sqrt{\mu_1}(1-\sqrt{\mu_1}) - \mu_2 - \mu_3$	$\begin{bmatrix} * & 0 & 0 & 0 \\ 0 & * & 0 & * \\ * & 0 & * & * \\ 0 & 0 & 0 & * \end{bmatrix}$
15	$-2pq + p + q - \mu_2 - \mu_3$	$\mu_1 + \mu_3 - q$ $(1-p)q - \mu_3$ $\mu_4 - (1-p)(1-q)$	$\left(1-\sqrt{\mu_4}, 1-\sqrt{\mu_4}\right)$ $\left(\frac{\mu_1}{\mu_1+\mu_3}, \mu_1+\mu_3\right)$ $\left(\mu_1+\mu_2, \frac{\mu_3}{\mu_3+\mu_4}\right)$ $\left(\frac{\mu_2}{\mu_2+\mu_4}, \mu_1+\mu_3\right)$	$2\sqrt{\mu_4}(1-\sqrt{\mu_4}) - \mu_2 - \mu_3$ $\frac{\mu_1}{\mu_1+\mu_3} - \mu_1 - \mu_2$ $-\frac{\mu_3}{\mu_3+\mu_4} + \mu_1 + \mu_3$ $-\frac{\mu_2}{\mu_2+\mu_4} + \mu_1 + \mu_2$	$\begin{bmatrix} * & 0 & 0 & 0 \\ 0 & * & 0 & * \\ * & 0 & * & 0 \\ 0 & 0 & 0 & * \end{bmatrix}$

Table 2. Algebraic analysis of the Wasserstein distance function shown in Fig. 2.

Case	Objective Function	Feasible Region, $0 \leq * \leq 1$	Solution	Minimum Value	Subdivision
First affine piece					
2	$-p+q+\mu_2-\mu_3$	$(1-p)q+\mu_2-1$	$(1-\sqrt{1-\mu_2},\sqrt{1-\mu_2})$	$2\sqrt{1-\mu_2}+\mu_2-\mu_3-1$	$\begin{bmatrix} 0 & * & 0 & 0 \\ 0 & * & 0 & 0 \\ * & * & * & * \\ 0 & * & 0 & 0 \end{bmatrix}$
9	$-p+q+\mu_2-\mu_3$	$\mu_1+\mu_3-(1-p)q$ $q-\mu_1-\mu_3$ $(1-p)q-\mu_3$ $(1-p)(1-q)-\mu_4$	$\left(\frac{\mu_2}{\mu_2+\mu_4},\mu_1+\mu_3\right)$ $(1-\sqrt{\mu_3},\sqrt{\mu_3})$ $\left(\frac{\mu_1}{\mu_1+\mu_3},\mu_1+\mu_3\right)$ $\left(\mu_1+\mu_2,\frac{\mu_3}{\mu_3+\mu_4}\right)$	$-\frac{\mu_2}{\mu_2+\mu_4}+\mu_1+\mu_2$ $2\sqrt{\mu_3}+\mu_2-\mu_3-1$ $-\frac{\mu_1}{\mu_1+\mu_3}+\mu_1+\mu_2$ $\frac{\mu_3}{\mu_3+\mu_4}-\mu_1-\mu_3$	$\begin{bmatrix} * & * & 0 & 0 \\ 0 & * & 0 & 0 \\ 0 & * & 0 & * \\ 0 & * & 0 & * \end{bmatrix}$
13	$-p+q+\mu_2-\mu_3$	μ_2-p $(1-p)q-\mu_1-\mu_3$ $1-\mu_2-(1-p)q$	$(1-\sqrt{\mu_1+\mu_3},\sqrt{\mu_1+\mu_3})$ $\left(\mu_2,\frac{\mu_1+\mu_3}{\mu_1+\mu_3+\mu_4}\right)$	$2\sqrt{\mu_1+\mu_3}+\mu_2-\mu_3-1$ $\frac{\mu_1+\mu_3}{\mu_1+\mu_3+\mu_4}-\mu_3$	$\begin{bmatrix} 0 & * & 0 & 0 \\ 0 & * & 0 & 0 \\ * & 0 & * & * \\ 0 & * & 0 & * \end{bmatrix}$
14	$-p+q+\mu_2-\mu_3$	$p-\mu_2$ $\mu_2-p(1-q)$ $\mu_1+\mu_2-p$ $\mu_4-(1-p)(1-q)$	$\left(\mu_1+\mu_2,\frac{\mu_3}{\mu_3+\mu_4}\right)$ $\left(\mu_1+\mu_2,\frac{\mu_1}{\mu_1+\mu_2}\right)$ $\left(\frac{\mu_2}{\mu_2+\mu_4},\mu_1+\mu_3\right)$	$\frac{\mu_3}{\mu_3+\mu_4}-\mu_1-\mu_3$ $\frac{\mu_1}{\mu_1+\mu_2}-\mu_1-\mu_3$ $-\frac{\mu_2}{\mu_2+\mu_4}+\mu_1+\mu_2$	$\begin{bmatrix} * & * & 0 & 0 \\ 0 & 0 & 0 & * \\ 0 & * & * & * \\ 0 & 0 & 0 & * \end{bmatrix}$
Second affine piece					
1	$p-q-\mu_2+\mu_3$	$p(1-q)+\mu_3-1$	$(\sqrt{1-\mu_3},1-\sqrt{1-\mu_3})$	$2\sqrt{1-\mu_3}-\mu_2+\mu_3-1$	$\begin{bmatrix} 0 & 0 & * & 0 \\ * & * & * & * \\ 0 & 0 & * & 0 \\ 0 & 0 & * & 0 \end{bmatrix}$
16	$p-q-\mu_2+\mu_3$	$q-\mu_3$ $\mu_3-(1-p)q$ $\mu_1+\mu_3-q$ $\mu_4-(1-p)(1-q)$	$\left(\mu_1+\mu_2,\frac{\mu_3}{\mu_3+\mu_4}\right)$ $\left(\frac{\mu_1}{\mu_1+\mu_3},\mu_1+\mu_3\right)$ $\left(\frac{\mu_2}{\mu_2+\mu_4},\mu_1+\mu_3\right)$	$-\frac{\mu_3}{\mu_3+\mu_4}+\mu_1+\mu_3$ $\frac{\mu_1}{\mu_1+\mu_3}-\mu_1-\mu_2$ $\frac{\mu_2}{\mu_2+\mu_4}-\mu_1-\mu_2$	$\begin{bmatrix} * & 0 & * & 0 \\ * & * & 0 & 0 \\ 0 & 0 & * & 0 \\ 0 & 0 & 0 & * \end{bmatrix}$
19	$p-q-\mu_2+\mu_3$	$p(1-q)-\mu_1-\mu_2$ μ_3-q $1-\mu_3-p(1-q)$	$(\sqrt{\mu_1+\mu_2},1-\sqrt{\mu_1+\mu_2})$ $\left(\frac{\mu_1+\mu_2}{\mu_1+\mu_2+\mu_4},\mu_3\right)$	$2\sqrt{\mu_1+\mu_2}-\mu_2+\mu_3-1$ $\frac{\mu_1+\mu_2}{\mu_1+\mu_2+\mu_4}-\mu_2$	$\begin{bmatrix} 0 & 0 & * & 0 \\ * & * & 0 & * \\ 0 & 0 & * & 0 \\ 0 & 0 & * & * \end{bmatrix}$
20	$p-q-\mu_2+\mu_3$	$\mu_1+\mu_2-p(1-q)$ $p(1-q)-\mu_2$ $p-\mu_1-\mu_2$ $(1-p)(1-q)-\mu_4$	$\left(\mu_1+\mu_2,\frac{\mu_3}{\mu_3+\mu_4}\right)$ $(\sqrt{\mu_2},1-\sqrt{\mu_2})$ $\left(\mu_1+\mu_2,\frac{\mu_1}{\mu_1+\mu_2}\right)$ $\left(\frac{\mu_2}{\mu_2+\mu_4},\mu_1+\mu_3\right)$	$-\frac{\mu_3}{\mu_3+\mu_4}+\mu_1+\mu_3$ $2\sqrt{\mu_2}-\mu_2+\mu_3-1$ $-\frac{\mu_1}{\mu_1+\mu_2}+\mu_1+\mu_3$ $\frac{\mu_2}{\mu_2+\mu_4}-\mu_1-\mu_2$	$\begin{bmatrix} * & 0 & * & 0 \\ * & * & 0 & 0 \\ 0 & 0 & * & 0 \\ 0 & 0 & * & * \end{bmatrix}$
Third affine piece					
3	$-p-q+\mu_1-\mu_4+1$	$p(1-q)-\mu_2$ $\mu_1+\mu_2-p$ $\mu_3-(1-p)q$	$\left(\mu_1+\mu_2,\frac{\mu_1}{\mu_1+\mu_2}\right)$ $\left(\mu_1+\mu_2,\frac{\mu_3}{\mu_3+\mu_4}\right)$	$-\frac{\mu_1}{\mu_1+\mu_2}+\mu_1+\mu_3$ $-\frac{\mu_3}{\mu_3+\mu_4}+\mu_1+\mu_3$	$\begin{bmatrix} * & 0 & 0 & 0 \\ * & * & 0 & 0 \\ 0 & 0 & 0 & 0 \\ 0 & * & 0 & * \end{bmatrix}$
4	$-p-q+\mu_1-\mu_4+1$	$p(1-q)-\mu_2$ $(1-p)q-\mu_3$ $(1-p)(1-q)-\mu_4$	$\left(\gamma^+,1-\frac{\mu_2}{\gamma^+}\right)$ $(1-\sqrt{\mu_4},1-\sqrt{\mu_4})$ $\left(\frac{\mu_2}{\mu_2+\mu_4},\mu_1+\mu_3\right)$ $\left(\mu_1+\mu_2,\frac{\mu_3}{\mu_3+\mu_4}\right)$	$-\gamma^++\frac{\mu_2}{\gamma^+}+\mu_1-\mu_4$ $2\sqrt{\mu_4}+\mu_1-\mu_4-1$ $-\frac{\mu_2}{\mu_2+\mu_4}+\mu_1+\mu_2$ $-\frac{\mu_3}{\mu_3+\mu_4}+\mu_1+\mu_3$	$\begin{bmatrix} * & 0 & 0 & 0 \\ * & * & 0 & 0 \\ * & 0 & * & 0 \\ * & 0 & 0 & * \end{bmatrix}$
5	$-p-q+\mu_1-\mu_4+1$	$(1-p)q-\mu_3$ $\mu_1+\mu_3-q$ $\mu_2-p(1-q)$	$\left(\frac{\mu_1}{\mu_1+\mu_3},\mu_1+\mu_3\right)$ $\left(\frac{\mu_2}{\mu_2+\mu_4},\mu_1+\mu_3\right)$	$-\frac{\mu_1}{\mu_1+\mu_3}+\mu_1+\mu_2$ $-\frac{\mu_2}{\mu_2+\mu_4}+\mu_1+\mu_2$	$\begin{bmatrix} 0 & * & 0 & 0 \\ 0 & * & 0 & 0 \\ * & * & 0 & * \end{bmatrix}$
17	$-p-q+\mu_1-\mu_4+1$	μ_1-pq $\mu_2-p(1-q)$ $\mu_3-(1-p)q$	$\left(\gamma^-,\frac{\mu_3}{1-\gamma^-}\right)$ $\left(\frac{\mu_1}{\mu_1+\mu_3},\mu_1+\mu_3\right)$ $\left(\mu_1+\mu_2,\frac{\mu_1}{\mu_1+\mu_2}\right)$	$-\gamma^--\frac{\mu_3}{1-\gamma^-}+\mu_1-\mu_4+1$ $-\frac{\mu_1}{\mu_1+\mu_3}+\mu_1+\mu_2$ $-\frac{\mu_1}{\mu_1+\mu_2}+\mu_1+\mu_3$	$\begin{bmatrix} * & 0 & 0 & 0 \\ 0 & 0 & 0 & 0 \\ 0 & 0 & * & 0 \\ * & * & * & * \end{bmatrix}$
Fourth affine piece					
6	$p+q-\mu_1+\mu_4-1$	$\mu_2-p(1-q)$ $p-\mu_1-\mu_2$ $(1-p)q-\mu_3$	$\left(\mu_1+\mu_2,\frac{\mu_3}{\mu_3+\mu_4}\right)$ $\left(\mu_1+\mu_2,\frac{\mu_1}{\mu_1+\mu_2}\right)$	$\frac{\mu_3}{\mu_3+\mu_4}-\mu_1-\mu_3$ $\frac{\mu_1}{\mu_1+\mu_2}-\mu_1-\mu_3$	$\begin{bmatrix} * & * & 0 & * \\ 0 & 0 & * & 0 \\ 0 & 0 & 0 & 0 \\ 0 & 0 & 0 & * \end{bmatrix}$
7	$p+q-\mu_1+\mu_4-1$	$\mu_2-p(1-q)$ $\mu_3-(1-p)q$ $\mu_4-(1-p)(1-q)$	$\left(\gamma^+,\frac{\mu_3}{1-\gamma^+}\right)$ $\left(\frac{\mu_2}{\mu_2+\mu_4},\mu_1+\mu_3\right)$ $\left(\mu_1+\mu_2,\frac{\mu_3}{\mu_3+\mu_4}\right)$	$\gamma^++\frac{\mu_3}{1-\gamma^+}-\mu_1+\mu_4-1$ $\frac{\mu_2}{\mu_2+\mu_4}-\mu_1-\mu_2$ $\frac{\mu_3}{\mu_3+\mu_4}-\mu_1-\mu_3$	$\begin{bmatrix} * & * & * & * \\ 0 & * & 0 & 0 \\ 0 & 0 & * & 0 \\ 0 & 0 & 0 & * \end{bmatrix}$
8	$p+q-\mu_1+\mu_4-1$	$\mu_3-(1-p)q$ $q-\mu_1-\mu_3$ $p(1-q)-\mu_2$	$\left(\frac{\mu_2}{\mu_2+\mu_4},\mu_1+\mu_3\right)$ $\left(\frac{\mu_1}{\mu_1+\mu_3},\mu_1+\mu_3\right)$	$\frac{\mu_2}{\mu_2+\mu_4}-\mu_1-\mu_2$ $\frac{\mu_1}{\mu_1+\mu_3}-\mu_1-\mu_2$	$\begin{bmatrix} * & 0 & * & * \\ 0 & * & 0 & 0 \\ 0 & 0 & * & 0 \end{bmatrix}$
11	$p+q-\mu_1+\mu_4-1$	$pq-\mu_1$ $p(1-q)-\mu_2$ $(1-p)q-\mu_3$	$\left(\gamma^-,\frac{\mu_3}{1-\gamma^-}\right)$ $(\sqrt{\mu_1},\sqrt{\mu_1})$ $\left(\mu_1+\mu_2,\frac{\mu_1}{\mu_1+\mu_2}\right)$ $\left(\frac{\mu_1}{\mu_1+\mu_3},\mu_1+\mu_3\right)$	$\gamma^-+\frac{\mu_3}{1-\gamma^-}-\mu_1+\mu_4-1$ $2\sqrt{\mu_1}-\mu_1+\mu_4-1$ $\frac{\mu_1}{\mu_1+\mu_2}-\mu_1-\mu_3$ $\frac{\mu_1}{\mu_1+\mu_3}-\mu_1-\mu_2$	$\begin{bmatrix} * & 0 & 0 & 0 \\ 0 & * & 0 & 0 \\ 0 & 0 & * & 0 \end{bmatrix}$
$\gamma^+ := (1+m_2-m_3)/2+\sqrt{(1+m_2-m_3)^2/4-m_2}$ and $\gamma^- := (1+m_2-m_3)/2-\sqrt{(1+m_2-m_3)^2/4-m_2}$					

We analyze all 20 components of $\mathrm{in}_d(I_A)$ in this manner, and we record the result in the first two columns of Tables 1 and 2. The rightmost column gives the support of the corresponding vertex of the transportation polytope. For instance, the matrix in the first row of the table shows the support of $\tilde{\pi}_\sigma$ in (12).

The third column of the table contains all candidates for the optimal point ν^* expressed as an algebraic function in the four coordinates of μ. Each of the 20 cases has one or several candidate solutions, listed in the third and forth columns of the table. Which of the candidates is the actual solution can be determined in terms of further case distinctions on μ, which we omit in the table. The smallest solution among all cases for a given μ is the desired Wasserstein distance $W(\mu, \nu^*) = W(\mu, \mathcal{M})$. Note that these expressions involve a square root, so the Wasserstein degree of the independence surface $\mathcal{M}$ equals two, as predicted by Corollary 1.

6   Conclusion

In this paper, we developed mathematical foundations for computing the Wasserstein distance between a point and an algebraic variety in a probability simplex. Our next goal is to develop a practical algorithm that scales beyond toy problems. We also plan to answer the questions raised in the introduction, such as characterizing scenarios when the minimizer is unique.

Acknowledgments. GM has received funding from the European Research Council (ERC) under the European Union's Horizon 2020 research and innovation programme (grant no 757983).

References

1. Allman, E., et al.: Maximum likelihood estimation of the latent class model through model boundary decomposition. J. Algebraic Stat. **34**, 51–84 (2019)
2. Arjovsky, M., Chintala, S., Bottou, L.: Wasserstein GAN. arXiv:1701.07875
3. Bassetti, F., Bodini, A., Regazzini, E.: On minimum Kantorovich distance estimators. Stat. Probab. Lett. **76**, 1298–1302 (2006)
4. Bernton, E., Jacob, P., Gerber, M., Robert, C.: On parameter estimation with the Wasserstein distance. Inf. Infer. J. IMA **8**(4), 657–676 (2019)
5. Cuturi, M.: Sinkhorn distances: lightspeed computation of optimal transport. In: Advances in Neural Information Processing Systems, Proceedings NIPS 2013, pp. 2292–2300 (2013)
6. De Loera, J.A., Rambau, J., Santos, F.: Triangulations: Structures for Algorithms and Applications, Algorithms and Computation in Mathematics, vol. 25. Springer-Verlag, Heidelberg (2010). https://doi.org/10.1007/978-3-642-12971-1
7. Duarte, E., Marigliano, O., Sturmfels, B.: Discrete statistical models with rational maximum likelihood estimator. arXiv:1903.06110
8. Kulas, K., Joswig, M.: Tropical and ordinary convexity combined. Adv. Geom. **10**, 333–352 (2010)
9. Lasserre, J.: An Introduction to Polynomial and Semi-Algebraic Optimization, Texts in Applied Mathematics. Cambridge University Press, Cambridge (2015)

10. Miller, E., Sturmfels, B.: Combinatorial Commutative Algebra. Graduate Texts in Mathematics, vol. 227. Springer-Verlag, New York (2005). https://doi.org/10.1007/b138602

11. Pele, O., Werman, M.: Fast and robust earth mover's distances. In: 2009 IEEE 12th International Conference on Computer Vision, pp. 460–467, September 2009

12. Peyre, G., Cuturi, M.: Computational optimal transport. Found. Trends Mach. Learn. **11**, 355–607 (2019)

13. Rostalski, P., Sturmfels, B.: Dualities in convex algebraic geometry. Rendiconti di Matematica **30**, 285–327 (2010)

14. Seigal, A., Montúfar, G.: Mixtures and products in two graphical models. J. Alg. Stat. **9**, 1–20 (2018)

15. Sturmfels, B.: Gröbner Bases and Convex Polytopes. University Lecture Series, vol. 8. American Mathematical Society, Providence (1996)

16. Sullivant, S.: Algebraic Statistics. Graduate Studies in Math. American Mathematical Society, Providence (2018)

17. Villani, C.: Optimal Transport: Old and New. Grundlehren Series, vol. 338. Springer Verlag, Heidelberg (2008). https://doi.org/10.1007/978-3-540-71050-9

18. Weed, J., Bach, F.: Sharp asymptotic and finite-sample rates of convergence of empirical measures in Wasserstein distance. arXiv:1707.00087

SFV-CNN: Deep Text Sentiment Classification with Scenario Feature Representation

Haoliang Zhang[1,3]($\boxtimes$), Hongbo Xu[1], Jinqiao Shi[2], Tingwen Liu[1], and Jing Ya[1]($\boxtimes$)

[1] Institute of Information Engineering CAS, Beijing 100093, China
evercristee@163.com, yajing@iie.ac.cn
[2] Beijing University of Posts and Telecommunications, Beijing, China
[3] School of Cyber Security, University of Chinese Academy of Sciences, Beijing, China

Abstract. In this paper, we present a deep learning approach to represent the scenario-related features of a sentence for text classification, and also demonstrate an interesting application which shows the nearest scenarios for a sentence. In order to improve the performance of text classification, it is necessary to make them be aware of the scenario switching at the background of the texts. We propose a CNN based sentiment analysis model named SFV-CNN for sentence classification. The proposed model can be improved by assigning suitable window for each scenario corpus in scenario word embedding training. Our experiments demonstrate that SFV-CNN brings an improvement in accuracy and also shows more obvious advantages when test on datasets across scenarios.

Keywords: Text classification · Sentiment analysis · Deep learning · Natural language processing · Word embedding

1 Introduction

Word embedding is used to transform text data from abstract symbols to a set of word vectors that contain a certain implicit semantic information [1, 2]. With the help of this method, it becomes possible for us to make use of CNN model in training text data just like processing images. Kim [3] introduced CNN method into the field of natural language processing, proposed the TextCNN model and achieved remarkable experimental results. Zhang [4] made full experiments on sensitivity of TextCNN model. Zhang [5] and Marinho [6] researched on text classification based on CNN Method at Character Level. Johnson [13] and Conneau [14] have discussed the effect of increasing network depth on improving the text categorization ability of CNN network. And there were many excellent works focused on combining CNN with RNN, LSTM, Attention mechanism and other methods to explore their improving effect [7–12, 15–17]. It is also a practical way to improve CNN based text classification model by making the word embedding carry more semantic information [24, 25]. A large number of studies have proved that word embedding, as a text information representation method, can reflect some grammatical features to a certain extent, such as co-occurrence relation of words.

D. Slamanig et al. (Eds.): MACIS 2019, LNCS 11989, pp. 382–394, 2020.
https://doi.org/10.1007/978-3-030-43120-4_30

Therefore, we can conclude that word embedding plays an important role in enabling CNN model to capture the semantic features from texts directly just like extracting image features.

However, the word embeddings also have their inherent limitations that some important semantic information hidden in text can be intuitively perceived by human, but it is difficult to be captured and quantified by word embeddings. For example, in different scenarios, the same words may contain different semantics, and the similar semantics can also be expressed in different language styles. The changes of sentence style caused by the scenario switching can hardly be noticed by a single word embedding, because the differences among scenarios are invisible to it.

In the training process of word embedding, the grammatical and semantic relations among words are all mined within the limited range of the training corpus which is collected from a specified scenario such as wiki, news, etc. In other words, each word embedding can only be bound to a single Scenario and can only see inside of the binding scenario, so the word embedding is destined to be blind of the differences among Scenarios, no matter how big the corpus is and no matter how many different kinds of sentences in it. Some research work has begun to focus on studying the fusion effect of cross-domain word embeddings [19–21, 23–26] and transfer learning on multi-datasets [18, 22].

We believe that studying the features of sentence styles in specific scenarios is of great significance to improving text classifiers, and especially to enhancing text sentiment classification effectiveness. To achieve this goal, we propose a Scenario Feature Vector (SFV) model to quantify the scenario-related implicit expression features, demonstrate its ability to distinguish the scenario-related differences among texts. Finally, we use this model to improve the CNN based classifier and propose SFV-CNN model for sentiment classification. Experiments show that this method can bring about improvement in accuracy of text sentiment classification. And we find that our model has more outstanding performance in the task of sentiment classification on cross-domain data.

2   Method

2.1   Expressive Grammar Plane Concept

Sentence grammar is the basic way in which it is structured and in different scenarios texts are structured differently about the grammatical styles and expressive habits. For example, "I forgot to list the coffee." and "Coffee not listed." have similar meanings while the first sentence is probably a sentence in one's blog, and the second sentence is more like a short message on Twitter. Essentially, this formal difference reflects the implicit changes that take place at the background of grammar, which are influenced by many indescribable factors, such as the object of expression, the application environment and the interactive limits. We define the indescribable characteristics of this grammatical style and habit as Grammatical Expressive Feature (GEF).

To facilitate our discussion, we propose a concept of Expressive Grammar Plane (EGP) on which place we project the GEF of sentences. To demonstrate the concept, we might as well assume that the X-axis represents the degree of how it is relevant to some professional fields, and the Y-axis of the plane represents the degree of normalization of sentences. If we make the texts from several typical scenarios distribute in this plane, we may see the condition similar to Fig. 1(a) based on our intuitive experience.

First, we suppose that the texts from the same scenario will be centrally distributed in a certain specific region on EGP. This is mainly due to the usage of the same vocabulary and idioms in specific fields, as well as similar grammatical requirements and other constraints to the texts from the same Scenario. Therefore, we draw different regions to represent the centralized distribution of the text corpora from different scenarios and show their relative positions. These regions have not only overlapping areas, but also independent parts that make them can be distinguished.

Second, there should not be strict boundaries among the scenario text's regions on EGP, mainly because the commonly used sentences are frequently used in all scenarios, take "Thank you very much!" for example.

Last, in a certain period of time, the relative relationship among these corpora from different scenarios is relatively stable, because the emergence of new words and the change of expression habits are relatively slow processes.

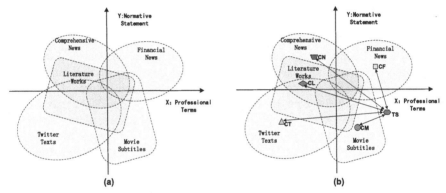

Fig. 1. (a) Expressive Grammar Plane. Texts from different scenarios are distributed on the this conceptual plane formed by X-axis and Y-axis. For demonstration, we suppose X-axis represents the usage of professional terms and Y-axis represents the normalization degree. (b) Position of Target Sentence (TS) can be evaluated by its distances to the grammatical gravity center (GGC) of regions corresponding to Scenario Corpora.

Based on the above description, we can conclude that although the sentences' distribution is very complex, if the scenario corpus is regarded as a whole, the grammatical features of scenario corpora can be steadily distinguished at the top level. We assume that there is a grammatical gravity center (GGC) for each region on EGP, which represents the overall GEF of scenario corpus. In this way, the grammatical differences of scenario corpora can be measured relatively by their GGC distances. Therefore, we can also regard EGP as a platform for grammatical relational computing.

2.2 Scenario Features Vectors

It is a hard mission to quantify and measure the grammatical features of text directly. For example, we can distinguish the GEF differences between the two sentences, "GitHub makes it easy to scale back on context switching." and "dude u totally poached my joke", that whose content is more professional and whose expression style is less formal, but we can hardly evaluate the degree of their differences accurately with numbers. While, the previous discussion of Expressive Grammar Plane (EGP) may provide a practicable inspiration for the solution.

If the GGC of each Scenario Text Set is regarded as an anchor point on EGP, shown in Fig. 1(b), then we can use the distance between the target sentence (TS) and the GGC of scenario corpus (CF, CN, CL, CT, CM) as one of the indicators of the target text's position on EGP, that is, the GEF of the target text. We arrange the target text's Scenario Distances in a certain order to form a vector $\{d_{TS,CF}, d_{TS,CM}, d_{TS,CT}, d_{TS,CN}, d_{TS,CL}\}$, and we define this vector as the Scenario Feature Vector (SFV) of the target text.

Scenario Similarity Computing Method. Scenario Feature Vectors (SFV) provides a framework for the quantitative representation of grammatical features. For the computation of SFV elements, we propose a Scenario Similarity Computing Method (SSCM) based on word embedding model.

Suppose S is a scenario such as twitter or wiki, C_s is the text corpus from S, $E_s \in R^d$ is the word embedding trained by C_s, we define E_s as Scenario Word Embedding (SWE). $W = \{w_1, w_2 \ldots w_n\}$ is the target text made up of words w_i ($i \in [1,n]$). Set w_i ($i \in [1,n]$) as central word and we use Context(w_i) to represent the context of w_i. Context(w_i) and w_i form the local text L_i

$$L_i = w_i \cup Context(w_i) \tag{1}$$

Given w_i and context(w_i), we calculate their co-occurrence probability y_i by making use of E_s, and we define y_i as Scenario Grammatical Similarity (SGS).

$$y_i = p(w_i|Context(w_i), E_s) \tag{2}$$

Suppose that Word Embedding contains the unique grammatical features of different scenarios, therefore, y_i can vary for the same L_i as compute with different E_s. The bigger y_i is, L_i is more similar to current scenario S, meanwhile the shorter Scenario Distance is on EGP.

We employ CBOW model based on Negtive Sampling to train Scenario Word Embeddings, by using word2vec toolkit. The training objective of the model is to maximize the g(w) function, and the expression is as follows:

$$g(w) = \sigma\left(x_w^T \theta^w\right) \prod_{u \in NEG(w)} \left[1 - \sigma\left(x_w^T \theta^u\right)\right] \tag{3}$$

Where $\sigma\left(x_w^T \theta^w\right)$ denotes the probability of w occurring as central word when given the context Context(w), and $\sigma\left(x_w^T \theta^u\right)$ denotes the probability that the central word is not w when the context is Context(w). The result of maximizing g(w) is that words conform to the grammatical habits of S achieve greater chances of appearing in the position of the central word, and contrarily minimizing the probability of words that do not conform. We quote the first half of g(x), and define it as the scenario similarity of the local text L, the expression is as follows:

$$y = p(w|Context(w), E_s), \sigma\left(x_w^T \theta_s^w\right) \tag{4}$$

where $\theta_s^w \in R^m$ (m is the size of the word vector) is the auxiliary vector for word w, x_w is the sum of word vectors of the words in Context(w).

$$x_w = \sum_{u \in context(w)} v(u) \tag{5}$$

where v(u) denotes the word vector of u, which is one of Context(w).

The target text $W = \{w_1, w_2 \ldots w_n\}$ derives its local texts $\{L_1, L_2 \ldots L_n\}$ as the input of the computation, therefore, we define the Scenarios Similarity of W as the average of $\{y_1, y_2 \ldots y_n\}$, represented as y.

$$y = \sum_{i=1}^{n} y_i / n \tag{6}$$

Usage of Scenario Feature Vector. We choose 5 typical corpora as Scenarios Corpus, including financial news (FN), movie subtitles (MS), twitter (TW), comprehensive news (NW) and literary works (LR), and use the CBOW model based on Negative Sampling to train the scenario word embeddings corresponding to the above five scenario corpora. Afterwards, we generate scenario feature vectors for the target text, as shown in Fig. 2.

Scenario Feature Vectors of Sentence

WORD	FN	MV	TW	NW	LR
stocks	0.6006	0.0878	0.2742	0.5641	0.1539
closed	0.3061	0.0585	0.1132	0.2984	0.1319
sharply	0.4960	0.0868	0.3393	0.3351	0.1079
lower	0.4450	0.0125	0.0946	0.3086	0.1151
on	0.0276	0.0309	0.0252	0.0282	0.0604
tuesday	0.0354	0.0335	0.0572	0.0706	0.0000
falling	0.2489	0.0041	0.0531	0.1268	0.0798
for	0.0348	0.0507	0.0334	0.0401	0.0270
a	0.0596	0.0523	0.0420	0.0441	0.0628
second	0.2522	0.1742	0.1806	0.1804	0.4134
day	0.3676	0.2264	0.2341	0.2717	0.3206
SENTENCE :	0.2613	0.0743	0.1315	0.2062	0.1339

WORD	FN	MS	TW	NW	LR
stocks	0.6006	0.0878	0.2742	0.5641	0.1539
closed	0.3061	0.0585	0.1132	0.2984	0.1319
sharply	0.4960	0.0868	0.3393	0.3351	0.1079
Lower	0.4450	0.0125	0.0946	0.3086	0.1151
on	0.0276	0.0309	0.0252	0.0282	0.0604
tuesday	0.0354	0.0335	0.0572	0.0706	0.0000
falling	0.2489	0.0041	0.0531	0.1268	0.0798
for	0.0348	0.0507	0.0334	0.0401	0.0270
a	0.0596	0.0523	0.0420	0.0441	0.0628
second	0.2522	0.1742	0.1806	0.1804	0.4134
day	0.3676	0.2264	0.2341	0.2717	0.3206
SENTENCE	0.2613	0.0743	0.1315	0.2062	0.1339

Scenario Feature Vectors of Sentence

WORD	FN	MV	TW	NW	LR
on	0.1299	0.0802	0.0731	0.0733	0.0722
the	0.0668	0.0727	0.0612	0.0432	0.0608
sky	0.1494	0.0700	0.2861	0.3278	0.3269
a	0.0475	0.0177	0.0187	0.0140	0.0183
brooding	0.0000	0.0433	0.0645	0.1535	0.1198
gloom	0.0881	0.0324	0.0259	0.1733	0.4378
in	0.0489	0.0423	0.0307	0.0179	0.0292
sunshine	0.2595	0.0383	0.0890	0.3292	0.4336
a	0.1061	0.0391	0.0436	0.0302	0.0408
lurid	0.0000	0.1053	0.0450	0.3102	0.6691
glare	0.2839	0.1315	0.1255	0.2170	0.7085
SENTENCE :	0.1073	0.0612	0.0785	0.1536	0.2652

WORD	FN	MS	TW	NW	LR
on	0.1299	0.0802	0.0731	0.0733	0.0722
the	0.0668	0.0727	0.0612	0.0432	0.0608
sky	0.1494	0.0700	0.2861	0.3278	0.3269
a	0.0475	0.0177	0.0187	0.0140	0.0183
brooding	0	0.0433	0.0645	0.1535	0.1198
gloom	0.0881	0.0324	0.0259	0.1733	0.4378
in	0.0489	0.0423	0.0307	0.0179	0.0292
sunshine	0.2595	0.0383	0.0890	0.3292	0.4336
a	0.1061	0.0391	0.0436	0.0302	0.0408
lurid	0	0.1053	0.045	0.3102	0.6691
glare	0.2839	0.1315	0.1255	0.2170	0.7085
SENTENCE	0.1073	0.0612	0.0785	0.1536	0.2652

Fig. 2. An application of SFV model for finding the nearest scenario for a sentence. The left column is the interface of the application, the right column is the heat map for the results. First sentence is judged to be nearest to financial news scenario with a similarity of 0.2613, and the second phrase is nearest to literature scenario with a similarity of 0.2652.

According to Fig. 2, we can find some interesting effects of the SFV based application, we input a short phrase such as "stocks closed sharply lower on Tuesday falling for a second day" in the command line, and get a Sentence SFV {0.2613, 0.0743, 0.1315, 0.2062, 0.1339}, which means the sentence has the highest similarity of 0.2613 to FN (financial) scenario, followed by NW (news) scenarios with 0.2062, and the local text of "stock" has the highest similarity value 0.6006. This result is consistent with our subjective judgment, because the discussion of stock-related content should be closer to financial scenarios according to our common sense. Think, people should seldom discuss stock related contents in life without the relevant scenarios of the financial business.

In addition, we input a short sentence "on the sky a brooding gloom in sun-shine a lurid glare", and get a SFV result {0.1073, 0.0612, 0.0785, 0.1536, 0.2652}, according to this result, LR (Literature) is the nearest scenario with a similarity of 0.2652, and the local text corresponding to "Lurid" and "glare" contributes most to the result in sentences. This result also conforms to our experience and common sense well.

2.3 Text Sentiment Classification

The expressive style of sentences in practice is flexible, it can vary with the change of scenarios diversely. In order to better understand the text semantics, people need to grasp the context information related to the scenario. Similarly, the machine also needs to know the context information related to the scenario. For this purpose, we propose a CNN based text classification model enhanced by Scenario Feature Vector (SFV-CNN), and apply it to text sentiment classification tasks. In this way, the text categorization model can learn the text features at a deeper level under the condition of perceiving the context features. The structure of the model is shown in Fig. 3.

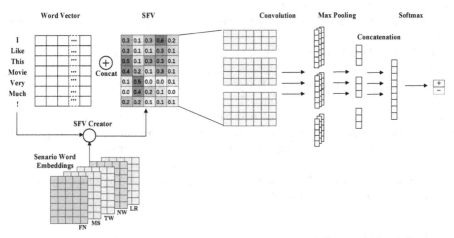

Fig. 3. Structure of SFV-CNN model. The input of the model consists of two parts, the first part is the word vectors queried from the pretrained word embeddings and the second part is the SFV vectors calculated through the SFV Creator module by use of Scenario Word Embeddings.

We use the pretrained word vectors from word2vec as the generic word embedding, and train another five Scenario Word Embeddings(SWE) separately with 5 typical scenario corpora extracted from open datasets, listed as {FN, MS, TW, NW, LR}. SFV vectors for input sentences are generated at the run time through the SFV Creator module. A new SFV vector will be generated for the same word as its context changes, and this characteristic is the key point that make the classifier understand the switching of scenario. Our experiments show that the classification ability of convolution neural network has been improved.

3   Experiments

3.1   Data Set

We used a total of 8 public datasets and the word2vec vectors. Among them, 5 datasets are used for unsupervised training of Scenario Word Embedding (SWE), 1 for supervised training of CNN based Sentiment Classification Model, and 2 for validating the classification effect across datasets.

SWE Training Datasets: (1) **Financial News**: United States Financial News articles from news publishers such as Bloomberg.com, CNBC.com, reuters.com, wsj.com, fortune.com [29]. (2) **OpenSubtitles**: This is a collection of English dialogue text extracted from the subtitles of movies. [30, 31] (3) **Twitter Corpus**: The corpus of Twitter (700k lines) which is shared by [32], odd lines are tweets, even lines are corresponding response tweets. (4) **Comprehensive News**: We used the News Crawl: articles from 2012 corpus, monolingual language training data of English which is provided by WMT 2014 [33]. The article texts are extracted from various online news publications. (5) **Literature Corpus**: Classic Literature in ASCII, which contains many English literary works stored in ASCII text files, is an open dataset from Kaggle [34]. Each book is stored in a single ASCII text file. There are fictional and non-fictional 2 kinds of works, and we used the fictional files for training.

MR: Sentence polarity dataset which we used for sentiment classifier model training [27], and it is also used by [1, 4].

Pretrained Word Vectors: Commonly used word2vec vectors trained on Google News dataset which is about 100 billion words [28].

Cross-Validation Datasets: (1) **Kaggle Movie Review**: This dataset Contains the sentimental sentences from the Rotten Tomatoes dataset and is provided by Kaggle net [35]. We extracted the sentences with Sentiment Tag 0 and 1 as negative, 3 and 4 as positive, and used the modified dataset for validating our model's adaptability across datasets. (2) **Sentiment140**: It is a popular dataset used for sentiment analysis of tweet [36]. There are three types of polarity label in the dataset (negative, neutral and positive), and we extracted the negative and positive sentences out to form a new dataset for validating our model's adaptability across scenarios.

3.2   Experimental Setup

For we mainly aim at verify the promoting effectiveness brought by SFV on the CNN based text classification models, we didn't pay more attention to fine-tune the parameter settings of CNN. The general settings of convolutional neural network of each investigated model in our experiments are as shown in Table 1.

The specific settings of each model are described as follows.

CNN-Rand: The model uses random numbers to initialize word vectors, which will be optimized during the model training process. The dimension of word vectors is set to 300, same as the other models.

Table 1. General settings for CNN based model.

Description	Values
Input word vector	Google word2vec
Dimension of word vector	300
Filter region size	(3,4,5)
Feature maps	128
Activation function	ReLU
Pooling	Max_pooling
Dropout rate	0.5
L2 norm constraint	0

CNN-WV: The 300-dimensional word embedding trained by Word2vec is used for transfer input words into word vectors, and the words which are not covered by word embedding are initialized by random numbers and fine-tuned during the classifier training process.

SFV-CNN-WV(n): SFV-CNN model that use pretrained word vectors to initialize input words. SFV Creator module will generate SFV vector for each word and concatenate it to the tail of word vector to form a 305-dimensional vector as input data. The word vectors will be fine-tuned in the training process, while the SFV vectors remain unchanged. The "n" of the model name represents the window size setting of SWE.

SFV-CNN-Rand(n): SFV-CNN model that initialize the input word vectors with random numbers and concatenate word vector and SFV vector together to form a 305-dimensional vector for classifier training. The word vectors are fine-tuned and the SFV vectors remain unchanged during the training process.

In addition, to verify if the window size setting for training SWE has a great influence on the Feature Capturing ability of SFV model, we conducted an experiment to compare the SFV-CNN models using SWE that trained with different window size, and the optional settings of window size include (3, 4, 5, 6, 7, 8, 9).

3.3 Results and Discussion

The first two experiments aim at validating the improvement effects brought by SFV, including the improvement of training efficiency, the improvement of accuracy, and the enhancement of adaptability on cross-scenario dataset. The purpose of the last experiment is to explore the influence brought about by window size setting of SWE on the SFV-CNN model.

Classifier Training on MR Dataset. We trained the four categorization models mentioned above with MR data. Among them, we set a default window size of SWE to 6 for SFV-CNN-Rand and SFV-CNN-WV model, so if we didn't mention explicitly the window size setting is always 6 for SFV-CNN model. (1) SFV-CNN models learn more quickly. Compared with CNN-Rand, as shown in Fig. 4(a), SFV-CNN-Rand has a steeper

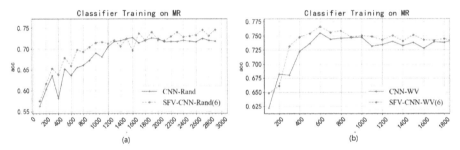

Fig. 4. Training curves for CNN based sentiment classification models. (a) Comparing the training process of our model and CNN based model under the same condition with randomly initialized word vectors. (b) Comparing the training process of our model and CNN based model with pretrained word vectors.

training curve at the begging and achieves a quite good accuracy at about 1300 steps, while CNN-Rand at about 1600 steps. SFV-CNN-WV achieves a fairly good accuracy at about 500 steps, 100 steps less than CNN-WV, as shown in Fig. 4(b). (2) SFV-CNN models are more accurate. As the accuracy scores listed in Table 2, we find that SFV-CNN models achieve higher accuracy scores with both randomly initialized word vectors or pretrained word vectors. The experiments demonstrate that SFV can bring about obvious improvements to CNN based sentiment classification models.

Validation Across Dataset and Scenario. We saved the trained classification models for each 300 steps in the training process of all these models, and conducted cross-dataset and cross-scenario sentiment classification experiments for the purpose of comparing their adaptiveness. We verify models by another open dataset, Kaggle MR, which is from the same scenario as MR belongs to. In addition, Twitter sentiment analysis datasets are selected to verify the cross-scenario adaptability of our model (Fig. 5).

Compared with CNN-Rand, the accuracy of SFV-CNN-Rand(6) is quite higher in the tests with both Kaggle-MR and Sentiment140 datasets. Especially in the test with sentiment140, our model is 3.31% points higher than that of CNN-Rand, as listed in Table 2; the accuracy of SFV-CNN-WV(6) is higher than that of CNN-WV by 0.26% points with Kaggle MR data, and 1.51% points higher when test with Sentiment 140 data set. The results prove that our model can obviously improve the adaptability of CNN based sentiment classification model.

The Influence of SWE's Window Size. In the above experiments, we use the SWE trained with default window size settings. Considering the sensitivity of this parameter to the ability of extracting textual features, we conducted this experiment for the purpose of finding the best SWE window size. With the optional window settings {3,4,5,6,7,8,9}, we train 7 SWE for SFV Creator, generate the SFV vectors and train classifiers correspondingly. The resulting accuracy of SFV-CNN models with different windows are shown in Table 2.

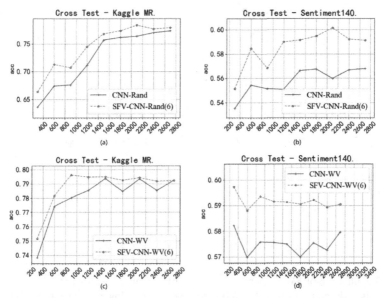

Fig. 5. Validate our model across dataset and Scenario. (a) (b) our model with randomly initialized word vectors compared to the corresponding CNN based model on adaptability. (c) (d) our model with pretrained word vectors compared to the corresponding CNN based model on adaptability.

Table 2. The results of our model obtained by evaluating on different data sets.

Model	MR	Kaggle MR	Sentiment140
CNN-Rand	72.70	77.34	56.83
SFV-CNN-Rand(6)	74.58	78.80	60.14
CNN-WV	75.52	79.37	58.23
SFV-CNN-WV(6)	76.64	79.63	59.74
SFV-CNN-Rand(3)	74.11	78.10	59.02
SFV-CNN-Rand(4)	73.36	78.68	59.38
SFV-CNN-Rand(5)	75.61	78.77	58.95
SFV-CNN-Rand(7)	74.20	79.33	59.19
SFV-CNN-Rand(8)	72.98	78.97	58.69
SFV-CNN-Rand(9)	74.30	78.63	56.46
SFV-CNN-WV(3)	74.95	79.92	59.17
SFV-CNN-WV(4)	75.42	79.50	58.87
SFV-CNN-WV(5)	75.52	79.52	59.82
SFV-CNN-WV(7)	75.80	79.39	59.23
SFV-CNN-WV(8)	76.98	79.56	59.65
SFV-CNN-WV(9)	75.33	79.79	59.97

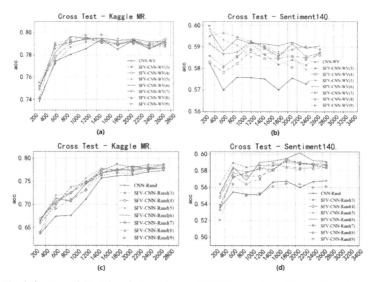

Fig. 6. The influence of SWE's window size. (a) (b) Our model with word vectors randomly initialized show different effects with SWE's window size changing in range of (3,4,5,6,7,8,9) on the Kaggle MR dataset and especially Sentment140 dataset. (c) (d) Our models initialized with pretrained word vectors show different effects with SWE's window size changing on the validation on Kaggle MR and Sentment140 datasets.

As shown in Fig. 6, differences brought by window size settings are not obvious for the tests on Kaggle MR dataset, but they are comparatively obvious when tested on sentiment140 dataset. The experimental results show that the change of window size has an important influence on the effectiveness of sentiment classification models, but it does not show obvious linear relations with the window setting. This may because that the corpus of specific scenario may have its own fitting window size, if we set the same window size for all SWEs, then the gains and losses offset each other to varying degrees, so it turns out to be somewhat chaotic and irregular in figures. In any case, it has been proved that windows have a great impact on classifiers.

4   Conclusions

In the present work we have proposed the method of quantifying grammatical features of corpus binding to a specific scenario, and apply it to sentiment classification model, and proposed SFV-CNN model. Our results demonstrate the improvement effect brought about by SFV, including the efficiency of model training, the accuracy of model and the adaptability in scenario transferring. Although SFV model is proved to be an effective method for extracting scenario related features, and fitting for enhancing CNN network, we aware that SFV's ability can be further improved. We have some choices for improvement. (1) Study on the optimal parameters for training of SWE. (2) Try to calculate SFV with more advanced word vector models such as BERT, because these models are proved to be so smart that they may have a deeper understanding of human language. (3) Well reprocessed corpora have great potential to enlarge SFV's ability to quantify semantic information.

Acknowledgments. This work was supported by The National Key Research and Development Program of China 2016YFB0801003.

References

1. Mikolov, T., et al.: Distributed representations of words and phrases and their compositionality. In: Burges, C.J.C., et al. (eds.) Advances in Neural Information Processing Systems 26, pp. 3111–3119. Curran Associates, Inc., Harrahs and Harveys, Lake Tahoe (2013)
2. Le, Q.V., Mikolov, T.: Distributed representations of sentences and documents. In: ICML, pp. 1188–1196 (2014)
3. Kim, Y.: Convolutional neural networks for sentence classification. http://arxiv.org/abs/1408. 5882 (2014)
4. Zhang, Y., Wallace, B.: A sensitivity analysis of (and practitioners' guide to) convolutional neural networks for sentence classification. arXiv preprint arXiv:1510.03820 (2015)
5. Zhang, X., et al.: Character-level convolutional networks for text classification. http://arxiv. org/abs/1509.01626 (2015)
6. Marinho, W., et al.: A compact encoding for efficient character-level deep text classification. In: IJCNN, pp. 1–8. IEEE (2018)
7. Er, M.J., et al.: Attention pooling-based convolutional neural network for sentence modelling. Inf. Sci. **373**, 388–403 (2016)
8. Socher, R., et al.: Recursive deep models for semantic compositionality over a sentiment treebank. In: Proceedings of the 2013 Conference on Empirical Methods in Natural Language Processing, pp. 1631–1642 (2013)
9. She, X., Zhang, D.: Text classification based on hybrid CNN-LSTM hybrid model. In: ISCID, no. 2, pp. 185–189. IEEE (2018)
10. Chen, N., Wang, P.: Advanced combined LSTM-CNN model for twitter sentiment analysis. In: CCIS, pp. 684–687. IEEE (2018)
11. Pontes, E.L., et al.: Predicting the semantic textual similarity with siamese CNN and LSTM. CoRR. abs/1810.10641 (2018)
12. Wang, J., et al.: Dimensional sentiment analysis using a regional CNN-LSTM model. In: ACL, no. 2. The Association for Computer Linguistics (2016)
13. Johnson, R., Zhang, T.: Convolutional neural networks for text categorization: shallow word-level vs. deep character-level. ArXiv, abs/1609.00718 (2016)
14. Conneau, A., et al.: Very deep convolutional networks for natural language processing. CoRR. abs/1606.01781 (2016)
15. Guo, L., Zhang, D., Wang, L., Wang, H., Cui, B.: CRAN: a hybrid CNN-RNN attention-based model for text classification. In: Trujillo, J.C., et al. (eds.) ER 2018. LNCS, vol. 11157, pp. 571–585. Springer, Cham (2018). https://doi.org/10.1007/978-3-030-00847-5_42
16. Liang, B., et al.: Context-aware embedding for targeted aspect-based sentiment analysis. In: Korhonen, A. et al. (eds.) ACL, no. 1, pp. 4678–4683. Association for Computational Linguistics (2019)
17. Zhang, L., et al.: Deep learning for sentiment analysis : a survey. arXiv:1801.07883 cs, stat. (2018)

18. Dong, X., de Melo, G.: A Helping hand: transfer learning for deep sentiment analysis. In: Gurevych, I., Miyao, Y. (eds.) ACL, no. 1, pp. 2524–2534. Association for Computational Linguistics (2018)
19. Sarma, P.K., et al.: Domain adapted word embeddings for improved sentiment classification. In: Gurevych, I., Miyao, Y. (eds.) ACL, no. 2, pp. 37–42. Association for Computational Linguistics (2018)
20. Wu, F., et al.: Active sentiment domain adaptation. In: Barzilay, R., Kan, M.-Y. (eds.) ACL, no. 1, pp. 1701–1711. Association for Computational Linguistics (2017)
21. Helmy, A.A., et al.: An innovative word encoding method for text classification using convolutional neural network. CoRR. abs/1903.04146 (2019)
22. Barnes, J., et al.: Bilingual sentiment embeddings: joint projection of sentiment across languages. In: Gurevych, I., Miyao, Y. (eds.) ACL, no. 1, pp. 2483–2493. Association for Computational Linguistics (2018)
23. Shi, B., et al.: Learning domain-sensitive and sentiment-aware word embeddings. CoRR. abs/1805.03801 (2018)
24. Wang, P., et al.: Semantic expansion using word embedding clustering and convolutional neural network for improving short text classification. Neurocomputing **174**, 806–814 (2016)
25. Gultepe, E., et al.: Latent semantic analysis boosted convolutional neural networks for document classification. In: BESC, pp. 93–98. IEEE (2018)
26. Johnson, R., Zhang, T.: Effective use of word order for text categorization with convolutional neural networks. In: Mihalcea, R., et al. (ed.) HLT-NAACL, pp. 103–112. The Association for Computational Linguistics (2015)
27. Pang, B., Lee, L.: Seeing stars: exploiting class relationships for sentiment categorization with respect to rating scales. In: ACL, pp. 115–124 (2005)
28. Pretrained word embedding download page. https://code.google.com/p/word2vec/. Accessed 10 Sep 2019
29. Financial articles webpage. https://www.kaggle.com/jeet2016/us-financial-news-articles. Accessed 20 Sep 2019
30. Opensubtitles Homepage. http://www.opensubtitles.org. Accessed 20 Sep 2019
31. Lison, P., Tiedemann, J.: OpenSubtitles2016: extracting large parallel corpora from movie and TV subtitles. In: Calzolari, N., et al. (ed.) LREC. European Language Resources Association (ELRA) (2016)
32. Twitter Corpus download page. https://github.com/Marsan-Ma-zz/chat_corpus. Accessed 20 Sep 2019
33. News articles download page. http://www.statmt.org/wmt14/training-monolingual-news-crawl. Accessed 20 Sep 2019
34. Literature works corpus download page. https://www.kaggle.com/mylesoneill/classic-literature-in-ascii. Accessed 20 Sep 2019
35. Movie Review corpus download page. https://www.kaggle.com/c/sentiment-analysis-on-movie-reviews/data. Accessed 20 Sep 2019
36. Go, A. et al.: Twitter sentiment classification using distant supervision. Processing, pp. 1–6 (2009)

Reinforcement Learning Based Interactive Agent for Personalized Mathematical Skill Enhancement

Muhammad Zubair Islam[ID], Kashif Mehmood[ID], and Hyung Seok Kim[(✉)]

Department of Information and Communication Engineering, Sejong University,
Seoul 05006, Republic of Korea
m.zubair_islam@outlook.com, kashif.mehmood224@gmail.com,
hyungkim@sejong.ac.kr

Abstract. Traditional intelligent systems recommend a teaching sequence to individual students without monitoring their ongoing learning attitude. It causes frustrations for students to learn a new skill and move them away from their target learning goal. As a step to make the best teaching strategy, in this paper a Personalized Skill-Based Math Recommender (PSBMR) framework has been proposed to automatically recommend pedagogical instructions based on a student's learning progress over time. The PSBMR utilizes an adversarial bandit in contrast to the classic multi-armed bandit (MAB) problem to estimate the student's ability and recommend the task as per his skill level. However, this paper proposes an online learning approach to model a student concept learning profile and used the Exp3 algorithm for optimal task selection. To verify the framework, simulated students with different behavioral complexity have been modeled using the Q-matrix approach based on item response theory. The simulation study demonstrates the effectiveness of this framework to act fairly with different groups of students to acquire the necessary skills to learn basic mathematics.

Keywords: Skill monitoring tool · Adversarial bandit · Recommendation system · Personalized learning model · Simulated learner

1 Introduction

With the increase in the world population, it is common in schools, colleges, and universities to teach several hundreds of students only with one teacher and thus quality education becomes one of the main problems. This teaching scenario makes personalized teaching even more challenging and eventually, students fail to reach their full potential due to lack of personalized tasks as per their competency level. So, for a teacher, it is highly difficult to give special assistance to each student at an individual level due to limited time and learning resources. A substantial amount of scientific research [1] proves that personalized teaching is a promising approach to increase the learning potential of students. There is a strong need to develop an effective automatic teaching system that can be used to accommodate the needs of students. The need to develop such a

© Springer Nature Switzerland AG 2020
D. Slamanig et al. (Eds.): MACIS 2019, LNCS 11989, pp. 395–407, 2020.
https://doi.org/10.1007/978-3-030-43120-4_31

system has become a reality with the expansion of existing algorithms offered for the scheduling problem. With the development of the competency-based intelligent system, it is possible to study the personal academic attributes from a student's knowledge and make recommendations proportional to proficiency level. The recent studies [2] in competency-based recommender systems cover collaborative filtering, content-based and hybrid approaches, the using student's academic characteristics (performance, experience, degree of achievement) history. However, utilizing these techniques in a real-time learning environment without historical traces of new students raise some issues like the cold-start problem [2, 3]. In [2, 4] authors present a hybrid technique using the ontological domain representation for adaptive learning to avoid these disadvantages. However, it is also based on the student's historical parameters such as schooling, learning, and proficiency level.

In order to overcome the existing challenges, a PSBMR framework has been proposed. It measures the student's competency level in real-time using an online estimation approach [5] to recommend the task as per the student's attribute. The framework estimates the student skill level based on their responses to the proposed task, without using any other information or student's academic historical data. To estimate the knowledge attributes of a student, the PSBMR makes a concept learning skill profile for each individual student. Therefore, the use of this system enables learners to learn at their own pace. In contrast, the existing systems such as [6, 7] use expert algorithm Exp4 and Fuzzy linguistic approach respectively. This paper utilizes the Exp3 algorithm [8] to recommend the optimal tasks to the student. The Exp4 and Exp3 are the exponential-weighting algorithms used for handling exploration and exploration dilemma [8] in bandit problems [5]. A bandit problem is like a game between the learner and the environment in which the learner plays an action and the environment then gives the reward to the learner against played action. The main contribution of this work can be summarized as follows.

- The student's behavior has been modeled as a student profile to understand the student's existing academic skills in real-time.
- The task and student skill relationships are characterized by M-matrix. It is utilized to update and estimate the student profile information.
- An exponential-weight algorithm Exp3 is utilized successfully to recommend the optimal task with fast running time and the ability to handle large numbers of actions as compared to the Exp4.
- The accuracy analysis of the proposed PSBMR using the population of simulated students is performed.

The structure of the remaining paper is organized as follows. Section 2 conducts an overview of the related work. In Sect. 3, a system model for recommendation strategy is formulated, and system modules overview is performed. Section 4 presents the proposed schema for the student profile model and the technique to acquire that profile. An algorithm for choosing an optimal task for PSMBR is proposed. Section 5 demonstrates the experiment to assess the performance of the proposed framework using virtual students' implementation. Finally, Sect. 6 concludes this paper and describes future work.

2 Related Works

There are various approaches found in the available literature to create competency-based recommendation systems [2]. It is difficult to apply traditional recommender approaches in many educational scenarios where the learning potential of the student changes over time. Several studies have focused on these challenges, along with studies on knowledge and the performance factor tracing approach [9]. These psychometric measurement approaches such as multidimensional item response theory (MIRT) [10] and diagnostic classification models (DCM) [11] are being used to assess the student's behavior by taking into account the student's knowledge. These models are mostly used for measuring latent variables of the students for binary responses in adaptive recommendation systems. The basic components of a competency-based recommender system are described along with a method to model student competency. Different approaches have been proposed to create a student model in various works including [7, 12]. In [7] a fuzzy linguistic approach is used to express the achievement of the student in each competency and also shows that it is insufficient to express student competency using numeric values. In contrast to the [7], the proposed model utilized a continuous number between [0,1] to represent students' acquired skill levels (e.g., 0.3 means 30% acquired and 0 means nothing acquired) as a student profile. Similarly, [12] propose a mathematical framework based on psychometric assessment of the learner by using a Markov decision process (MDP) for the adaptive learning system. The proposed model in [12], recommends the learning material to the learner associated with each skill level using different optimal strategies like c-μ rule and Gittins index. The author formulates the recommendation strategies as a stochastic problem by assuming the fixed procedures in the learning process, whereas, the PSBMR formulate it as a non-stochastic problem and not used fixed learning procedures.

In [6] a tutoring model for optimal teaching sequence for individual students using MAB is proposed. It uses expert knowledge for exploration and exploitation trade-off to find a possible set of actions for a personalized task selection. It utilizes the expert algorithm Exp4 to select an optimistic action for each activity. However, it is difficult to apply the Exp4 within a real-time environment when numerous actions and experts corresponding to each activity are used [5, 8]. The work in [13] also used expert advice to employ teaching strategies for personalized instructions. In contrast, the proposed approach in this paper utilizes the Exp3 algorithm which does not rely on expert advice for the optimization of the task corresponding to each action for personalization.

3 System Model

The proposed PSBMR framework consists of three components for a recommendation strategy, including student profile module, learning measurement module and task selection strategy module as shown in Fig. 1.

3.1 Student Profile Module

The student profile module is used to manage the student's real-time learning outcomes such as competency and learning level and thus student characteristics store into the

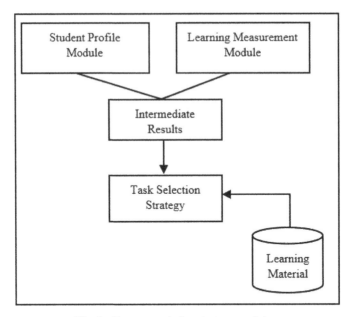

Fig. 1. Recommendation strategy modules.

M-matrix using the Q-matrix representation. In the Q-matrix representation, the matrix $Q_{(M \times N)}$ is an M by N dimension matrix with entries, each row represents the number of questions and each column is equal to the required competencies for the corresponding questions respectively [14]. The Q-matrix is used to understand student performance by representing the relationship between questions and their required concepts. In this context, the proposed framework creates an $M_{(A \times C)}$ matrix similar to the Q-matrix whereas it represents a student progress level into numeric values. The $M_{(A \times C)}$ matrix represents the relationship between the task set A and competency set C as shown in Fig. 2. Each task from $A = (a_1, a_2, a_3 \ldots \ldots \ldots, a_m)$ has a relation with competency set $C = (c_1^a, c_2^a, c_3^a \ldots \ldots \ldots, c_n^a)$ and the numeric values corresponding to each task and competency pair represent the estimated skill level (ESL) of a student. Therefore, a student profile is represented by the skill level of each task a_i, belonging to the competency level c_j acquired by the student. For example, the cell (a_1, c_2) shows that a student has achieved 20% for a task a_1 with difficulty as well as competency level c_2. The profile module utilizes a student answer (right or wrong) corresponding to the task to store the student's learning progress.

3.2 Learning Measurement Module

The learning measurement module is used to understand and estimate the student learning skills as well as characterize the learning material. This module defines a set of values required to acquire a specific task associated with proficiency levels. This module represents the required skill level (RSL) to solve each task and it is indexed by (a_i, c_j) as represented in Fig. 2. Here (1) is used to define independent RSL to each

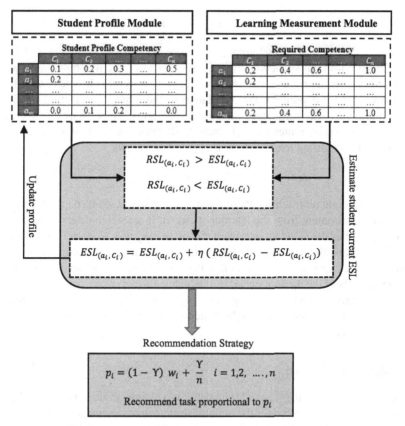

Fig. 2. A systematic overview of the PSBMR framework.

task in the learning material automatically and the parametric value $RSL_{(a_i,C_j)}$ shows the required competency level against the proposed task which is used to estimate the student's current ESL.

$$RSL_{(a_i,C_j)} = \left(\frac{T_j \times (T_i + 1)}{len(A)} \right) - \left(\frac{T_j \times T_i}{len(A)} \right), \tag{1}$$

where T_j is the required conceptual understanding of the chosen task as a column number, T_i is the proposed task row number in the matrix dimension and $len(A)$ is the total number of tasks in the learning material. When the task is recommended by the recommendation strategy to a student, then it receives the right or wrong answer from a student as a response. The learning module helps to estimate the student performance through the corresponding parameter value $RSL_{(a_i,C_j)}$ of the proposed task. The learning module assesses the student at every task and assigns the reward to each individual student as per his progress in learning material. The reward function is a core component of the learning module and its computational interface is depicted in (2) and (3) respectively

$$R_{(a_i,C_j)} = max\{(RSL_{(a_i,C_j)} - ESL_{(a_i,C_j)}), 0\}, \tag{2}$$

$$R_{(a_i,C_j)} = min\{(RSL_{(a_i,C_j)} - ESL_{(a_i,C_j)}), 0\}, \tag{3}$$

where $R_{(a_i,C_j)}$ is the reward belonging to the progress of the student if a task a_i is recommended with competency c_j. Hence, Eqs. (2) and (3) cover both cases when a student's answer either right or wrong. Here (2) is used to measure the reward for the right answer and (3) is used for the calculation of the wrong answer to the proposed task. With the describing parametric values in (1), (2) and (3) it is possible to update and estimate the learning attributes for each student. Therefore, the ESL of a student for any recommended task during learning time is updated as follows

$$ESL_{(a_i,C_j)} = ESL_{(a_i,C_j)} + \eta R_{(a_i,C_j)}, \tag{4}$$

where η is the weight representing the student learning speed, which depends on the student's skill improvement from the learning material. It is used to determine the reward preference, smaller the value more reward means an easy task to learn. The best setting fitting value means the recommendation of the task for a better outcome in the learning process. The expression (4) shows that the performance of the student depends on receiving rewards of the recommended task which is further used to motivate as well as increase the learning proficiency. Therefore, (4) shows that a student can get progress only if he gives the correct answer to the difficult task rather than his previous estimated competency level.

3.3   Task Selection Model

The core component of the proposed framework is a recommendation strategy. The task selection module is used to select the optimal strategy to recommend the task based on ESL as per the student profile attributes. However, the selection of an optimal task from the task pool A is very challenging due to the dimensionality problem [15]. In order to overcome these challenges, the PSBMR framework is formulated under the adversarial bandit problem and the Exp3 algorithm is utilized for optimization and dimensionality reduction. Figure 2, shows an overview of the recommendation strategy on how to estimate and update the student profile.

4   Proposed Methodology

4.1   Concept of Adversarial Bandits

The concept of the multi-armed bandit is used for algorithms that make decisions under uncertainty over time. Mostly, it is investigated in stochastic scheduling and decision-making problems. In bandit problem, the player decides which action to take to play in a sequence to get maximum reward. In the MAB model, an agent selects any arm from $(K \geq 1)$ at each time step $(t \in T)$ to play and gain some reward $(x \in \mathbb{R})$ with a certain probability $P \in [0, 1]$. The MAB is an efficient online estimation approach allowing the strategy to train the algorithm in online fashion when it receives new data points to find the optimal action to make the best decision [16]. A variant of the MAB framework with a non-statistical assumption about the reward generation process rather than predefined fixed distribution is known as an adversarial or non-statistical bandit [17, 18]. In the adversarial model, rewards are selected by an adversary as per-play with unbounded computational power from a bounded range. This problem is related to the unknown repeated matrix game in learning to play problem. In the proposed framework, the student's progress only depends on the ESL after learning the specific mathematical task. It also has random distributions for task selection to train the student on multiple mathematical tasks. It represents the non-stochastic environment of the proposed framework. Therefore, the proposed PSBMR has been formulated under the non-stochastic bandit problem to train the students under online fashion. Another interesting point to build the PSBMR using adversarial bandit is that task recommendation only depends on the existing knowledge attributes of the student, not on academic historical traces.

4.2 Adversarial Bandit Solution Using the Exp3 Algorithm

For the adversarial bandit problem, the Exp3 is a very simple and powerful algorithm. It is based on the *Hedge* algorithm [19] which is used to solve exploration and exploitation trade-off in the MAB problem. It can be described as follows

$$\mathbf{E}\big[\hat{x}_i(t)|i_1,\ldots,i_{t-1}\big] = \mathbf{E}\bigg[p_i \cdot \frac{x_i}{p_i(t)} + (1 - p_i(t)) \cdot 0\bigg]. \tag{5}$$

The Exp3 draws an action i_t from $(1 \leq i \leq K)$ in each round t with probability distribution p and obtain an actual reward $x_i(t)$. Here in (5), $\mathbf{E}\big[\hat{x}_i(t)|i_1,\ldots,i_{t-1}\big]$ represents the estimated reward corresponding action i_t taken in each round t. The reward estimation process $\mathbf{E}\big[p_i.\frac{x_i}{p_i(t)} + (1 - p_i(t)).0\big]$ shows that for the observed arms the Exp3 estimates the reward by dividing the actual reward with probability and assign zero rewards to the unobserved arms. The probability p_i is the mixture of uniform and weighted distribution probability for each arm defined as

$$p_{i,t} = (1 - \Upsilon)w_{i,t} + \frac{\Upsilon}{K} \quad i = 1, 2, \ldots, K. \tag{6}$$

where $\Upsilon \in [0, 1]$ is an exploration parameter for action i. It is used to estimate the time taken by the algorithm for decision-making to select an action i. $(1 - \Upsilon)$ is equal to the remaining time taken by an algorithm to calculate the weight w_i distribution of the actual reward $x(t)$. w_i is the exponential weight of the action taken by the corresponding arm of the bandit described as follows

$$w_{i(t+1)} = w_i(t).exp^{\left(\frac{\Upsilon.\hat{x}(t)}{p_i(t).K}\right)}. \tag{7}$$

The Exp3 algorithm consists of two parts, and one is used for the calculation of uniform probability to pull the good arm and the second part is used to keep a list of weights for each pulled arm in each round and update it. In the PSBMR, the Exp3 selects a task from set A as per drawn action according to the probability distribution p. The Exp3 generates the pulled arm reward $\hat{R}$ by dividing the actual gain to the probability of the chosen action, which is equal to the estimated reward as in (5). The process of estimating the expected rewards for the Exp3 in PSBMR can be defined as follows

$$\hat{R}_j(i) = \begin{cases} \frac{R_j(i)}{p_j(i)}, & j = i \\ 0, & j \neq i \end{cases}, \tag{8}$$

where $R_j(i)$ indicates the student reward $R_{(a_i,C_j)}$ as described in (2) and (3). $p_j(i)$ is known as the uniform probability distribution of the proposed task. The pseudo-code of the PSBMR framework for the selection and recommendation of the optimal task is depicted in Fig. 3.

Algorithm 1: Exp3 algorithm for PSBMR Framework
Exploration parameter: $\Upsilon \in [0,1]$ called egalitarianism
Initialization of weight : w_i uniformly
Initialization of ESL

1 **While** *studying* **do**
2   **for** i = 1,2, . . ., n **do**
3          $w_i = \frac{w_i}{\sum_{j=1}^{n} w_j}$
4          $p_i = (1 - \Upsilon) \, w_i + \frac{\Upsilon}{K} \quad i = 1,2, \dots, K$
5   Select task proportional p_i
6   Recommend Task $A = (a_1, a_2, a_3 \dots \dots \dots, a_m)$
7   Reward Generation process
8   **If** answer = 1
9              $R_{(a_i,C_j)} = \left\{ max \left(RSL_{(a_i,C_j)} - ESL_{(a_i,C_j)} \right), 0 \right\}$
10   **else**
11             $R_{(a_i,C_j)} = \left\{ min \left(RSL_{(a_i,C_j)} - ESL_{(a_i,C_j)} \right), 0 \right\}$
12   Update Estimated Competence
               $ESL_{(a_i,C_j)} = ESL_{(a_i,C_j)} + \eta \, R_{(a_i,C_j)}$
13   **for** j = 1,2, . . ., n **do**
14             $\hat{R}_j(i) = \begin{cases} \frac{R_j(i)}{p_j(i)}, & j = i \\ 0, & j \neq i \end{cases}$
15             $w_{j(i+1)} = w_j(i) * exp^{\left(\Upsilon * \hat{R}_j(i) / p_i \right)}$
16   **end**
17   **end**
18 **end**

Fig. 3. Pseudo-code for the PSBMR framework.

5   Performance Evaluation and Results

This section demonstrates the performance of the proposed MAB based framework using simulated students' implementation [20, 21]. There are different approaches to create a simulation environment to train students [22]. In this study, a student model is developed to create virtual student profiles with different learning complexity. Each virtual student has been created with different aptitude levels for each learning skill attribute. For this experiment, two groups of simulated students (above and below average), 100 mathematical tasks and 6 skill level attributes for each task have been used. Each of the competencies has a value between 0 and 1. Specifically, the simulation depicted in this section is aimed to evaluate the correctness of the proposed model and compare it with the baseline model designed by an educational expert for this problem.

5.1   Simulation Environment

In order to evaluate the proposed framework, two different groups of virtual students Q-students and P-students were created to check how well the framework is capable of recommending a task for each individual student. The population of above-average students Q have different learning rates and different maximum competency levels for each task set for all students. While the population of below-average students P have a limitation to use the task set and have specific competency levels for the task for each student. Both groups of simulated students have been recommended with the same learning material, basic algebra questions. For the modeling of simulated students, success probability for each task and a good schema for ESL evaluation is needed. The success probability to solve the tasks is modeled using the concept of Item Response Theory (IRT) [22] can be written as follows

$$p(success) = \prod_{i,j=1}^{m} \left[\frac{\delta(a)}{1 + \exp\left(\alpha + \beta\left(RSL_{(a_i,C_j)} - ESL_{(a_i,C_j)}\right)\right)} \right]^{1/m}, \quad (9)$$

where α and β are the constants that allow tuning the success probability. α and β describes the probability of success to the student competency as per required competency of the task and the rate of the drop or rise in the probability of success respectively. The function δ is used to define two different groups of students Q and P. For the population of Q-students $\delta(a) = 1$, meaning that these students have the capability to solve all the tasks. The $\delta(a) = 0$ for P-students, represent limited skills to solve the tasks. The evolution schema for the profile is very simple the student just makes progress when he gets success in any task and his estimated competency level in the profile model is lower than the parametric value RSL as described in (4).

5.2   Baseline Method

To compare the performance of the PSBMR with the expert teacher teaching sequence a predefined expert sequence (PES) has been designed as a baseline method. The PES is designed by a mathematics subject expert to perform the recommendation according to predefined expert advice. The expert teacher defined a sequence of tasks according to the difficulty level. A student starts with the first task and moves to the next target as per the defined sequence after succeeding in the proposed task. To design expert sequence, the idea has been borrowed from [6] along with the numerical setting of the difficulty levels for each task.

5.3 Experimental Results

The combination of graphs and empirical values are used to show that the proposed framework is capable of providing the optimal incremental tasks to train the student according to the student's learning attributes. An experimental study has been conducted to examine the learning progress of the student in the knowledge competency against the proposed task for PSBMR and PES model, as depicted in Fig. 4. For the PES, a task is chosen for the next recommendation as per defined hierarchy chain. In Fig. 4, it can be seen that there is no progress in the student's knowledge attributes during complex levels after a certain point by using the PES. The reason for this problem is that the PES imposed the same skill hierarchy structure for each individual student without estimating a student's existing understanding level. To cope with this challenge, the proposed schema develops a relationship between learning material and students' skill level to understand their learning concept level. So, that each task is a pair of question and skill attribute which has been recommended to the student by the PSBMR. In this way, the PSBMR not only selects the optimal task but also proposes the task by considering the student specific competency level. Therefore, the PSBMR optimal design based on the learning material and the student's profile has outperformed the PES strategy, with more competencies for each task for each individual student.

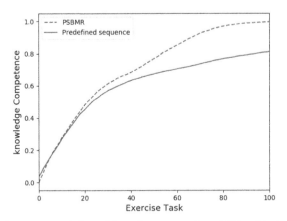

Fig. 4. Automatic recommendation of the task as per student skill level by PSBMR and PES.

Figure 5, illustrates the number of failures during task recommendations made by PSBMR and PES for the population of P and Q students. The experiment has been carried out with the following parametric values. Case-I: above-average students ($\delta(a) = 1$, T = 50 rounds over 100 iterations, Number of students = 50, Number of tasks = 100).

Case-II: below-average students ($\delta(a) = 0$, T = 50 rounds over 100 iterations, Number of students = 50, Number of tasks = 100). The rating parameters have been used with the following values $\alpha = 1.38$, $\beta = 8$. The parameters are used to define the different ability levels of simulated students. At the initial, the proposed framework had understood the pattern for estimating the unknown parameters of the student profile model. It is the reason for the high number of failures at the beginning of the results.

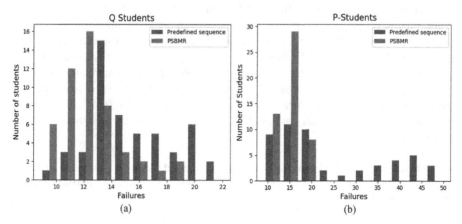

Fig. 5. The average number of failures for above and below average students.

In the first scenario, as depicted in Fig. 5(a), the average number of errors made by PSBMR and PES for Q students are shown. Based on the simulation results, it can be seen that the proposed model is capable of adapting that learning material which meets the ongoing learning skills, therefore, students make fewer errors during the learning process. As the number of task recommendations increased, the error of the PSBMR decreased and it made the framework more adaptive for each individual student. In the second scenario, as shown in Fig. 5(b), the analysis shows that the PSBMR slightly improved the learning progress of the P rather than Q students. The specific reason for this improvement can be explained as the above-average students already had a good opportunity to solve the task while the below-average students had limited ability to accomplish the tasks. This analysis makes a sense, that PSBMR recommended the task according to the ESL of the students and potentially motivates the below-average student to learn. Whereas, the PES had recommended the task according to the predefined hierarchy of difficulties for all students. Hence, the PES created frustration for students and the students did not perform better in the learning environment. Thus, the overall experimental study shows that the proposed framework is useful to train the students who are getting bored and discouraged with mathematical studies. The proposed model not only recommends the trusted and optimal task using the individual profile model as well as helping the students to motivate by removing their disappointment towards mathematics. These findings show that the student's skill level progression in the learning subject could be improved by providing an optimal personalized learning task.

6   Conclusion and Future Works

In this paper, a Personalized Skill-Based Math Recommendation (PSBMR) framework is proposed in which a non-stochastic problem known as an adversarial bandit problem along with the exponential-weight Exp3 is analyzed. The aim of this recommender system is to automatically estimate the current skill level of the student and recommend the task similar to the expert sequence design to motivate and maximize the student learning outcomes. The proposed framework is illustrated as a set of optimal action values proportional to the distribution probability and estimated reward functions obtained by the implementation of the multi-armed bandit concept. Simulation studies validate the optimistic nature of the PSBMR framework to recommend the mathematical task. As future work, it would be a direction to explore more advanced techniques to model the student skills and learning material in order to compare with expert sequence and baseline methods. A second direction is to evaluate the proposed system in the real-life educational scenario with real students to check the ability of the system in real-life applications.

Acknowledgment. This work was supported by the National Research Foundation of Korea (NRF) grant funded by the Korea government (MSIT) (No. 2019R1A4A1023746, No. 2019R1F1A1060799).

References

1. Miliband, D.: Personalised Learning: Building a New Relationship with Schools, Speech to the North of England Education Conference, Belfast, January 2004
2. Yago, H., Clemente, J., Rodriguez, D.: Competence-based recommender systems: a systematic literature review. Behav. Inf. Technol. **37**(10–11), 958–977 (2018)
3. Ricci, F., Rokach, L., Shapira, B.: Introduction to Recommender Systems Handbook. In: Ricci, F., Rokach, L., Shapira, B., Kantor, P. (eds.) Recommender Systems Handbook. Springer, Boston (2011). https://doi.org/10.1007/978-0-387-85820-3_1
4. Zaphiris, P., Ioannou, A. (eds.): LCT 2015. LNCS, vol. 9192. Springer, Cham (2015). https://doi.org/10.1007/978-3-319-20609-7
5. Slivkins, A.: Introduction to Multi-Armed Bandits, April 2019. arXiv:1904.07272v3
6. Clement, B., Roy, D., Oudeyer, P.-Y., Lopes, M.: Online optimization of teaching sequences with multi-armed bandits. In: 7th International Conference on Education Data Mining, London, UK (2014)
7. Serrano-Guerrero, J., Romero, F.P., Olivas, J.A.: Hiperion: a fuzzy approach for recommending educational activities based on the acquisition of competencies. Inf. Sci. (NY) **248**, 114–129 (2013)
8. Auer, P., Cesa-Bianchi, N., Freund, Y., Schapire, R.E.: The nonstochastic multiarmed bandit problem. SIAM J. Comput. **32**(1), 48–77 (2003)
9. Wang, Y., Heffernan, N.: Extending knowledge tracing to allow partial credit: using continuous versus binary nodes. In: Lane, H.C., Yacef, K., Mostow, J., Pavlik, P. (eds.) AIED 2013. LNCS (LNAI), vol. 7926, pp. 181–188. Springer, Heidelberg (2013). https://doi.org/10.1007/978-3-642-39112-5_19
10. Reckase, M.D.: Multidimensional Item Response Theory Models. Springer, New York (2009). https://doi.org/10.1007/978-0-387-89976-3

11. Rupp, A.A., Templin, J.L.: Unique characteristics of diagnostic classification models: a comprehensive review of the current state-of-the-art. Meas. Interdiscip. Res. Perspect. **6**(4), 219–262 (2008)
12. Chen, Y., Li, X., Liu, J., Ying, Z.: Recommendation system for adaptive learning. Appl. Psychol. Meas. **42**(1), 24–41 (2018)
13. Koedinger, K.R., Brunskill, E., Baker, R.S., Mclaughlin, E.A., Stamper, J.: New potentials for data-driven intelligent tutoring system development and optimization. AI Mag. **34**(3), 27–41 (2013)
14. Leighton, J.P., Gierl, M.J., Hunka, S.M.: The attribute hierarchy method for cognitive assessment: a variation on Tatsuoka's rule-space approach. J. Educ. Meas. **41**(3), 205–237 (2004)
15. Nino-Mora, J.: Stochastic scheduling. In: Floudas, C.A., Pardalos, P.M. (Eds.) Encyclopedia of Optimization, pp. 3818–3824 (2009)
16. Burtini, G., Loeppky, J., Lawrence, R.: A Survey of Online Experiment Design with the Stochastic Multi-Armed Bandit. arXiv:1510.00757v4
17. Auer, P., Cesa-Bianchi, N., Freund, Y., Schapire, R.E.: Gambling in a rigged casino: the adversarial multi-armed bandit problem. In: Proceedings of Annual Symposium Foundations of Computer Science, Milwaukee, WI, pp. 322–331 (1995)
18. Lattimore, T., Szepesvá, C.: Bandit Algorithms. Cambridge University Press, Cambridge (2018). Draft of 28th July, Revision 1016
19. Schapire, R.E., Freund, Y.: A decision-theoretic generalization of on-line learning and an application to boosting. J. Comput. Syst. Sci. **55**, 119–139 (1997)
20. Mertz, Jr., J.: Using a simulated student for instructional design. In: Proceedings of the Seventh World Conference on Artificial Intelligence in Education (1995)
21. Beck, J.: Modeling the student with reinforcement learning. Paper presented at the 6th Annual Conference on User Modelling, Sardina, Italy (1997)
22. Meneghetti, D.D.R., Junior, P.TA.: Application and Simulation of Computerized Adaptive Tests Through the Package catsim (2017). arXiv:1707.03012v2

Common Vector Approach Based Image Gradients Computation for Edge Detection

Sahin Isik$^{(\boxtimes)}$ ⬤ and Kemal Ozkan ⬤

Computer Engineering Department, Eskisehir Osmangazi University, 26480 Eskişehir, Turkey
sahini@ogu.edu.tr

Abstract. In this study, the concept of Common Vector Approach (CVA) is adopted for image gradients computation in terms of revealing edge maps stated on images. Firstly, noise stated on image is smoothed by Gaussian filtering, secondly gradient map computation using CVA is carried out, then the angle and direction maps are obtained from the gradient map and lastly peak points are selected and a smart routing procedure is performed to linking them. With an unusual methodology, the derivatives of image through vertical and horizontal directions have obtained by utilizing the CVA, which is the crucial step and gained the novelty to this work. To compare results objectively, we have judged the performance with respect to a comparison metric called ROC Curve analysis. As a contribution to the edge detection area, CVA-ED presents satisfactory results and edge maps produced can be used in the tasks of object tracking, motion estimation and image retrieval.

Keywords: Edge detector · Common vector approach · Edge gradient map

1 Introduction

In computer vision applications, edge detection is a process to determine the typical characteristics of objects situated on image. The main purpose of edge detection methods is to identify where the brightness of the local area changes significantly. Since impacts of edge detection affect the success on object tracking [1], motion estimation [2], image retrieval [3] tasks, studies on edge detection constitute a crowded set in the literature of image processing. It means that if the edges of an image identified accurately, then the outline of an object such as geometrical shape information and other basic properties can be simultaneously measured with respect to segmented regions.

Although there is not yet certain rule index to judge performance of edge detection, but in general, it is agreed that edge segments should be thin, well-localized, non-jittered and one-pixel wide for a good edge detector. Beside, a proposed method on edge detection should be preserving all detailed information such as sharp discontinuities, junctions, corners etc., of the processed image. For this purpose, a vast number of edge detectors used gray level variation of pixels and employs the first and second order derivatives of image. However, since the derivative process is sensitive to noise due to sources including electronic, semantic and digitization/quantization impacts, so a jagged form

© Springer Nature Switzerland AG 2020
D. Slamanig et al. (Eds.): MACIS 2019, LNCS 11989, pp. 408–421, 2020.
https://doi.org/10.1007/978-3-030-43120-4_32

can be occurring in obtained edge map. To alleviate such problems, firstly, the image is generally smoothed with a filter. However, sharp discontinuities corresponding to edges can be also suppressed because of this smoothing. Moreover, each one of the most widely used differential operators such as Sobel [4], Roberts [5], Prewitt [6] and Scharr [7] are useful particular tasks since the edge segments obtained by them are in a different form.

In the literature, numerous studies on edge detection methods [3, 8–14] have been proposed in the past decades. The history of edge detection methods begins with the study of Marr-Hildreth [15], called the zero crossing of Laplacian edge detector, which is a gradient based operator and considers the step difference in the intensity of the image as representing with the second derivative by the zero crossing. As an interesting point of this operator, with endorsement from biological vision systems, paved the way of new approaches in field of edge detection. Therefore, the most of gradient based edge detectors had been inspired by study of Marr-Hildreth. Unfortunately, the drawbacks of developed operators are returning false edges corresponding to the local minima of the gradient magnitude and giving poor localization at curved edges. Afterwards, Canny [16] introduced a new approach in which an objective function is developed to be optimized for accurate edge localization and minimizing multiple responses to a single edge. Due to advantages such as fast, reliable, robust and generic, it is widely accepted as an optimal edge detector in industry. However, because of the parameters, Canny's algorithm is more sensitive to weak edges, resulting in a corrupted edge map and accuracy is not satisfactory. Therefore, some algorithms [11, 17, 18] have been proposed to improve the localization and quality of edge maps obtained by the Canny operator.

Because of the short-sight ideas of the traditional edge detectors such as discontinues in edge segments, thick edges and being resistless to noise, recently, some new studies in this area emerged. In [10], a new edge segment detection algorithm was presented, so called Edge Drawing (ED), in which, firstly a set of anchor edge points is computed in processed image and then connecting these anchor points by drawing edges between them. This algorithm generates the high quality edge maps and runs faster than the widely known fastest edge detection algorithms. However, since this edge detector connects anchor points by one pixel one, the edges obtained from this algorithm is sensitive to noise and natural deformations can be occurred in an image. Also in [8], a different algorithm has been presented based on the hybrid of gradients and zero crossings obtained by convolving the image with the corresponding operators. In related algorithm the least squares support vector machine (LS-SVM) with a typical and most frequently experimented kernel function, called the Gaussian radial basis function, is carried out to reveal edge maps. In a different approach, Discrete Cosine Transform (DCT) method [9] applied to extract edges from local features. A DCT basis images was used to generate gradient basis images corresponding to different operators including "Roberts", "Prewitt" and "Sobel". Subsequently, edge map of each block were calculated using DCT coefficients. However, edges obtained from last two methods are not thin and seem sensitive to noise. Recently, a different paradigm of unsupervised edge detection [12] is introduced by improving the hysteresis thresholding that utilized on the computational edge detection approach of Canny. As a contribution of work, a new post-processing step of non-maximal suppression has presented in order to attain the good edge points and ignore the other ones. Moreover, the author emphasized that it is possible to drop the idea of using the two-threshold hysteresis that operated on Canny algorithm, by using only a unique predefined threshold.

Technically, the performance of an edge detection algorithm depends on the utilized gradient extraction technique or thresholding scheme. One can say that a developed method, which depends on only utilized thresholding scheme, may be not suitable for all image types. With a different point of view, we have focused on the gradient extraction by employing a new gradient extraction technique. With this aim, we have developed a new edge detection method based on the common vector approach (CVA), called CVA-ED, in where, firstly the noise situated on image is smoothed by Gaussian filtering in order to reduce noise, secondly gradient map computation by using CVA is carried out, then the angle and direction maps are obtained from the gradient map and lastly peak points are selected and a smart routing procedure is performed to linking them. Contrary to ordinary, the derivatives of image through vertical and horizontal directions are obtained by utilizing the CVA, which is gained the novelty to this paper. When the proposed methods on the edge detection area are considered, this idea is a new approach in terms of edge detection procedure. Our experimental results on numerous experiments indicate that CVA-ED, as a new edge detection method, exhibits great performance and the obtained edge maps can be used with a purpose of image matching or compression task.

The remaining parts of the paper are arranged as follows. Section 2 reviews CVA method and presents how this method adopted to our work. Also smoothing with Gaussian filter is mentioned in Subsect. 2.1, the gradient map computation by using CVA is introduced in Subsect. 2.2, the procedure for extraction of peak points is explained in Subsect. 2.3 and the connection of peak points by the smart routing procedure is illustrated on Subsect. 2.4. Moreover, the Sect. 3 displays our edge detection results compared with the results of [12]. Finally, a conclusion is touched in Sect. 4.

2 Common Vector Approach Based Edge Detection

Essentially, the stages involved in CVA-ED method can be summarized with following steps:

i.   Suppression of noise by Gaussian filtering.
ii.   Gradient map computation using CVA.
iii. Extraction of peak points.
iv.   Connection of peak points by the smart routing procedure.

2.1 Suppression of Noise by Gaussian Filtering

The main idea behind the image smoothing is reducing or removing the details including noise, outliers and sharp transitions, while preserving the more meaningful information containing in the image. For this purpose, the gray image should be smoothed with Gaussian low-pass filter in order to suppress noise. For all results given in this paper, a rotationally symmetric Gaussian low pass filter of size 5 with standard deviation 1 is carried out as a pre-processing step of edge detection.

2.2 Gradient Map Computation Using CVA

The CVA is a subspace-based recognition method that gives satisfactory results in voice and pattern recognition tasks [19]. Let suppose that we have given m samples in R^n corresponding to the any class (image blocks, $\{\vec{a}_i\}_{i=1}^m$). It is possible to represent each $\vec{a}_i$ vector as the sum of

$$\vec{a}_i = \vec{a}_{com} + \vec{a}_{i,diff}. \qquad (1)$$

A common vector ($\vec{a}_{com}$) is what is left when the differences between feature vectors are removed from class members, and it is invariant throughout the class, whereas $\vec{a}_{i,diff}$ is called the remaining vector, represent the particular trends of this particular sample. There are two cases in CVA where the number of vectors is either sufficient or insufficient, which are the cases as the feature vector dimension is lower or higher than the number of vectors, respectively.

The sufficient data case occurs when the dimension of the vectors in a class (i) is less than the number of vectors. In this case, the common vector of a class can be obtained by calculating the eigenvalues and eigenvectors of covariance matrix related to the given class [20]. Moreover, in sufficient data case, the eigenvectors span the entire difference space and an indifference subspace does not exist. Therefore, the common vector becomes $\vec{0}$. In this case, $\vec{a}_{com}$ can be estimated by ordering the eigenvalues of $n \times n$ covariance matrix and the smallest k eigenvalues build the indifference subspace $(B^{\perp})$. Since indifference subspace $B^{\perp}$ and difference subspace (B) are orthogonal, the remained $(n - k)$ eigenvalues can be used to construct the difference vector by projection of the data onto the difference subspace (B). In this study, the blocks of the processed image are gathered to provide the insufficient data case due to $\vec{a}_{com}$ does not exist in sufficient data case.

Let the vector dimension and the number of vectors are defined as n and m, respectively, in an R^n vector space. To obtain the common and difference vectors of a block as formed from the neighborhoods of the i, j point, the required steps are explained with following steps.

$$block = \begin{bmatrix} I_{i-2,j-2} & I_{i-2,j-1} & I_{i-2,j} & I_{i-2,j+1} & I_{i-2,j+2} \\ I_{i-1,j-2} & I_{i-1,j-1} & I_{i-1,j} & I_{i-1,j+1} & I_{i-1,j+2} \\ I_{i,j-2} & I_{i,j-1} & I_{i,j} & I_{i,j+1} & I_{i,j+2} \\ I_{i+1,j-2} & I_{i+1,j-1} & I_{i+1,j} & I_{i+1,j+1} & I_{i+1,j+2} \\ I_{i+2,j-2} & I_{i+2,j-1} & I_{i+2,j} & I_{i+2,j+1} & I_{i+2,j+2} \end{bmatrix} \qquad (2)$$

In Eq. (2), a block is given in order to explain the common vector procedure. To compute the common and difference gradients in vertical direction, each column is considered as a vector. If our aim is to obtain horizontal gradient map, then each row vector is processed. Suppose that, we want to construct common and difference vectors in vertical direction. It clearly appears in the Eq. (2) that the dimension of a vector (n) is 5 and the number of vectors (m) is 5. Therefore, the insufficient data case is occurred as n >= m.

$$\vec{a}_1 = \left[I_{i-2,j-2} \ I_{i-1,j-2} \ I_{i,j-2} \ I_{i+1,j-2} \ I_{i+2,j-2} \right]^T$$
$$\vec{a}_2 = \left[I_{i-2,j-1} \ I_{i-1,j-1} \ I_{i,j-1} \ I_{i+1,j-1} \ I_{i+2,j-1} \right]^T$$
$$\vec{a}_3 = \left[I_{i-2,j} \ I_{i-1,j} \ I_{i,j} \ I_{i+1,j} \ I_{i+2,j} \right]^T \quad (3)$$
$$\vec{a}_4 = \left[I_{i-2,j+1} \ I_{i-1,j+1} \ I_{i,j+1} \ I_{i+1,j+1} \ I_{i+2,j+1} \right]^T$$
$$\vec{a}_5 = \left[I_{i-2,j+2} \ I_{i-1,j+2} \ I_{i,j+2} \ I_{i+1,j+2} \ I_{i+2,j+2} \right]^T$$

In insufficient data case, the common vector of a class can be obtained either with eigenvalues and eigenvectors of the covariance matrix related to vectors or Gram-Schmidt orthogonalization procedure. In this study, we have carried out the Gram-Schmidt orthogonalization procedure to obtain the common and difference vectors. An example is illustrated for explaining the calculation process of common and difference gradient maps through the horizontal direction by taking the 5 row vectors given in Eq. (2).

To make data as normalized and centered on the coordinate system, the mean is subtracted from each vectors. Thus, each vectors have zero mean ($a_i = a_i - \frac{1}{5} \sum_{i=1}^{5} a_i$). Then the difference set is constructed by subtracting each vectors ($a_i (i = 1, \dots, m)$) from a predefined reference vector, which is shown in Eq. (4). Also, it has been noted that selection of any vector as reference, produce the same common vector and does not affect expected result [19].

$$\vec{b}_1 = \vec{a}_2 - \vec{a}_1, \ \vec{b}_2 = \vec{a}_3 - \vec{a}_1, \ \vec{b}_3 = \vec{a}_4 - \vec{a}_1, \ \vec{b}_4 = \vec{a}_5 - \vec{a}_1 \quad (4)$$

Let B denotes the subspace spanned by vectors of difference set; $B = span\left\{ \vec{b}_1, \vec{b}_2, \vec{b}_3, \vec{b}_4 \right\}$ refers to the difference space of given vectors. As shown in Eq. (5), the orthonormal space is formed with Gram-Schmidt orthogonalization procedure by using the basis vectors of difference space;

$$\vec{d}_1 = \vec{b}_1, \ \vec{z}_1 = \frac{\vec{d}_1}{\left\| \vec{d}_1 \right\|}$$
$$\vec{d}_2 = \vec{b}_2 - <\vec{b}_2, \vec{z}_1> \vec{z}_1, \ \vec{z}_2 = \frac{\vec{d}_2}{\left\| \vec{d}_2 \right\|}$$
$$\vec{d}_3 = \vec{b}_3 - <\vec{b}_3, \vec{z}_1> \vec{z}_1 - <\vec{b}_3, \vec{z}_2> \vec{z}_2, \ \vec{z}_3 = \frac{\vec{d}_3}{\left\| \vec{d}_3 \right\|} \quad (5)$$
$$\vec{d}_4 = \vec{b}_4 - <\vec{b}_4, \vec{z}_1> \vec{z}_1 - <\vec{b}_4, \vec{z}_2> \vec{z}_2 - <\vec{b}_4, \vec{z}_3> \vec{z}_3, \ \vec{z}_4 = \frac{\vec{d}_4}{\left\| \vec{d}_4 \right\|}$$

Finally, the projection of any given vector $\vec{a}_i$ into the obtained orthonormal basis vectors can be computed as shown in Eq. (6). The projection process indicates the difference vector of related processed block.

$$\vec{a}_{diff} = \sum_{k=1}^{m-1} <\vec{a}_i, \vec{z}_k> \vec{z}_k. \quad (6)$$

To compute the horizontal gradient map, the procedure that is given from Eq. (3) to Eq. (6) is repeated for all 5 × 5 blocks in a processed image. Likewise, the same procedure is illustrated to extract vertical gradient map. Finally the horizontal and vertical gradient maps are combined to attain a single gradient map, which exhibits the edge strength in a given image. In case of gradient map computation over blocks, the overlapping with weighted averaging is considered by taking the recommendation in [21] into account.

The utilized algorithm for gradient extraction can be summarized as;

Algorithm: Single gradient map with CVA
1. *Take 5x5 blocks are taken from the input image*
2. *Subtracting reference one from each vectors in order to obtain the difference set.*
3. *Obtaining orthonormal vectors spanning difference subspace of processed block by applying Gram–Schmidt orthogonalization process onto the difference set of that block.*
4. *Calculating the difference vector for processed block by taking projection of a randomly chosen vector onto the basis returned from Gram–Schmidt orthogonalization process of that block.*
5. *Repeat, the same procedure to obtain a single gradient map*
6. *by utilizing the concept of CVA onto all remaining blocks.*

2.3 Extraction of Peak Points

While the derivative of image in horizontal direction indicates horizontal gradient image, the derivative of image in vertical direction refers to vertical gradient image. In this study, the CVA is employed to acquire derivative of the image through horizontal and vertical direction. This is the most crucial step for proposed algorithm and considered as a novel aspect in edge detection area. Similar to traditional methods, the gradient of image is computed by sum of absolute value of horizontal and vertical gradient image. Once the vertical and horizontal gradients obtained, the angle image, which contains the directions of each pixels in edge segment, is computed by dividing vertical gradients with horizontal one. Based upon the angle image, the direction image is computed in order for the computation of further steps including computation of peak points and connecting them with a smart routing algorithm.

To extract peak points for edge linking, the non-maximal suppression (NMS) method [16] has been carried out for the suppression of the local non-maxima of the magnitude image, in the study of Canny. After applied NMS, only the most powerful edge segments retained and rest of them has suppressed. This is resulting in a single pixel wide edge segment, which is often necessary for application of subsequent algorithms. However, the non-maximum suppression also generates quite a large number of edges. To overcome this problem, Canny proposed the two-threshold hysteresis in order to filter the output of the edge response.

Similarly, in this study, an optimal thresholding criteria has performed in order to suppress some pixels, which are less than a specified threshold, as a new NMS procedure.

Technically, the threshold value is determined by using the Otsu's method [22] which is named after its inventor Nobuyuki Otsu in 1979. It is an optimal threshold method as aimed in finding the optimal value for the global thresholding. Otsu's method selects the threshold value by minimizing the intra-class variance or maximizing inter-class variance. The main idea under the algorithm is that the image to be thresholded contains bi-modal histogram (two groups of pixels). With this respect, the optimum threshold for two groups of pixels has computed with Otsu's methodology. Then, foreground and background has separated by the thresholding operation. Hence, their combined spread (intra-class variance) becomes minimal. The ratio of between class variance dividing by global variance gives the threshold value. Moreover, the detailed information about the implementation of the algorithm can be analyzed in referred paper [22]. To obtain the binary image map, the edge pixels are determined with rules given in Eq. (7).

$$\forall (i, j),\ I(i, j) = no\_edge\ if \begin{cases} I(i, j) < Otsu\_Threshold \\ I(i, j) < I(i + n_x, j + n_y) \\ I(i, j) < I(i - n_x, j - n_y) \end{cases} \tag{7}$$

$$\forall (i, j),\ I(i, j) = edge \qquad otherwise$$

where $-1 \leq n_i \leq 1, i = x, y$. In the Eq. (7), sign of n_x and n_y is taken according to the edge direction at the pixel location.

Fig. 1. Connection proceeding; (a) East-West, (b) North-South, (c) Northeast-Southwest and (d) Northwest-Southeast.

2.4 Connection of Peak Points

To improve the quality of the edge maps produced by traditional edge detectors, many post-processing techniques [17, 23–25] have been proposed. As noted in previous steps, the obtained gradient image, peak points and directions of pixels should be carried out in the operation of connecting pixel points. To bind the pixels one by one, a smart connection procedure is performed by inspiring from the idea behind the ED algorithm [10]. As demonstrated in Fig. 1, four directions have considered as *east/west, northeast/southwest, north/south* and *northwest/southeast*, respectively. The smart routing process works as follows; initially a peak point is chosen and direction of the edge passing through this peak point is determined. In case of linking peak points, four directions are considered as;

i.   If the direction of edge falls east or west held on the peak point, the algorithm starts the connection process by proceeding to the east and to the west (refer to Fig. 1(a).

ii. If the direction of edge is north or south held on the peak point, the algorithm starts the connection process by proceeding to the north and to the south (refer to Fig. 1(b).

iii. If the direction of edge is northwest or southeast held on the peak point, the algorithm starts the connection process by proceeding to the northwest and to the southeast (refer to Fig. 1(c).

iv. If the direction of edge is northeast or southwest held on the peak point, the algorithm starts the connection process by proceeding to the northeast and to the southwest (refer to Fig. 1(d).

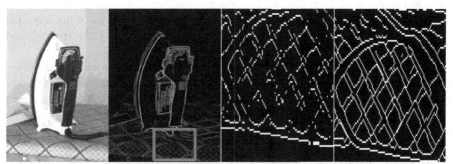

(a) Original image,   (b) *CVA-ED's edge map* (c) *Peak points.* (d) Linking peaks

Fig. 2. An illustration of the result of smart routing procedure.

During a move, only three immediate neighbors are considered and the one having the maximum gradient value is picked. We move out the edge area when two distinct conditions are occurred:

i. The gradient value of the current pixel is less than a predetermined threshold value,
ii. Or, in case of the algorithm hits a previous picked point.

To demonstrate how proposed method extracts the well-localized and non-jittered edges in a given region; the proposed algorithm is executed on a real and the result is given in Fig. 2. A part of *iron* image is cropped to give insights into the performance of CVA-ED, which includes the thin and continuous edges. It is clearly seen that the CVA-ED reveals the well linked lines and preserve the structural form of edge map and robust to extract well localized and contiguous edges.

Unlike the traditional methods which compute the gradient map by taking the derivative as pixel by pixel without considering the edge status of neighboring pixels (vectors), in the context of CVA-ED method, the neighbor pixels have taking into account in terms of overcome the noise coming from different sources. For this purpose, the CVA is carried out for each block, so that the edges are stored in the difference vector, $\bar{a}_{diff}$, after obtaining the common features from the original data of block. One can clearly observe that the quality of edge segments obtained from connection of peak points give satisfactory results.

3   Experiments and Results

Although an abundance of work on different methodology for edge detection has been proposed, but a solid method for the performance evaluation of developed edge detectors have not been generally accepted in the edge detection literature. The possible strategies to measure the performance of edge detectors would be probability of false alarms, probability of missed edges, errors of estimation in the edge angle, localization errors, the tolerance to distorted edges, corners and junctions. In a referred study [26], the published methods for edge detector performance evaluation was categorized into two fields including "theoretical" and "analytical"; ground truth required or not required. However, some proposed methods are subjective and there is not an objective method in terms performance evaluation of edge detectors.

Moreover, the human evaluation on edge detector output is widely accepted by researchers. In addition to this, in a variety of proposed edge comparison methods, visual rating which reports the perceived quality of the edges for identifying an object are employed to construct ground truth images which consist of different segments with different numbers specified by reader subjectively. Furthermore, it is generally accepted that different edge detectors would be better suitable for different tasks.

Although a crowded set of gradient extraction methods appears on edge detection literature, but making a comparison with widely known ones can be considered as sufficient in terms usability and effectiveness. In case of performance evaluation, we have judged the performance of proposed method versus to Ray edge detector. The reason for comparing our results with only Ray edge detector can be attributed to the page limitation, where there is no enough space to expand the comparison stage.

3.1   Performance Evaluation in Terms of Objective Measures

In this stage, we have focused on the evaluation of the performance for the proposed edge detector by judging the performance with respect to a comparison metric called ROC Curve analysis. For all of the experiments, 60 real images have carried out to investigate the capability of CVA for edge detection. To judge the performance, outputs of proposed method are presented and a direct comparison with a recently published [12], Ray's edge detector method, is commented by using the public available database of ROC curves which contains 60 real images, 50 of general objects and 10 of aerial scenes and their manually specified ground-truth segmentation data. The real images and their specified ground truths are available in the website of [27]. Noting that all experiments given in this work were executed on same hardware (Intel core i5-2400 with 3.10 GHz CPU and 4 GB memory) with software implemented on the Matlab.

As mentioned above the ROC curve [23] for a given edge detector can be constructed from true positive and false positive edge pixels which are obtained from the comparison of detected edge pixels to the specified ground truth. The segmented regions and edge pixels included in ground truth (GT) have distinct meanings, i.e. the black pixels represent the edges, and gray represents no edges and white represents "don't care". The areas, which are not classified neither by edge detector nor by GT, are called "don't care" regions. If a detector reports an edge pixel within a specified tolerance or a pixel is reported as edge by either of the two, it is counted as a true positive (TP). If an edge

Table 1. Obtained objective values from 10 aerial images.

Image	Ray's edge		Proposed	
	AR	F	AR	F
Average	1.38	0.28	1.72	0.37

Table 2. Obtained objective values from 50 object images.

Image	Ray's edge		Proposed	
	AR	F	AR	F
Average	1.13	0.20	1.39	0.31

pixel is decided as "edge" in GT, but coincide with no edge in edge detector it is counted as a false positive (FP). No any count is taken for "don't care" regions, which are marked with white color. The points that are non-edge in GT and also coincide with non-edge points in the detector output are called true negatives (TN). Points that were decided "non-edge" in GT and coincide with edge detector are reported as false negatives (FN). In case of TP and FN, the percent rate is derived in terms of total edge pixels in GT, for the FP and TN, the percent rate is obtained in terms of corresponding total pixels in the image. The accuracy rate of the edge detector is computed by comparing the rates of TP and TN against to the rates of FP and FN. The accuracy rate (AR) is computed by using the following formula.

$$AR = (TP + TN)/(FP + FN) \qquad (9)$$

Additionally, the *F-measure (F)* values of both proposed and compared methods are obtained. Generally, F is used to crosscheck the accuracy by considering the precision p and recall r to compute score. In the context of F, while *1* indicates high score, the zero (0) refers to worst score. The traditional F is the harmonic mean of *precision* and *recall* and computed by the formula of;

$$F = 2x \, (precision \, x \, recall)/(precission + recall) \qquad (10)$$

By observing the average F values, we can see that the proposed method is superior to other method. For the objects images, the obtained best F values are *0.43, 0.42* and *0.41* from the images, *43, 138 and Pitcher*, respectively. Moreover, in case of aerial images, the best F value is reaching *0.52* for the *Series* image. Moreover, while with the proposed method it is able to obtain the overall F values as *0.37* and *0.31*, but the other method yields *0.28* and *0.20*, for the real and aerial images, given in Tables 1 and 2, respectively. The results unveil the evidence of good quality for proposed method.

In addition to F-measure, the goodness of the edge detector has analyzed by using the accuracy rate, where the high value for TP and TN against the low value for FP and FN indicates the robustness of an edge detector. The quantitative measurements and success rates obtained to compare the edge maps, which are presented in Tables 1 and 2. For each table, the result in the last row indicates the average value of each given column. The highest accuracy rates are achieved with valuable scores of *2.12*, *2.05* and *1.95* which means that the performance proposed edge detector is more better than [12] when conducting on the such images, namely *110*, *Block* and *126*, respectively. Moreover, when the overall accuracy rates are considered in case of aerial images, for the best performance, while the proposed method yields *1.72* success rate, but the compared one exhibits *1.38* accuracy rate which indicates the evidence of superiority of the proposed method against Ray's edge detector.

3.2 Performance Evaluation in Terms of Subjective Measures

Moreover, some subjective outputs are presented in Fig. 3 in terms of visually evaluation. A synthetic image, named *103,* and an aerial image, named *Mainbuilding,* are used in case performance comparison. The edge maps of compared edge detector are taken from the original paper of the Ray's edge detector [12]. While the third row presents the edge maps obtained from the compared edge detector, the fourth row shows the edge maps of proposed edge detector. At a glance, it can be seen that the some details of images are hidden in edge maps of Ray's edge detector. On the other hand, CVA-ED's edge maps are very clean consisting of contiguous, well localized and more apparent. By inspecting the Fig. 3, we can see that the proposed edge detector can produce more clear lines.

Furthermore, the performance of our edge detector could be improved by illustrating a post-processing procedure, i.e. eliminating some edge segments that lower than a minimum length. However, the parameter tuning is not needed for the proposed method and it is aimed at as a parameter free edge detector. We believes that the success rate of proposed method comes from modification of some crucial steps such as the smoothing the noise with an ideal Gaussian filter (5×5), employing the CVA in terms of the gradient computation, which is the vital point and gaining the novelty to this experiment and lastly the idea under the ED method [10] is employed to link the edge pixel as dot to dot procedure after a smart thinning procedure. Hence, thin, contiguous and well-localized edges are obtained as shown in Fig. 3.

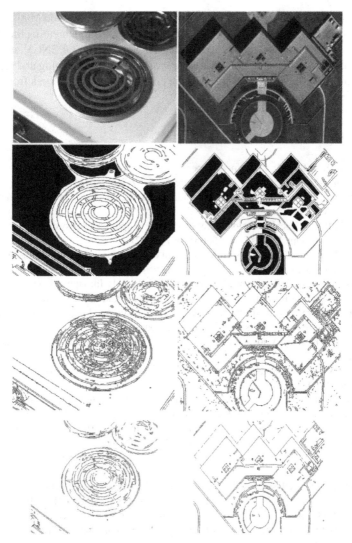

Fig. 3. The visual performance comparison with Ray's edge detector (Third row) and CVA-ED (Fourth row).

4   Conclusion

We have proposed a new image's gradients computation method by using the CVA, which has been used in the literature for some pattern recognition tasks. It is obviously seen that the proposed method presents the better edge map results than the Ray's work [12] when conducting on given images. Moreover, the value of threshold directly affects the performance of edge detector, it would be better if we used the local adaptive thresholding. However, for optimal threshold selection, Otsu's method is able to extract the well-localized and continuous edges. Although the elapsed running time for the

CVA-ED method is between 8 and 9 s without optimizing the executed Matlab code, but this duration will be improved by investigating new ways as implementing the algorithm on new platforms with different programming languages, i.e., OPENCV and C++, or after optimizing the Matlab code. As a future work, an image matching and registration method is aimed to develop by extracting the corner points and features from the edge maps responded from CVA-ED.

References

1. Kim, T., Lee, S., Paik, J.: Combined shape and feature-based video analysis and its application to non-rigid object tracking. IET Image Process. **5**, 87–100 (2011)
2. Paul, A., Wu, J., Yang, J.-F., Jeong, J.: Gradient-based edge detection for motion estimation in H. 264/AVC. IET Image Process. **5**, 323–327 (2011)
3. Singh, C., Pooja: Local and global features based image retrieval system using orthogonal radial moments. Opt. Lasers Eng. **50**, 655–667 (2012)
4. Sobel, I., Feldman, G.: A 3x3 isotropic gradient operator for image processing. A Talk at the Stanford Artificial Project, pp. 271–272 (1968)
5. Roberts, L.: Machine perception of three dimensional solids. In: Tippet, J., et al. (eds.) Optical and Electro-Optical Information Processing. MIT Press, Cambridge (1965)
6. Prewitt, J.M.: Object Enhancement and Extraction. Academic Press, New York (1970)
7. Scharr, H.: Optimal operators in digital image processing (2000)
8. Peng, W., Qichao, C.: A novel SVM-based edge detection method. Phys. Procedia **24**, 2075–2082 (2012)
9. Qian, Z., Wang, W., Qiao, T.: An edge detection method in DCT domain. Procedia Eng. **29**, 344–348 (2012)
10. Topal, C., Akinlar, C.: Edge drawing: a combined real-time edge and segment detector. J. Vis. Commun. Image Represent. **23**, 862–872 (2012)
11. Li, B., Söderström, U., Ur Réhman, S., Li, H.: Restricted hysteresis reduce redundancy in edge detection. J. Signal Inf. Process. **4**, 158–163 (2013)
12. Ray, K.: Unsupervised edge detection and noise detection from a single image. Pattern Recogn. **46**, 2067–2077 (2013)
13. Flores-Vidal, P.A., Olaso, P., Gómez, D., Guada, C.: A new edge detection method based on global evaluation using fuzzy clustering. Soft. Comput. **23**, 1809–1821 (2019)
14. Kimia, B.B., Li, X., Guo, Y., Tamrakar, A.: Differential geometry in edge detection: accurate estimation of position, orientation and curvature. IEEE Trans. Pattern Anal. Mach. Intell. **41**, 1573–1586 (2018)
15. Marr, D., Hildreth, E.: Theory of edge detection. Proc. R. Soc. Lond. B Biol. Sci. **207**, 187–217 (1980)
16. Canny, J.: A computational approach to edge detection. IEEE Trans. Pattern Anal. Mach. Intell. **8**, 679–698 (1986)
17. Wong, Y.-P., Soh, V.C.-M., Ban, K.-W., Bau, Y.-T.: Improved canny edges using ant colony optimization. In: 2008 Fifth International Conference on Computer Graphics, Imaging and Visualisation, CGIV 2008, pp. 197–202. IEEE (2008)
18. Bernal, J.: Linking Canny edge pixels with pseudo-watershed lines (2010)
19. Gulmezoglu, M.B., Dzhafarov, V., Keskin, M., Barkana, A.: A novel approach to isolated word recognition. IEEE Trans. Speech Audio Process. **7**, 620–628 (1999)
20. Gülmezoğlu, M.B., Dzhafarov, V., Edizkan, R., Barkana, A.: The common vector approach and its comparison with other subspace methods in case of sufficient data. Comput. Speech Lang. **21**, 266–281 (2007)

21. Dabov, K., Foi, A., Katkovnik, V., Egiazarian, K.: Image denoising by sparse 3-D transform-domain collaborative filtering. IEEE Trans. Image Process. **16**, 2080–2095 (2007)
22. Otsu, N.: A threshold selection method from gray-level histograms. IEEE Trans. Syst. Man Cybern. **9**, 62–66 (1979)
23. Shih, F.Y., Cheng, S.: Adaptive mathematical morphology for edge linking. Inf. Sci. **167**, 9–21 (2004)
24. Jevtić, A., Melgar, I., Andina, D.: Ant based edge linking algorithm. In: 2009 35th Annual Conference of IEEE Industrial Electronics, IECON 2009, pp. 3353–3358. IEEE (2009)
25. Rahebi, J., Elmi, Z., Shayan, K.: Digital image edge detection using an ant colony optimization based on genetic algorithm. In: 2010 IEEE Conference on Cybernetics and Intelligent Systems (CIS), pp. 145–149. IEEE (2010)
26. Heath, M.D., Sarkar, S., Sanocki, T., Bowyer, K.W.: A robust visual method for assessing the relative performance of edge-detection algorithms. IEEE Trans. Pattern Anal. Mach. Intell. **19**, 1338–1359 (1997)
27. Bowyer, K., Kranenburg, C., Dougherty, S.: Edge detector evaluation using empirical ROC curves. Comput. Vis. Image Underst. **84**, 77–103 (2001)

Optimizing Query Perturbations to Enhance Shape Retrieval

Bilal Mokhtari[1]([✉]), Kamal Eddine Melkemi[2], Dominique Michelucci[3], and Sebti Foufou[3,4]

[1] Laboratory of Applied Mathematics LMA, University of Biskra, BP 145 RP, 07000 Biskra, Algeria
`bilal.mokhtari@univ-biskra.dz`
[2] Department of Computer Science, University of Batna 2, 05000 Batna, Algeria
[3] Laboratoire d'Informatique de Bourgogne, EA 7534, Université de Bourgogne, BP 47870, 21078 Dijon Cedex, France
[4] New York University of Abu Dhabi, P.O. Box 129188, Abu Dhabi, UAE

Abstract. 3D Shape retrieval algorithms use shape descriptors to identify shapes in a database that are the most similar to a given key shape, called the query. Many shape descriptors are known but none is perfect. Therefore, the common approach in building 3D Shape retrieval tools is to combine several descriptors with some fusion rule. This article proposes an orthogonal approach. The query is improved with a Genetic Algorithm. The latter makes evolve a population of perturbed copies of the query, called clones. The best clone is the closest to its closest shapes in the database, for a given shape descriptor. Experimental results show that improving the query also improves the precision and completeness of shape retrieval output. This article shows evidence for several shape descriptors. Moreover, the method is simple and massively parallel.

Keywords: Computer vision · 3D Shape matching and recognition · Shape Retrieval · Shape Descriptors · Cloning · Genetic Algorithms

1 Introduction

Shape Retrieval computes which shapes in a database resemble the most to a given key shape Q, called the query [41]. Shapes are polyhedra with triangular faces. Output should be accurate (no false positive) and complete (no omitted solution). Basically, the shape retrieval algorithm computes off-line a shape descriptor, intuitively a signature or a feature vector, for each shape in the database. They do not depend on queries. It also computes on-line the shape descriptor of the query Q. Each shape descriptor induces a dissimilarity measure, or distance for short. For example, if the shape descriptor is an histogram, the dissimilarity measure can be the Chi-squared distance, the Kullback-Leibler divergence, the Hellinger distance, etc. Then, the algorithm computes this induced distance between Q and each shape in the database. Finally, the algorithm outputs the m (we use $m = 11$) shapes with the smallest dissimilarity to the query Q.

© Springer Nature Switzerland AG 2020
D. Slamanig et al. (Eds.): MACIS 2019, LNCS 11989, pp. 422–437, 2020.
https://doi.org/10.1007/978-3-030-43120-4_33

Several shape descriptors have already been proposed in the literature, but none achieves satisfying retrieval results with all kinds of shapes [8,10,12,14, 19,20]. The classical approach to solve this issue is to combine several shape descriptors using some fusion rules [1,4,6,25,27].

This article proposes to solve the problem by improving the query shape itself. Our approach is therefore orthogonal to the classical approaches, which use only one query at a time, and (a fusion of) many shape descriptors.

To improve the query, we propose a genetic algorithm (GA) [21,23,33,36] called GA-SR: Genetic Algorithm for 3D Shape Retrieval. GA-SR makes evolve a population of perturbed copies of the query shape. Perturbed copies are called clones. The fittest clone Q^* is the clone the closest to its m closest shapes $M(Q^*, D)$ in the database, for a given shape descriptor and its induced distance D. The m closest shapes to Q are the m closest shapes to the fittest clone.

All shapes in the database, query Q and its clones are (generically non convex) polyhedra with triangular faces. Q and all its clones share the same topology, *i.e.*, the same incidence relations between vertices, edges and faces. The sole difference between Q and any one of its clones is that the 3D coordinates of some vertices of Q are weakly perturbed. The perturbation is small enough, in order for the query and its clones to have similar appearance for the human eye.

Improving the query also improves the precision (no false positive) and completeness (no forgotten solution) of shape retrieval, regardless of the used shape descriptor and its induced dissimilarity measure. This article shows evidences for several shape descriptors: VND (Vertex Normal Descriptor), DMC (Discrete Mean Curvature), LSD (Local Shape Descriptor), and TD (Temperature Distribution).

Shapes in the database are usually classified into several classes or clusters to facilitate the work of classical shape retrieval methods [1,5,6]. In opposite, GA-SR does not need to know the class of shapes in the database. This information is only needed for measuring and comparing performances of GA-SR [19,23,26, 27,32,38,42].

The rest of this paper is organized as follows. Section 2 presents the background. Section 3 details GA-SR. Section 4 presents experimental results. Section 5 concludes.

2   Background and Principles

2.1   Improvement of Shape Retrieval

Several efforts have already been conducted to improve shape retrieval [37,43, 50]. Most of improvement methods are based on fusion of shape descriptors and their related dissimilarity measures. Chahooki et al. [6] proposed a method to fuse contour and region-based features for improving the retrieval precision. Akgül et al. proposed a fusion-based learning algorithm [1], which combines dissimilarity measures operating on different shape features. It computes their optimal combination by minimizing the empirical ranking risk criterion. Other fusion methods exist [7,27].

Improving pre-existing shape descriptors also improves shape retrieval output. For example, Bronstein and Kokkinos [3] present a scale-invariant version of the heat kernel descriptor previously proposed by Sun et al. [42]. Ling and Jacobs [28] aim to make the shape context descriptor by Sun et al. invariant to articulation: they replace the Euclidean distance by the inner (also called geodesic) distance to build a shape context descriptor. Other methods [51] for improving shape retrieval associate the database with a graph whose nodes are the database shapes. Therefore, the distance between shapes is defined as the length of the geodesic path in the graph associated to the database. A learning method permits to improve dissimilarity measure using graph transduction.

The concept of perturbation has been used as a successful strategy to improve many algorithms [15,16,22,29,46,49].

For example, Thompson and Flynn [46] extract the iris from an image by finding circular boundaries that approximate the circle surrounding the iris. A perturbation is performed by changing the values of one or more parameters of the method.

Stochastic arithmetic is another field which uses random perturbations to improve the robustness of numerical computations [49].

In Computational Geometry, small random perturbations of geometric data remove all degeneracies such as, in 2D, three collinear points or four co-cyclic points [15,16,22]. Perturbation greatly simplifies geometric algorithms, because only a small number of generic cases needs to be considered, while the number of degenerate cases increases exponentially with the geometric dimension of the problem.

In Stochastic Resonance, perturbation enhances the transmission of information and the detection of low signals [11,40].

In Machine Learning, several works [13,24] recently showed that noisy computations improve associative memories.

More recently, a face recognition system [30] is enhanced by using landmark perturbation technique that sweeps more landmarks, which improves faces comparison.

Yin et al. [52] establish connections between evolutionary algorithms and stochastic approximations.

In this wake, Vaira and Kurasova [48] use a genetic algorithm based on random insertion heuristics for the vehicle routing problem with constraints.

Ernest et al. [17] use GA and Genetic Fuzzy trees to compute deterministic fuzzy controllers, for autonomous training and control of squadron of unmanned combat aerial vehicles.

GAs have been used for solving complex optimization problems [34,35]. GAs have been also used as a powerful strategy to improve the precision in Information Retrieval Systems [18] and in Web Retrieved Documents [45].

GAs have been also used in Computer Vision and Graphics for measuring similarity of visual data, and in CBIR (Content-Based Image Retrieval). Syam and Rao [44] propose a GA-based similarity measure for CBIR: the GA integrates distinct image features in order to find images that are most similar to a given query image. Aparna [2] proposes a GA-based CBIR method to merge similarity scores: it computes the adequate weight associated with each similarity measure. Chan and King [7] combine different shape features: a GA computes suitable weights for considered features.

Several fitness functions have been used for information retrieval involving GAs. Thada and Jaglan [45] give a comparative study of similarity coefficients used to find the best fitness function, in order to find the most relevant text documents for a set of given keywords. Fan et al. [18] computes the best fitness function with a GA for information retrieval.

2.2 Shape Descriptors

Shape descriptor represents an essential ingredient for measuring the similarity of shapes. For a polyhedric shape with vertices V, it consists in calculating a signature for some of its vertices. It can be for all vertices in V, or for a strict subset of V refereed to by feature vertices.

Several researches have been conducted to propose discriminant shape descriptors [8,10,12,14,20].

We have considered several shape descriptors selected from different categories, such as Vertex Normal-based Descriptor VND [47], Local Shape Descriptor LSD [26], Temperature Distribution TD [19], and a Discrete Mean Curvature DMC [32]. The GA-SR methods based on these descriptors are referred as GA-VND, GA-LSD, GA-TD, and GA-DMC, respectively. We have selected these descriptors for their simplicity and efficiency. GA-SR improves all these shape descriptors, in terms of recall-precision curves. Other descriptors can be used.

The Vertex Normal-Based Descriptor (VND). The VND [47] descriptor is simple and fast. It considers the normal vector at vertices. The normal vector $\overrightarrow{N}$ at a vertex v is the average of normal vectors in the 1-star of the vertex:

$$\overrightarrow{N}(v) = \frac{1}{l}\sum \alpha_f \overrightarrow{N_f} \tag{1}$$

where l is the number of faces surrounding the vertex v, and α_f is the ratio area of the face f to the total area of the 1-star. The normal vector $\overrightarrow{N_f}$ of a face f with three points p_1, p_2 and p_3, is given by:

$$\overrightarrow{N_f} = (p_2 - p_1) \times (p_3 - p_1) \tag{2}$$

$p_i = (x_i, y_i, z_i)$, $i = 1, 2, 3$, and $\times$ stands for the cross product. The orientation of $\overrightarrow{N_f}$ does not matter. Let F be the subset of feature vertices $n(F) = 3000$. Then the descriptor VND of a vertex v in F is given by:

$$VND(v) = \frac{\|\overrightarrow{N}(v)\|_2}{\sum_{v' \in F} \|\overrightarrow{N}(v')\|_2} \tag{3}$$

Discrete Mean Curvature (DMC). The Discrete Mean Curvature [32] of a vertex v is given by:

$$DMC(v) = \frac{1}{4} \sum_{i=1}^{d} l_i(\pi - \beta_i) \tag{4}$$

where d is the degree of vertex v, β_i the internal dihedral angle (in radians) between two consecutive faces around the vertex v, and l_i the length of the edge common to those faces.

Local Shape Distribution (LSD). The LSD descriptor [26] extracts n random vertices ($n = 3000$), and characterizes each sample vertex v in terms of Euclidean distances to all other points belonging to its neighborhood. The neighborhood is a spherical region centered at point v. The LSD descriptor associates to each region a histogram of Euclidean distances between the point v and points in its neighborhood.

To compute the similarity between two shapes A and B, a complete bipartite graph g is built as follows: the first set of vertices of g is given by the regions of A, the second set is given by the regions of B. The cost of an edge (a, b) between two regions in g is the Chi-squared distance between the a histogram and the b histogram. By definition, the distance between A and B is the smallest cost of perfect matchings in g. This method does not only compute a distance between two shapes A and B, but it also matches regions in A with regions in B.

The Temperature Distribution (TD). The temperature distribution [42] simulates the heat diffusion process on the surface of a model, which starts at a vertex, and goes through other vertices over time.

The temperature distribution descriptor [19] of a vertex is represented as the average of temperatures measured on all vertices in the surface of the model, after applying a unit heat at that vertex. The average temperature for a vertex v, at heat dissipation time t, is given by:

$$TD(v) = \frac{1}{n-1} \sum_{w, w \neq v} \sum_i e^{-\lambda_i t} \phi_i(v) . \phi_i(w) \tag{5}$$

where n is the number of vertices (usually $n \approx 3400$), $t = 50$ is a constant, and λ_i is the ith eigenvalue (sorted in decreasing order) of the Laplacian of the underlying graph of the mesh, and ϕ_i its ith eigenvector. In practice, only few eigenvectors are used, four in our experiments.

The distribution of the average temperature values is then represented by means of a histogram. The distance between two shapes is the L_2-norm computed from their histograms.

TD descriptor is invariant to isometric transformations like pose changes, and robust against noise and geometric textures like bumps. However, TD is improved by GA-SR.

2.3 Shape Similarity and Statistical Distances

There are many statistical distances to calculate dissimilarity between two shapes represented as distributions (histograms): Kullback-Leibler divergence, Hellinger distance, Bhattacharyya distance, Chi-squared distance, L_n norm, etc. In this work, we have used some of these distances to measure the dissimilarity of shapes based on each of the used descriptors. We have used Chi-squared distance [39] for VND, LSD and DMC, and L_2-norm for TD [9], in accordance to their experiments. Note that the number of drawers of histograms of compared shapes is $b \approx \sqrt{n}$ ($b \approx 50$). In the rest of this paper, $D(A, B)$ refers to the distance between two shapes A and B.

3   GA-SR: Genetic Algorithm for Shape Retrieval

3.1 Notations and Definitions

All shapes *i.e.,* the query, its clones, and shapes in the database, are polyhedra with triangular faces. A polyhedron is represented with a geometric part V and a topologic part F. V is an array of the 3D coordinates of vertices of the polyhedron: $V_i = (x_i, y_i, z_i) \in \mathbb{R}^3$. Coordinates are floating point numbers. F is an array of triangular faces: $F_k = (a_k \in \mathbb{N}, b_k \in \mathbb{N}, c_k \in \mathbb{N})$, where a_k, b_k, c_k are the indices in array V of the vertices composing face F_k. a_k, b_k, c_k are typically ordered counterclockwise, seen from outside the polyhedron. For convenience, a scaling normalization is applied to all polyhedra, so that the sum of all triangles areas equals one (one square meter, say).

A query and all its clones have the same topologic part F. However, the geometric parts are different. Let $Q = \text{shape}(V, F)$ be the query shape. Let $Q' = \text{shape}(V', F)$ be a clone of Q. The geometric part V' of Q' is defined as:

$$V' := V + P, \quad ||P||_\infty \le \epsilon, \quad ||P||_0 = \mu = \lceil \rho n \rceil \tag{6}$$

where $P_i = (x_i, y_i, z_i) \in [-\epsilon, \epsilon]^3$ is a perturbation vector, $\epsilon \in \mathbb{R}^+$ the noise threshold, n the number of vertices. P is the unknown of our problem.

$||P||_\infty \le \epsilon$ is imposed to guarantee the perturbation is small. This constraint is compatible with GA cross-over. Typically, ϵ is between 0.002 (2 mm) and 0.06 (6 cm). The optimal values of ϵ for VND, TD, DMC, and LSD are respectively 0.0074, 0.0022, 0.0562, and 0,0005.

Moreover, we impose that P is sparse. Let ρ be the probability for a vertex to be ϵ-perturbed. In practice, $\rho = 1/4$. The number of perturbed vertices is $\mu = \lceil \rho n \rceil$, with n the number of vertices. The number of perturbed vertices is the same for all clones of a query. This constraint is sometimes written $||v||_0 = \mu$, where $|.|_0$ is a pseudonorm *i.e.,* $||v||_0$ is the number of non zero coordinates of v.

Clones are not re-normalized. It is assumed that the perturbation size is less than the Least Feature Size of shapes, so perturbations do not introduce self-intersections or other geometric inconsistencies.

Let $M(Q, D)$ or $M(Q)$ be the set of the $m = 11$ shapes in the database which are the closest to Q, according to the dissimilarity measure D.

Let q be a shape, typically a clone of Q or Q itself. Its fitness $f(q)$ or $f(q, D)$ is the averaged distance between q and shapes in $M(q, D)$ defined as:

$$f(q, D) := (1/m) \sum_{b \in M(q,D)} D(q, b) \tag{7}$$

We are looking for the perturbation P such that the clone $q = \text{shape}(V+P, F)$ is the closest to its m most similar shapes in the database, *i.e.*, such that $f(q, D)$ is minimal. For convenience, pose $g(P) := f(\text{shape}(V + P, F), D)$. Then the problem becomes: find the optimal or a good enough perturbation X which minimizes $g(X)$ with $||X||_\infty \leq \epsilon$, $||X||_0 = \mu$:

$$X^* = \text{argmin } g(X), \quad ||X||_\infty \leq \epsilon, \quad ||X||_0 = \mu \tag{8}$$

3.2 Sensitivity to Perturbations and Discretization Artifacts

Shape descriptors are very sensitive to noise, *i.e.*, small random perturbations and artifacts due to discretization. This sensitivity, which can be seen as a shortcoming of shape descriptors, is illustrated in Fig. 1: it shows for several shape descriptors the distance curves between a model David1 and clones of a model David2. David1 and David2 are two statues of David, in different poses. Let V_2, F_2 be the geometry and the faces of David2. Let P_2 be the normalized direction of some perturbation vector: $||P_2||_\infty = 1$ for simplicity. Each curve in Fig. 1 shows the curve $d(t) = D(\text{David1}, \text{shape}(V_2 + tP_2, F_2))$, with t sampled in $[0., 0.1]$. t is on the horizontal axis, and $d(t)$ on the vertical axis. $d(0)$ is not zero: it is the distance between David1 and David2. It depends on the used descriptor. $d(t)$ quickly falls below $d(0)$ for tiny values of t in $[0, 0.006]$, then slowly increases until $t = 0.01$ or 0.07 depending on the used shape descriptor, and finally quickly increases. For $t \in (0., 0.01]$ or $(0., 0.07]$ depending on the used shape descriptor, all $d(t)$ are below $d(0)$ for this random perturbation direction P_2. These distance curves are rough or noisy. This is due to discretization artifacts. When it is possible, increasing n, the number of samples or feature vertices, and thus increasing the ratio n/b yield to smoother curves. Anyway, this noise does not jeopardize GA-SR, so it is useless to try to reduce it.

These features can be reproduced and more easily understood in the much simpler context of 1D shapes, see Fig. 2. A 1D shape is a continuous and derivable function from $[0, 1]$ to $[0, 1]$, for convenience. For discretization, the interval $[0, 1]$ is divided into n intervals, with $n = 250$ or 5000 in Fig. 2. Each function f is discretized with a vector F such that $F[i] = f(i/n)$, $i \in [0, n]$. The distance between two 1D shapes f_1 with vector F_1 and f_2 with vector F_2 is the Chi-squared distance between their histograms $H(F_1)$ and $H(F_2)$ with $b = 50$ buckets per histogram (this is the value used in references). The example in Fig. 2 uses 1D shapes $f_1(x) = L(a_1, x)$ and $f_2(x) = L(a_2, x)$, where $L(a, x) = ax(1 - x)$ is the Logistic map, and $a_1 = 0.7$ and $a_2 = 0.75$. Visually, f_1 and f_2, or their respective vectors F_1 and F_2, are very close, but the Chi-squared distance between their histograms is 0.16 or 0.17 (the possible maximal value in 1). Figure 2 shows that

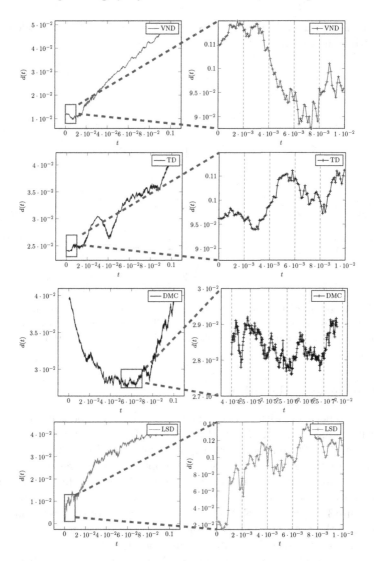

Fig. 1. The impact of perturbation parameter t (horizontal axis) applied on the clones of David2 model, using VND, TD and DMC descriptors in term of distance (vertical axis) to David1 model. Distances are computed between a model David1 and clones of David2. Values of graphs of the left column are picked with a step of 0.0001 in the interval $[0, 0.1]$, and those of the right column are picked with the step in the intervals $[0, 0.01]$, $[0, 0.01]$, $[0.04, 0.07]$, and $[0, 0.01]$.

clones of F_2 are closer to F_1. Each point (x, y) of a curve in Fig. 2 is $(x = t, y = \chi^2(H(F_1), H(F_2(t))))$ and $F_2(t) = F_2 + tP_2$, where P_2 is a random perturbation vector. Three random perturbation vectors P_2 were tried. For all of them, some clone of F_2 is better than the query F_2, *i.e.*, closer to F_1. With $n = 5000$, curves are smoother than with $n = 250$.

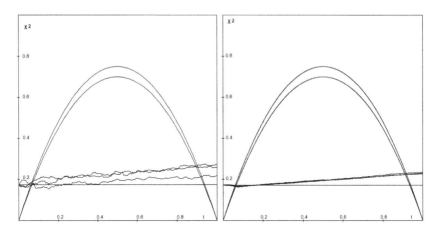

Fig. 2. Distance curves between F_1 and clones of F_2. Left: $n = 250$ samples. Right: $n = 5000$ samples. The height of the horizontal line is 0.17, the Chi-squared distance between F_1 and F_2 histograms. Many clones of F_2 are closer to F_1 than F_2 itself.

3.3   The Genetic Algorithm

GA-SR is a genetic algorithm. Let Q be the query, and D be the shape descriptor and its induced distance. GA-SR makes evolve a population P of $K = 15$ clones during $G = 20$ iterations, from generation P_0 to P_G. The first population P_0 contains Q and $K - 1$ mutants. Next populations are generated with GA operators: crossover and mutation. Equation (7) defines the fitness of a clone q: the closest to the set $M(q, D)$ of its most similar shapes in the database, the fittest. Figure 3 shows the evolution of the fitness value of the best clone at each generation of the GA. The best solution of the GA corresponds to the minimal fitness value, in this case at the fourteenth generation.

The genotype of a clone is an unsorted array of its μ perturbed vertices: (i, x_i, y_i, z_i), where i is the index of the perturbed vertex, and (x_i, y_i, z_i) the 3D coordinates of the vertex after perturbation. The tuple (i, x_i, y_i, z_i) is called a gene in the GA parlance. It is easy to obtain vertex coordinates of the clone from the vertex coordinates of the query and from the genotype.

Each clone in the first population P_0 is generated with mutations of the query. Let V, F be the geometric and topologic parts of Q. The μ genes of each clone are generated as follows. μ distinct vertex indices in $1, \ldots n$ are picked at random. Let i be one of these integers. Let $V_i = (x_i, y_i, z_i)$ be the 3D coordinates of vertex V_i of Q. The gene is $(i, x_i + \epsilon R(), y_i + \epsilon R(), z_i + \epsilon R())$ where function $R()$ returns a pseudorandom floating point value uniformly distributed in the interval $[-1, 1]$.

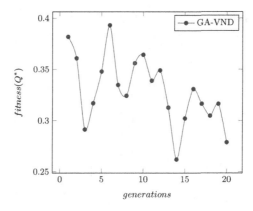

Fig. 3. Evolution of the fitness of the best clone among generations.

For each population P_g, $g = 0, \ldots G$, the fitness function (see Eq. 7) of every clone is computed. To renew the population, standard genetic operators are applied to selected parents to generate new clones: there is no elitism, so the curve in Fig. 3 is not monotonous. More precisely, $K/2$ pairs of clones are selected using the fitness-proportionate selection rule (also called the roulette-wheel selection). Each selected pair generates two new clones with a standard crossover operation between the two genotypes. These two new clones replace their parents in the next generation.

The standard crossover between a first genotype $G_1 = L_1 R_1$ (L for left, R for right) and a second genotype $G_2 = L_2 R_2$ gives two genotypes $L_1 R_2$ and $L_2 R_1$, where lengths of L_1, L_2 are equal. Any classical crossover operator can be used. Some vertices may be perturbed several times, without hindering GA-SR.

Mutation is an important operator in evolutionary algorithms. Each generated clone is subject to a post-mutation: with probability 0.01, each gene (i, x_i, y_i, z_i) is changed to $(j, x_j + \epsilon_x, y_j + \epsilon_y, z_j + \epsilon_z)$, where j is selected randomly and ϵ_x, ϵ_y, ϵ_z are pseudo random values uniformly distributed in $[-\epsilon, \epsilon]$.

4 Experiments

4.1 Databases Used

We used the databases TOSCA [53] and SHREC'11 [26]. TOSCA contains 148 3D models (eg. Cats, Centaurs, Dogs, Wolves, Horses, Lions, Gorillas, Sharks, Female and Male figures). The models are distributed into 10 categories including a variety of poses. SHREC'11 contains about 600 non-rigid 3D objects classified into different groups of models, each of which contains approximately the same number of models. In both databases, 3D models are represented as triangular meshes stored in ASCII files in .off format (Object File Format). The name of each file implicitly gives the class (*e.g.,* Cats, Dogs, etc), which permits measurement of performances of retrieval algorithms.

4.2 Tests and Results

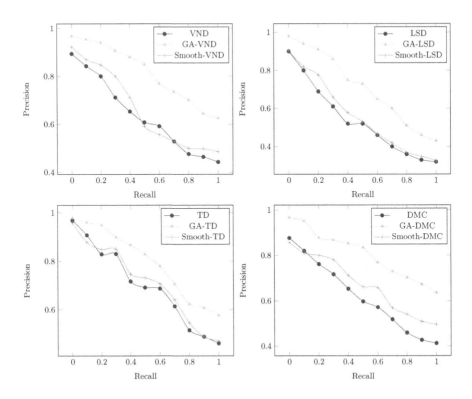

Fig. 4. Averaged 11-point precision-recall curves of random queries of the TOSCA database using original descriptors (blue), Smooth-based descriptors (violet) and the GA-based descriptor (red). (Color figure online)

The output quality of shape retrieval algorithms is measured with precision-recall curves. They account both for precision and completeness. They are drawn with the 11-point interpolated average precision algorithm by Manning et al. [31]. It is the reason why we use $m = 11$. The higher the precision-recall curve, the better the retrieval.

Figures 4 for TOSCA and Fig. 5 for SHREC'11 show the precision-recall curves of descriptors VND, LSD, DMC, and TD compared to their GA counterparts. Clearly, GA-SR significantly improves all these descriptors.

To show the effectiveness of our method, we compare it to the following existing methods: D2 [38], MDS−ZFDR [27], GPS [42], and GT [51]. Comparison results are illustrated in plots of Fig. 6. GA-SR shows better performance.

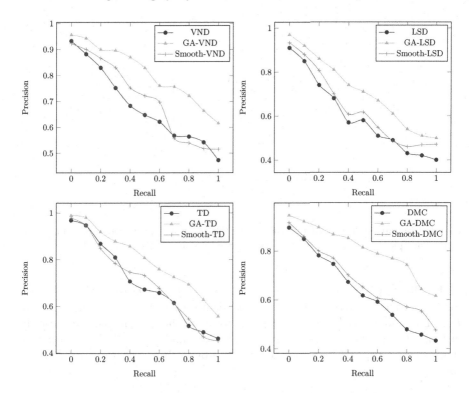

Fig. 5. Averaged 11-point precision-recall curves of random queries of the SHREC'11 database using original descriptors (blue), Smooth-based descriptors (violet) and the GA-based descriptor (red). (Color figure online)

Smoothing of a query shape is another possible way to improve shape retrieval. To smooth a shape, all its vertices are smoothed (without any constraint regarding the order). Let v be a vertex, let g be the barycentre of its neighbors. Then v', the corresponding smoothed vertex, is defined using (9).

$$v' := \mathrm{smooth}(v) := \alpha v + (1 - \alpha)g \tag{9}$$

where α is a parameter in $[0, 1]$. Then D', the smoothed distance for D in VND, LSD, DMC, TD, is (10):

$$D' := D(\mathrm{smooth}(A), \mathrm{smooth}(B)) \tag{10}$$

where smooth(.) is the smoothing operator. Smoothing reduces noise and irregularities, so intuitively, we expect smoothing to reduce distances: $D'(A, B) \leq D(A, B)$. Smoothing is simple and fast, in particular faster than GA-SR. Figures 4 for TOSCA and Fig. 5 for SHREC'11 show the precision-recall curves for VND, LSD, DMC, TD, their smoothed counterparts, and their GA counterparts. Clearly, cloning achieves better retrieval results than smoothing.

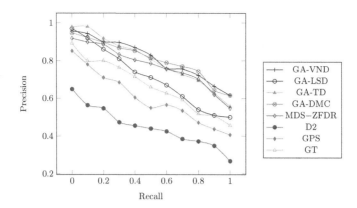

Fig. 6. Comparing results of GA-SR with other methods proposed in the literature. Curves are plotted according to results reported on the SHREC'11 database.

Finally, GA-SR is compatible with fusion: let $D_1, \ldots D_s$ be s shape descriptors and their induced distances. Then define their fusion distance D with: $D(A, B) := \min(D_1(A, B), \ldots D_s(A, B))$ (or any other fusion rule), and use this distance D with GA-SR. We compared GA-SR and SR with this min-merged shape descriptor, and here too, GA-SR improves SR. No figure is provided for conciseness.

5   Conclusion

Shape descriptors are very sensitive to small perturbations. This shortcoming is also an opportunity for improving shape retrieval. GA-SR achieves better results than previous classical retrieval methods, and better results than smoothing. Other shape descriptors are easily taken into account. GA-SR is simple and massively parallel. It needs no machine learning, no deep learning, no supervision.

References

1. Akgül, C.B., Sankur, B., Yemez, Y., Schmitt, F.: Similarity score fusion by ranking risk minimization for 3D object retrieval. In: Proceedings of the 1st Eurographics Conference on 3D Object Retrieval, pp. 41–48. Eurographics Association (2008)
2. Aparna, K.: Retrieval of digital images based on multi-feature similarity using genetic algorithm. Int. J. Eng. Res. Appl. (IJERA) **3**(4), 1486–1499 (2013)
3. Bronstein, M.M., Kokkinos, I.: Scale-invariant heat kernel signatures for non-rigid shape recognition. In: 2010 IEEE Conference on Computer Vision and Pattern Recognition (CVPR), pp. 1704–1711. IEEE (2010)
4. Bu, S., Cheng, S., Liu, Z., Han, J.: Multimodal feature fusion for 3D shape recognition and retrieval. IEEE MultiMedia **21**(4), 38–46 (2014)

5. Carneiro, G., Chan, A.B., Moreno, P.J., Vasconcelos, N.: Supervised learning of semantic classes for image annotation and retrieval. IEEE Trans. Pattern Anal. Mach. Intell. **29**(3), 394–410 (2007)

6. Chahooki, M., Charkari, N.M.: Shape retrieval based on manifold learning by fusion of dissimilarity measures. IET Image Process. **6**(4), 327–336 (2012)

7. Chan, D.Y.-M., King, I.: Genetic algorithm for weights assignment in dissimilarity function for trademark retrieval. In: Huijsmans, D.P., Smeulders, A.W.M. (eds.) VISUAL 1999. LNCS, vol. 1614, pp. 557–565. Springer, Heidelberg (1999). https://doi.org/10.1007/3-540-48762-X_69

8. Chang, M.C., Kimia, B.B.: Measuring 3D shape similarity by graph-based matching of the medial scaffolds. Comput. Vis. Image Underst. **115**(5), 707–720 (2011). Special issue on 3D Imaging and Modelling

9. Chang, S.K., Wong, Y.: Ln norm optimal histogram matching and application to similarity retrieval. Comput. Graph. Image Process. **13**(4), 361–371 (1980)

10. Chao, M.W., Lin, C.H., Chang, C.C., Lee, T.Y.: A graph-based shape matching scheme for 3D articulated objects. Comput. Animation Virtual Worlds **22**(2–3), 295–305 (2011)

11. Chapeau-Blondeau, F., Rousseau, D.: Raising the noise to improve performance in optimal processing. J. Stat. Mech. Theory Exp. **2009**(01), P01003 (2009)

12. Chen, D.Y., Ouhyoung, M.: A 3D object retrieval system based on multi-resolution Reeb graph. In: Computer Graphics Workshop, pp. 16–20 (2002)

13. Chen, H., Varshney, L.R., Varshney, P.K.: Noise-enhanced information systems. Proc. IEEE **102**(10), 1607–1621 (2014)

14. Coifman, R., et al.: Geometric diffusions as a tool for harmonic analysis and structure definition of data: diffusion maps. Proc. Nat. Acad. Sci. **102**(21), 7426–7431 (2005)

15. De Berg, M., Van Kreveld, M., Overmars, M., Schwarzkopf, O.C.: Computational Geometry. Springer, Heidelberg (2000). https://doi.org/10.1007/978-3-662-04245-8

16. Emiris, I.Z., Canny, J.F.: A general approach to removing degeneracies. SIAM J. Comput. **24**(3), 650–664 (1995)

17. Ernest, N., Cohen, K., Kivelevitch, E., Schumacher, C., Casbeer, D.: Genetic fuzzy trees and their application towards autonomous training and control of a squadron of unmanned combat aerial vehicles. Unmanned Syst. **3**(03), 185–204 (2015)

18. Fan, W., Gordon, M.D., Pathak, P.: A generic ranking function discovery framework by genetic programming for information retrieval. Inf. Process. Manage. **40**(4), 587–602 (2004)

19. Fang, Y., Sun, M., Ramani, K.: Temperature distribution descriptor for robust 3D shape retrieval. In: 2011 IEEE Computer Society Conference on Computer Vision and Pattern Recognition Workshops (CVPRW), pp. 9–16. IEEE (2011)

20. Gal, R., Cohen-Or, D.: Salient geometric features for partial shape matching and similarity. ACM Trans. Graph. **25**(1), 130–150 (2006)

21. Goldberg, D.E., et al.: Genetic Algorithms in Search Optimization and Machine Learning, vol. 412. Addison-Wesley, Reading (1989)

22. Hoffmann, P.H.C.M., Revol, W.L.N. (eds.): Reliable Implementation of Real Number Algorithms: Theory and Practice. Springer, Heidelberg (2008). https://doi.org/10.1007/978-3-540-85521-7

23. Holland, J.H.: Adaptation in Natural and Artificial Systems: An Introductory Analysis with Applications to Biology, Control, and Artificial Intelligence. University of Michigan Press, Ann Arbor (1975)

24. Karbasi, A., Salavati, A.H., Shokrollahi, A., Varshney, L.R.: Noise facilitation in associative memories of exponential capacity. Neural Comput. **26**(11), 2493–2526 (2014)
25. Kittler, J., Hatef, M., Duin, R.P., Matas, J.: On combining classifiers. IEEE Trans. Pattern Anal. Mach. Intell. **20**(3), 226–239 (1998)
26. Li, B., et al.: SHREC'12 track: generic 3D shape retrieval. In: 3DOR, pp. 119–126 (2012)
27. Li, B., Godil, A., Johan, H.: Hybrid shape descriptor and meta similarity generation for non-rigid and partial 3D model retrieval. Multimedia Tools Appl. **72**(2), 1531–1560 (2014)
28. Ling, H., Jacobs, D.W.: Using the inner-distance for classification of articulated shapes. In: 2005 IEEE Computer Society Conference on Computer Vision and Pattern Recognition, CVPR 2005, vol. 2, pp. 719–726. IEEE (2005)
29. Luo, J., Gu, F.: An adaptive niching-based evolutionary algorithm for optimizing multi-modal function. Int. J. Pattern Recognit. Artif. Intell. **30**(03), 1659007 (2016)
30. Lv, J.J., Cheng, C., Tian, G.D., Zhou, X.D., Zhou, X.: Landmark perturbation-based data augmentation for unconstrained face recognition. Signal Process. Image Commun. **47**, 465–475 (2016)
31. Manning, C.D., Raghavan, P., Schütze, H., et al.: Introduction to Information Retrieval, vol. 1. Cambridge University Press, Cambridge (2008)
32. Meyer, M., Desbrun, M., Schröder, P., Barr, A.H.: Discrete differential-geometry operators for triangulated 2-manifolds. In: Hege, H.C., Polthier, K. (eds.) Visualization and Mathematics III, pp. 35–57. Springer, Heidelberg (2003). https://doi.org/10.1007/978-3-662-05105-4_2
33. Miranda, V., Ranito, J., Proenca, L.M.: Genetic algorithms in optimal multistage distribution network planning. IEEE Trans. Power Syst. **9**(4), 1927–1933 (1994)
34. Misevičius, A.: Experiments with hybrid genetic algorithm for the grey pattern problem. Informatica **17**(2), 237–258 (2006)
35. Misevičius, A., Rubliauskas, D.: Testing of hybrid genetic algorithms for structured quadratic assignment problems. Informatica **20**(2), 255–272 (2009)
36. Mitchell, M.: An Introduction to Genetic Algorithms. MIT Press, Cambridge (1998)
37. Mohamad, M.S., Deris, S., Illias, R.M.: A hybrid of genetic algorithm and support vector machine for features selection and classification of gene expression microarray. Int. J. Comput. Intell. Appl. **5**(01), 91–107 (2005)
38. Osada, R., Funkhouser, T., Chazelle, B., Dobkin, D.: Shape distributions. ACM Trans. Graph. (TOG) **21**(4), 807–832 (2002)
39. Pele, O., Werman, M.: The Quadratic-Chi histogram distance family. In: Daniilidis, K., Maragos, P., Paragios, N. (eds.) ECCV 2010. LNCS, vol. 6312, pp. 749–762. Springer, Heidelberg (2010). https://doi.org/10.1007/978-3-642-15552-9_54
40. Rousseau, D., Anand, G., Chapeau-Blondeau, F.: Noise-enhanced nonlinear detector to improve signal detection in non-Gaussian noise. Signal Process. **86**(11), 3456–3465 (2006)
41. Safar, M.H., Shahabi, C.: Shape Analysis and Retrieval of Multimedia Objects, vol. 23. Springer, Boston (2003). https://doi.org/10.1007/978-1-4615-0349-1
42. Sun, J., Ovsjanikov, M., Guibas, L.: A concise and provably informative multi-scale signature based on heat diffusion. In: Computer Graphics Forum, pp. 1383–1392. Wiley Online Library (2009)
43. Super, B.J.: Retrieval from shape databases using chance probability functions and fixed correspondence. Int. J. Pattern Recognit. Artif. Intell. **20**(08), 1117–1137 (2006)

44. Syam, B., Rao, Y.: An effective similarity measure via genetic algorithm for content based image retrieval with extensive features. Int. Arab J. Inf. Technol. (IAJIT) **10**(2), 143–151 (2013)

45. Thada, V., Jaglan, V.: Comparison of Jaccard, Dice, Cosine similarity coefficient to find best fitness value for web retrieved documents using genetic algorithm. Int. J. Innov. Eng. Technol. **2**(4), 202–205 (2013)

46. Thompson, J., Flynn, P.: A segmentation perturbation method for improved iris recognition. In: 2010 Fourth IEEE International Conference on Biometrics: Theory Applications and Systems (BTAS), pp. 1–8, September 2010

47. Thürrner, G., Wüthrich, C.A.: Computing vertex normals from polygonal facets. J. Graph. Tools **3**(1), 43–46 (1998)

48. Vaira, G., Kurasova, O.: Genetic algorithm for VRP with constraints based on feasible insertion. Informatica **25**(1), 155–184 (2014)

49. Vignes, J.: A stochastic arithmetic for reliable scientific computation. Math. Comput. Simul. **35**(3), 233–261 (1993)

50. Wong, W.T., Shih, F.Y., Su, T.F.: Shape-based image retrieval using two-level similarity measures. Int. J. Pattern Recognit. Artif. Intell. **21**(06), 995–1015 (2007)

51. Yang, X., Bai, X., Latecki, L.J., Tu, Z.: Improving shape retrieval by learning graph transduction. In: Forsyth, D., Torr, P., Zisserman, A. (eds.) ECCV 2008. LNCS, vol. 5305, pp. 788–801. Springer, Heidelberg (2008). https://doi.org/10.1007/978-3-540-88693-8_58

52. Yin, G., Rudolph, G., Schwefel, H.P.: Establishing connections between evolutionary algorithms and stochastic approximation. Informatica **6**(1), 93–117 (1995)

53. Young, S., Adelstein, B., Ellis, S.: Calculus of nonrigid surfaces for geometry and texture manipulation. IEEE Trans. Visual Comput. Graphics **13**(5), 902–913 (2007)

Authorship Attribution by Functional Discriminant Analysis

Chahrazed Kettaf$^{(\boxtimes)}$ and Abderrahmane Yousfate

Laboratoire de Mathématiques, Djillali Liabes University, Sidi Bel Abbes, Algeria
kchahrazed8@gmail.com, yousfate@univ-sba.dz

Abstract. Recognizing the author of a given text is a very difficult task that relies on several complicated and correlated criterias. For this purpose, several classification methods are used (neuronal network, discriminant analysis, SVM...). But a good representation of the text that keeps the maximum of the stylistic information is very important and has a considerable influence on the result. In this paper, we will tackle the problem of the authorship attribution for very long texts using the discriminant analysis extended to the functional case after presenting the texts as elements of a separable Hilbert space.

Keywords: Authorship attribution · Textmining · Big textual data · Discriminant analysis · Funtional classification · Functional data analysis

1 Introduction

Textual data is the major part of the data on computers (information, course, article, contract, CV,...), Textmining aims to explore this data to extract or retrieve information, synthesize, translate... which induces, as in datamining, the use of new statistical tools including data analysis, neural networks, SVM...

Before applying any of these tools, the textual data must go through a preprocessing step. This step is essential and greatly influences the quality of the results. Moreover, it depends on several factors including: the format of the texts (article, CV, book,... etc), the language and the nature of the problem (thematic classification, search for information, identification of the author,... etc.), which makes it too difficult. The authorship attribution is very important in several fields (crime, plagiarism [21],...), It depends on the preferred vocabulary of the author, the length of the sentences, the genre of the author's works (theater, poem, narrative text...). In order to identify the style of the author, we must have a method that keeps the maximum of this author's stylistic information. In this article, we present a new method of representation of texts by functions. By applying the functional discriminant analysis (FDA) to these functional data, we will show that this method is adapted to the problem of identification of the author and gives extremely interesting results.

D. Slamanig et al. (Eds.): MACIS 2019, LNCS 11989, pp. 438–449, 2020.
https://doi.org/10.1007/978-3-030-43120-4_34

2   Related Work

Previous author attribution studies have proposed various descriptors to quantify the style of writing, called style markers, under different criterias [6,7].

The current functions of text representation for stylistic purposes focus mainly on lexical, character, syntactic, semantic and application-specific requirements for measuring them. The vast majority of the author's attribution studies are (at least partially) based on lexical characteristics to represent the style.

The simplest approach to represent texts is the vectors of words frequencies. The text is thus considered as a set of words each having a frequency of occurrence without taking into account the contextual information.

However, there is a significant difference in the classification of text based on the style: the importance of function words (determiners, conjunctions, prepositions, pronouns, auxiliary verbs, modals, qualifiers and question words) wich are among the best elements of discrimination between authors, because they are used in a largely unconscious by the authors and are independent of the subject. Thus, they are able to capture the stylistic choices of the author through different themes [8,9].

2.1   Character Features

According to this family of measures, a text is considered as a simple succession of characters. Thus, various levels of character measures can be defined, including alphabetic characters, numbers, uppercase and lowercase characters, punctuation characters,... [2,10]. This type of information is readily available for any natural language and any corpus. It has been proven very useful for quantifying the writing style [5].

Kjell [3] used the bigrams and trigrams characters to discriminate the "Federalist Papers". Forsyth and Holmes [11] found that bigrams and the variable length character n-grams yield better results than lexical descriptors in several text categorization tasks, including author attribution. Similarly, a comparative study of different lexical descriptors and characters on the same evaluation corpus [5] showed that the n-gram characters were the most effective measures.

This method is well adapted to long and short corpus, Posadas-Durán, Gómez-Adorno and Sidorov [18] use it for short corpus.

2.2   Lexical Features

The simplest lexical representation of texts has been introduced in the context of the vector model, it is called "bag of words". Texts are simply transformed into vectors, each component of which represents a term. C term is in general:

– the lexical stem of the word rather than the entire word. Several algorithms have been proposed to substitute words by their root; one of the best known is Porter's algorithm.

– The lemma: for that we use the grammatical analysis to replace the verbs by their infinitive form and the nouns by their singular form. Lemmatization is therefore more complicated to implement than the search for stems, since it requires a syntactic analysis of the texts. An efficient algorithm based on decision trees, named TreeTagger [19], has been developed for the English, French, German and Italian languages.

A recent study by Akimushkin, Amancio, Oliveira is based on lexical descriptors (words) to author attribution problem [20].

2.3   Syntaxic Features

The idea is that authors tend to use similar syntactic structures unconsciously. Therefore, syntactic information is considered to be more reliable for identifying author with respect to lexical information. The extraction of the syntax measure is a language dependent procedure. Baayen, van Halteren, and Tweedie [12] were the first to use syntactical measures for author attribution. Similarly, Gamon [13] uses the output of a parser to measure the frequencies of the rewriting of syntactic rules.

2.4   Semantic Features

The NLP (Natural Language Processing) tools are successfully applied to low-level tasks, seen previously, unfortunately, more complex tasks such as full parsing, semantic analysis and pragmatic analysis can not yet be satisfactorily addressed by these tools. As a result, very few attempts have been made to exploit high-level descriptors for stylistic purposes. Gamon [13] used a tool to produce semantic dependency graphs.

McCarthy, Lewis, Dufty and McNamara [14] describe another approach for extracting semantic measures. Based on WordNet [15], they estimated the information of synonyms and hyperonyms of words, as well as the identification of causal verbs.

A very important method of exploiting semantic information has been described by Argamon et al. [17]. Inspired by the theory of Systemic Functional Grammar (SFG), they defined a set of features that associate certain words or phrases with semantic information.

2.5   Application-Specific Features

Application-specific metrics can be defined to better represent the nuances of style in a given text domain such as e-mail messages and online forum posts.

Structural measures include the use of greetings and goodbyes in messages, types of signatures, the length of the paragraph,... [2,10].

Although these measures need to be defined manually, they can be very effective when it comes to certain types of texts.

2.6   Selection and Extraction of Features

Attribution studies of the author often combine several types of descriptors such as lexical descriptors and characters. In general, the descriptors chosen by these methods are examined individually on the basis of the authors' discrimination against a given corpus [16]. However, some descriptors that seem unimportant when examined independently may be useful in combination with other descriptors. Descriptors should be carefully selected according to the authors' universal properties to avoid dependence on a specific learning corpus.

3   Functional Discriminant Analysis (FDA)

Like most standard statistical models, discriminant analysis has been extended to the functional case. Preda [1], proposes an application of the PLS regression ((Partial Least Square) as part of the discriminant analysis. A penalization approach of the covariance operator has been successfully applied to Discriminant Analysis by Hastie et al. [4]. In what follows, we present the principle of functional discriminant analysis.

Data

We have a set of individuals (n random functions) $X = \{X_t\}_{t \in T}$ also noted $X(.,t) \in L^2(\Omega)$ which means that $\int_\Omega X_t(w)^2 \mathbb{P}(dw) < \infty$, and a qualitative variable Y, with q modalities (variable to explain).

The variable Y generates a partition $\{\Omega_i, i = 1, ..., q\}$ of the *Omega* set of individuals.

Goal. As in the multidimensional case, functional discriminant analysis has two goals:

Description: Search among all the possible PCAs one whose graphical representation of individuals discriminates at best the q classes generated by the variable Y.

Decision: Assign a new individual to one of the q modalities of the variable Y.

Let $\Gamma(s,t) = cov(X_s, X_t) = \int_\Omega X_t(\omega) X_s(\omega) Pd(\omega)$

The application Γ is called covariance operator.

Let V be the total covariance matrix such that $V_{s,t} = cov(X_s, X_t)$.

The functional expectation of the set *Omega* is $\widetilde{X} \in L^2(\Omega)$ where

$$\widetilde{X}(\omega) = \int_\Omega X_t(\omega) d\mathbb{P}(\omega)$$

Let $V_{g,t,s}$ be the inter-class covariance matrix such that $V_{g,t,s} = cov(\widetilde{X_t}, \widetilde{X_s}) = \int_\Omega \widetilde{X_t}(\omega)\widetilde{X_s}(\omega) Pd(\omega)$.

The solution is to maximize inter-class variance and minimize intra-class variance:

$$\begin{cases} \max_{\|u\|_V} \langle V^{-1}V_g u, u \rangle_V \\ \langle u, u \rangle_V = 1 \end{cases}$$

Using schwartz inequality, we obtain:

$$\langle V^{-1}V_g u, u \rangle_{H'} \leq \|V^{-1}V_g\| \|u\|$$

The maximum is reached for $V^{-1}V_g u = \lambda u$.

u_1 is the argument of the solution (1), which is the eigenvector of $V^{-1}V_g$ associated with the highest eigenvalue λ_1. the other arguments of the solution are obtained by iteration with V orthonormality.

4   Functional Representation of Texts

4.1   Preliminary Processing of Texts

- **Segmentation**: this task consists of dividing the text into a series of tokens (words), for that we used as separators: the space character, the different punctuation characters and the line break.
- **Lemmatization**: each word in the text is replaced by its lemma using Tree-tagger software.
- **Elimination of unnecessary punctuation characters**: all punctuation characters have been removed except those at the beginning and the end of the sentence: dot '.', Dash '-', exclamation point '!' and the question mark '?', however no function word is eliminated.

4.2   Transformation of Texts into Functions

- **Quantifying words:** at the beginning the code of a word is equal to its frequency of appearance in all the learning texts.

$$C(M) = \sum_{i=1}^{n} F(M)_{T_i}$$

where $C(M)$ is the code of the word M, n is the number of learning texts and $F(M)_{T_i}$ is the number of appearance of the word M in the text T_i.

It may be that after this coding, k different words have the same code if their frequencies of appearance in the learning texts are equal, to remedy this problem, we must *distinguish* the codes by adding a weak term to differentiate them.

Let g_η be the set of words whose frequency of occurrence of each word $M_i \in g_\eta$ $i = 1...k$ equals η with k the cardinality of g_η.

$$C(M_i) = \eta + (i-1)/(2.k)$$

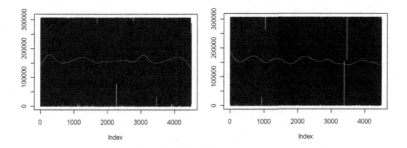

Fig. 1. Curves representing two different texts

Note that the union of sets g_η is a partition of the set of words.

Since the set of word codes will represent the values that the functions take (stochastic processes), they must be correctly positioned on the y-axis. We have, then, transformed the set of codes obtained in the previous step C into a new set C' so that the highest codes are positioned at the center of the interval $[y_{min}, y_{max}]$, such that $y_{min} \leq max\ X_i(t)_{\{i=1...n\}} \leq y_{max}$, thus forming a Gaussian distribution,

- **Representation by transition between words:** after having correctly coded the words of the learning texts, each of these texts T_i is represented by a curve taking the code values of its words in the order of their appearance in the text.

$$X_i(l) = C_i(M_l)\ pour\ l = 0, ..., L(T_i) - 1$$

where $L(T_i)$ is the length (number of words) of the text T_i and $C_i(M_l)$ the code of the lth word in the text T_i. The Fig. 1 represents the curves obtained by a b-spline smoothing of two different texts.

4.3   Advantages

This representation is very interesting and is well adapted to the problem of identifying the author by the fact that:

- it keeps the order of the words in the sentence, so it keeps contextual information unlike the wordbag representation.
- it keeps the preferred vocabulary information (a word with the code y is very used by the author if the curve $X(t)$ takes the value y quite often).
- does not require a reduction of the dimension, since the texts are represented as elements of the hilbert space (infinite dimension).
- takes into account the length of the sentences: since we have kept the punctuation characters of beginning and end of sentence ('-', '.', '?', '!'), then if a curve often takes one of the values of the codes of these words, it means that this text contains short sentences and vice versa.

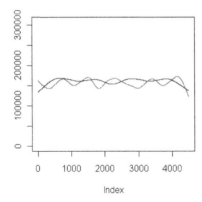

Fig. 2. Curves representing two texts: narrative in blue and theatrical in red (Color figure online)

- it keeps the genre of the text information, for example in a theatrical text, the sentences are quite short, begin with a dash '-' and end with one of the characters ('.', '?', '!'), on the other hand in a narrative text, we usually find long sentences, the dash '-', '?' and '!' are rare. For example, we chose a part of the book "Il ne faut jurer de rien" written by "ALFRED DE MUSSET" and part of the book "L'Éranger" of "ALBERT CAMUS", the first text is theatrical, the second is narrative. The Fig. 2 presents the curves of these two texts. It is clear that a graphical representation of these two texts by curves contains a big part of discrimination.

5   Experimentation

5.1   Presentation of the Used Corpus

The Table 1 presents the set of frensh texts constituting the corpus of learning and test used, the number of authors is 09, the total number of texts is 561, the size of each text is 4500 words. The Texts are downloaded from Gutenberg library.

5.2   Results

After representing the 561 texts of the array (Table 1) by functions (belong to separable Hilbert space), we chose the statistical method: functional discriminant analysis to discriminate them and assign an anonymous text to its author.

Table 1. Table presenting the corpus used.

Author	Work title	Size of the plain text (MB)	Number of texts
ALEXANDRE DUMAS	La dame aux camelias, la dame de monsoreau v.3, la reine margot tome i, la tulipe noire, le capitaine pamphile, le vicomte de bragelonne t3, les quarante-cinq - tome ii	3.79	119
ALFERD DE MUSSET	CROISILLES, EMMELINE, FREDERIC et BERNERETTE, histoire dun merle blanc, il ne faut jurer de rien, la mouche, le fils du titien, les caprices de MARIANNE, les deux maîtresses, l'habit vert, mimi pinson, PIERRE et CAMILLE, premières poésies, secret de JAVOTTE	1.65	62
ANDRE GIDE	ISABELLE, la porte etroite, la symphonie pastorale, les caves du vatican, les nourritures terrestres, l'immoraliste, si le grain ne meurt	1.92	63
ALBERT CAMUS	CALIGULA, l'etranger, l'homme révolte, la peste	1.41	34
GUSTAVE FLAUBERT	éducation sentimentale, Hérodias, la légende de saint julien l'hospitalier, madame BOVARY, mémoires d'un fou, SALAMMBO, un coeur simple, la tentation de saint ANTOINE, le candidat	2.66	76
HONORE DE BALZAC	la bourse, le chef d'oeuvre inconnu, le contrat de mariage, l'élixir de longue vie, l'Enver de l'histoire contemporaine, les chouans, mémoires de deux jeune maries, peau de chagrin, père GORIOT, recherche de l'absolu, traite des excitants modernes, une passion dans le désert	3.69	80
JEAN JACK ROUSSEAU	discours sur l'économie politique, discours sur l'inégalité, les confessions, le promeneur solitaire	3.37	14
VICTOR HUGO	CLAUDE GUEUX, l'ane, le dernier jour d'un condamne, les châtiments, les contemplations, les feuilles d'automne, les misérables - tome II - cosette, les orientales, les rayons et les ombres, notre dame de paris, quatre-vingt-treize	3.90	89
VOLTAIRE	CANDIDE, la pucelle d'Orléans, le monde comme il va, lettres philosophiques, l'homme aux quarante écus, Micromégas	0.69	24

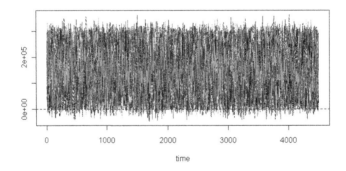

Fig. 3. Graphical representation of functions with decomposition into 560 b-spline basis functions

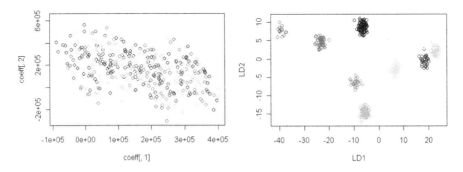

Fig. 4. Result of the LDA on the b-spline basis functions decomposition coefficients

5.3   Decomposition into Basis Functions

To implement the FDA, we proceeded to decomposition into basic functions. The Fig. 3 shows all the learning functions in the form of decomposition into b-spline basis functions.

5.4   Application of the Discriminant Analysis on the Coefficients

The Fig. 4 shows the result of the discrimination on the obtained coefficients of the decomposition.

5.5   Result of the Classic LDA on the Same Texts

To show the interest of passing from the finite dimension to the infinite dimension, we applied the LDA (Linear Discriminant Analysis) on the same texts using finite dimension vectors representation (see the Fig. 5, The number of variables is equal to 1000).

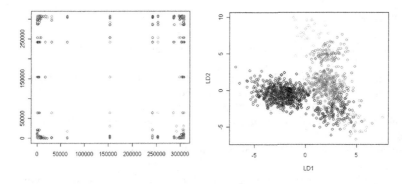

Fig. 5. Result of the discrimination using LDA in finite dimension

5.6 Evaluation of the Classification

Table 2 shows the evaluation of the classification quality of the discriminant analysis in the functional and multi-dimensional case on the same learning and test data.

For this we used 14-cross validation. The set of functions representing the texts is thus divided into 14 subsets. 13 subsets are used for learning while the 14th is the test subset. By iterrating 14 times. We chose 14-cross validation because all classes have a cardinality greater than or equal to 14.

Table 2. Comparative table of evaluation of the classification using FDA and LDA methods by 14-cross validation

	Correctly assigned	Incorrectly assigned
FDA	87.14%	12.86%
LDA	7.011%	92.989%

6 Conclusion

LDA is a simple and highly effective supervised classification method. The extension of this method to the functional case makes it possible to exploit the information presented by a functional random variable along its entire trajectory. Despite its simplicity, the method of text representation by transition between words has the advantage of keeping a lot of stylistic information and not requiring a prior reduction of the dimension. The results of the experiment, showed that the application of functional LDA to the identification of the author made it possible to discriminate perfectly the classes, in addition this approach gave a very interesting results when the assignment of new texts.

References

1. Preda, C.: L'approche PLS pour l'analyse de données fonctionnelles. Bull. Soc. Sci. Méd. **2**, 171–185 (2006)
2. Zheng, R., Li, J., Chen, H., Huang, Z.: A framework for authorship identification of online messages: writing style features and classification techniques. J. Am. Soc. Inf. Sci. Technol. **57**, 378–393 (2006)
3. Kjell, B.: Discrimination of authorship using visualization. Inf. Process. Manage. **30**, 141–150 (1994)
4. Hastie, T., Ruja, A., Tibshirani, R.: Penalized discriminant analysis. Ann. Stat. **23**, 73–102 (1995)
5. Grieve, J.: Quantitative authorship attribution: an evaluation of techniques. Lit. Linguist. Comput. **22**, 251–270 (2007)
6. Holmes, D.I.: Authorship attribution. Comput. Humanit. **28**, 87–106 (1994)
7. Stamatatos, E., Fakotakis, N., Kokkinakis, G.: Automatic text categorization in terms of genre and author. Comput. Linguist. **26**, 471–495 (2000)
8. Burrows, J.F.: Word patterns and story shapes: the statistical analysis of narrative style. Lit. Linguist. Comput. **2**, 61–70 (1987)
9. Argamon, S., Levitan, S.: Measuring the usefulness of function words for authorship attribution. In: Proceedings of the Joint Conference of the Association for Computers and the Humanities and the Association for Literary and Linguistic Computing (2005)
10. de Vel, O., Anderson, A., Corney, M., Mohay, G.: Mining e-mail content for author identification forensics. ACM SIGMOD Rec. **30**, 55–64 (2001)
11. Forsyth, R., Holmes, D.: Feature-finding for text classification. Lit. Linguist. Comput. **11**, 163–174 (1996)
12. Baayen, R., van Halteren, H., Tweedie, F.: Outside the cave of shadows: using syntactic annotation to enhance authorship attribution. Lit. Linguist. Comput. **11**, 121–132 (1996)
13. Gamon, M.: Linguistic correlates of style: authorship classification with deep linguistic analysis features. In: Proceedings of the 20th International Conference on Computational Linguistics (2004)
14. McCarthy, P.M., Lewis, G.A., Dufty, D.F., McNamara, D.S.: Analyzing writing styles with Coh-Metrix. In: Proceedings of the Florida Artificial Intelligence Research Society International Conference (2006)
15. Fellbaum, C.: WordNet: An Electronic Lexical Database. MIT Press, Cambridge (1998)
16. Forman, G.: An extensive empirical study of feature selection metrics for text classification. J. Mach. Learn. Res. **3**, 1289–1305 (2003)
17. Argamon, S., Whitelaw, C., Chase, P., Hota, S.R., Garg, N., Levitan, S.: Stylistic text classification using functional lexical features. J. Am. Soc. Inform. Sci. Technol. **58**, 802–822 (2007)
18. Posadas-Durán, J.P., Gómez-Adorno, H., Sidorov, G.: Application of the distributed document representation in the authorship attribution task for small corpora. Soft Comput. **21**, 627–639 (2017). American Society for Information Science and Technology
19. Schmidt, H.: Probabilistic part-of-speech tagging using decision trees (1994)

20. Akimushkin, C., Amancio, D.R., Oliveira, O.N.: On the role of words in the network structure of texts: application to authorship attribution. Phys. A (2017). https://doi.org/10.1016/j.physa.2017.12.054
21. Stamatatos, E., Koppel, M.: Plagiarism and authorship analysis: introduction to the special issue (2016). http://www.jstor.org/stable/41486024

Tools and Software Track

An Overview of Geometry Plus Simulation Modules

Angelos Mantzaflaris[(⊠)]

Inria Sophia Antipolis - Méditerranée, Université Côte d'Azur, Nice, France
`angelos.mantzaflaris@inria.fr`

Abstract. We give an overview of the open-source library "G+Smo". G+Smo is a C++ library that brings together mathematical tools for geometric design and numerical simulation. It implements the relatively new paradigm of isogeometric analysis, which suggests the use of a unified framework in the design and analysis pipeline. G+Smo is an object-oriented, cross-platform, fully templated library and follows the generic programming principle, with a focus on both efficiency and ease of use. The library aims at providing access to high quality, open-source software to the community of numerical simulation and beyond.

Keywords: C++ · B-splines · NURBS · Isogeometric analysis · Geometric design

1 Introducing Isogeometric Analysis

Isogeometric analysis suggests the use of common spline basis for geometric modeling and finite element analysis of a given model. Engineering process begins with designers encapsulating their concepts on the computer using Computer Aided Design (CAD) software. With the finite element analysis (FEM) technology, the CAD model must be first meshed before any numerical simulation can be performed on it. The mesh generation results in a piecewise linear approximation of the original model, that both increases the data volume and introduces unwanted approximation errors in the geometric description. Moreover, the mesh generation phase involves tedious manual work for producing finite element meshes that are suitable for analysis. Last but not least, the results of the analysis should be projected back to the original CAD model, for allowing the designer to update it accordingly. The data transfer between models suitable for design (CAD) and analysis (FEM) poses a severe problem in industry today.

To address these issues Hughes and co-workers introduced in [4] an analysis framework which is based on NURBS (Non-Uniform Rational B-Splines), which is standard technology employed in CAD systems. They propose to keep the exact CAD geometry by NURBS surfaces, and apply numerical simulation using the same NURBS as basis functions for the analysis [2].

D. Slamanig et al. (Eds.): MACIS 2019, LNCS 11989, pp. 453–456, 2020.
https://doi.org/10.1007/978-3-030-43120-4_35

2   General Concept of the Library

The design and software development philosophy of G+Smo was triggered by the observation that simply augmenting existing libraries for geometric design or numerical analysis will not succeed, since the implementation of isogeometric methods requires a fundamentally new approach to unlock the full set of benefits of the isogeometric paradigm.

The goal of G+Smo is to realize the seamless integration of Finite Element Analysis (FEA) and Computer-aided design (CAD) with open-source code from and to the isogeometric analysis community.

Three general guidelines have been set for the development process. Firstly, we promote both efficiency and ease of use; secondly, we ensure code quality and cross-platform compatibility and, thirdly, we always explore new strategies better suited for isogeometric analysis before adopting FEA practices.

G+Smo is an object-oriented, templated C++ library, that implements a generic concept for IgA, based on abstract classes for geometry map, discretization basis, assemblers, solvers etc. It makes use of object polymorphism and inheritance techniques in order to support a variety of different discretization bases, namely B-spline, Bernstein, NURBS bases, hierarchical and truncated hierarchical B-spline bases of arbitrary polynomial order. The library was first announced in the extended abstract [5], a more detailed description is contained in [6].

3   Open-Source License

The library is licensed under the Mozilla Public License v2.0 (MPL). The MPL is a simple copy-left license. The MPL's "file-level" copy-left is designed to encourage contributors to share their modifications with the library, while still allowing them to combine the library with code under other licenses (open or proprietary) with minimal restrictions. The latest revision can be downloaded from GitHub: https://github.com/gismo.

4   Modules

The library is partitioned into smaller entities, called modules. Examples of available modules include the dimension-independent NURBS module, the data fitting and solid segmentation module, the PDE discretization module and the adaptive spline module, based on hierarchical splines of arbitrary dimension and polynomial degree.

At the present point, the B-spline, Bernstein, NURBS bases, and also the truncated and non-truncated hierarchical B-spline bases (cf. [3]) are in a stable and functional state, for general dimension. From the simulation side, isogeometric simulation algorithms with the Galerkin approach, notably over non-conforming multi-patch physical domains are implemented and in a stable state. A module treating a large class of linear and non-linear elasticity problems was also released recently [7]. A module dedicated to isogeometric multigrid solvers on multipatch domains is also available [8].

Another interesting module is the assembler module based on *expression templates*, which are a powerful C++ technique for the development of user-friendly and versatile code. They have the advantage of bringing the implementation of a numerical method much closer to the mathematical notation. This allows the easy formation of isogeometric Galerkin matrices using a high-level language, without sacrificing efficiency.

The library takes advantage of modern, efficient move semantics, which are available in recent C++ versions, for more efficiency and cleaner code.

5   Code Management and Documentation

In order to deliver a successful library, quality of the code needs to be assured. To this end, the CMAKE cross-platform compilation system is used in the library, which allows for seamless installation of the library on different platforms. Moreover Jenkins and a CDASH testing servers are in place for executing regular nightly builds, memory checks and unit-tests on different platforms (Windows, Linux and MacOSX). Documentation is done mostly in the form of in-source text, using Doxygen as well as in the Wiki pages http://gs.jku.at/gismo. The documentation is currently available online in the form of HTML pages to all the users, and is updated regularly, as we actively continue adding new material, see https://gismo.github.io.

6   Third-Party Dependencies

Apart from the C++ Standard Library, we use open-source third party software for common tasks, for example the Eigen (http://eigen.tuxfamily.org) linear algebra library for (sparse) matrices and linear algebra computations, as well as tools for argument parsing, input and output of XML formatted files, the openNurbs library for reading Rhino's 3DM commercial CAD format (https://www.rhino3d.com/opennurbs), Trilinos (https://trilinos.github.io) for high performance solvers etc.

7   Plugins and Extensions

Concerning plugins, visualization and user interfaces, we developed a plug-in of G+Smo for the Axl geometric modeling platform ([1], http://axl.inria.fr. This provides an interactive GUI (graphical user interface) that allows the user to manipulate multi-patch geometries by their control points and work with complicated CAD models. Nevertheless, our primary approach to visualization is based on the well-established Paraview software (https://www.paraview.org/)and has been enhanced, e.g., with trimmed surface visualization, gradient fields and absolute error plots, etc. Another connection has been setup with Siemens'

Parasolid geometric CAD kernel (https://www.plm.automation.siemens.com), which is widely used in industry. Currently, this allows us to exchange data and to use the advanced modeling utilities available in Parasolid from within G+Smo.

Acknowledgement. G+Smo is jointly developed by several contributors at the Johannes Kepler University, at the RICAM Institute of the Austrian Academy of Sciences (in the frame of the Austrian Science Fund NFN project S117) at INRIA and at the Department of Applied Mathematics of TU Delft (The Netherlands). More contributions have been made by developers from other institutions. The full list of contributors is available at https://github.com/gismo/gismo/wiki/About--G-Smo.

References

1. Christoforou, E., Mantzaflaris, A., Mourrain, B., Wintz, J.: Axl, a geometric modeler for semi-algebraic shapes. In: Davenport, J.H., Kauers, M., Labahn, G., Urban, J. (eds.) ICMS 2018. LNCS, vol. 10931, pp. 128–136. Springer, Cham (2018). https://doi.org/10.1007/978-3-319-96418-8_16
2. Cottrell, J., Hughes, T., Bazilevs, Y.: Isogeometric Analysis: Toward Integration of CAD and FEA. Wiley, Chichester (2009)
3. Giannelli, C., Juettler, B., Kleiss, S.K., Mantzaflaris, A., Simeon, B., Speh, J.: THB-splines: an effective mathematical technology for adaptive refinement in geometric design and isogeometric analysis. Comput. Methods Appl. Mech. Eng. **299**, 337–365 (2016). http://dx.doi.org/10.1016/j.cma.2015.11.002
4. Hughes, T., Cottrell, J., Bazilevs, Y.: Isogeometric analysis: CAD, finite elements, NURBS, exact geometry and mesh refinement. Comput. Methods Appl. Mech. Eng. **194**(39–41), 4135–4195 (2005). http://dx.doi.org/10.1016/j.cma.2004.10.008
5. Juettler, B., Langer, U., Mantzaflaris, A., Moore, S., Zulehner, W.: Geometry + simulation modules: Implementing isogeometric analysis. Proc. Appl. Math. Mech. **14**(1), 961–962 (2014). http://dx.doi.org/10.1002/pamm.201410461
6. Langer, U., Mantzaflaris, A., Moore, S.E., Toulopoulos, I.: Multipatch discontinuous galerkin isogeometric analysis. In: Jüttler, B., Simeon, B. (eds.) Isogeometric Analysis and Applications 2014. LNCSE, vol. 107, pp. 1–32. Springer, Cham (2015). https://doi.org/10.1007/978-3-319-23315-4_1
7. Shamanskiy, A., Simeon, B.: Isogeometric simulation of thermal expansion for twin screw compressors. IOP Conf. Ser.: Mater. Sci. Eng. **425**, 012031 (2018). https://doi.org/10.1088/1757-899x/425/1/012031
8. Takacs, S.: Fast multigrid solvers for conforming and non-conforming multi-patch isogeometric analysis, arXiv preprint https://arxiv.org/abs/1902.01818 (2019)

DD-Finite Functions Implemented in Sage

Antonio Jiménez-Pastor$^{(\boxtimes)}$

Johannes Kepler University, Altenbergerstr. 69, 4020 Linz, Austria
ajpastor@risc.uni-linz.ac.at
https://www.dk-compmath.jku.at/people/antonio

Abstract. We present here the Sage package dd_functions which provides symbolic features to work with DD-finite functions, a natural extension of the class of holonomic or D-finite functions, on the computer. Closure properties, composition of DD-finite functions and sequence extraction are key features of this package. All these operations reduce the problem to linear algebra computations where classical division-free algorithms are used.

Keywords: Holonomic · D-finite · Generating functions · Closure properties · Formal power series

1 Introduction

A formal power series $f(x) = \sum_{n \geq 0} a_n x^n$ is called D-finite, if it satisfies a linear differential equation with polynomial coefficients [7,11]. Many generating functions of combinatorial sequences are of this type as well as the most commonly used special functions [1,2,9]. These objects can be represented in finite terms using a defining differential equation and sufficiently many initial values.

We recently extended this class to a wider set of formal power series that satisfy linear differential equations with D-finite coefficients, and call them DD-finite [4]. There were several implementations of D-finite functions in several computer algebra systems [6,8,10], but here we present the first implementation for DD-finite functions with a Sage package including all the closure properties and operations that have been proven for DD-finite functions.

The finiteness of the representation for D-finite functions allows to finitely represent DD-finite functions, with a defining differential equation and initial conditions. A difference between our package and other implementations is that ours handle mainly particular solutions of differential equations, making the initial conditions an important point of our implementation. This leads to two different steps while manipulating DD-finite functions: computing the final differential equation and computing the initial data required for the specific object. More details on the structure we used can be found in Sect. 2.

This author was funded by the Austrian Science Fund (FWF): W1214-N15, project DK15.

D. Slamanig et al. (Eds.): MACIS 2019, LNCS 11989, pp. 457–462, 2020.
https://doi.org/10.1007/978-3-030-43120-4_36

The package `ore_algebra` [6] has implemented the differential operators computations for D-finite functions and the package presented here uses it by default. If the user desires to avoid the package `ore_algebra`, or it is not installed, a different implementation [3] is automatically provided. Such implementation is used by default with DD-finite functions.

At the time of writing, the package described in this document is still under construction and has not been added to the official Sage distribution. Readers who want to try it are invited to download and install the current version from the public git distribution using the command

```
sage -pip [--user] install git+https://github.com/Antonio-JP/dd_functions.git
```

and are encouraged to send bug reports, feature requests or other comments. If the user has not a Sage installation available, it is still possible to test a demo of the package, using Binder, from the personal webpage of the author.

Once installed, the user can start using the package typing the following command on Sage:

`sage: from ajpastor.dd_functions import *`

All the features of the package (structures for our objects, arithmetic operations over them, built-in examples, tests, etc.) can be used without further configuration. More details on how to build the structures and the examples available can be found in Sects. 2 and 3.

2 Data Structure

Although the main goal of our package is to manipulate DD-finite functions, the theory can be stated in a more general setting that is covered by our implementation as well. For further details, see [4].

Definition 1. *Let R be a non-trivial differential subring of $K[[x]]$ and $R[\partial]$ the ring of linear differential operators over R. We call $f \in K[[x]]$ differentially definable over R if there is a non-zero operator $\mathcal{A} \in R[\partial]$ that annihilates f, i.e., $\mathcal{A} \cdot f = 0$. By $D(R)$ we denote the set of all $f \in K[[x]]$ that are differentially definable over R.*

Differentially definable functions are implemented using the *Parent-Element* model of Sage and our Parent structure is included in the *categories* framework, allowing the user to have an extended coercion system. With this, the user have a more natural management of the differentially definable functions.

2.1 The Parent Structure: *DDRing*

The structure `DDRing` describes the sets of differentially definable functions. Thus, they are built from any ring structure in Sage, R. Following the *Parent-Element* model of Sage, `DDRing` is a parent class that will include as element any differentially definable function over R.

To create a `DDRing`, the user has to provide the base ring R for the coefficients of the differential operators with a derivation. There are several optional inputs that allow more flexibility and a wider use of the structure, such as different derivations or adding parameters. For more information, use `DDRing?`.

This structure is only used to create differentially definable functions and to cast other elements (if possible). It is also used to compute a common parent (`pushout`) for two different elements, which allows the user to compute more easily with these objects.

Several `DDRing`s are defined by default:

- `DFinite`: the ring of D-finite functions ($R = \mathbb{Q}[x]$),
- `DDFinite`: the ring of DD-finite functions ($R = D(\mathbb{Q}[x])$),
- `DFiniteP`: the ring of D-finite functions with a parameter `P` ($R = \mathbb{Q}(P)[x]$).
- `DFiniteI`: the ring of D-finite over the Gaussian rationals ($R = \mathbb{Q}(i)[x]$).

2.2   The Element Structure: *DDFunction*

The class `DDFunction` is the Element class of `DDRing` and represents a differentially definable function within a particular `DDRing`.

A `DDFunction` is always included in a `DDRing` so, for creating a new one, the method `element` from the corresponding `DDRing` must be used. Any `DDFunction` is represented using a linear differential equation and some initial values, so the user must provide two lists to create a `DDFunction`: a list $[r_0, \ldots, r_d]$ with coefficients for the differential equation and a list $[a_0, \ldots, a_n]$ with initial values for the function:

$$([r_0, \ldots, r_d], [a_0, \ldots, a_n]) \mapsto \begin{cases} r_d f^{(d)}(x) + \cdots + r_0 f(x) = 0, \\ f(0) = a_0, \ldots, f^{(n)}(0) = a_n. \end{cases}$$

The method `element` also checks that the coefficients r_i are in the appropriate ring R and that the initial conditions a_i are elements of the field where the function $f(x)$ is considered and are valid for the equation defined by the list $[r_0, \ldots, r_d]$.

If not enough initial conditions a_i are provided, the object will represent all the solutions to the given differential equation with the given initial conditions. This makes the creation of objects flexible but some features (such as composition) are not available unless enough initial data is provided.

3   Selected Methods and Main Features

3.1   Utility Methods

There are several methods that allow the user to extract information or manipulate a `DDFunction`.

- `equation`: the differential operator that defines the function.
- `getInitialValue`: the value of the nth derivative of the function at $x = 0$.

- `getSequenceElement`: the value of the nth coefficient of the formal power series.
- `getOrder`: the order of the differential operator defining the function.
- `min_coefficient`: the first non-zero coefficient of the formal power series.
- `zero_extraction`: the order of the formal power series and the `DDFunction` defined after factoring the maximal power of x possible.
- `change_init_values`: returns a new `DDFunction` with the same differential equation and some different initial conditions given as parameters.

3.2 Operational Methods

Differentially definable functions satisfy several closure properties. They are closed under addition, multiplication, differentiation and integration [4]. The user can also compute the division [4] and composition [5] whenever those operations are defined as formal power series.

- **Addition**: `add` or simply `+`
- **Multiplication**: `mult` or simply `*`
- **Difference**: `sub` or simply `-`
- **Division**: `div` or simply `/`
- **Exponentiation**: `pow` or simply `^`
- **Derivation**: `derivative`.
- **Integration**: `integrate`. The value at $x = 0$ can be specified. By default, it is 0.
- **Composition**: `compose` or the magic method `__call__`.

The implementations are based on linear algebra computations. Namely, computing nullspaces for matrices where the coefficients are elements of the base ring R. These linear algebra operations are usually carried by standard algorithms on Sage, although for DD-finite functions, i.e., when the coefficients of the matrices are D-finite functions, we have implemented a specialized version of Bareiss' algorithm. For further details, see [3].

Several algebraic properties has also been proven (see [5,7]) for `DDFunctions` and they can also be computed with our package using the following methods:

- `DAlgebraic`: given an algebraic equation of $f(x)$ over a field F, it computes its representation as an element of $D(F)$.
- `diff_to_diffalg`: given a `DDFunction`, computes a differentially algebraic equation (i.e., a non-linear differential equation).

3.3 Built-in Functions of the Package

Aiming for a more user friendly interface, we provide a set of examples of D-finite and DD-finite functions that can be easily called and built from Sage once the package is loaded. Some of these functions are elementary as the exponential function (`Exp`) or trigonometric functions (`Sin`, `Cos`, `Tan`); some are special functions as the Bessel functions (`BesselD`), hypergeometric $_pF_q$ functions, solutions to the Riccati differential equation (`RiccatiD`), the Mathieu functions (`MathieuD`) or some generating functions for combinatorial sequences (`FibonacciD`, `CatalanD`).

All implemented functions can be checked in Sage by typing `ddExamples?`. This command displays the documentation for the `ddExamples` file where all these built-in functions are explained with extended details. Also, check the online documentation for a web view of the built-in examples and all the details of the implementation.

3.4 Proving Identities with the Package

With the built-in examples and the operations included on the package the user can check and prove equality of expressions (using either the Python syntax `==` or the method `equals` of the `DDFunction` class).

We offer two examples proving some trigonometric identities to show how to use the package to prove identities:

```
sage: x = DFiniteI.variables()[0]# Getting 'x'
sage: I = DFIniteI.base().base().gens()[0]# Getting 'I'
sage: (Exp(I*x) + Exp(-I*x))/2 == Cos(x)
True
```

Similar syntax can be used also with DD-finite elements:

```
sage: t = Tan(x)# Getting the 'tan(x)' function
sage: %time t.derivative() == 1 + t^2
CPU times: user 2.82 s, sys: 117 ms, total: 2.94 s
Wall time: 2.78 s
True
```

4   Conclusion

The Sage package `dd_functions` provides a functional-focused implementation of differentially definable functions for arbitrary differential rings. In particular, it can be used for working with regular D-finite functions and DD-finite functions.

This package can be used to prove symbolically identities between DD-finite functions and to obtain and handle combinatorial sequences that are out of the D-finite scope.

Closure properties, such as addition, multiplication and exponentiation can be performed by the package, as well as other operations such as composition or division. It also provides several methods to convert algebraic functions to `DDFunction` and to compute a non-linear equation with polynomial coefficients for any differentially definable function.

Further work will be carried into this package adding more features, applying the standard Sage conventions for names through the whole package and, in last stage, pushing the package into the official Sage distribution. Readers can check the public GitHub project and documentation to get information about the new changes and features.

References

1. Andrews, G.E., Askey, R., Roy, R.: Special Functions. Encyclopedia of Mathematics and Its Applications. Cambridge University Press, Cambridge (1999)
2. Olver, F.W.J., et al. (eds.): NIST Digital Library of Mathematical Functions. Release 1.0.16 of 2017–09-18. http://dlmf.nist.gov/
3. Jiménez-Pastor, A., Pillwein, V.: Algorithmic arithmetics with DD-finite functions. In: Carlos, A. (ed.) Proceedings of the 2018 ACM on International Symposium on Symbolic and Algebraic Computation, ISSAC 2018, pp. 231–237. ACM, New York (2018). https://doi.org/10.1145/3208976.3209009
4. Jiménez-Pastor, A., Pillwein, V.: A computable extension for D-finite functions: DD-finite functions. J. Symb. Comput. **94**, 90–104 (2019). https://doi.org/10.1016/j.jsc.2018.07.002
5. Jiménez-Pastor, A., Pillwein, V., Singer, M.F.: Some structural results on D^n-finite functions. Technical report, Doctoral Program Computational Mathematics, Preprint series (2019, submitted to journal)
6. Kauers, M., Jaroschek, M., Johansson, F.: Ore polynomials in Sage. In: Gutierrez, J., Schicho, J., Weimann, M. (eds.) Computer Algebra and Polynomials. Lecture Notes in Computer Science, pp. 105–125. Springer, Cham (2014). https://doi.org/10.1007/978-3-319-15081-9_6
7. Kauers, M., Paule, P.: The Concrete Tetrahedron: Symbolic Sums, Recurrence Equations, Generating Functions, Asymptotic Estimates, 1st edn. Springer, Vienna (2011). https://doi.org/10.1007/978-3-7091-0445-3
8. Koutschan, C.: Advanced applications of the holonomic systems approach. Ph.D. thesis, RISC-Linz, Johannes Kepler University (2009). http://www.risc.uni-linz.ac.at/research/combinat/software/HolonomicFunctions/
9. Rainville, E.D.: Special Functions, 1st edn. Chelsea Publishing Co., Bronx (1971)
10. Salvy, B., Zimmermann, P.: GFUN: a Maple package for the manipulation of generating and holonomic functions in one variable. ACM Trans. Math. Softw. **20**(2), 163–177 (1994)
11. Stanley, R.P.: Differentiably finite power series. Eur. J. Comb. **1**(2), 175–188 (1980). https://doi.org/10.1016/S0195-6698(80)80051-5

Author Index

Ablinger, Jakob 42
Akoglu, Tulay Ayyildiz 3
Alexandersson, Per 333
Alnajjarine, Nour 288
Ay, Nihat 357

Bayrakci, Alp Arslan 187
Borges-Quintana, Mijail 218
Borges-Trenard, Miguel Ángel 218
Büyükçolak, Yasemin 280

Camargos Couto, Ana C. 80
Carlini, Luca 357
Çelik, Türkü Özlüm 364
Cenk, Murat 202
Corless, Robert M. 80

Diatta, Sény 16
Drămnesc, Isabela 153

England, Matthew 341
Esirci, Fatma Nur 187

Florescu, Dorian 341
Foufou, Sebti 422
Fukasaku, Ryoya 10

Görgen, Christiane 357
Gözüpek, Didem 280
Güvel, Muhammet Selçuk 324

Huang, Bo 169

Imbach, Rémi 122
Isik, Sahin 408
Islam, Muhammad Zubair 395

Jamneshan, Asgar 364
Jebelean, Tudor 153
Jeffrey, David J. 80
Jiménez-Pastor, Antonio 457

Kampel, Ludwig 313
Kettaf, Chahrazed 438
Kim, Hyung Seok 395

Lai, Jiahua 138
Lavrauw, Michel 288
Ledoux, Viviane 35
Levin, Alexander 64
Linder, David 80
Liu, Tingwen 382
Luan, Qi 89, 105

Mantzaflaris, Angelos 453
Martínez-Moro, Edgar 218
Mehmood, Kashif 395
Mehta, Sanyam 234
Melkemi, Kamal Eddine 422
Michelucci, Dominique 422
Mokhtari, Bilal 422
Montúfar, Guido 364
Moreno Maza, Marc 80
Moroz, Guillaume 16, 35
Mou, Chenqi 138

Nabeshima, Katsusuke 10, 48
Niu, Wei 169

Özkahya, Lale 324
Ozkan, Kemal 408
Özkan, Sibel 280

Pan, Victor Y. 89, 105, 122
Pouget, Marc 16

Restadh, Petter 333

Saraswat, Vishal 234
Sato, Yosuke 10
Sekigawa, Hiroshi 10

Sevim, Taha   324
Shi, Jinqiao   382
Simos, Dimitris E.   313
Sturmfels, Bernd   364
Svadlenka, John   89
Szanto, Agnes   3

Tajima, Shinichi   48
Tarsissi, Lama   295
Torres-Guerrero, Gustavo   218

Ulu, Metin Evrim   202
Uncu, Ali Kemal   273

Venturello, Lorenzo   364
Vuillon, Laurent   295

Wagner, Michael   313

Xu, Hongbo   382

Ya, Jing   382
Yıldırım, Hamdi Murat   249
Yousfate, Abderrahmane   438

Zhang, Haoliang   382
Zhao, Liang   89

Printed in the United States
By Bookmasters